Man and Machines

Scientiam non dedit natura semina scientiae nobis dedit
"Nature has given us not knowledge itself, but the seeds thereof."
Seneca

The Joy of Knowledge Encyclopaedia is affectionately
dedicated to the memory of John Beazley 1932–1977,
Book Designer, Publisher, and Co-Founder of the
publishing house of Mitchell Beazley Limited, by all
his many friends and colleagues in the company.

The Joy of Knowledge Library

General Editor: James Mitchell
With an overall preface by Lord Butler, Master of Trinity College,
University of Cambridge

Science and The Universe	Introduced by Sir Alan Cottrell, Master of Jesus College, University of Cambridge; and Sir Bernard Lovell, Professor of Radio Astronomy, University of Manchester
The Physical Earth	Introduced by Dr William Nierenberg, Director, Scripps Institution of Oceanography, University of California
The Natural World	Introduced by Dr Gwynne Vevers, Assistant Director of Science, the Zoological Society of London
Man and Society	Introduced by Dr Alex Comfort, Professor of Pathology, Irvine Medical School, University of California, Santa Barbara
History and Culture 1 **History and Culture** 2	Introduced by Dr Christopher Hill, Master of Balliol College, University of Oxford
Man and Machines	Introduced by Sir Jack Callard, former Chairman of Imperial Chemical Industries Limited
The Modern World	
Fact Index A–K	
Fact Index L–Z	

Man and Machines

Introduced by Sir Jack Callard

former Chairman of Imperial Chemical Industries Limited

MITCHELL BEAZLEY

The Joy of Knowledge Library

Editorial Director	Frank Wallis
Creative Director	Ed Day
Project Director	Harold Bull

Volume editors	
Science and The Universe	John Clark
	Lawrence Clarke
The Natural World	Ruth Binney
The Physical Earth	Erik Abranson
	Dougal Dixon
Man and Society	Max Monsarrat
History and Culture 1 & 2	John Tusa
	Roger Hearn
Time Chart	Jane Kenrick
Man and Machines	John Clark
Fact Index	Stephen Elliott
	Stanley Schindler
	John Clark

Art Director	Rod Stribley
Production Editor	Helen Yeomans
Assistant to the Project Director	Graham Darlow
Associate Art Director	Anthony Cobb
Art Buyer	Ted McCausland
Co-editions Manager	Averil Macintyre
Printing Manager	Bob Towell
Information Consultant	Jeremy Weston

Sub-Editors	Don Binney
	Arthur Butterfield
	Charyn Jones
	Jenny Mulherin
	Shiva Naipaul
	David Sharp
	Jack Tresidder
Proof-Readers	Jeff Groman
	Anthony Livesey
Researchers	Peter Furtado
	Malcolm Hart
	Peter Kilkenny
	Ann Kramer
	Lloyd Lindo
	Heather Maisner
	Valerie Nicholson
	Elizabeth Peadon
	John Smallwood
	Jim Somerville

Senior Designer	Sally Smallwood
Designers	Rosamund Briggs
	Mike Brown
	Lynn Cawley
	Nigel Chapman
	Pauline Faulks
	Nicole Fothergill
	Juanita Grout
	Ingrid Jacob
	Carole Johnson
	Chrissie Lloyd
	Aean Pinheiro
	Andrew Sutterby
Senior Picture Researchers	Jenny Golden
	Kate Parish
Picture Researchers	Phyllida Holbeach
	Philippa Lewis
	Caroline Lucas
	Ann Usborne

Assistant to the Editorial Director	Judy Garlick
Assistant to the Section Editors	Sandra Creese
Editorial Assistants	Joyce Evison
	Miranda Grinling
Production Controllers	Jeremy Albutt
	John Olive
	Anthony Bonsels
Production Assistants	Nick Rochez
	John Swan

Major contributors and advisers
to The Joy of Knowledge Library

Fabian Acker CEng, MIEE, MIMarE;
Professor H.C. Allen MC; Leonard Amey
OBE; Neil Ardley BSc; Professor H.R.V.
Arnstein DSc, PhD, FIBiol; Russell Ash
BA(Dunelm), FRAI; Norman Ashford
PhD, CEng, MICE, MASCE, MCIT;
Professor Robert Ashton; B.W. Atkinson
BSc, PhD; Anthony Atmore BA;
Professor Philip S. Bagwell BSc(Econ),
PhD; Peter Ball MA; Edwin Banks
MIOP; Professor Michael Banton; Dulan
Barber; Harry Barrett; Professor J.P.
Barron MA, DPhil, FSA; Professor W.G.
Beasley FBA; Alan Bender PhD, MSc,
DIC, ARCS; Lionel Bender BSc; Israel
Berkovitch PhD, FRIC, MIChemE;
David Berry MA; M.L. Bierbrier PhD;
A.T.E. Binsted FBBI (Dipl); David
Black; Maurice E.F. Block BA,
PhD(Cantab); Richard H. Bomback BSc
(London), FRPS; Basil Booth
BSc(Hons), PhD, FGS, FRGS; J. Harry
Bowen MA(Cantab), PhD(London);
Mary Briggs MPS, FLS; John Brodrick
BSc (Econ); J.M. Bruce ISO, MA,
FRHistS, MRAeS; Professor D.A.
Bullough MA, FSA, FRHistS; Tony
Buzan BA(Hons) UBC; Dr Alan R.
Cane; Dr J.G. de Casparis; Dr Jeremy
Catto MA; Denis Chamberlain; E.W.
Chanter MA; Professor Colin Cherry
DSc(Eng), MIEE; A.H. Christie MA,
FRAI, FRAS; Dr Anthony W. Clare
MPhil(London), MB, BCh, MRCPI,
MRCPsych; Sonia Cole; John R. Collis
MA, PhD; Professor Gordon Connell-
Smith BA, PhD, FRHistS; Dr A.H. Cook
FRS; Professor A.H. Cook FRS; J.A.L.
Cooke MA, DPhil; R.W. Cooke BSc,
CEng, MICE; B.K. Cooper; Penelope J.
Corfield MA; Robin Cormack MA, PhD,
FSA; Nona Coxhead; Patricia Crone BA,
PhD; Geoffrey P. Crow BSc(Eng), MICE,
MIMunE, MInstHE, DIPTE; J.G.
Crowther; Professor R.B. Cundall FRIC;
Noel Currer-Briggs MA, FSG;
Christopher Cviic BA(Zagreb),
BSc(Econ, London); Gordon Daniels
BSc(Econ, London), DPhil(Oxon);
George Darby BA; G.J. Darwin; Dr
David Delvin; Robin Denselow BA;
Professor Bernard L. Diamond; John
Dickson; Paul Dinnage MA; M.L.
Dockrill BSc(Econ), MA, PhD; Patricia
Dodd BA; James Dowdall; Anne Dowson
MA(Cantab); Peter M. Driver BSc, PhD,
MIBiol; Rev Professor C.W. Dugmore
DD; Herbert L. Edlin BSc, Dip in
Forestry; Pamela Egan MA(Oxon);
Major S.R. Elliot CD, BComm; Professor
H.J. Eysenck PhD, DSc; Dr Peter
Fenwick BA, MB, BChir, DPM,
MRCPsych; Jim Flegg BSc, PhD, ARCS,
MBOU; Andrew M. Fleming MA;
Professor Antony Flew MA(Oxon),
DLitt(Keele); Wyn K. Ford FRHistS;
Paul Freeman DSc(London); G.E. Fussell
DLitt, FRHistS; Kenneth W. Gatland
FRAS, FBIS; Norman Gelb BA; John
Gilbert BA(Hons, London); Professor
A.C. Gimson; John Glaves-Smith BA;
David Glen; Professor S.J. Goldsack BSc,
PhD, FINSTP, FBCS; Richard Gombrich
MA, DPhil; A.F. Gomm; Professor A.
Goodwin MA; William Gould
BA(Wales); Professor J.R. Gray;
Christopher Green PhD; Bill Gunston;
Professor A. Rupert Hall LittD; Richard
Halsey BA(Hons, UEA); Lynette K.
Hamblin BSc; Norman Hammond;
Professor Thomas G. Harding PhD;
Richard Harris; Dr Randall P. Harrison;
Cyril Hart MA, PhD, FRICS, FIFor;
Anthony P. Harvey; Nigel Hawkes
BA(Oxon); F.P. Heath; Peter
Hebblethwaite MA(Oxon), LicTheol;
Frances Mary Heidensohn BA; Dr Alan
Hill MC, FRCP; Robert Hillenbrand MA,
DPhil; Professor F.H. Hinsley; Dr
Richard Hitchcock; Dorothy
Hollingsworth OBE, BSc, FRIC, FIBiol,
FIFST, SRD; H.P. Hope BSc (Hons,
Agric); Antony Hopkins CBE, FRCM,
LRAM, FRSA; Brian Hook; Peter

Howell BPhil, MA(Oxon); Brigadier K.
Hunt; Peter Hurst BDS, FDS, LDS,
RSCEd, MSc(London); Anthony Hyman
MA, PhD; Professor R.S. Illingworth
MD, FRCP, DPH, DCH; Oliver Impey
MA, DPhil; D.E.G. Irvine PhD; L.M.
Irvine BSc; Anne Jamieson cand
mag(Copenhagen), MSc(London);
Michael A. Janson BSc; Professor P.A.
Jewell BSc(Agric), MA, PhD, FIBiol;
Hugh Johnson; Commander I.E.
Johnston RN; I.P. Jolliffe BSc, MSc, PhD,
CompICE, FGS; Dr D.E.H. Jones ARCS,
FCS; R.H. Jones PhD, BSc, CEng, MICE,
FGS, MASCE; Hugh Kay; Dr Janet Kear;
Sam Keen; D.R.C. Kempe BSc, DPhil,
FGS; Alan Kendall MA(Cantab);
Michael Kenward; John R. King
BSc(Eng), DIC, CEng, MIProdE; D.G.
King-Hele FRS; Professor J.F. Kirkaldy
DSc; Malcolm Kitch; Michael Kitson
MA; B.C. Lamb BSc, PhD; Nick Landon;
Major J.C. Larminie QDG, Retd; Diana
Leat BSc(Econ), PhD; Roger Lewin BSc,
PhD; Harold K. Lipset; Norman
Longmate MA(Oxon); John Lowry;
Kenneth E. Lowther MA; Diana Lucas
BA(Hons); Keith Lye BA, FRGS; Dr
Peter Lyon; Dr Martin McCauley; Sean
McConville BSc; D.F.M. McGregor BSc,
PhD(Edin); Jean Macqueen PhD;
William Baird MacQuitty MA(Hons),
FRGS, FRPS; Jonathan Martin MA; Rev
Canon E.L. Mascall DD; Christopher
Maynard MSc, DTh; Professor A.J.
Meadows; J.S.G. Miller MA, DPhil, BM,
BCh; Alaric Millington BSc, DipEd,
FIMA; Peter L. Moldon; Patrick Moore
OBE; Robin Mowat MA, DPhil; J.
Michael Mullin BSc; Alistair Munroe
BSc, ARCS; Professor Jacob Needleman;
Professor Donald M. Nicol MA, PhD;
Gerald Norris; Caroline E. Oakman
BA(Hons, Chinese); S. O'Connell
MA(Cantab), MInstP; Michael Overman;
Di Owen BSc; A.R.D. Pagden MA,
FRHistS; Professor E.J. Pagel PhD; Carol
Parker BA(Econ), MA(Internat. Aff.);
Derek Parker; Julia Parker DFAstrolS;
Dr Stanley Parker; Dr Colin Murray
Parkes MD, FRC(Psych), DPM;
Professor Geoffrey Parrinder MA, PhD,
DD(London), DLitt(Lancaster); Moira
Paterson; Walter C. Patterson MSc; Sir
John H. Peel KCVO, MA, DM, FRCP,
FRCS, FRCOG; D.J. Penn; Basil Peters
MA. MInstP, FBIS; D.L. Phillips FRCR,
MRCOG; B.T. Pickering PhD, DSc; John
Picton; Susan Pinkus; Dr C.S. Pitcher
MA, DM, FRCPath; Alfred Plaut
FRCPsych; A.S. Playfair MRCS, LRCP,
DObstRCOG; Dr Antony Polonsky;
Joyce Pope BA; B.L. Potter NDA,
MRAC, CertEd; Paulette Pratt; Antony
Preston; Frank J. Pycroft; Margaret
Quass; Dr John Reckless; Trevor Reese
BA, PhD, FRHistS; Derek A. Reid BSc,
PhD; Clyde Reynolds BSc; John Rivers;
Peter Roberts; Colin A. Ronan MSc,
FRAS; Professor Richard Rose
BA(Johns Hopkins), DPhil(Oxon);
Harold Rosenthal; T.G. Rosenthal
MA(Cantab); Anne Ross MA,
MA(Hons, Celtic Studies),
PhD(Archaeol and Celtic Studies, Edin);
Georgina Russell MA; Dr Charles
Rycroft BA(Cantab), MB(London),
FRCPsych; Susan Saunders MSc(Econ);
Robert Schell PhD; Anil Seal MA,
PhD(Cantab); Michael Sedgwick
MA(Oxon); Martin Seymour-Smith
BA(Oxon), MA(Oxon); Professor John
Shearman; Dr Martin Sherwood; A.C.
Simpson BSc; Nigel Sitwell; Dr Alan
Sked; Julie and Kenneth Slavin FRGS,
FRAI; Alec Xavier Snobel BSc(Econ);
Terry Snow BA, ATCL; Rodney Steel;
Charles S. Steinger MA, PhD; Geoffrey
Stern BSc(Econ); Maryanne Stevens
BA(Cantab), MA(London); John
Stevenson DPhil, MA; J. Stidworthy MA;
D. Michael Stoddart BSc, PhD; Bernard
Stonehouse DPhil, MA, BSc, MInstBiol;
Anthony Storr FRCP, FRCPsych;
Richard Storry; Professor John Taylor;
John W.R. Taylor FRHistS, MRAeS,
FSLAET; R.B. Taylor BSc(Hons,
Microbiol); J. David Thomas MA, PhD;
Harvey Tilker PhD; Don Tills PhD,
MPhil, MIBiol, FIMLS; Jon Tinker; M.
Tregear MA; R.W. Trender; David

Trump MA, PhD, FSA; M.F. Tuke PhD;
Christopher Tunney MA; Laurence
Urdang Associates (authentication and
fact check); Sally Walters BSc;
Christopher Wardle; Dr D. Washbrook;
David Watkins; George Watkins MSc;
J.W.N. Watkins; Anthony J. Watts; Dr
Geoff Watts; Melvyn Westlake; Anthony
White MA(Oxon), MAPhil(Columbia);
P.J.S. Whitmore MBE, PhD; Professor
G.R. Wilkinson; Rev H.A. Williams CR;
Christopher Wilson BA; Professor David
M. Wilson; John B. Wilson BSc, PhD,
FGS, FLS; Philip Windsor BA,
DPhil(Oxon); Professor M.J. Wise; Roy
Wolfe BSc(Econ), MSc; Dr David
Woodings MA, MRCP, MRCPath;
Bernard Yallop PhD, BSc, ARCS, FRAS
Professor John Yudkin MA, MD,
PhD(Cantab), FRIC, FIBiol, FRCP.

The General Editor wishes particularly
to thank the following for all their
support:
Nicolas Bentley
Bill Borchard
Adrianne Bowles
Yves Boisseau
Irv Braun
Theo Bremer
the late Dr Jacob Bronowski
Sir Humphrey Browne
Barry and Helen Cayne
Peter Chubb
William Clark
Sanford and Dorothy Cobb
Alex and Jane Comfort
Jack and Sharlie Davison
Manfred Denneler
Stephen Elliott
Stephen Feldman
Orsola Fenghi
Dr Leo van Grunsven
Jan van Gulden
Graham Hearn
the late Raimund von
  Hofmansthal
Dr Antonio Houaiss
the late Sir Julian Huxley
Alan Isaacs
Julie Lansdowne
Andrew Leithead
Richard Levin
Oscar Lewenstein
The Rt Hon Selwyn Lloyd
Warren Lynch
Simon macLachlan
George Manina
Stuart Marks
Bruce Marshall
Francis Mildner
Bill and Christine Mitchell
Janice Mitchell
Patrick Moore
Mari Pijnenborg
the late Donna Dorita
  de Sa Putch
Tony Ruth
Dr Jonas Salk
Stanley Schindler
Guy Schoeller
Tony Schulte
Dr E. F. Schumacher
Christopher Scott
Anthony Storr
Hannu Tarmio
Ludovico Terzi
Ion Trewin
Egil Tveteras
Russ Voisin
Nat Wartels
Hiroshi Watanabe
Adrian Webster
Jeremy Westwood
Harry Williams
the dedicated staff of MB
Encyclopaedias who created this
Library and of MB Multimedia
who made the IVR Artwork Bank.

Man and Machines/Contents

Keystone

Lord Butler, Master of Trinity College,
Cambridge, knocks on the great door of
the college during his installation
ceremony on October 7, 1965

Preface

I do not think any other group of publishers could be credited with producing so comprehensive and modern an encyclopaedia as this. It is quite original in form and content. A fine team of writers has been enlisted to provide the contents. No library or place of reference would be complete without this modern encyclopaedia, which should also be a treasure in private hands.

The production of an encyclopaedia is often an example that a particular literary, scientific and philosophic civilization is thriving and groping towards further knowledge. This was certainly so when Diderot published his famous encyclopaedia in the eighteenth century. Since science and technology were then not so far developed, his is a very different production from this. It depended to a certain extent on contributions from Rousseau and Voltaire and its publication created a school of adherents known as the encyclopaedists.

In modern times excellent encyclopaedias have been produced, but I think there is none which has the wealth of illustrations which is such a feature of these volumes. I was particularly struck by the section on astronomy, where the illustrations are vivid and unusual. This is only one example of illustrations in the work being, I would almost say, staggering in their originality.

I think it is probable that many responsible schools will have sets, since the publishers have carefully related much of the contents of the encyclopaedia to school and college courses. Parents on occasion feel that it is necessary to supplement school teaching at home, and this encyclopaedia would be invaluable in replying to the queries of adolescents which parents often find awkward to answer. The "two-page-spread" system, where text and explanatory diagrams are integrated into attractive units which relate to one another, makes this encyclopaedia different from others and all the more easy to study.

The whole encyclopaedia will literally be a revelation in the sphere of human and humane knowledge.

Butler

Master of Trinity College,
Cambridge

The Structure of the Library

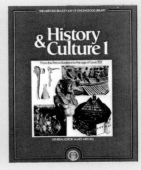

Science and The Universe

The growth of science
Mathematics
Atomic theory
Statics and dynamics
Heat, light and sound
Electricity
Chemistry
Techniques of astronomy
The Solar System
Stars and star maps
Galaxies
Man in space

The Physical Earth

Structure of the Earth
The Earth in perspective
Weather
Seas and oceans
Geology
Earth's resources
Agriculture
Cultivated plants
Flesh, fish and fowl

The Natural World

How life began
Plants
Animals
Insects
Fish
Amphibians and reptiles
Birds
Mammals
Prehistoric animals and
plants
Animals and their habitats
Conservation

Man and Society

Evolution of man
How your body works
Illness and health
Mental health
Human development
Man and his gods
Communications
Politics
Law
Work and play
Economics

History and Culture

Volume 1 From the first
civilizations to the age of
Louis XIV

The art of prehistory
Classical Greece
India, China and Japan
Barbarian invasions
The crusades
Age of exploration
The Renaissance
The English revolution

Man and Machines is a book of popular general knowledge about technology. It is a self-contained book with its own index and its own internal system of cross-references to help you to build up a rounded picture of technology.

Man and Machines is one volume in Mitchell Beazley's intended ten-volume library of individual books we have entitled *The Joy of Knowledge Library*—a library which, when complete, will form a comprehensive encyclopaedia.

For a new generation brought up with television, words alone are no longer enough—and so we intend to make the *Library* a new sort of pictorial encyclopaedia for a visually oriented age, a new "family bible" of knowledge which will find acceptance in every home.

Seven other colour volumes in the *Library* are planned to be *Science and The Universe, The Physical Earth, Man and Society, History and Culture* (two volumes), *The Natural World*, and *The Modern World*. *The Modern World* will be arranged alphabetically: the other volumes will be organized by topic and will provide a comprehensive store of general knowledge.

The last two volumes in the *Library* will provide a different service. Split up for convenience into A-K and L-Z references, these volumes will be a fact index to the whole work. They will provide factual information of all kinds on peoples, places and things through approximately 25,000 mostly short entries listed in alphabetical order. The entries in the A-Z volumes also act as a comprehensive index to the other eight volumes, thus turning the whole *Library* into a rounded *Encyclopaedia*, which is not only a comprehensive guide to general knowledge in volumes 1–7 but which now also provides access to specific information as well in *The Modern World* and the fact index volumes 9 and 10.

Access to knowledge
Whether you are a systematic reader or an unrepentant browser, my aim as General Editor has been to assemble all the facts you really ought to know into a coherent and logical plan that makes it possible to build up a comprehensive general knowledge of the subject.

Depending on your needs or motives as a reader in search of knowledge, you can find things out from *Man and Machines* in four or more ways: for example, you can

simply browse pleasurably about in its pages haphazardly (and that's my way!) or you can browse in a more organized fashion if you use our "See Also" treasure hunt system of connections referring you from spread to spread. Or you can gather specific facts by using the index. Yet again, you can set yourself the solid task of finding out literally everything in the book in logical order by reading it from cover to cover from page 1 and to page 255: in this the Contents List (page 7) is there to guide you.

Our basic purpose in organizing the volumes in *The Joy of Knowledge Library* into two elements—the three volumes of A-Z factual information and the seven volumes of general knowledge—was functional. We devised it this way to make it easier to gather the two different sorts of information—simple facts and wider general knowledge, respectively—in appropriate ways.

The functions of an encyclopaedia
An encyclopaedia (the Greek word means "teaching in a circle" or, as we might say, the provision of a *rounded* picture of knowledge) has to perform these two distinct functions for two sorts of users, each seeking information of different sorts.

First, many readers want simple factual answers to simple factual questions, such as "What is uranium?" They may be intrigued to learn that it is a dense metal, and that a piece of uranium the size of a matchbox weighs nearly a pound and can provide as much energy as 1,500 tons of coal. Such direct and simple facts are best supplied by a short entry and in the *Library* they will be found in the two A-Z *Fact Index* volumes.

But secondly, for the user looking for in-depth knowledge on a subject or on a series of subjects—such as "How does a nuclear reactor work?"—short alphabetical entries alone are inevitably bitty and disjointed. What do you look up first—"reactor"? "uranium"? "atomic energy"? "nuclear power"? "atom"?— and do you have to read all the entries or only some? You normally have to look up *lots* of entries in a purely alphabetical encyclopaedia to get a comprehensive answer to such wide-ranging questions. Yet comprehensive answers are what general knowledge is all about.

A long article or linked series of longer articles,

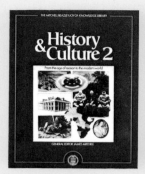

History and Culture

Volume 2 From the Age
of Reason to the
modern world

Neoclassicism
Colonizing Australasia
World War I
Ireland and independence
Twenties and the
  depression
World War II
Hollywood

Man and Machines

The growth of
  technology
Materials and techniques
Power
Machines
Transport
Weapons
Engineering
Communications
Industrial chemistry
Domestic engineering

The Modern World

Almanack
Countries of the world
Atlas
Gazetteer

Fact Index A-K

The first of two volumes
containing 25,000 mostly
short factual entries
on people, places and
things in A-Z order. The
Fact Index also acts as
an index to the eight
colour volumes. In
this volume, everything
from Aachen to Kyzyl.

Fact Index L-Z

The second of the A-Z
volumes that turn the
Library into a complete
encyclopaedia. Like the
first, it acts as an
index to the eight
colour volumes. In this
volume, everything from
Ernest Laas to Zyrardow.

organized by related subjects, is clearly much more
helpful to the person wanting such comprehensive answers.
That is why we have adopted a logical, so-called *thematic*
organisation of knowledge, with a clear system of
connections relating topics to one another, for teaching
general knowledge in *Man and Machines* and the six
other general knowledge volumes in the *Library*.

The spread system
The basic unit of all the general knowledge books is the
"spread"—a nickname for the two-page units that
comprise the working contents of all these books. The
spread is the heart of our approach to explaining things.

Every spread in *Man and Machines* tells a story—almost
always a self-contained story—a story on aluminium and
its uses, for example (pages 34 to 35) or how computers
work (pages 106 to 107) or photography (pages 218 to 219)
or how cosmetics and perfumes are made (pages 246 to
247). The spreads on these subjects all work to the same
discipline, which is to tell you all you need to know in two
facing pages of text and pictures. The discipline of having
to get in all the essential and relevant facts in this
comparatively short space actually makes for better
results—text that has to get to the point without any waffle,
pictures and diagrams that illustrate the essential points in
a clear and coherent fashion, captions that really work and
explain the point of the pictures.

The spread system is a strict discipline but once you get
used to it, I hope you'll ask yourself why you ever thought
general knowledge could be communicated in any other way.

The structure of the spread system will also, I hope,
prove reassuring when you venture out from the things you
do know about into the unknown areas you don't know,
but want to find out about. "Well, if they treat the story
of how technology began (pages 20 to 21) like that, then
they will probably treat the story of the way cars work
(pages 130 to 131) in much the same way." There are many
virtues in being systematic. You will start to feel at home
in all sorts of unlikely areas of knowledge with the spread
system to guide you. The spreads are, in a sense, the
building blocks of knowledge. Like the various circuits
and components that go to make up a computer, they are
systematically "programmed" to help you to learn more

easily and to remember better. Each spread has a main
article of 850 words summarising the subject. The article
is illustrated by an average of ten pictures and diagrams,
the captions of which both complement *and* supplement
the information in the article (so please read the captions,
incidentally, or you may miss something!). Each spread,
too, has a "key" picture or diagram in the top right-hand
corner. The purpose of the key picture is twofold: it
summarises the story of the spread visually and it is
intended to act as a memory stimulator to help you to
recall all the integrated facts and pictures on a given subject.

Finally, each spread has a box of connections headed
"See Also" and, sometimes, "Read First". These are
cross-reference suggestions to other connecting spreads.
The "Read Firsts" normally appear only on spreads with
particularly complicated subjects and indicate that you
might like to learn to swim a little in the elementary
principles of a subject before being dropped in the deep end.

The "See Alsos" are the treasure hunt feature of *The Joy
of Knowledge* system and I hope you'll find them helpful
and, indeed, fun to use. They are also essential if you want
to build up a comprehensive general knowledge. If the
spreads are individual components, the "See Alsos" are the
circuit diagram that tells you how to fit them together into
a computer that stores all general knowledge.

Level of readership
The level for which we have created *The Joy of Knowledge
Library* is intended to be a universal one. Some aspects of
knowledge are more complicated than others and so
readers will find that the level varies in different parts of
the *Library* and indeed in different parts of this volume,
Man and Machines. This is quite deliberate: *The Joy of
Knowledge Library* is a library for all the family.

Some younger people should be able to enjoy and to
absorb most of the spread in this volume on classic cars,
for example, from as young as ten or eleven onwards—
but the level has been set primarily for adults and older
children who will need some basic knowledge to make sense
of the pages on information retrieval or computers, for
example.

Whatever their level, the greatest and the bestselling
popular encyclopaedias of the past have always had one

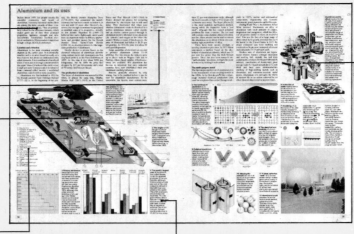

Main text Here you will find an 850-word summary of the subject.

Connections "Read Firsts" and "See Alsos" direct you to spreads that supply essential background information about the subject.

Illustrations Cutaway artwork, diagrams, brilliant paintings or photographs that convey essential detail, re-create the reality of art or highlight contemporary living.

Annotation Hard-working labels that identify elements in an illustration or act as keys to descriptions contained in the captions.

Captions Detailed information that supplements and complements the main text and describes the scene or object in the illustration.

Key The illustration and caption that sum up the theme of the spread and act as a recall system.

A typical spread Text and pictures are integrated in the presentation of comprehensive general knowledge on the subject.

---

thing in common—simplicity. The ability to make even complicated subjects clear, to distil, to extract the simple principles from behind the complicated formulae, the gift of getting to the heart of things: these are the elements that make popular encyclopaedias really useful to the people who read them. I hope we have followed that precept throughout the *Library*: if so our level will be found to be truly universal.

Philosophy of the Library
The aim of *all* the books—general knowledge and *Fact Index* volumes—in the *Library* is to make knowledge more readily available to everyone, and to make it fun. This is not new in encyclopaedias. The great classics enlightened whole generations of readers with essential information, popularly presented and positively inspired. Equally, some works in the past seem to have been extensions of an educational system that believed that unless knowledge was painfully acquired it couldn't be good for you, would be inevitably superficial, and wouldn't stick. Many of us know in our own lives the boredom and disinterest generated by such an approach at school, and most of us have seen it too in certain types of adult books. Such an approach locks up knowledge, not liberates it.

The great educators have been the men and women who have enthralled their listeners or readers by the self-evident passion they themselves have felt for their subjects. Their joy is natural and infectious. We remember what they say and cherish it for ever. The philosophy of *The Joy of Knowledge Library* is one that precisely mirrors that enthusiasm. We aim to seduce you with our pictures, absorb you with our text, entertain you with the multitude of facts we have marshalled for your pleasure—yes, *pleasure*. Why not pleasure?

There are three uses of knowledge: education (things you ought to know because they are important); pleasure (things which are intriguing or entertaining in themselves); application (things we can do with our knowledge for the world at large).

As far as education is concerned there are certain elementary facts we need to learn in our schooldays. The *Library*, with its vast store of information, is primarily designed to have an educational function—to inform, to be a constant companion and to guide everyone through school and college.

But most facts, except to the student or specialist (and these books are not only for students and specialists, they are for everyone), aren't vital to know at all. You don't *need* to know them.

But discovering them can be a source of endless pleasure and delight, nonetheless, like learning the pleasures of food or wine or love or travel. Who wouldn't give a king's ransom to know when man really became man and stopped being an ape? Who wouldn't have loved to have spent a day at the feet of Leonardo or to have met the historical Jesus or to have been there when Stephenson's *Rocket* first moved? The excitement of discovering new things is like meeting new people—it is one of the great pleasures of life.

There is always the chance, too, that some of the things you find out in these pages may inspire you with a lifelong passion to apply your knowledge in an area which really interests you. My friend Patrick Moore, the astronomer, who first suggested we publish this *Library* and wrote much of the astronomy section in our volume on *Science and The Universe*, once told me that he became an astronomer through the thrill he experienced on first reading an encyclopaedia of astronomy called *The Splendour of the Heavens*, published when he was a boy. Revelation is the reward of encyclopaedists. Our job, my job, is to remind you always that the joy of knowledge knows no boundaries and can work miracles.

In an age when we are increasingly creators (and less creatures) of our world, the people who *know*, who have a sense of proportion, a sense of balance, above all perhaps a sense of insight (the inner as well as the outer eye) in the application of their knowledge, are the most valuable people on earth. They, and they alone, will have the capacity to save this earth as a happy and a habitable planet for all its creatures. For the true joy of knowledge lies not only in its acquisition and its enjoyment, but in its wise and living application in the service of the world.

Thus the Latin tag "Scientiam non dedit natura, semina scientiae nobis dedit" on the first page of this book. It translates as "Nature has given us not knowledge itself, but the seeds thereof." It is, in the end, up to each of us to make the most of what we find in these pages.

General Editor's Introduction
The Structure of this Book

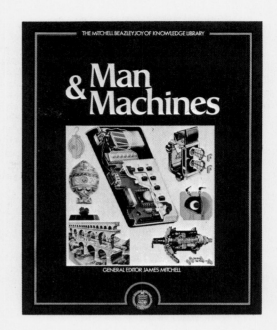

Man and Machines is a book that provides general information about technology—the practical application of science to everyday needs. It includes everything we think most interesting and relevant about machines, mechanisms, inventions and engineering in all their aspects. Our intention has been to present the basic information in words and pictures in such a way that it makes sense and tells a logical, comprehensible and coherent story rather than appears as an arbitrary and therefore meaningless jumble. The following pages explain the plan to which, with this in mind, we have created *Man and Machines*.

Before itemizing the contents of *Man and Machines* I'm going to assume that you—just like me when I began planning the book—are coming to this subject more as a "know-nothing" than as a "know-all". Incidentally, knowing nothing can be a great advantage as a reader—or as an editor, as I discovered in the early days of selecting topics for this book. If you admit to knowing nothing, but want to extend your knowledge, you ask awkward questions all the time. I spent much of my time as General Editor of this *Library* asking acknowledged experts awkward questions and refusing to be fobbed off with complicated answers I could not understand. As a result, *Man and Machines*, like every other book in this *Library*, has been through the sieve of my personal ignorance in its attempt to inform simply and understandably.

If you are totally new to the subject, I suggest that you start with Sir Jack Callard's introduction on pages 16 to 19. He discusses the advent of technology in an historical setting and explains how it is that the subject has arranged itself into various logical categories and how, while many inventions and developments were the children of necessity, a significant proportion were leaps forward made by men of vision and genius.

You will find out about man's need to build shelter from the elements, about his need to clothe and feed himself and his family, about his ability to extend the power of his own muscles and harness natural forces to "work" for him, how he made the best use of the minerals and materials he discovered, how he learned to get about more efficiently than his own two feet, and how he learned to communicate with his fellow men over distances beyond the range of ordinary human speech.

Technology today

One of the most fascinating (and to some people, most frightening) aspects of technology is the ever-increasing speed at which new ideas and techniques are put into practice. Tens of thousands of years passed while man struggled to improve something as simple as the stone axe; thousands of years passed as he strove to discover a means of transport better than the horse and cart; hundreds of years went by before the firearm was reliable enough to replace the bow. But then technology got into its stride. In seventy years man went from his first powered flight of sixty yards to a quarter-million-mile journey to the Moon. In the course of one six-year war he turned propeller-driven aircraft into jets and then into rockets. Today, aircraft go straight from the drawing board into production—the idea of building a prototype to see if the design is sound has been abandoned as unnecessary.

The capabilities of "big" technology can now be taken for granted. The question is not whether we can get a man to the Moon, but how technology is to be applied in the everyday interests of mankind. One of the most important aspects of raising a people's standard of living is the introduction of "small" technology. *Man and Machines* devotes several spreads to looking at ways in which, by using local resources and simple tools and techniques, developing countries can improve their living standards. For example, "Building with local resources" (pages 50 to 51) describes how bamboo can be used to reinforce concrete; "Small technology in the home" (pages 52 to 53) shows how waste products such as sugar cane waste and coconut fibre can be turned into excellent roofing materials without sophisticated technology.

That, then, is the outline of how this book was put together. If you do not want to begin on the first page and read through to the end, but prefer to plunge straight into the "facts", I suggest you study eight spreads in the book before anything else (see panel on page 14). These are "Read First" spreads. They will give the basic facts about the key areas of technology and provide you with a framework of essential information in its proper perspective. Once you have digested these spreads you can build up a more comprehensive general knowledge by proceeding to explore the rest of the book.

Man and Machines, like most volumes in *The Joy of Knowledge Library*, tackles its subject topically on a two-page spread basis. Though the spreads are self-contained, you may find some of them easier to understand if you read certain basic spreads first. Those spreads are illustrated here. They are "scene-setters" that will give you an understanding of the fundamentals of technology. With them as background, the rest of the spreads in *Man and Machines* can be more readily understood. The eight spreads are:

Plan of the book

There are broadly ten sections, or blocks of spreads, in *Man and Machines*. The divisions between them are not obvious in the text because we did not want to spoil the continuity of the book. They are:

The Growth of Technology

Here, in twelve pages, we review the history of the subject, from "Early technology" (pages 20 to 21) to "The Steel Age" (pages 30 to 31). Technology began in the Stone Age, when men first fashioned crude tools and weapons from flint and bone. But those materials did not satisfy his needs—and so the story progresses naturally through his discovery of bronze and iron to the coming of steel.

Materials and techniques

Modern man uses an enormous amount of steel—but he also uses a wide range of other metals and materials, some specially created. Spacecraft heat shields, for example, are made of ceramics designed to vaporize— in a controlled way—at temperatures well above the melting points of the metals they protect. This section, from "Metals and their uses" (pages 32 to 33) to "Making paper" (pages 60 to 61), discusses both the materials and the uses to which they are put.

Power

Machines without power are merely pieces of furniture. In twelve spreads we discuss basic engines and the energies—coal, oil, petrol, natural gas, wind, water, geothermal, solar and nuclear—that power them. We do not overlook conservation—proper insulation, for example, can save thirty per cent of the heat loss from an ordinary home, and we devote a spread to this topic.

Machines

Most machines involve mechanical linkages, gears, pulleys, wheels and so on—a pair of scissors and an aircraft loading jack both work on the same principles. This section of *Man and Machines* discusses the principles and then looks at machines for weighing, measuring, lifting and moving heavy loads. There is a sports stadium in Hawaii that can be changed from a football pitch to a baseball diamond by moving the stands—this section tells you how it is done.

Transport

Getting from point A to point B in a shorter and shorter space of time, and in greater and greater comfort, has been one of man's principal concerns. This section describes the evolution of water transport from rafts to hovercraft, of land transport from sleds to monorails, of air transport from balloons to Concorde.

Weapons

It is perhaps deplorable, but nevertheless true, that a considerable number of technological advances have come through war. It is certain that these days defence is as much a matter of technology as of manpower. *Man and Machines* looks at early weapons, traces the development of firearms, artillery, armoured fighting vehicles, fighting ships and military aircraft, and ends the section by looking at what may be the warfare of the future—nuclear, chemical and biological.

Engineering

Man's engineering works rank among his major achievements: a tunnel twelve miles long and 7,000 feet below the summit of a mountain; bridges with single spans of more than 4,000 feet; canal systems more than 1,000 miles long; a dam that holds more than 200,000 million cubic yards of water. From "Road building" (pages 182 to 183) to "Sewage treatment" (pages 202 to 203) the section assembles the wonders of the world.

Communications

Today's means of communication include the printed word, photography, and tele-communications that range from the telegraph through radio and television to video recording. In seventeen spreads, from "History of printing" (pages 204 to 205) to "Radar and sonar" (pages 236 to 237) the section looks at how men talk to one another.

Industrial chemistry

This section discusses the applications of chemistry to industry. Chemical engineering provides us with soaps and detergents, fireworks and explosives, cosmetics and perfumes. Detergents are new; but cosmetics are as old as the first woman who decided she would try to improve on nature.

Domestic engineering

In this section you will find the mechanisms you use every day but see only from the outside—zip fasteners, ball pens, cigarette lighters, refrigerators, vacuum cleaners, washing

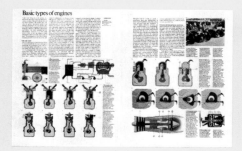

Basic types of engines

Levers and wedges

Carts, coaches and carriages

Road building

History of printing

Chemical engineering

machines, gas meters and lawnmowers. They seem to be simple, but they include some of man's most ingenious conceptions and call for the application of some of his most skilled techniques. If you ever wondered how a sewing machine makes a stitch, this is the place to find out.

Man and Machines may not turn you into a "know-all". Indeed, considering the derogatory overtones that word has, I hope it doesn't! But if we have done our job properly (and I hope we have) it ought to open your eyes to some aspects of our lives that we tend to take for granted but are in fact extraordinary. There ought really to be no argument between the "technologists" and the "anti-technologists", between the "small technologists" and the "big technologists". Technology, as a wit once said about sex, is here to stay. Our concern ought not to be whether it is used, or on what scale, but that we use it for the right ends.

Man and Machines

Sir Jack Callard,

former Chairman of Imperial Chemical Industries Limited

Science is concerned with the exploration of nature, the understanding of natural phenomena and the constant extension of human knowledge of all that occurs around us. But it is through the application of knowledge gained from these explorations that full benefit in practical form can be brought to bear upon everyday life.

Pure science must eventually be applied and, as knowledge increases, it can be used beneficially only if it is applied in intelligent and productive ways. Technology is the systematic application of various branches of knowledge to practical tasks. One type of technology – such as that involved in the design and construction of a supersonic aircraft – may have developed from the highly organized application of information discovered by scientists of many disciplines and indeed from non-scientific knowledge as well: all technologists find that their work is a constant challenge to their organizational ability and skill. It can be that the practical achievement of any technological objective falls short through inadequate basic understanding – by a limitation of knowledge in one branch of science – and so stimulates the pure scientist, by further exploration and experimentation, to push out again the boundary of knowledge to encompass the required field and so bring into range the practical objective in an economic way. Early steam engines were crude because the principle behind them was not fully understood and because materials and the skills of metallurgists were limited. Engineering ability had to improve within only a few years from the comparative crudity of the blacksmith to the precision and high tolerances of the instrument maker.

Scientific discovery constantly makes possible further advances in application leading to constant change and increased sophistication of technology. Conversely, the problems encountered by the technologist in applying available scientific information continually stimulate, challenge and guide the scientist. For, by discovering limitations in application and by displaying where greater knowledge is needed for advancement in technology, the technologist assists the scientist in the organization of further exploration. For example, the principle of the transistor had been known for several years before sufficiently pure semiconductors were developed to make the idea a reality.

In many applications of science, the pure and the applied move closely together, each aiding and developing the skills and endeavours of the other. Similarly, when a period of rapid advance in one science occurs (as when, for example, after a period of relative stagnation, the whole new branch of knowledge called atomic physics broke upon the world and men

came to understand the possibilities of fission and fusion and the effects of achieving them in a controlled way) great stimulation is given to other pure scientists to make progress in their understanding.

Scientific frontiers are constantly being extended and technologists are constantly under challenge to put new knowledge – such as the advances that have accompanied the conquest of space – to practical use. A world with a rapidly increasing population requires a rapidly increasing supply of food. This fundamental need not only stimulates the farmer to increase production and to improve his efficiency of production, but also directs the work of the botanist to select and improve strains of grain or of grass to give higher yields. Similarly, the agronomist is stimulated to develop methods of agriculture suited to a particular geographical area, the biologist to develop animal species more suited to food producing, the chemist to discover products that eliminate the ravages of insects or bacteria upon crops, or to synthesize proteins from minerals, and the engineer to develop machines to simplify agricultural work or to improve irrigation. It is perhaps man's need to survive – sometimes, paradoxically, at the expense of his fellows – and his wish to master his environment that provide the greatest stimuli to technological advance and invention.

Modern technologists can apply their expertise to solving some of the problems of developing countries – the so-called Third World. Local materials and labour can be used to fulfil many of the urgent needs, for example to provide better roads, homes, cooking facilities, water supply and sewage disposal, so improving communications, living standards and the health of the community.

Growth of technology

It was from the stimulus of the need to survive that the earliest technology derived. The few and scattered men of the Old Stone Age developed little that could be classified as a conquering of their environment, but they needed to fill their stomachs and so must have made some study of the changes of climate, season or soil, and their effect upon the ability to gather food. Susceptible to cold, they frequented caves, improvised shelters and made use of fire. Even the lighting of fire by creating friction between two pieces of wood may perhaps have derived from the observation of dry boughs being rubbed together by the wind. The creation and reduction of friction was certainly an early motivator, particularly in the need for transport. The realization that it was easier to drag a load than to carry it; the experience that on soft ground the load was easier to drag if it was first placed on twigs or boughs; that some things would slip more easily across a surface than others; and that if the bearers rolled across the surface the force required to move the load was further reduced were all observations that started the development of transport systems. From sliding – and Stone Age carvings on cave walls depict man using skis – to rolling and so to the use of the wheel, perhaps the greatest invention of the third millennium, to the cart and the use of an animal to pull it, so the sequence ran to horsepower.

To create fire, to restrict it, to quench it with water and to note that the application of its heat could change the physical form of natural things – could melt them – were simple observations necessary to the development of other than stone implements for tools and weapons and enabled man to increase his control over his natural environment. The development of his knowledge of metalworking was highly significant to the furtherance of his conquest of his environment – the dawn, perhaps, of the material side of civilization as we know it.

Man's need for food which led to husbandry of nature, both animal and vegetable, was a third spur to the application of simple observations: that selected seeds planted on selected sites, or tillage, could increase food supply and simplify its collection; and the herding of animals increased their usefulness to man. Stock-keeping, nomadic in nature, implied new, if temporary, settlements as new sites were chosen and itself made men conscious of new wants. Once the need for food was satisfied, time became more freely available for the trading of surpluses and the development of new crafts and skills.

These examples of the early applications of man's observation of natural phenomena to the solution of his practical problems and for the betterment of his existence show that technology, simple in the extreme though such examples are, had its beginnings at least 20,000 years ago. All the sophisticated materials, machines, communications and manufactures enjoyed today have resulted from thousands of years of experiment and invention. Modern materials became available because of the need for better performance starting from implements or weapons of stone and flint and bone, through copper to bronze to iron, to the development of smelting, to alloying, casting, and forging and steelmaking.

And so the streams developed –

– From the use of random stone, through mud bricks to fired bricks, to fused materials, to glass, from simple clay pot-

Entertainment without effort – technology brings a celluloid actor to a wheeled audience.

tery (was the potter's wheel the earliest machine?) to china, porcelain and modern ceramic materials.

– From hides and hair to wool, to the spinning machine and threads, to woven fabrics, to the shuttle and the loom and a nigh infinite range of textiles.

– From wind power harnessed by sail, to water power, from heat to steam, the steam engine and the turbine, from the application of magnetism to the generation of electricity and the electric motor; the steady succession of electrical devices, to electronics, the computer, control mechanisms and automation.

– From quarrying to mining and mineral extraction from the surface, to oil-well drilling to depths of 1,000 metres, the processing of oil and methods of using it, to the internal combustion engine.

– From the dugout canoe and the raft, to the use of sail, to the fabricated wooden ship, to the steamship, to the 350,000-tonne tanker; from the balloon to the aeroplane, to the supersonic aircraft or remote-controlled flight.

– From the drum and the smoke signal to the signal lamp to the telegraph and telephone. From radio and radar to television.

The catalogue of development of applied science is endless whatever particular field we consider. It may also be timeless. Progress in application is a result of many branches of science and the arts which interrelate and which provide mutual stimuli and mutual development. Progress in any one so often requires progress in another and the direction of development in any period of time may be strongly influenced by the emphasis created by the priority of the need. The shortage of a natural material places emphasis on the need for a substitute and man's increasing ability to synthesize has opened up vast new areas for technological progress. Aided by the steady advancement of instruments of measurement and analysis, the chemist's knowledge of the constitution and structure of molecules is now highly advanced and enables molecules to be synthesized to provide designed properties. In this way, there are now materials that are at least adequate substitutes and at best have a better combination of properties than those they displaced. So industries are born and the demand for natural materials can be moderated in relation to their abundance, location and cost.

The economics of attaining an objective is also a force constantly creating change and stimulating development; and social forces increasingly have their influence. The world's consumption of timber far exceeds, today, the rate at which timber is being produced and the factors of cost and availability encourage the development of more versatile substitutes such as plastics that can be produced at lower cost and greater convenience.

The development of the plastics industry – using oil as raw material, which advances in other technologies had permitted men to extract in even larger quantities – lessened the need for timber and greatly increased the consumption of oil, which in turn displaced coal. The discovery by chemists of methods by which molecules of simple hydrocarbons could be polymerized – that is, that many molecules could be made to join together to create new materials with better physical properties than others – was a major discovery of this century. The key to the production of polyethylene (polythene) was the highly speculative experiment conducted by Imperial Chemical Industries between 1935 and 1939 in which ethylene was subjected to pressure up to ten times greater than had previously been applied. A waxy solid was the result. To do the experiment at all and, later, to develop a production process, required special engineering techniques in design and construction of large-scale chemical manufacturing equipment capable of generating and withstanding pressures of up to 2,000 atmospheres.

Speculative experimentation by organic chemists and chemical engineers had discovered a new plastic material with

Dwarfed by a stamp – tiny integrated circuits from a computer's memory.

an interesting and exploitable range of properties of which one stood out – its electrical insulation or dielectric properties. The need to make the new product in commercial quantities provided stimulus and challenge to men of other scientific disciplines – the mechanical engineer, the metallurgist, the instrument engineer – at the same time as chemists were refining the process using catalysts.

The dielectric properties of polyethylene provided an answer to the problems of those seeking to develop radar, particularly airborne radar in which equipment had to be robust, compact and light as well as being capable of working at very high voltages and very high frequencies. No other known material could meet these conditions and so the discovery of polyethylene permitted further development in other branches of technology, particularly radar and telegraphy but also, in conjunction with other thermoplastics, a wide range of processes and machines for the production of sheets, films and mouldings. This is but one example of the effect of a single discovery through observation upon a range of other scientific disciplines.

Significance for the future

It is undeniable that the physical resources of the world are limited and that the supply of the necessary materials for industrial use is diminishing. It is also fair to predict that the demand for the products of industry, so vital to living standards, will continue to increase. Although the more sophisticated products surround all who live in industrialized countries, there are many hundreds of millions of people in underdeveloped countries who have as yet had little opportunity to enjoy their advantages or even to sample them. In addition, the population of the world is increasing rapidly and so the potential demand for goods is likely to grow still further. It is probable that man's continuing and increasing need for food, warmth, shelter, clothing and health will remain the chief basic stimuli and that the efforts of the pure and applied scientists will continue mainly to be directed to the satisfaction of such needs. Increased demand for goods and services and the requirement of warmth lead to increased demand for energy as well as for basic materials.

As the stocks of known mineral resources become exhausted the search for new sources intensifies and men will be forced to explore less accessible areas whether they be at greater depths below the surface of the land or beneath the seas. It is some comfort to remember that there are still vast areas of land surface that have not yet been thoroughly explored in geological terms and that two-thirds of the earth's surface is water – areas that have scarcely been explored at all! But to discover and win further sources of iron, coal, oil, copper, tin and so forth will require extensions of present technologies to the extraction of minerals from under the sea, which may require the development of under-sea residences for both men and mechanisms. Meanwhile other technologists continue to devise man-made materials as substitutes for natural ones in short supply.

Additional food needed to feed an increasing population will accentuate the advance of techniques of agronomy. It is likely that, apart from the cropping for food of additional land (enabled by improved techniques for irrigation and cultivation, increasing the frequency and yield of crops), emphasis will be given to increased chemical, botanical, and biological applications to increase the rapidity of growth and reduce the deprivations of plant and animal diseases. Further means may be developed to harvest and to increase the food-generating capacity of the seas, and supplies from chemical synthesis of proteins will increase.

The essential requirements of energy will encourage techniques for obtaining it by means other than the combustion of fossil fuels. The possible supply from nuclear sources seems infinite and the limitation on its use is one of available technology not excluding the techniques necessary to protect all forms of life from radiation. Energy from the sun whether by direct absorption of heat or indirectly through the harnessing of the movement of sea waves is clearly practicable if economic problems can be solved.

The desire for health, general wellbeing, and the relief of suffering that is within all men will motivate continuing effort in the medical and pharmaceutical fields. Chemotherapy is still in its infancy and medical science – both curative and preventive – is advancing fast. Improved techniques of surgery, and greater use of transplants, may well contribute to greater longevity which in turn will tend to increase demand for everything else.

Some of the advances of modern technology have left in their wake disadvantages, chief of which are disruption of community life, noise and pollution. Factory machines, road vehicles and jet aircraft generate noise, sometimes enough to be a hazard to health. Designers strive to reduce noise from such sources without impairing the efficiency of the machines. Chemical and nuclear waste products – whether from factories or lorries and cars – are another potential health hazard and a threat to the survival of some forms of wildlife. Again technologists can devise processes to reduce or even eliminate such pollution. For example a lead compound (a potential pollutant) is added to petrol to improve its octane rating in car engines. The lead can be eliminated and an equally good but more expensive fuel produced, although it will be used only if motorists are prepared to pay more for their petrol.

Whereas it is from the increasing material needs of an increasing population that the main spur to technological advance will continue to come, it is of man's nature to be curious, to struggle to understand and to improve his knowledge. Already great voyages of exploration into space have been made and samples of the surface of the Moon have been brought back to Earth. Photographic evidence of the nature of the surface and of the atmosphere of more distant terrestrial bodies has been obtained as a result of the development of transportation, control and photographic techniques that were undreamt of only a generation ago. The appetite to know more will stay strong and the skills available constantly increase in extent and complexity.

Maybe one day man will learn to organize them in such a way that he will discover whether the Earth on which he lives is the only terrestrial body among so many millions in the universe that is capable of sustaining life in the form he recognizes. There have already been manned landings on the Moon and unmanned spacecraft have conducted scientific experiments on the inhospitable surface of Mars. Spaceprobes have been sent to "fly" past the outer planets, to the limits of the Solar System. Man and his machines will surely one day travel beyond the Milky Way and to other galaxies.

Power house – the home of turbine generators at a hydroelectric station.

Early technology

Perhaps the first event in the history of technology was the making of a crude stone hand axe more than a million years ago. Another achievement of prehistoric technology was the control of fire for hardening the points of wooden tools and weapons. By the end of the Old Stone Age, some tens of thousands of years ago, man-made objects of stone, horn, bone and wood reached high standards.

Advances in Neolithic technology

It was not until the New Stone Age, or Neolithic, however, when man first settled down in one place, that there arose a need for a technology extending beyond the manufacture of tools, weapons and garments.

From about 5000 BC, technological development began to increase rapidly. In the thousand years that followed, an urban civilization arose in Mesopotamia. Ploughs, first with blades of stone and later of bronze, sledges for transport and buildings of baked mud were among the major developments.

By 3500 BC a number of city states were established between the Tigris and Euphrates rivers. The potter's wheel and the solid cartwheel were introduced not long after. The Sumerians also made inventories of their property in the first written records – cuneiform carvings on clay.

Mesopotamian architecture reached its peak in the ziggurats, great stepped buildings made from clay bricks laid in a herringbone pattern. The Egyptians however had earlier built monuments that were even larger, but in stone. The Great Pyramid of Cheops still stands, 148m (485ft) high and covering five hectares (13 acres) yet accurate in design and construction to a tiny fraction of a centimetre. One of the greatest feats of engineering, it was built without the use of the wheel.

Far from the Middle East, other civilizations began to emerge with their own special technological achievements. In ancient China, silk was being made as early as 2000 BC, although it did not reach the West until the sixth century BC. The Indus civilization, c. 2300 to 1750 BC, built the northern Indian cities of Harappa and Mohenjodaro, which had vast complexes of houses, granaries and wells, complete with the earliest corporate drainage and sewage system.

Europe, Africa (excluding Egypt) and great tracts of Asia, however, experienced only patchy technological developments at that time. South America generated an elaborate civilization of its own, but not for another 2,000 years. The Mayan people (in Mexico) remained in a Stone Age, although they developed the most accurate of calendars and perfected an arithmetic that contained a symbol for zero for the first time.

Other parts of the world failed to develop an advanced technology: until the arrival of the white settlers, Australian Aborigines continued as food gatherers, relying on nature to supply all the necessities of life.

The mastery of materials

Ancient technology, therefore, developed in fits and starts in limited areas of the world. Surprisingly the Hellenic Greeks, in many ways the most talented of all ancient peoples, did not make great advances in technology. Although they enjoyed philosophical, including scientific, speculation they held inventions and manufacturing in low esteem. One thousand years before the Greeks,

1 A
Flint sickle Flint set in a wooden handle 19th-century cradle scythe 19th-century scythe

Bronze sickle Roman sickle Medieval sickle Medieval scythe Hainault scythe

B

1 Farming tools [A] for cutting and reaping form an unbroken sequence down the ages. The Egyptians used flint-edged wooden sickles at first and later bronze ones. By the Middle Ages the sickle had developed a curve and was set in a wooden handle. Medieval harvesters [B] used either short-handled sickles or scythes. Later developments include the Hainault scythe still in use today.

3 A

B

2 Prehistoric Celts were the first to fit horses with shoes, but heavy iron horseshoes were not common in Europe in medieval times. Horseshoes of various designs are associated with certain cultures and they achieved their final form only in the late 19th century.

1 Frankish
2 Syrian
3 Saxon
4 Roman hipposandal
5 Celtic
6 Moorish from Algeria 12th-century
7 Roman
8 16th-century French

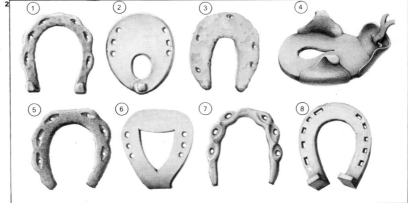

2

3 Greek metalsmiths [A] fashioned delicate ornaments from silver as early as 700 BC. This silver, later part of the vast wealth of democratic Athens, was mined using entirely slave labour. In Roman times, mining extended throughout the empire and Roman smiths [B] were busy forming themselves into powerful guilds, the fabri. Much of the smith's equipment, except for the open forge, would be familiar to a European smith of today.

Hittite smiths in Mesopotamia had worked in iron, a technology that they handed on to many later civilizations [Key].

Egyptian tombs were rich in objects made from pottery, wood, glass, ivory, copper and bronze and, later, iron. During Roman times mining activities extended throughout Europe and beyond and bronze and iron were used in large quantities for armour and engines of war. After the rise of Islam in AD 622, the Arabs became the custodians of scientific ideas but, like the Greeks, they made few technological innovations.

Apart from clocks and stained-glass windows of cathedrals, Europe had contributed little new technology for hundreds of years. But the fourteenth century saw the start of the Renaissance, which released scientific thought from religious shackles and in doing so promoted technology.

Renaissance technology

By 1450, Johannes Gutenberg's printing press was active in Germany, European canals were busy with trade and early blast furnaces were making iron. In the sixteenth century mining was improved by mechanized drainage and ventilation systems [6]. This was also the age of Galileo's telescope and many other scientific instruments [7, 8].

The real breakthrough – the large-scale generation and control of energy – had yet to arrive. When it did, in the eighteenth century, technology advanced on many fronts. Windmills had been in use for a thousand years and waterwheels for half that time, but neither of these inventions advanced industry generally. Then three developments stimulated the Industrial Revolution; these were steam power, coal mining and iron smelting. The application of steam power required the engines of Thomas Newcomen (1663–1729) and the availability of coal to heat their boilers. Coal was also employed in the early 1700s by Abraham Darby (c. 1678–1717) to make coke, and this new fuel was used by him in blast furnaces to produce a plentiful supply of iron. Later the steam engine, invented by James Watt (1736–1819), introduced powered transport, and sulphuric acid manufacture by the lead chamber process established the beginnings of the chemical industry.

The control of fire is fundamental to technology. As early as 3000 BC, copper and tin ores were smelted in charcoal fires, refined and alloyed into bronze. Two thousand years later, iron objects were being made in Egypt and elsewhere as the result of a technique handed down by Hittite smiths. Iron could not be melted over a fire, even with the aid of bellows such as these, but it could be made red hot and then hammered into ornaments and implements.

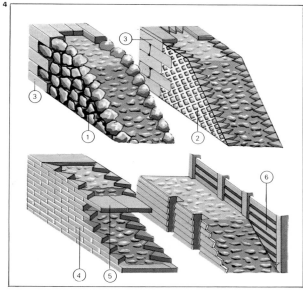

4 Roman wall-building included small stones laid in mortar, opus incertum [1]. Opus reticulatum [2] was a later diagonal pattern. These walls had outer vertical angles finished with stone quoins [3]. Opus testaceum had a brick facing [4] often strengthened by bonded tiles [5]. Walls of foundations were cast in a timber framework [6], removed when the mortar had hardened.

5 Medieval stone masons, most of whom were employed in the building of churches and cathedrals, were among the elite of workmen and formed themselves into guilds, as did the Roman smiths before them.

1 Carving decorated stonework with hammer and punch
2 Measuring out a design on stone with compass and rule
3 Cutting stone with a frame saw
4 Shaping a moulding with pitching tool

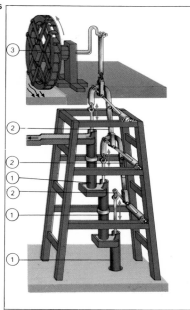

6 Mines became deeper and more efficient when it was possible to pump out floodwater. In this 16th-century mine, a series of lift pumps [1], similar to a village pump, were linked by levers [2] and powered by an undershot waterwheel [3] driven by a nearby natural stream or one diverted for the purpose.

7 An accurate thermometer invented in the 1760s by J. A. Deluc had a brass plate [1] that could be moved along the scale [2] by a screw [3], which acted like a micrometer to give precise readings. As the liquid inside became warmer, it expanded into the overflow vessel [4]. The whole cumbersome device was supported in a carrier [5].

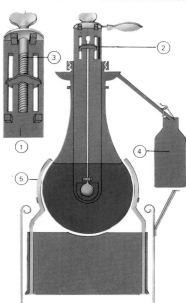

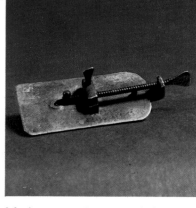

8 A microscope made by the Dutch scientist Anton van Leeuwenhoek (1632–1723) is an example of 17th-century precision engineering. It had a single lens mounted between two plates and ground so accurately that it gave a clearer magnification than other compound microscopes made at that time. The specimen was mounted on the point of the longer of the two focusing screws.

Modern technology

The great events that directed science and technology on to the paths they pursue today were the profound social changes that occurred in the eighteenth century. One of these events was the French Revolution; another the Industrial Revolution.

The age of revolution

The French Revolution stressed the need for the rational conduct of human affairs. One effect was to give a strong impetus to calculation and mathematics. This French development inspired the English mathematician Charles Babbage (1791–1871) to build the first "computer-type" calculating machine.

The Industrial Revolution [4] posed new problems. Its central features were the extensive use of coal and iron and the development of the steam engine by Thomas Newcomen (1633–1729) and James Watt (1736–1819) [2]. Industry was thus presented with the prospect of unlimited power, and the problem of the nature of power and energy was brought to the forefront of science. The Industrial Revolution created a new demand for a knowledge of the properties of materials

and a fresh motive for the development of chemistry. It made mankind aware of the possibility of fundamental change.

The Industrial Revolution stimulated a search into the principles of motion associated with other phenomena such as electricity. Michael Faraday (1791–1867) conceived the idea of the electromagnetic field to explain electrical motions. He also demonstrated the principles of the dynamo and the electric motor. With the coming of power stations, electricity could be generated and distributed almost anywhere to factories and even individual homes. The new industries required fresh sources of new materials and so were responsible for an expansion in overseas trade. Other countries became markets for manufactured goods.

A few pioneering scientists concentrated on working out the laws of thermodynamics – the branch of physics concerned with processes involving heat changes. Sadi Carnot (1796–1832) gave a mathematical theory to the cycles of heat changes in the steam engine, as determined experimentally by James Watt. James Joule (1818–89) mea-

sured the mechanical equivalent of heat (how much heat is generated by a given amount of mechanical energy, and vice versa) and how various other forms of energy are interrelated. From this the concept of the conservation of energy was developed – the basic idea of physics that energy can be neither created nor destroyed, merely converted from one form into another.

James Clerk-Maxwell (1831–79) gave expression to Faraday's ideas of lines of force acting in space in his equations of the electromagnetic field. He showed that light waves are a form of electromagnetic waves and deduced that other forms, different from light waves, might exist. One such form, now used in radio, was discovered by Heinrich Hertz (1857–94) in about 1887.

The new chemistry, which Antoine Lavoisier (1743–94) had taken the chief part in founding, was developed especially in Germany by Justus von Liebig (1803–73), who had received his early chemical training in Paris. He was followed by August Hofmann (1818–92), who worked for a long period in England. Hofmann had a pupil in his London

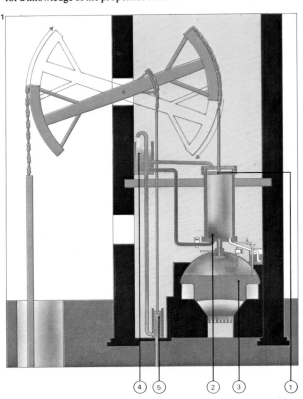

1 Thomas Newcomen's steam engine of 1712 was a great advance on Savery's earlier machine, although it was also only a pumping engine. The piston [1] was forced up by steam pressure and the weight of the pump mechanism. A cold-water spray [2] condensed the steam, creating a vacuum which "sucked" the piston down. A boiler [3] provided steam and a tank [4], fed by a secondary pump [5], the cold water.

2 James Watt's engine of the 1770s also condensed the steam with water but in a separate condenser [1]. This technique, and that of admitting steam to both sides of the piston [2], greatly increased the efficiency of the steam engine. Pushing the piston in both directions made it double-acting. Watt soon adapted it to produce rotary motion, used for machines other than pumps.

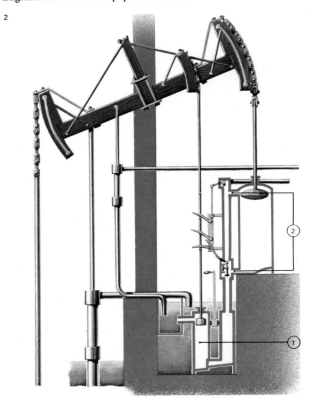

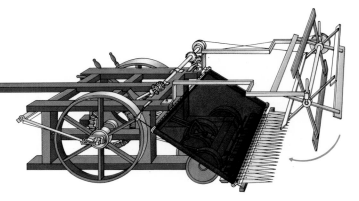

3 Agriculture was also revolutionized by new machines. A reaper invented by the Scotsman Patrick Bell in 1828 was pushed by two horses and could cut an acre of corn in an hour. A large bevel gear on the main axle drove a high-speed shaft that worked crank rods and an oscillating cutter bar. A belt on the same shaft turned the blades of a collector to push the corn on to the blades. A side-ways moving belt moved the cut corn to the side of the reaping machine.

laboratory, William Perkin (1838–1907), who at the age of 18 discovered the first synthetic chemical dye, mauve, which led to the foundation of aniline dye manufacture, at first based on coal tar. This was the start of the modern chemical industry.

Modern science

Following Hertz's discovery of long electromagnetic waves, Wilhelm Roentgen (1845–1923) discovered short electromagnetic waves in 1895 and called them X-rays. They came as a surprise to physicists. Searches were immediately made for other new rays, and in 1896 Antoine Becquerel (1852–1908) discovered that uranium emitted a new kind of radiation. Pursuing this discovery Marie Curie (1867–1934) and her husband Pierre (1859–1906) showed in 1897 that far more powerful sources of this kind of radiation existed in the new elements that they named polonium and radium. While these researches were progressing, other scientists were investigating how electricity is conducted in gases. This culminated in the discovery of the electron by Joseph (J. J.)

Thomson (1856–1940) in 1897, which led to the development of the cathode ray tube, valves and the whole science of electronics.

Developments in atomic theory provided scientists with new instruments of investigation. Ernest Rutherford (1871–1937) utilized radioactivity to discover the structure of the atom, and how atoms could be transmuted, opening the way to the release – for good or evil – of atomic energy.

Molecules of life

Lawrence Bragg (1890–1971), working with his father William (1862–1942) explained how the structure of molecules can be determined by means of X-rays. Working in Bragg's laboratory, James Watson (1928–   ) and Francis Crick (1916–   ) discovered in 1953 the double-helix structure of the molecule of DNA (deoxyribonucleic acid), upon which one of the fundamental processes of life, the transmission of hereditary qualities, depends. Such knowledge has made "genetic engineering" possible and led, for example, to new strains of crop plants to help feed the world's ever growing population.

The first cast-iron bridge, built over the River Severn at Coalbrookdale, England, in 1779, represented the beginnings of iron as a constructional material. The key to cheap iron was the development of blast furnaces and a smelting process using coke instead of coal. This was introduced from 1709, also at Coalbrookdale, by the Englishman Abraham Darby (1678–1717).

4 The Industrial Revolution began in Georgian and Victorian Britain. Based on coal and iron, and powered by steam, the world's first industrial society could by the mid-nineteenth century claim to be "the workshop of the world" and stage a Great Exhibition of Trade and Industry at the Crystal Palace in London's Hyde Park in 1851. One of the products of this industrialization, and the one that probably most affected the way of life of the people, was the factory town. By 1851 in Britain, more people lived in towns and cities than in country areas. Mills often dominated the urban landscape. Their steam-powered machinery demanded that men, women and even children work for 72 hours a week, often in dangerous and unhealthy conditions. Textile mills produced cotton and woollen cloth and yarn to be exported throughout the world. Shown here [left] are carding machines and spinning jennies. Iron foundries made everything from pots and pans to great iron beams and girders and, of course, the machines themselves. But neither machines nor products would have been any use without an efficient transport system. Improvement of natural waterways and then the digging of hundreds of kilometres of canals [right] provided a transport network linking all the major industrial centres. The itinerant navvies (short for navigators) dug the canals, at first largely by hand using picks, shovels and wheelbarrows. With the coming of the railways in the nineteenth century the navvies took on the task of making cuttings and building huge earth embankments. The railways completed the transport system.

The Stone Age

According to the most recent discoveries man's first ancestors, the Australopithecines, lived about three million years ago. These creatures, only 1.2m (about 4ft) tall, were distinguished from apes mainly by their anatomical features and because they walked upright. If they employed tools then these were probably simple pieces of stick as used today by chimpanzees. Many Australopithecine remains have been found in southern and eastern Africa [3].

We do not know how or when man discovered the use of fire. We are sure, however, that by the time of the lower Palaeolithic (early Old Stone Age) about 500,000 years ago man not only made fire safely but also hardened wood in it to make tools and weapons. He also made stone hand axes – crude but hard-wearing and effective tools. The earliest stone tools date from about 1.8 million years ago [Key].

The Stone Age in Europe
In Europe a series of ice ages continued until only 10,000 years ago when the ice sheets finally began to retreat northwards. Human fossils from the first half of the Stone Age are rare, but it seems likely that the type of man called *Homo erectus* gradually gave way to, or evolved into, a number of later types of which only one, *Homo sapiens* or modern man, has survived. Human artefacts of this long period of the middle Palaeolithic include improved types of stone flake tools. These are associated with the Levalloisian culture of the middle Palaeolithic, located around Levallois-Perret in France.

About 30,000 years ago the last type of man recognizably different from *Homo sapiens* (although he is sometimes classified within this species) became extinct. This was Neanderthal man who had a stooping posture, heavy brow ridges and a sloping forehead, but whose brain was as large as ours. He is associated with the Mousterian culture of the middle Palaeolithic, located in the region of the River Vézère in southwest France. He made fire and elegant hand axes and buried his dead with funeral rites.

The rest of the Palaeolithic is the story of prehistoric cultures of *Homo sapiens*: the Aurignacian of the upper Palaeolithic, located at Haute Garonne, France; Gravettian of the upper Palaeolithic, located on the Dordogne, France; Solutrean of the upper Palaeolithic, located at Saône-et-Loire, France; and Magdalenian cave dwellers of the upper Palaeolithic, located in the region of the River Vézère, France.

Sophisticated artefacts
Weapons of these cultures were often finely made [Key C, D]. The Solutreans were particularly sophisticated toolmakers. But men were also becoming creative in a totally new way: they began to make carvings representing humans and cave paintings of animals.

The carvings were dumpy "Venuses" that possibly had a magical significance [1]. The cave paintings of Lascaux in France and Altamira in Spain [4], both of the Magdalenian culture, probably served magical purposes connected with hunting.

Men of these cultures also made simple stone vessels including lamps [2A]. They used flint, bones and antlers as materials for tools and they probably worked in wood and

1 This "Venus", carved from a mammoth bone, is from the Gravettian culture of the Old Stone Age. It was one of 43 figures found at Kostienki in southern Russia. It is a fertility symbol and refers to the fertility of both woman and the earth.

2 From the late Old Stone Age to the time of his settling down the Neolithic or New Stone Age man worked in a number of materials and made many kinds of objects. The cave painters made small household objects such as the stone lamps [A] which date from 15,000 BC. The painted pebbles [B] of the Mesolithic Azilian culture date from 10,000 BC and may be toys or magical objects. Also Mesolithic, but from the Maglemosian culture of 8000–5000 BC, are decorative animals carved in amber [C]. Neolithic tools were often elaborately manufactured as in this flint-bladed bone-handled sickle of about 5000 BC [D]. Antlers were used as picks and ox shoulder blades as shovels [F]. Stone tools were used until later times in Britain. A good example is the stone-headed axe [E].

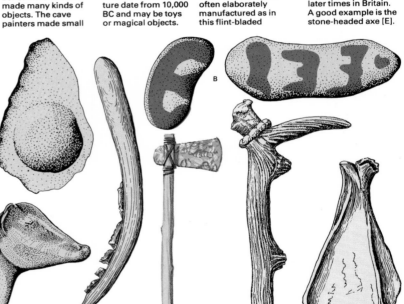

3 Man probably first evolved in Africa. Fossils resembling men rather than apes have been dug up in the Olduvai Gorge of Tanzania by Louis Leakey in 1959. Some date from 1.8 million years ago.

4 Man was an artist 20,000 years ago – but his paintings also carried a magical significance. This example is from the cave paintings at Altamira, Santander in Spain, which showed various animals, and were the first to be recognized as Palaeolithic.

leather as well, although these materials have not survived. In a later culture the Azilian people (located at Ariège, southwest France) of the Mesolithic – the Middle Stone Age – made objects such as painted pebbles [2B].

The Mesolithic, lasting from 10,000 to 7000 BC in the Near East and later in Europe, was a time of great changes. The end of the Ice Age freed men to wander: indeed, it forced them to, because the game animals spread out from local areas and became more difficult to hunt. In the milder European climate forests sprang up. The Maglemosian culture (located at Maglemose, Denmark) associated with these forests produced some carvings that were purely decorative [2C].

Man – settled and nomadic
From 7000 to 5000 BC, in the equable climates of Turkey and Mesopotamia, tribes that had been nomadic began to settle down in the first villages and raise animals and grow crops for food. This was the beginning of the Neolithic, or New Stone Age. In the fertile plain between the Tigris and Euphrates rivers, this village life eventually gave rise

to the first great city civilization, Sumeria.

In the Near East the Stone Age lasted only until 3500 BC after which, as men learned how to smelt metals, it gave way to the Bronze Age. In more backward northern Europe the Neolithic persisted until about 2000 BC. Intermediates were such village cultures as that at Habasesti in Romania [5].

The Neolithic was also a time of great migrations of peoples, particularly those eastwards and westwards from southwest Russia. These people spoke the language from which many Eastern and nearly all European languages have since descended.

In the West the Neolithic wanderers left behind many weapons and tools of stone and bone [2D, E, F] as well as clay vessels and toys. More impressively, they left behind, from the Mediterranean to north Europe, massive stone blocks arranged in lines and circles erected for astronomical or perhaps religious purposes. These stones (megaliths), such as Stonehenge, are the last great relics of the Stone Age and were probably used to calculate the times of sunrise and sunset at various seasons of the year.

KEY

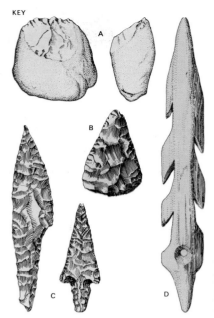

The earliest tools of ancient man were crude stone hand axes some of which (from east Africa) [A] are nearly two million years old. Much later, about 50,000 years ago, Neanderthal man made stone hand axes of a more sculptured kind [B]. These were shaped by being hammered carefully with another stone. Late in the Old Stone Age, about 18,000 BC, Solutrean hunters were making very elegant scrapers and arrowheads [C]. By about 15,000 BC *Homo sapiens* was also an expert fisherman who carved bone harpoons [D]. This was the time of Cro-Magnon man, who was physically indistinguishable from the various races of modern man.

5 At Habasesti, a site in Romania dating from 3000 BC, edged tools such as axes and hoes are made of stone and sickles are edged with flint. Thus the site (illustrated here) belongs to the New Stone Age. But ornaments and tools of copper are

also found. Perhaps these are copies of items imported from Anatolia to the east, which was already in the Bronze Age. Habasesti shell ornaments, with holes for stringing, are probably also copies but this time from the ornaments of visiting Mediterr-

anean peoples. Neolithic villages of the European forest-steppe zones were often larger than English villages 4,000 years later: Habasesti contains remains of 44 longhouses, each of which could house 500 people. But these were probably built and

occupied over a long period by a smaller number of people. The longhouses had walls of clay on a framework of tree branches. Roofs were reed and thatch. House furnishings in these villages were often surprisingly sophisticated. The pottery vessels of

Habasesti bear a distinctive S-shaped design. They were painted on red and white bases and outlined by channelled grooves and white, black and red paint. They were fired, together with many small animal and human clay figures, in top-

draught kilns. The Habasesti settlers were farmers who raised wheat and other grains, and vetches, and kept cattle, goats, pigs and dogs. The remains of roe deer and wild pigs show, however, that the settlers also hunted for food. Their lives

were busy and prosperous because the settled life allowed them to collect a surplus of food, weapons and ornaments. Life at Habasesti also had its dark side: defensive ditches dug around the settlement show that attack was likely.

Fire and bronze

Man first used metal about 6,000 years ago in an era from which the beginning of Western civilization can be dated. Earlier the stones he had picked up, perhaps first to throw at an attacker, had been refined into the sharp tools and weapons of the Neolithic (New Stone Age). But now a new material was at hand – the malleable copper which men first, perhaps, picked up as the native (pure) metal and beat into the required shape. When the pure metal became scarce, copper was smelted from the ores containing it. Soon man was to find that the rather soft copper could be alloyed, by mixing with tin, into the harder, more useful bronze. A new range of skills was now made possible.

Increasing use of metals

The longest Copper Age, that of ancient Egypt, extended before the first dynasty in about 5000 BC until about 3700 BC, after which bronze was used. But bronze objects had been made earlier still, in the city states of the Fertile Crescent of Mesopotamia where the earliest of all known city dwellers, the mysterious Sumerians, were also the first

to use copper. They probably obtained it by trading with miners in Asia Minor.

Europe in general saw the introduction of the Bronze Age about 1,000 years later than did Asia Minor [Key, 1, 5], although some copper artefacts discovered in eastern parts of Europe date from times as ancient as those of Sumeria [4].

Beyond the deserts and mountains of central Asia, the great civilization of China was pursuing its own course, producing ceramics and intricate jade and bronze vessels from 1500 BC onwards. But some ancient societies, such as those beyond the oceans, in Central America, had no metals at all at this time.

The first civilizations were controlled by the priests in theocratic Sumeria, and the civil servants and administrators of aristocratic Egypt. They also created the first slaves, war captives whom it was more profitable to put to work than to kill. Thus there was a need for both utility and luxury goods, including those made of metals, and a labour force existed that released artisans to do skilled work in shaping tools and ornaments.

The largest class, however, remained the peasant farmers. They also needed metals, in the form of ploughs, hoes, axes and other edge tools which in Neolithic times had been made of wood and stone. In times of war, many peasants were recruited as soldiers and, as at other times in history, peaceful metal-working was adapted to the production of bronze helmets, battleaxes, spearheads, daggers and later swords.

Bronze Age achievements

Huge sacred buildings were constructed by slaves both in Mesopotamia and Egypt. In Mesopotamia, pyramidal ziggurats were used for astrological observations and as temples. In Egypt, the pyramids were the greatest and grandest tombs ever built. Interred with each Egyptian luminary were all kinds of domestic and luxury objects to be taken into the next world. The tomb of the pharaoh Tutankhamen, the most intact yet discovered, contained a wealth of luxury articles made of bronze, gold, silver, ivory and glass. If a relatively unimportant 18-year-old pharaoh deserved all this, then it is probable that even

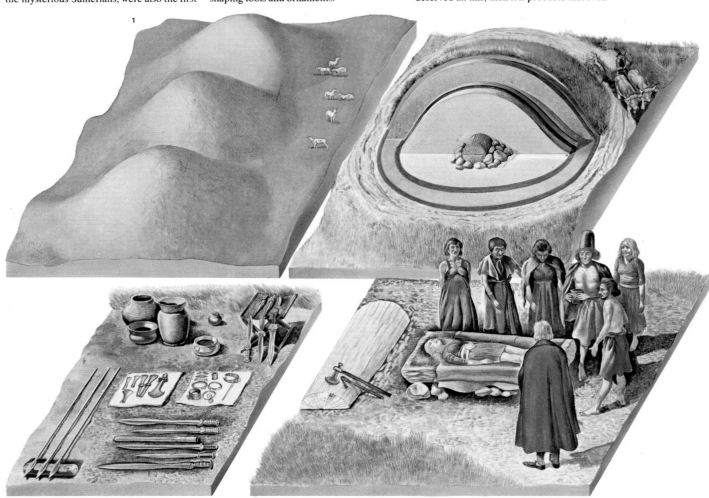

1 The ancients left their messages for posterity in the graves of their dead, whether great pyramids or simple turf burial mounds. This reconstruction depicts a burial ceremony at Egtved, in east Jutland, a few centuries after the Bronze Age culture had reached northwestern Europe, about 1500 BC. The corpse is to be interred in a round barrow, or tumulus grave mound. Such Bronze Age barrows and long barrows from the Neolithic (New Stone Age) are widespread in western Europe. The corpse, that of a rich young woman, was discovered in a barrow, in a coffin made from a split and hollowed tree trunk. The coffin was enclosed in a heap of stones and then surrounded by an enormous quantity of turf, enough to strip several fields. The dead woman wore a belt ornamented by a large bronze disc, bronze arm rings and an earring of bronze. The grave objects laid out for the ceremony are varied and plentiful, since they are those of a rich woman. They include weapons, jewellery and pottery. The weapons are typical of the period and culture; numbered among them are palstaves – narrow bladed axes with high flanges on either side – and small thrusting swords with hilts only large enough for a three-fingered grip. Decorations on the weapons echo those of Mycenaean Greece about 1500 BC. The background to the scene is an aspect of Bronze Age agriculture, a cart drawn by cattle. Seed could have been scattered in front of animals to be trodden in by their hoofs. The harvest, if it was of grain, would have been reaped with bronze-bladed sickles.

richer tombs once existed, containing still greater funeral hoards.

Egyptian civilization, with only a few major upsets, endured for more than three thousand years. In contrast, the civilizations in Mesopotamia came and went over the centuries. The Sumerians, who had invented writing and the wheel, disappeared and were replaced by the Akkadians by 2350 BC. The Akkadians left behind one of the great Bronze Age sculptures, that of their warrior-king Sargon I, which still survives. In their turn, the Akkadians were swept away by the Amorites, who gave way to the Hittites in 2000 BC. The Hittites came from Anatolia and were the greatest of ancient ironworkers and steelmakers. They belong not to the Bronze Age but to the succeeding Iron Age.

Technology of metals

Why copper and bronze should have been employed for so long to make axes, hammers and heavy weapons is at first glance odd, since iron ore was plentiful, and iron is harder and tougher than bronze.

The answer to why the Bronze Age lasted for more than 3,000 years partly lies in the technology that was available. Copper will melt at 1,083°C (1,981°F), whereas iron melts at 1,539°C (2,802°F), and cannot be melted for moulding without the use of some kind of force draught or blast furnace. Even the Hittites lacked such a furnace: they beat the metal out of roasted, but not melted, ore. True smelting of iron came even later.

Bronze could be strengthened by the addition of small amounts of zinc, antimony and other elements that improved its hardness and toughness. These elements were at first present in ancient bronze as impurities, but later were probably added deliberately. At first, the mixtures of copper and tin or their ores [6] were made haphazardly. But later, as skills improved, they were mixed in definite proportions for different bronzes.

The Sumerians also hardened bronze by hammering, and made nails from it. They made wire, sheet and castings (in gold and silver, also) in clay moulds by the *cire perdue*, or lost wax process, which continued to be used for thousands of years (as in the Benin sculptures of West Africa).

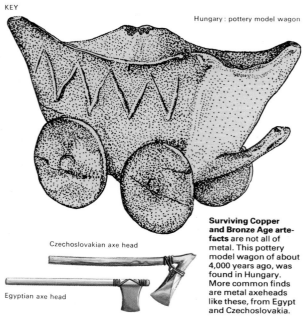

Hungary : pottery model wagon

Czechoslovakian axe head

Egyptian axe head

Surviving Copper and Bronze Age artefacts are not all of metal. This pottery model wagon of about 4,000 years ago, was found in Hungary. More common finds are metal axeheads like these, from Egypt and Czechoslovakia.

2 Unable to multiply or divide, the ancient Egyptians achieved a wonder of practical mathematics when they built the Great Pyramid. Its base covers an area of 5 hectares (13 acres) yet differs from a perfect square by only 1.5cm (0.6in). The stone blocks of which the pyramids were built weigh up to 1,000 tonnes and were moved into position without the use of either rollers or wheels.

3 Some Egyptian tombs have pictures carved in relief in rock. Others have paintings executed in water-based paints, to which were often added honey, gum or egg white to give substance to the colour. The pictures themselves provide a record of the ways of life in various classes of Egyptian society. Officialdom, as in all civilizations, thrived; here a dignitary is attended by many servants.

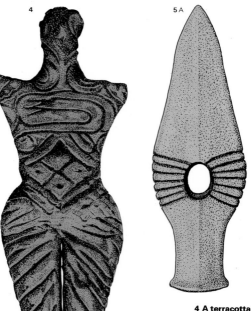

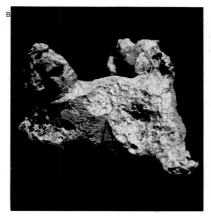

4 A terracotta figurine from the Copper Age of Romania dates from before 3500 BC, and gives some idea of the spread of culture along the River Danube during the time of the earliest metal ages of Sumeria and Egypt.

5 Across the Black Sea, north of the Fertile Crescent, traces of the Copper and Bronze Ages have been found in the Soviet Union. This axe (top [A], side [B]) was found at Fatyanovo and dates from the early Bronze Age, about 1000 BC.

6 Bronze alloys are made primarily from two metals, tin and copper. In the metallurgy of the Bronze Age bronzes could have been made – and probably were – in a number of ways. Tin occurs mainly in nature as an oxide of tin called cassiterite [A], which is often found in rivers or lakes as an alluvial deposit. Copper also occurs as many kinds of ores, as well as the "ready-to-use" native metal [B]. In some regions, a Copper Age preceded the Bronze Age.

The Iron Age

Man might first have encountered iron as a gift from the heavens: iron meteorites could have supplied the metal for weapons and ornaments made in about 3000 BC or even earlier. The manufacture of iron tools had to wait another thousand years until the introduction of iron smelting – the extraction of the often impure metal from iron ores. Iron's high melting temperature means that it cannot be melted directly over a wood or coal fire. Some kind of forced-draught, or blast, furnace must be employed to attain the required temperature of 1,539°C (2,802°F).

Primitive smelting ovens did produce a malleable iron of fair quality and were used for several centuries in different countries. Iron was of fundamental importance to a number of civilizations and by the first century AD the Noricans, Parthians and Chinese were able to produce mild steel.

The methods of the Hittite smiths

Iron in the form of bog ore is quite plentiful, widespread and easy to get at. To make iron objects the ore is first smelted to make a "bloom" of iron. It must then be heated and hammered to remove the slag, then quenched in water. This process, repeated several times, makes iron of adequate strength. The technique was used widely by Hittite ironsmiths, the first true technologists of the Iron Age, in the second millennium BC [2].

These Hittites closely guarded their smelting method but after the destruction of the Hittite Empire in 1200 BC the smiths were scattered, so that other tribes and nations benefited from their skill in making tough metal tools. At this time a kind of steel – iron containing 1.5 per cent or less of carbon – had also been made and was used for tools and weapons needing a sharp cutting edge. An early Iron Age spearhead or sickle [3] was limited by lack of a socket, which could be made only by casting the metal in a mould. A few socketed iron weapons have been found dating from as early as the second millennium BC but their sockets are made from gold or copper (copper melts at 1,083°C [1,981°F] and gold at 1,064°C [1,947°F], so these metals could be melted over a fire for casting). Iron swords of this period were stronger and less brittle than bronze ones.

More peaceful early Iron Age objects include the tongs, hammers and anvils used by the ironsmiths themselves, and iron nails, a great improvement on bronze nails that were of only limited strength.

The first iron ploughshares, from Palestine, date from about 1100 BC but the Greeks of the sixth century BC, do not seem to have used iron for this purpose and iron-shod ploughs became common only in Roman times. By 600 BC the Catalans of northeastern Spain were able to soften, but not melt, iron in a type of furnace having a forced draught provided by bellows.

Romans and their use of iron

If the Hittites were the first ironsmiths, then the Romans were the first mining engineers. Sites of Roman metal mines range from southern Scotland to southern Spain in the west to Romania in the east.

As would be expected, Roman iron and steel were put to many military uses. The famous short-sword, the nine-foot spear and the cuirass inherited from the Greeks were all made largely of iron. Later Roman soldiers

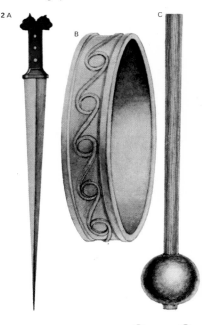

1 Iron ores are very common and widespread, second only to those of aluminium. But early Iron Age man would not have used the richer ores such as haematite and magnetite, which can contain as much as 60% iron. These ores must be mined. He would more probably have first exploited bog iron ore which, as its name implies, occurs freely in marshy ground. It is formed by the decomposition of other iron minerals and is finally precipitated in the water by the action of bacteria. It needs only to be sieved out to be used. However, by 850 BC several other iron ores were being worked in Europe, and this map shows the regions occupied by various peoples who worked iron ore.

Thracians
Greeks
Illyrians
Etruscans
Sicans
Italics
Ligurians

0   1,000km

2 Earliest of man-made iron objects were ornaments and small weapons, dating from before 2500 BC. The Hittite dagger-sword [A] was made somewhat later. Its blade would have been beaten out from heated iron ore. This produces a relatively crude form of iron but the Hittite smiths eventually became the masters of iron technology. They influenced European ironwork, such as this collar [B]. By 500 BC, in Styria and Carinthia, central Europe, iron-smiths were making a kind of steel by hammering charcoal (a form of carbon) into heated iron. The iron-headed mace [C] is a late example of primitive iron technology from Chieti, Italy. It dates from about 600 BC.

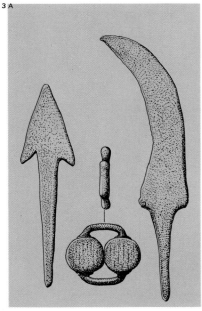

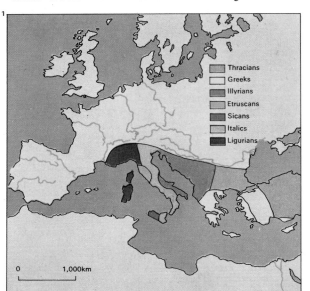

Turf
Chalk
Chalk parapets
Clay
Earth

Unexcavated

3 Maiden Castle is a large Iron Age fort in Britain. The site, a saddleback hill [B], was occupied from Neolithic times (3rd millennium BC) and by 100 BC had grown to a town-sized fortress with six phases of ramparts (1-6), the earliest dating to about 500 BC. The iron arrowhead and small sickle and the bronze belt buckle [A] are objects that were made by a Wessex Iron Age people who inhabited Maiden Castle from about 100 BC until AD 70; they were known at the time as the Durotriges. These Celts also left behind them many artefacts made of iron, including rings, axes and "currency bars" (shaped like swords) that were used for trading. The castle was attacked by the Romans under Vespasian in AD 43 and a cemetery of the defenders has been found. The castle continued to be occupied until about AD 70, when it was replaced by the nearby Roman town of Dorchester.

also used iron long-swords and throwing spears and shot iron darts from their catapults. Iron-headed battering-rams had been used since the time of the Assyrians; a Roman improvement was the *terebra* which bored into battered walls and gates.

In architecture the Greeks had made the most imaginative use of iron, in wrought-iron beams. In the construction of the Parthenon, such beams acted as cantilevers which held up the heaviest statues on the pediment. The Romans later made use of T-shaped iron girders, as in the Baths of Caracalla where they help to support a dome 36m (118ft) across.

Little is known about mining and metallurgy in the early Middle Ages except that Saxon miners were at work in the Harz mountains before AD 1000. Knowledge of later medieval mining technology is much greater, due largely to a single man, the Saxon physician usually known as Agricola. His great book *De re metallica* (1556) details methods and machinery and demonstrates that mining was a profitable industry.

Centuries before Agricola's time, the iron industry was lively but was what today would be called an armaments industry. From the time of Charlemagne (AD 742–814) onwards, iron armour was invented afresh, very little being copied from the Romans. Later there evolved a totally new iron technology – the manufacture of iron firearms [5].

Iron smelting and casting

The final great discovery of the Iron Age was that iron could be melted and cast in moulds. The first blast furnaces were developed as enlarged versions of medieval *Stückofen*. They used first charcoal, then coal, both as fuel and as a necessary ingredient (carbon) for the smelting process. By 1711, Abraham Darby (1678–1717) in Coalbrookdale, England, began to use coke to smelt large quantities of ores to make good-quality cast iron suitable for forging as well as for casting in sand moulds. Previously, cast utensils, such as pots, could be made only from expensive metals such as brass. Soon every home in the land had cast-iron cooking pots and pans. Iron was used for making bridges and railways. The Industrial Revolution and the age of steel had begun.

Iron Age peoples occupied hill-forts in many parts of Europe often fortifying them, as here at Maiden Castle, England, with a series of concentric earth ramparts.

4 After the fall of the Roman Empire, metalworking in Europe declined, surviving mainly in the elegant iron swords made by the Burgundian and Frankish smiths, which had patterns formed by strips of welded iron. The horsemen of Charlemagne, king of the Franks in the 8th century AD, were clothed in heavy iron armour of a design that owed little to the Romans: the Germanic conquerors of the Roman Empire had apparently despised the thought of copying armour of a defeated foe. Chain mail came later still as an import from the East. This feudal knight wears up to 50kg (110lb) of mail and plate. Mail was made from rings of iron wire riveted or welded together and shaped to cover the arms, feet and head. By the 14th century, plate armour, originally a reinforcement for chain mail, was beginning to take its place. Ironsmiths were by this time adept at shaping accurately the joints necessary for plate armour and were also skilled at producing decorative metal inlays.

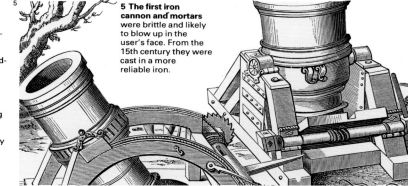

5 The first iron cannon and mortars were brittle and likely to blow up in the user's face. From the 15th century they were cast in a more reliable iron.

6 Coke smelting was first introduced by Abraham Darby in Coalbrookdale, England (shown here) in the early 18th century. It replaced the use of charcoal, which was dependent on failing supplies of wood, and greatly accelerated England's rapid industrial development.

The Steel Age

Ten thousand years ago the technology of our ancestors was based upon the use of stone, three thousand years ago upon bronze, and two thousand years ago upon iron. Thus we refer to the Stone Age, the Bronze Age and the Iron Age. There is no corresponding period that archaeologists can specify as an age based upon steel. Steel has been made in large quantities for only about 120 years, during which time it has become vital to our civilization. But the origins of steel lie in the remote past.

The first steelmakers

Plain or mild steel, as used in a host of applications, is an alloy of iron with a little carbon. More particularly, mild steel contains between 0.15 and 0.25 per cent carbon, partly in the form of iron carbide or cementite. Steel is much harder than pure iron and less brittle than cast iron (which contains more carbon than steel does); its strength, toughness and springiness account for its great usefulness.

The first makers of iron tools and weapons on a large scale were the Hittites,

and it was men of a subject tribe of the Hittites, the Chalybes, who first made steel in about 1400 BC. Chalybean steelsmiths in Asia Minor employed a cementation process; that is, they hammered hot but still unmolten iron together with charcoal until the iron became steel. During the hammering, carbon from the charcoal diffused into the iron, forming cementite – hence the name.

Moulding and hardening

Variations on the cementation process continued in use for more than a thousand years, spreading from the Middle East into Europe and India. But is was in southern India that iron was melted for the first time and steel cast into moulds. The Romans imported this steel in the shape of small, rounded cakes. They thought that it originated in China and called it "Seric" (meaning Chinese) iron.

In Europe steel was sometimes made directly from an ore with the correct carbon content, but this was rare. More often steel resulted from the cementation process and, from the eighth century AD onwards, cementation steel began to be exported from

the iron-rich parts of central Europe known as Styria and Carinthia. This steel had been further hardened by quenching it from red heat in water. It had taken a long time for this tempering technique to be learned, probably because the earlier known metals copper and bronze, quenched in this way, become softer.

By the fifteenth century early printers were using steel punches in the manufacture of their type moulds and, midway through the seventeenth century, tempered steel coach springs first added a little comfort to travel on Europe's rutted and pot-hole-strewn roads. Such examples show a thorough appreciation of the possibilities of steel but these were not to be fully realized until cheap steel became available with the coming of the Bessemer furnace in the 1850s.

The Bessemer process

A great problem with early steel was the presence in it of slag waste from the ore, which made it difficult to manufacture large steel objects without structural weaknesses. This problem was solved by the mid-eighteenth century as the result of the work of Torbern

1 **A complex modern steelmaking plant** uses coal as the initial fuel. Iron and limestone [1] are brought in by train or ship and are conveyed [2] from storage to a blast furnace. But some ore and limestone are first heated with coke [3]. Coal from a coal store [4] is converted into coke in a coking plant [5]. All these materials are then fed to the blast furnace [6] in which iron reacts with carbon from the coke and the other materials to form a layer of molten slag. A pump-house [7] supplies the furnace air, which is preheated in heat exchangers [8] by hot gases [9] from the blast furnace. Too much carbon is absorbed by the iron in the blast furnace and this excess is removed in a second basic oxygen furnace [10] containing an oxygen lance [11] fed from large spherical tanks [12]. Coke and limestone are also used in the refining process. Gas from the oxygen furnace is taken off [13] for cleaning before being discharged into the atmosphere. The refined steel, containing between 0.15 and 0.25 per cent carbon, is poured from the oxygen furnace [14] and cast into ingots ready for working [15].

Bergmann (1735–84), a Swedish metallurgist, and Benjamin Huntsman (1704–76), a British steelmaker. "Swedish" steel had a controlled carbon content and was free from slag. But it was still quite expensive. In 1850 the entire output of steel in Britain was only 60,000 tonnes. Twenty years later, Bessemer furnaces, or "converters", made steel at an average rate of one tonne per minute and great batches of steel were being made as cheaply as cast iron.

The secret of the Bessemer process – developed by Henry Bessemer (1813–98) from the invention of a bankrupt Kentucky steelmaker, William Kelly – was that excess carbon could be oxidized (combined with oxygen) by forcing bubbles of air through a mass of molten iron. Moreover, the carbon burned to carbon dioxide in the blast of air, so acting as a fuel. Once started, the process continued without the addition of more fuel coal and was thus extremely economical.

Within five years the Bessemer process [3] had a rival in the open-hearth method of steelmaking [2] in which iron, iron ore and scrap steel are melted in such proportions as

to drive off most of the carbon and oxygen as carbon monoxide. This inflammable gas is then burned "regeneratively" to pre-heat the blast. This efficient operation was, by 1900, producing more steel than even the Bessemer process.

The twentieth century saw further revolutions in steelmaking, notably the development of continuous casting of steel [4] and special steels such as those for making machine tools and turbine blades. The newer steels include stainless steels, which contain chromium, nickel and sometimes molybdenum, and are made in an electric furnace [5]. They are used mainly for cutlery, kitchen utensils and chemical plant. The Bessemer process is being replaced by the basic oxygen process [3B].

By the end of the nineteenth century world production of steel was already more than 28 million tonnes. In the years since, it has received further boosts from the two world wars, which demanded steel for armaments, and from the development of the internal combustion engine, which has made every motorist the owner of a tonne of steel.

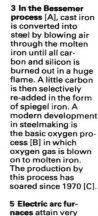

Steelmaking is often hot, tough work, as the faces of these Japanese steelworkers show. In the 19th century men whose job it was to tap or release the molten steel from the furnaces had no protective clothing, and the intense heat could kill them. Today automatic control protects workers.

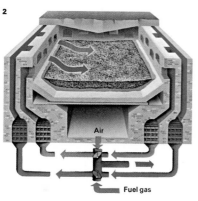

2 The Siemens-Martin open-hearth process, in which streams of air and fuel gas are fed alternately on to the furnace contents, has been used to make most of the steel during this century, but is being superseded by electric and basic oxygen methods. It uses the gases from the molten charge to preheat the air blast and so economizes on fuel. An alkaline lining is used if the ore contains phosphorus.

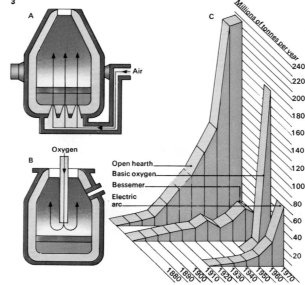

3 In the Bessemer process [A], cast iron is converted into steel by blowing air through the molten iron until all carbon and silicon is burned out in a huge flame. A little carbon is then selectively re-added in the form of spiegel iron. A modern development in steelmaking is the basic oxygen process [B] in which oxygen gas is blown on to molten iron. The production by this process has soared since 1970 [C].

5 Electric arc furnaces attain very high temperatures, with all oxygen excluded. For this reason, they are used to make steels containing oxidizable metals such as chromium and vanadium.

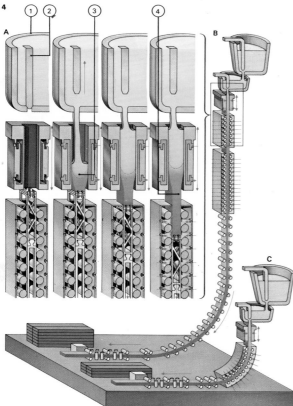

4 Continuous casting is a modern method of making steel in the form of bars and rods cheaply and quickly. A casting run [A] starts with molten steel being poured into a copper mould [1] which is closed by a plug [2] at its lower end. The steel is cooled by water in the mould and when it is nearly solid the plug is withdrawn; the moulded steel moves down [3] and is then further cooled by means of water sprays before being cut into the required lengths. While the plug is withdrawn molten steel is poured continuously [4]. The original casting process [B] had a straight, vertical cooling section, but a newer type [C] has a curved section that requires less space.

Metals and their uses

Iron and its alloys are dominant metals today and so it is customary to group all other metals into the category "non-ferrous". But non-ferrous metals have little in common apart from the negative attribute implied by the name. There are more than 20 non-ferrous metals of industrial importance and some of the chief ones – copper, silver, lead, tin and mercury – have been known since antiquity. Others such as titanium have found wide application only recently.

Gold, copper and silver
The legendary metal gold [1] occurs in nature almost exclusively in a native (free or uncombined) state. First found by chance, it was being mined by 3000 BC and was being chemically extracted from crushed quartz (by amalgamation with mercury) by 1000 BC. Gold is yellow and lustrous, outstandingly workable for drawing or hammering into shapes, and is resistant to corrosion and to chemical attack. For this reason it is widely used for costly ornaments and jewellery and as a means of storing wealth. The main producers are the USSR and South Africa.

The conventional test for gold is to drop hydrochloric or nitric acid separately on to the metal; pure gold resists them but dissolves in a mixture of the two, known as aqua regia. Other solvents for gold are chlorine water, alkaline cyanides and mercury. Gold is extremely soft and generally used only in alloys, usually with silver or copper. The number of "carats" indicates the number of parts of pure gold in 24 parts of the alloy (9 carat gold is $^9/_{24}$ pure gold and $^{15}/_{24}$ copper).

Copper was being smelted by about 4000 BC although it was probably found earlier than this in its native state. The camp fire was probably the original smelting furnace, as copper ore among the stones was reduced accidentally by wood charcoal in the fire to yield metallic copper. Nowadays copper is a major commercial metal, used at the rate of thousands of millions of tonnes a year.

Copper is easily worked and readily alloyed [5]. Only silver is a better conductor of heat and electricity. Copper is also readily joined by brazing, soldering or welding and is relatively resistant to corrosion. This combination of properties gives it widespread uses in electrical work or where heat is to be transferred quickly – as in radiators. Often it is used in its pure state, at other times in one of its many alloys such as brasses (with zinc) and bronzes (with tin).

Silver is another legendary metal, often mentioned in poetry and story. It is thought to have been discovered first as the free metal, then in chlorides and later from its occurrence with lead in galena ores. Silver's main uses are industrial and are based on its resistance to corrosion. It is also used widely, in the form of salts, for photographic emulsions, as well as for medals, decorative plate and other commemorative products. Silver's excellent electrical properties determine many of its applications and its value makes it suitable as a means of storing wealth.

Lead, tin and their alloys
Described by Shakespeare as "base" (compared with the "nobler" metals), lead [7] has a long history. Yet it is not found free in nature. It is thought to have been discovered by accidental smelting, much in the same way as copper. In this case, however, the ore was

1 Gold has always been a symbol of material wealth and, because of its unique qualities, has been used to create art treasures of dazzling beauty, such as this medieval goblet [A]. Complete freedom from corrosion also makes gold technically valuable. To minimize contact resistance, the terminals of electrical sub-assemblies such as a printed circuit board [B] may be thinly electro-plated with gold.

4 In a mercury switch a pool of mercury completes an electrical circuit. Terminals are fused into a glass bulb, which can be tilted. In the "on" position the mercury flows around the terminals: in the "off" position it flows away and breaks the connection.

2 Bronze is thought to have been the first metal alloy ever made, although it was probably produced originally by the accidental smelting of mixed ores of copper and tin. People in the Middle and Far East used it both for practical objects such as mirrors and for ornaments such as this Egyptian cat, which is thought to have been sculpted and cast in bronze at least 4,000 years ago.

3 A new alloy of aluminium containing 20% of tin was developed by tin researchers to improve the strength of white metal bearings and meet modern higher bearing loads. Here, an aluminium bearing is being fitted to a large diesel engine.

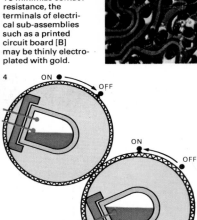

5 Alloys of copper find many uses in marine engineering. High-tensile brass, as that of ships' propellers, is based on Muntz metal containing about 40% of zinc. Other elements are often added to promote particular properties for various uses. To produce the range of high-tensile brasses, tin, aluminium, iron and manganese may be among the additives used; products are sometimes known as manganese bronzes although technically they are brasses.

galena (lead sulphide). Tumblers of lead have been found in remains dating from 3500 BC. The Romans used lead extensively for water pipes, known in English as "plumbing" – *plumbum* being the Latin name for lead – and it was later used as a roofing material. Lead has a low melting-point (327°C [620°F]), is easily cast and makes a number of valuable alloys – with tin for solders, with antimony for electric storage batteries and with several other metals for alloys for bearings and as sheathing for electric cables. An organic compound of lead is used in petrol as an anti-knock additive. In recent years there has been concern about the amount of lead (which is a poison) in the environment.

In antiquity, tin was used in the alloy bronze, but compositions were variable. Again it is thought that the first discovery was due to accidental smelting of mixed ores of copper and tin. Later the alloy was made deliberately from the component metals. Today its main use is as a coating on steel to make what is called tin-plate. Tin protects steel from corrosion, allows soldering to be carried out quickly and easily and is non-

toxic; thus tin-plate makes cans for food.

The second largest use for tin is in solders and there are also extensive uses in modern bronzes, metals for bearings [3] and as a minor alloying additive to control the structure of cast iron. Organic compounds of tin have wide applications as chemical stabilizers and as biocides for crop spraying; they do not contaminate the soil permanently because the chemicals break down to harmless inorganic compounds.

Mercury – the liquid metal

Mercury, the only metal that is liquid at ordinary temperatures, has long fascinated poets, who referred to it variously as "fluid silver" or later "quicksilver". It was probably known by about 500 BC and was prepared by treating cinnabar, its sulphide (also used directly as a red cosmetic), with vinegar, or simply by roasting it. The Romans used mercury to extract gold from ores. Today it is extensively used in scientific instruments – barometers, thermometers, micro-switches [4] and others – and as a compound in "initiators" for detonating explosives.

Coins are now used mainly for the lower values of money, but were once the only form of currency in international use. They need to be made of metal, such as cupro-nickel, that is fairly strong and yet resistant to corrosion and abrasion and at the same time readily worked into shape.

6 Nickel is a non-corrodable metal used with small amounts of alloying metals in vacuum tubes such as this output valve and in cathode ray tubes for television. But in most of its metallurgical uses it is alloyed – for example forming a constituent of steels for high strength and low corrosion, notably with chromium in stainless steel. Nickel also has many chemical applications.

7 A lead tank made in the 1880s shows no sign of corrosion, demonstrating how well this metal resists attack by ordinary water. Roman architects made extensive use of lead in their elaborate systems of water distribution and public baths. Very soft water attacks lead and, since an accumulation of the metal in the body is poisonous, lead pipes are now rarely used to carry drinking water.

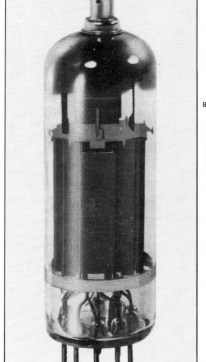

8 Chromium plating, here applied to car bumpers, improves durability as well as appearance. Chromium is a major alloying element in the making of stainless steels or other steels that need extra hardness and resistance to corrosion. Alloyed with nickel it is used for resistance wires for electrical heating elements in fires and industrial plants. But its main use is as a hard layer electroplated on to the surface of steel or plastic components.

9 Uranium, which is chemical element number 92, has the highest atomic weight of any naturally occurring metal. It or its compounds are used in atomic reactors or bombs because it is fissionable, with an atomic nucleus that breaks down when bombarded with neutrons, yielding further neutrons that can establish a chain reaction. Transuranium elements are also formed in reactor piles such as these and in bombs; they include the metal plutonium, probably the most dangerous of poisons.

Aluminium and its uses

Before about 1890, few people outside the scientific community had heard of aluminium. Yet today this metal and its alloys are among the most versatile of those commonly used by man [Key]. They are particularly valued by the aerospace industry, which makes good use of their three principal properties: lightness, strength and non-corrodibility. Aluminium's excellent electrical conductivity [8] also makes it important in high-voltage electrical conductors.

Location and extraction
Aluminium is the most abundant metallic element in the earth's crust. It is chemically reactive and so is never found naturally as a free element, but always in combination with other elements. It is a constituent of nearly all kinds of rock and its average concentration in the upper 16km (10 miles) of the earth's crust is eight per cent. Commercially it is extracted from a mineral called bauxite, a weathered aluminium oxide found in many countries.

Aluminium was first isolated in 1825 by the Danish scientist Hans Christian Oersted (1777–1851). At the beginning of the century the British chemist Humphry Davy (1778–1829) had postulated the metal's existence but had been unable to isolate it. It was not until 15 years after Oersted's discovery that enough aluminium was produced to establish some of its properties – notably its low density. Napoleon III (1808–73) believed that such a lightweight metal could have an important future in military applications and provided money for the French chemist Henri Sainte-Claire Deville (1818–81) to develop a process for the large-scale production of aluminium.

Deville devised a process based on the chemical reduction of aluminium chloride using sodium metal and aluminium was first displayed publicly at the Paris Exhibition of 1855. At this time it cost about $250 per kilogramme, but by 1886 the price had dropped by $15 per kilogramme and a total of 50 tonnes had been produced.

The production of aluminium
The future of aluminium was assured in 1886 when, at almost the same time, Charles Martin Hall (1863–1914) in the United States and Paul Héroult (1863–1914) in France devised the process for extracting aluminium by electrolysis that is still used today. They discovered that when pure alumina (aluminium oxide) is dissolved in a molten aluminium mineral called cryolite and an electric current passed through it, aluminium metal is liberated at one electrode (the cathode) and oxygen at the other (the anode). Within six years of this discovery, the price of aluminium had fallen to $1.30 per kilogramme. In 1976 the price was about 88 cents per kilogramme.

Large quantities of electricity are needed to produce aluminium (about 13–18 kWh/kg). For this reason the industry grew up in places such as Niagara Falls and in Norway, where cheap supplies of hydroelectricity are available. But aluminium has become so important that most industrial countries have now established their own aluminium smelting industries.

Bauxite, generally obtained by strip mining, has to be purified before it can be used for aluminium manufacture. To be acceptable, the bauxite must contain more

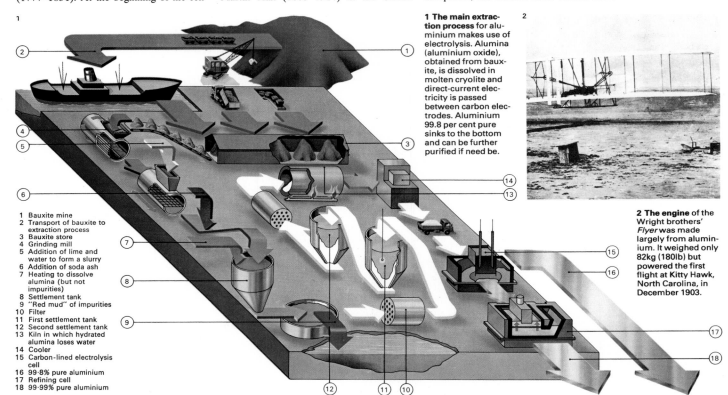

1 The main extraction process for aluminium makes use of electrolysis. Alumina (aluminium oxide), obtained from bauxite, is dissolved in molten cryolite and direct-current electricity is passed between carbon electrodes. Aluminium 99.8 per cent pure sinks to the bottom and can be further purified if need be.

1 Bauxite mine
2 Transport of bauxite to extraction process
3 Bauxite store
4 Grinding mill
5 Addition of lime and water to form a slurry
6 Addition of soda ash
7 Heating to dissolve alumina (but not impurities)
8 Settlement tank
9 "Red mud" of impurities
10 Filter
11 First settlement tank
12 Second settlement tank
13 Kiln in which hydrated alumina loses water
14 Cooler
15 Carbon-lined electrolysis cell
16 99.8% pure aluminium
17 Refining cell
18 99.99% pure aluminium

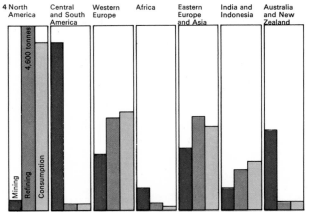

2 The engine of the Wright brothers' *Flyer* was made largely from aluminium. It weighed only 82kg (180lb) but powered the first flight at Kitty Hawk, North Carolina, in December 1903.

3 World primary aluminium production (million tonnes)

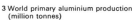

	5	6	7	8	9	10	11	12	13	14
1966										
1967										
1968										
1969										
1970										
1971										
1972										
1973										
1974										

3 Primary aluminium, extracted from ore and excluding that reclaimed from scrap, has been produced in steadily increasing quantities over the last few years. Production doubled between 1966 and 1974, although the price (at an all-time low of 31c per kg in the 1940s) slowly continues to rise as more and more uses are found for this versatile metal and its alloys. Recently, more effort has been made to collect scrap aluminium and re-use it.

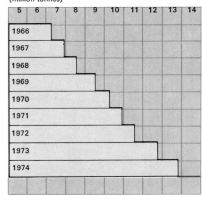

4 The world's largest user of aluminium, the United States, is not the greatest producer of the ore, bauxite, although more aluminium (66 per cent of the total output) is refined in the USA than anywhere in the world. The chief producers are Jamaica and Australia. This diagram shows where bauxite is mined and aluminium refined and used. It takes 4kg of bauxite to produce 1kg of aluminium. Current world production is 26,000 tonnes a day.

than 32 per cent aluminium oxide, although the level is usually as high as 45 to 55 per cent in bauxite used today. The Bayer process [1] is the most common purification method, during which the impurities form a "red mud", although this has caused pollution problems for some countries. The red mud still contains some alumina, mixed with silica, and the Alcoa process can be used to extract this aluminium or to obtain it from ores containing high proportions of silica.

There have been various attempts to develop alternative processes. In 1973 Alcoa announced a method based on the electrolysis of aluminium chloride, which uses 30 per cent less power than the Hall-Héroult process. There has also been research into "carbothermic" processes, in which the oxide is reduced by heating it with carbon.

The multi-purpose metal

The first major application of aluminium was in the manufacture of cast cooking utensils in the 1890s. In the first decade of this century, usage included electrical conductors (first used for telephone lines in the Chicago stock-

yards in 1897), marine and aeronautical components, engineering and scientific instruments, paint pigments and bottle caps.

During World War I, the German Alfred Wilm invented Duralumin, an alloy of aluminium with small amounts of copper, magnesium and manganese, which has physical properties similar to those of structural steel. This was the first of a large range of alloys which have greatly extended the use of the metal. In the United States in 1975 the major estimated uses were building and construction 26 per cent; transport, electrical and containers 15–16 per cent each.

Modern uses of aluminium are legion: metal window frames; coinage; saucepans and cooking foil; aircraft, bus and train components; catalysts for the petrochemicals industry; constituents of deodorants; giant storage tanks for liquid natural gas [11]; soft drink and beer cans; aluminothermic smelting of high-melting metals such as chromium; and corrosion-resistant metal paints. Aluminium now pervades the whole of modern life to an extent achieved by no other element discovered in modern times.

Aluminium and its alloys have a myriad of uses, from spacecraft to cooking foil. Many different elements may be added to aluminium to produce alloys with a remarkable range of properties and those properties may be further modified by heat treatment, ageing, solution treatment with various chemicals and mechanical working. Copper, magnesium and zinc are alloyed with aluminium to produce strong metals. Silicon, manganese, chromium and nickel are used in various combinations to make alloys for special components or to make the alloys suitable for casting, chemical treatment (such as anodizing) and certain machining operations.

5 The hardness of aluminium is increased by alloying and by heat treatment. When cooled rapidly from high temperatures [A], alloying element atoms are evenly dispersed. The alloy is strengthened by short-term heating, called ageing. At lower ageing temperatures [B] the alloying atoms cluster, making it hard. At higher temperatures the formation of intermetallic compounds [C] relieves strain, but makes soft alloys.

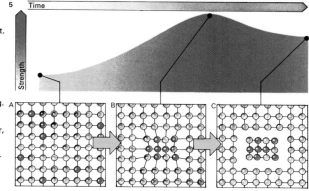

6 A thin film of oxide rapidly forms on aluminium on its exposure to air. Unlike rust on iron the oxide film is hard and non-porous, making aluminium non-corrodible but difficult to weld.

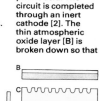

7 Anodizing [A] is the name of the electrolytic process for deliberately forming a relatively thick film of oxide on aluminium and its alloys. The article to be anodized [3] is made the anode (positive electrode) in an electrolyte [1], generally chromic or sulphuric acid. The circuit is completed through an inert cathode [2]. The thin atmospheric oxide layer [B] is broken down so that further oxidation, to form a thicker layer [C], can take place. Anodized aluminium can be dyed various colours, including metallic ones to make it resemble, say, brass or copper. The oxide layer is then sealed [D]. Letters and symbols can be made photographically. The process is more than merely a decorative finish and gives a tough surface that is much harder than the original base metal.

8 The electrical conductivity of aluminium (Al) is only about 60% of that of copper (Cu) but Al is lighter and cheaper. Here [A] weights and resistances are compared for 1,000m of cable – Cu [B] and Al [C].

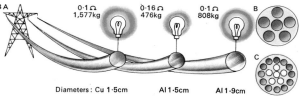

Diameters: Cu 1·5cm   Al 1·5cm   Al 1·9cm

9 Polished aluminium reflects much more light than iron and nearly as much as silver, making it an ideal material for mirrors for telescopes and other optical instruments.

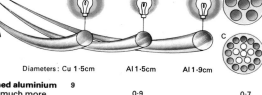

Silver   Aluminium   Iron

10 Alloying aluminium can rob it of some of its corrosion resistance. To overcome this drawback, thin sheets of pure aluminium can be pressure-welded to the outsides of alloy sheets.

11 A large, spherical "tank" at a nuclear power station makes good use of many of aluminium's properties: the tank is light, can be polished to reflect heat and is strong enough to withstand pressure.

12 Aluminium is soft because its atoms form a close-packed, cubic structure. This has many glide planes to let the atoms easily slide over each other. It is malleable and easily drawn out into wires.

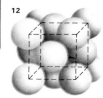

Tailor-made materials

The development of modern civilization has been closely connected with man's increasing ability to master his environment and adapt parts of it for his use. For centuries primitive man relied on the natural materials around him, such as stone and wood, to provide shelter and protection. Later he learned to modify these materials, producing such improvements as pottery, bricks and metals. But it has been only during the past century that materials science has emerged as a subject of real importance that enables the technologist to tailor materials able to withstand the extreme conditions often demanded by modern processes such as nuclear power generation or supersonic flight.

The importance of alloys

Alloys (combinations of a metal with one or more other elements) have played an important role in human development for thousands of years – a role recognized by terms such as the Bronze Age.

Bronze is generally a mixture of copper and tin, with the two components varying widely in proportion. By contrast, many of today's alloys are made largely of a single element, the properties of which are modified by the addition of precise quantities of carefully defined "impurities". From the now common stainless steels, made from iron mixed with small percentages of such elements as carbon, chromium and nickel, the technology of materials has advanced through the nickel superalloys, which retain their strength even when white-hot, to titanium alloys [2] which are essential to many jet engines.

Improvements since the 1950s

The molecular principles used in producing diamond substitutes (and improvements) such as boron nitride have spread throughout many industries since the 1950s. For instance, as, with attempts to achieve more speed or efficiency, the conditions under which machinery operates become more extreme, it has been necessary to develop not only new structural materials but also new lubricants. In many applications, conventional mineral oils have been replaced by specially tailored chemical lubricants. Some of these do more than merely operate effec-

tively at high temperatures and pressures. Some sulphur-containing lubricants, for example, are good for "running-in" new machinery. Even with modern machining techniques, surfaces are likely to be irregular and to produce "hot spots" during running, which explains why some have to be "run-in" cautiously. The sulphur-containing lubricants react with hot spots, removing them as soft metal sulphide particles, which are less damaging to the machinery than the small metal or metal oxide particles that are ground off in the presence of an unreactive lubricant.

Another field in which tailor-making plays a crucial role is replacement-part surgery [7]. The silicone polymers have made an outstanding contribution in this area because there is no interaction between them and body fluids and tissues. A replacement part for a heart valve, for example, has to be resistant to the corrosive effects of the litres of blood in which it will be bathed during its lifetime and the blood must not react abnormally to the presence of material foreign to the body. In biomedical technology, the trend is now towards the development of

CONNECTIONS

See also
44 Working with ceramics
54 Rubber and plastics
56 Fibres for fabrics
240 Soaps and detergents
244 Colour chemistry

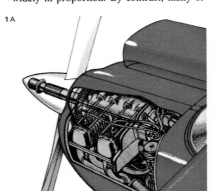
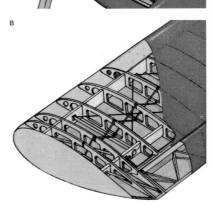

1 A light aeroplane such as this Bellanca Champion Scout would be impossible to make without the range of modern tailor-made materials. The propeller [A] is made from aluminium alloy and the wings [B] from aluminium ribs covered with the synthetic polymer Dacron, with glass-reinforced polyester wing-tips. The Scout has a fuselage of molybdenum steel tubing, also covered with Dacron. The transparent material in the windows is Perspex [C], and synthetic rubber is used for the tyres [D] which are supported on landing gear made from a type of high-tensile steel.

2 The Lockheed YF 12A broke four major world records on 1 May 1965 and still holds the official air speed record of just over 3,200 km/h (2,000 mph). This aeroplane is built largely of titanium, a metal that was not available in commercial quantities until after World War II. Titanium has the strength of steel but is only half the weight, hence its use in aircraft. It is the ninth most plentiful element in the earth's crust.

3 Development of new materials has been made possible by the increased range of scientific techniques available for studying the properties of different substances. This electron probe microanalyser sends a carefully focused beam of electrons onto a selected portion of the materials being studied. From the X-rays being emitted as a result, it is possible to identify elements present in the sample and analyse for impurities on surfaces.

materials similar to those evolved by nature.

In dealing with materials for biomedical use, high standards of purity are clearly necessary. The ability to produce ultra-pure materials has also been a major factor in the development of several other modern technologies – particularly solid-state electronics. Just as the properties of a metal can be altered by alloying it with some other element, so its properties may be altered by refining it from, say, 99 to 99.99 per cent purity. Without techniques such as zone refining (in which impurities are concentrated at one end of a rod by melting it) it would have been nearly impossible in the 1950s to prepare materials pure enough to make semiconductors and transistors.

The formation of composites

Another important aspect of materials science is the way in which several materials can be combined together to form composites. Often two or three different materials may be blended together in a precisely controlled manner to produce a new material with properties different from any of the individual components. Thus plastics reinforced with glass fibre are now used widely in automobile bodies and hulls for light boats, in which they replace older structural materials such as metal and wood.

Not all attempts to "compose" new materials have been so successful. A few years ago it was predicted that composites would soon replace leather for making shoes. Synthetic shoe materials are now available, but some companies lost large sums of money on research that failed to produce materials as good as natural leather. Another case in which technology over-reached itself occurred with the fan-blades for the Rolls-Royce RB2-11 turbine engine. In the original specification these were to be manufactured from a carbon-fibre reinforced composite [5]. Unfortunately this was not satisfactory in practice and more expensive, heavier titanium blades had to be used, contributing to the company's bankruptcy in 1971. Despite these failures it seems likely that as understanding of the behaviour of matter increases, technologists will continue to tailor new materials.

Coins were once made from pure, rare metals such as gold and silver. For years now they have been made from a mixture of cheaper metals, as used in this stamping mill at a mint. Sophisticated alloys and non-metallic, tailor-made materials play an ever-increasing role in modern technology.

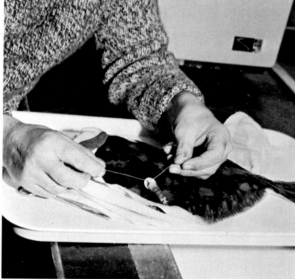

4 The miniaturization of electronic circuitry that followed the development of semiconductor materials has produced unexpected "spin-offs". For example, animals such as this plaice can now be fitted with small radio transmitters so that their migration patterns can be studied.

5 Carbon fibres, long whiskers of pure carbon produced under carefully controlled conditions, were heralded as the miracle material of the 1960s. It was believed that they could be used to reinforce plastics to make them as strong as metals. But they need more development before being used in some applications.

6 New materials enter into all areas of daily life, from the transistor radio to the instant bathroom. With the increasing cost of traditional building materials and, in many countries, a shortage of skilled labour, the building or renovating of houses is becoming more difficult and more expensive. However, with modern technology it is possible to mould an entire bathroom in plastics. It can then be lowered into place in a house where the walls are constructed of traditional materials that will provide the basic structural strength. After a few plumbing connections have been made the bathroom is ready to be used.

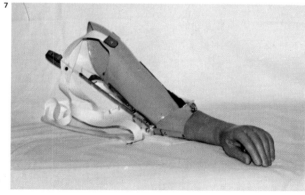

7 The most complex materials on earth are possibly those present in living tissues such as skin, bone, muscle and cartilage. To devise new synthetic materials to replace worn out or damaged parts of the human body is one of the greatest challenges facing materials scientists and technologists today. For internal use, rugged materials are required that will withstand the corrosive environment and yet produce no undesirable side reactions, such as inflammation. For external use, as in this prosthetic hand, the ideal is a combination of materials that will give not only the right appearance in terms of colour and texture, but also power and diversity of action, so totally imitating the natural limb.

Hand tools

Hand tools are "powered" by the muscles of the user. Craftsmen of almost every trade use a profusion of many kinds of tools. Modern materials and manufacturing methods have led to the development of new hand tools such as a tool that combines some of the functions of a plane and a file. But traditional forms can usually be recognized immediately in the modern version. In addition, today's consumer society has led to the development of special tools for the amateur – gardening tools are a good example [2]. Almost all craftsmen's tools have evolved from a limited number of standard types and may be grouped according to their principal function: hammering; cutting; splitting and shaping; piercing and boring; measuring and marking; grasping and holding; sharpening; and screw-based tools.

Hammering, cutting and piercing

The most familiar kind of hammer is the carpenter's claw-hammer whose primary function is to drive nails into wood and, when necessary, pull them out. But there are many other kinds of hammering tools [1] including mallets, sledgehammers and pestles. Blacksmiths, boilermakers, bricklayers, prospectors, woodcarvers [4], stonemasons, jewellers and chemists use a great variety of hammers. They vary in weight from a sledgehammer, almost too heavy to lift, to the jeweller's delicate instrument for embossing precious metals. Each type has its own special form for a particular function. The head of the tool may be hard enough to forge iron or soft enough to avoid damaging the wooden handle of another tool. One of the most important parts of the hammer is the handle, which must be carefully balanced for convenience and maximum efficiency in use.

Cutting, splitting and shaping tools may be classified according to the number of cutting edges they possess. Single-edged tools are probably the most numerous, and include razors, chisels, gouges, edging-tools and axes. Even an axe, which is generally regarded as an instrument for cutting down trees, and its smaller relation the hatchet, are available in many special forms. Knives and swords are made in endless variety from the sabre to the surgeon's scalpel. The carpenter's plane is an example of a specialized form of knife that is used for smoothing [3]. Another is the bricklayer's trowel, used for applying mortar.

Knives may have two edges, although each can function independently. The action of scissors and shears depends on two edges working in opposition. A file or saw has many cutting edges: it may have small teeth for cutting hard materials or coarser teeth for relatively soft ones. Saws also vary in the "set" of the teeth – that is, by the amount each is slightly bent to the left or right away from the saw blades. In the carver's riffler the cutting edges are reduced almost to points and scattered at random on the surface. And there are a vast number of cutting points on glasspaper and emery paper.

Cutting tools can be sharpened on artificial materials such as carborundum [8], but often natural stone is used: old paving stones made good sharpeners for cold chisels. A rotary stone can be used for sharpening flat chisels, but hand finishing is generally necessary. For curved-edge tools the stones must be chosen to match the required shape.

1 The hammer and the G-clamp are among the tools of the wheelwright. First he assembles the hub. Then he takes the spokes, which have been individually shaped by hand, and hammers them into place. The rim, which is made in sections, is fitted on to the spokes while the wheel is held in a vice. Finally a steel tyre is heated and fitted. As it cools it shrinks tightly into place to provide strength.

2 Cutting tools are a valuable aid to the gardener as well as to the carpenter. Garden shears operate on the scissors principle of two cutting edges working in opposition to each other. Lawn shears [A] are employed to cut the grass in awkward corners that would be inaccessible to a lawn-mower. Edging shears [B] are shaped differently for neatly trimming the grass on the edges of lawns grown on raised beds.

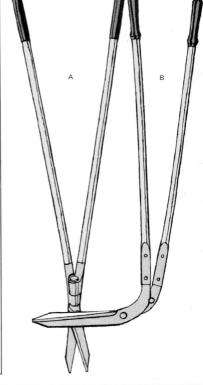

3 Woodworking tools have changed little over the centuries. This detail from the painting "Christ in the Carpenter's Shop" by the English Pre-Raphaelite artist John Millais (1829–96) shows a plane, pinchers and a frame or bow saw in which the blade is tensioned by a twisted cord across the top of the "bow". Other tools can be seen in the rack on the wall, including chisels, drills and a gimlet for boring small holes in wood; there is a vice in the foreground.

4 Grinling Gibbons (1648–1721) was a complete master with woodcarving tools. His delicate designs ranged over a diversity of subjects including fruits, flowers and shellfish. They can be admired today in a number of historic English houses and in churches and cathedrals, including Petworth House in West Sussex, Canterbury Cathedral and the chapel at Windsor Castle. This ornamentation is from the interior of Trinity College, Oxford.

Making holes in hard materials calls for specialized tools such as drills. Drills are generally made of hardened steel that is carefully shaped to deal with the material to be cut and the cutting speed. A drill can be held in a three-jaw chuck turned by a wheelbrace. A drill for boring timber is called a bit and is held in the two-jaw chuck of a carpenter's brace. Electrically powered drills are also in common use today.

Measuring and marking instruments

Science has devised instruments of remarkable precision for a wide variety of measurements, but many of those in everyday use are straightforward derivatives of prototypes used hundreds of years ago. A vertical is defined by a plumbline – a thin cord with a weight at one end. The horizontal is usually established with a spirit level, a transparent tube containing alcohol or other liquid with a bubble of air sealed into it. Compasses and dividers are used to draw circles and arcs; callipers are used to mark out or measure distances and micrometers are used to check small diameters accurately. For making larger measurements a flexible steel rule is more convenient. A set-square is used for marking and checking right-angles.

Grippers and screw-based tools

Pliers, tongs and tweezers are grasping instruments that handle objects more conveniently and more securely than the hand that manipulates them. They range from blacksmith's tongs for holding glowing hot iron bars to surgical forceps [7]. Workbenches and vices are commonly used by carpenters and metalworkers. The material that forms the face of a vice varies from hard steel to felt, according to the delicacy of the piece being worked on and the force applied. There are also other forms of clamp, including G-clamps [1], locking wrenches and simple foot-held straps.

Wood-screws and nuts and bolts are made in various shapes and sizes and of various materials, including plastics. Screwdrivers, spanners and wrenches [6] must fit their particular job. A tool that is too heavy may cause damage; too fine a tool will be ineffective and may itself be damaged.

Hand tools have been developed for a wide range of specialized jobs. The human hand is a versatile instrument, but unaided it has its limitations, particularly where power or precision is required. The gouging tool used so skilfully by this early engraver is, like all other hand tools, effectively an extension of his hand.

5 Cobblers' awls, seen in the bench rack, are typical piercing tools.

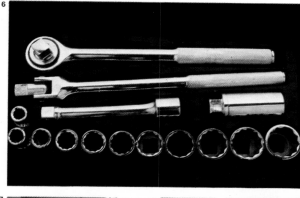

6 A set of socket spanners provides a range of sizes with a single handle, so making the set very compact. By means of a torque wrench they can be used to tighten nuts to a precise tension.

7 A modern operating theatre is equipped with a wide range of precision instruments made from durable, stainless materials. The surgeon's forceps are a kind of specialized gripping tool.

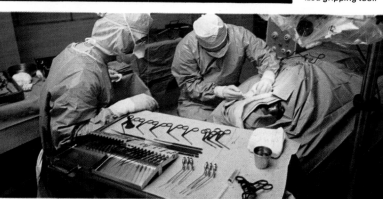

8 A hand grindstone is preferred for sharpening because there is little danger of overheating the tool. In many European cities travelling knife grinders can still be seen working in the streets.

9 A woodcarver uses a set of chisels and curved gouges with a wooden mallet to cut intricate designs or make sculptures in wood. The tools need to be extremely sharp; even big cuts are made using a series of carefully controlled small blows.

Hand-working metals

Most metals conduct heat and electricity well and many common metals can be mixed together to form alloys – substances with properties different from those of the parent metals but more useful in particular applications. Primitive man was not aware of many properties of metals, yet he appreciated their toughness and durability, their hardness and their ability to be worked into almost any desired shape. He also found that some metals retain an untarnishable lustre. Because of this, and their attractive appearance and rarity, these metals were used for money and personal adornment and became known as "precious".

Early steps in metallurgy

Civilized people of 6,000 years ago had learned to work with gold, copper, silver and probably tin. The earliest metallurgy stemmed from the discovery of native metals (which occurred naturally in their metallic state), but gradually man learned how to extract metals from minerals – metallic compounds – by heating them in a furnace, usually in the presence of charcoal. By about AD 1400 water-driven bellows could make a furnace so hot that even iron could be smelted.

Once metal could be melted it could be cast by pouring it as a white-hot liquid into a mould and leaving it to cool. Gradually men learned how to make more complicated moulds, designed to open so that the cast object could be extracted. But the most important early method of shaping metals was forging. The metal is not melted, but merely softened by being heated to a bright orange; then it is shaped by blows from a hammer. The smith [1] developed the anvil and an increasing range of shaped tools for hammering against hot metal [Key]. He learned to know the quality of metal by eye, to judge its exact temperature and to shape it accurately before it cooled. Medieval swordsmiths had to fashion blades that were hard (and therefore tended to be brittle) on the outside, to take a keen cutting edge, yet tough and supple inside.

Over the centuries a host of familiar objects gradually matured, differing from one region or country to another but each serving its purpose. Knives, ploughshares, horseshoes, door hinges, clasps, roasting spits, fire-irons and spurs were typical everyday objects made with no prior drawing or measurement but simply by skill and experience. The art of the jeweller also relied wholly on the craftsman's experience [5] and on his manual dexterity in fashioning objects that were both valuable and delicate.

It was in the making of jewellery that men first used the technique of soldering [7], in which a special alloy with a low melting-point is used to join other metal parts. It was also in jewellery that metals were first beaten out into thin wires, although during the Renaissance a way was found to "draw" wire by pulling it through a small hole.

Drawing and welding

The ability to be drawn is a characteristic of metals, stemming from their tensile strength and ductility (ability to change shape without breaking). Because the earliest metals were all fairly soft unless specially treated, for example as sword edges, they could be readily cut and pierced. This had to be done hot with large items, but small objects could

CONNECTIONS

See also
38 Hand tools
42 Metalworking machines
138 Small technology and transport

1 The smithy is centred around the open-hearth forge, where the temperature is kept high with bellows [1]. Anvils are pierced with a tool hole (hardie hole) [2] and punch hole [3]; the heavy body [4] resting on an elm block [5]. The anvil face [6] terminates in the beak [7]. A rake [8] lies near a tool rail [9]; a water trough [10] is used for quenching hot work. The smith [11] carefully positions work for the striker [12]. A floor mandrel [13] holds additional tools near the hearth [14]. Blacksmiths not only shod horses but made every kind of locally needed metal artefact for agriculture, engineering and even for the home.

2 In India delicate metalworking for decorative purposes reached a standard 1,000 years ago that has never been surpassed. Methods were devised for chasing and engraving soft metals, filling in the cavities with enamel, and even for inlaying one metal into a base of another metal of contrasting colour. Two of the best metals for decorative work are gold and brass, because they are highly malleable and can be deformed cold without splitting. This brass engraver in Jaipur (Rajasthan) is making a tray that will probably be sold in the West for hanging on a wall. In many countries such labour-intensive work would be economically impossible.

3 This selection of smith's products was typical of customer needs in Western countries in 1200–1850, during which period designs changed slowly. Door hinges [A] could be made on a small scale in brass for a decorative casket. An idle back [B] suspended a kettle over the hearth and could be tipped to pour the boiling water safely. A horseshoe [C] was an item made by the millions in distinct family designs that often reveal the place and time of manufacture. A lamp [D] could be pinned or screwed to a house wall and was an item on which the smith could demonstrate his skill with fine wrought ironwork. The 17th-century Kentish plough [E] is still wooden but incorporates vital iron components.

be shaped cold. The most delicate work of all was the beautiful technique of inlaying patterns of one metal into recesses in another metal of contrasting colour. Sometimes the inlaid pieces were secured by hammering the edges of the recesses. But in other cases soldering or brazing was employed, the latter using a brass (an alloy of copper and zinc), which was melted and run into the joints.

By 1800 the technique of welding metals using a hot gas flame was being developed and by the end of the century it had reached a high degree of perfection and been supplemented by welding with an electric arc. Welding, in which some of the metal is melted, exerts a profound effect on the metal's properties and much patient research was needed to find the best methods while simultaneously avoiding distortion caused by the severe heating.

The value of hand tools

Craftsmen using hand tools with the manual skill and judgment born of experience often provide the most efficient solution to many metalworking problems, especially where

tasks are essentially creative or the objects made to individual order.

The same is true in countless jobs in which metalworking plays a part, such as building construction, plumbing, farming, motor bodybuilding and repair, and in what have become known as do-it-yourself handyman tasks about the home. A special part is played by artists who work with metal, either in creating jewellery or in the larger forms grouped under the heading of sculpture [4].

Most of these tasks involve some form of joining, such as welding or soldering, and a few require a forge. All use hand tools such as shears, hacksaws and files to cut, shape and trim metal while cold. In nearly every case the precise dimensions are unimportant, provided that mating parts fit precisely. Even a watch-repairer normally works either by inserting a standard new part or making existing parts fit by eye and by feel. In modern manufacturing industry, however, the aims are entirely different. Mass-production of successive identical items usually calls for metalworking machines instead of slower human craftsmen.

KEY

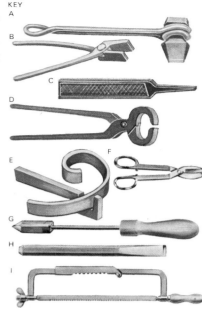

Tools for hand-working metals have changed little over the centuries. Cutting tools have either a single chopping blade (like a chisel), two shearing blades (like scissors) or many blades (as in a saw or file). Other tools were devised for gripping the hot metal while a workman bent or cut it. Blacksmith's tools shown here include a hot set [A] for cutting iron softened at red heat, plain tongs [B], horse rasp [C], hoof clippers [D] and a scrolling dog [E] for bending wrought iron. Also shown are tin snips [F] for cutting thin sheets, soldering iron (G), a cold chisel [H], which is used on cold metals as opposed to hot metals, and a hacksaw [I].

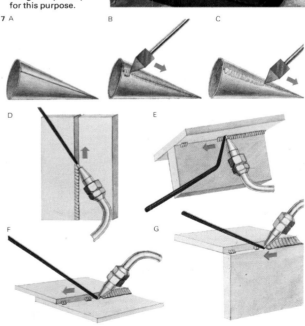

5 Jewellers fashion delicate designs in gold and silver by brazing previously shaped pieces together. A propane torch provides a fine needle flame; the delicate parts must be held with tweezers. Because precious metals are used, the jeweller must be careful to collect even the smallest pieces of waste material. Larger ornamental pieces, such as silver cups and bowls are fashioned from a flat sheet which is beaten with a round-headed mallet on a leather cushion until the required shape gradually emerges. Final shaping may be done on a series of anvils designed specially for this purpose.

4 Modern sculptors use oxy-acetylene welding gear to cut, join and shape metals. If it is made hot enough, the molten metal flows and the artist can control the effect to create new textures unobtainable in any other way. Such techniques also allow a sculptor to employ a wide range of materials. At one time, most metal sculptures were made by casting, using metals such as bronze. Today a sculptor can work with high-melting-point metals such as stainless steel. Like all welders he must wear dark goggles to protect his eyes from the glare and flying sparks.

6 Riveting is a method of joining metals without melting them. A hole is drilled through the parts to be joined [A] and a hot rivet passed through it [B]. The end of the rivet is beaten with a ball-pane hammer [C] until it flattens over, making a tight joint [D]. Large structures such as girders and plating on ships' hulls can also be joined using rivets and mechanical hammers.

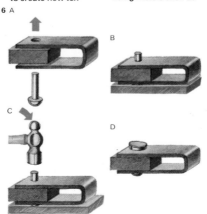

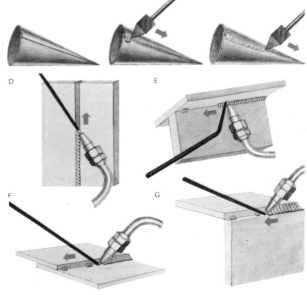

7 Soldering can be used for joining sheet metal, particularly copper and brass. In making this seam, the metal is first bent into shape [A]. The metal is treated with a flux (to clean the edges to be joined) and the two edges "tinned" with a thin film of solder [B]. Finally a hot soldering iron is passed along the joint to fuse the two tinned edges together [C]. There are various ways of joining sheet metal by welding. Butt welds can be used to join two plates in line [D] or at right-angles to each other [E]. A lap joint [F] is welded on each side and a corner joint [G] has a fillet of weld run along each seam.

Metalworking machines

Ever since Hiram of Tyre made the pillars of Solomon's temple in metal, men have continued to work metals for construction mainly by casting and forging. In casting, molten metal is poured into a mould; in forging, the metal – hot or cold – is beaten into shape with a hammer. These are still the two basic metalworking processes.

Casting and forging can be used to produce articles in which accuracy is not of prime importance. But when accuracy in a product is essential, then foundries supply metal in stock forms such as ingots, castings, sheet, bar and tube for subsequent machining to the required shapes.

Casting and forging

Traditional sand casting [1], such as a sculptor might use, has to be employed for bulky or irregular shapes, especially in iron and steel. Huge solid masses, such as a propeller for a giant tanker, could hardly be made in any other way. Even so, the finished casting needs to be machined or ground to the exact profile and smoothness.

Some small metal parts can be made by the repetitive process of die casting [2]. There are also special items, such as turbine blades for jet engines, which have to be made of metals that stay very strong even when extremely hot and casting is often the only economic way of making them even if they are quite small.

For many other items of all sizes, forging can be the answer. The modern equivalent of the blacksmith's hammer is the giant hydraulic press [Key]. Used for large masses of metal, it squeezes white-hot billets under a force of perhaps 50,000 tonnes. For small products die forging is used. The soft, hot "blank" is banged into shape between two precisely shaped dies which come together in perfect register.

A variation of forging is hot rolling. An ingot or "bloom" is passed, glowing with heat, between pairs of shaped rollers like a giant mangle. It may make dozens of passes, sometimes being reheated, as it grows longer and thinner. Eventually it will become an exact strip, an angle-section bar, a structural girder, a railway rail or, in modern plant, a seamless length of tube.

In other modern metalworking techniques basic metal stock, such as bar, rod or tube, is machined to the final required shape.

Using machine tools

Machine tools work with much greater power than a craftsman can exert with his hands. They therefore fashion metal parts faster and at a lower cost. But far more important is the fact that they exert rigid control over the workpiece and are cutting or shaping all the time. They can go on making the desired number of parts, each identical to the last, so that the machine-made parts are interchangeable. Each conforms exactly to an original drawing showing precise shapes, dimensions and tolerances. The tolerances indicate by how much a dimension can be slightly larger or smaller than the desired value, and are measured in micrometres or thousandths of an inch. Without modern machines a craftsman would be hard-pressed to maintain such accuracy and his rate of output would be very low indeed.

Machining a metal involves cutting it with a machine tool such as a milling machine

1 Sand casting is the most widely used casting method. A wooden pattern is made up from the original design, then packed with sand up to its largest cross-sectional area in a steel box [A]. The top part of the mould is assembled. It is clamped on to the bottom box and more sand rammed into it. A wooden runner and riser are fixed in position [B]. The sand core for a hollow in the casting is made. The sand for this is mixed with sodium silicate which forms "silica gel" when carbon dioxide is pumped through it. This "gel" has a syrupy consistency and binds the sand together. The mould is split and the pattern removed. The core is placed in position and the mould assembled [C]. Runner and riser are removed. Molten metal is poured into the dried mould through the conical-shaped runner. Displaced air escapes through the riser [D]. After cooling, the mould is split open and the casting removed [E]. The runner and riser are cut off and the sand core is knocked out. The finished casting [F] shows the hollow produced by the core.

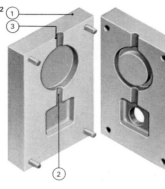

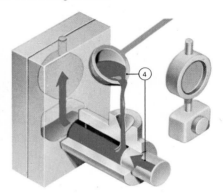

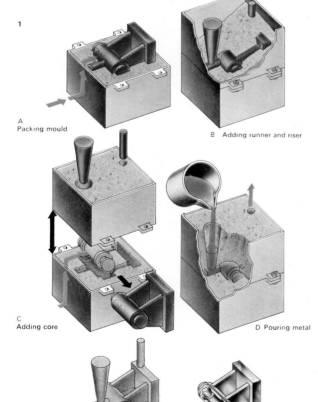

1

A
Packing mould

B Adding runner and riser

C
Adding core

D Pouring metal

E
Rough casting

F Finished casting

2 Die casting is a method of making large numbers of castings from non-ferrous metals. Common die castings, such as letter boxes and door knobs, are made from alloys of zinc. Instead of a sand mould [1A] which makes only one casting, a die or permanent mould [1] is used. This is made of cast iron with runners [2] and risers [3] like a sand mould, and it can be opened to get the casting out. For a hollow casting, the metal cores are either part of the die or retractable to release the casting. The metal may be cast by gravity or pressure from a plunger [4].

3 Press tools which consist of a die [1] and a counter die [2] are made of hard steel [A]. A piece of sheet metal [3] is put between the two and the press closes on it [B]. This method is used for forming sheet metal into various shapes, including simple bends and complex curves. The various sections of body panels of an automobile [C], for example, are generally pressings. Sharp curves may require several successive pressings, each with a tighter curve than the one before, on different machines. The different press operations such as blanking (cutting a piece to the right shape in the flat) or piercing (making holes in it) may be done on different presses or some may be combined, depending on the size and thickness of the material.

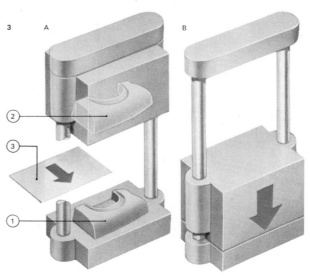

3 A

B

C

(miller), planer, router or lathe. Millers [4] hold the work still while rotating cutters pass over it. Planers draw large items past a fixed cutter, whereas a router resembles a drill which mills the surface instead of boring holes. Grinders [5] use another method of machining, but instead of having cutting tools of very hard metal (or sometimes even diamond) they use wheels with surfaces made up of millions of hard fragments, each of which takes a small cut. Lathes [6] rotate the workpiece while tools are arranged to cut it.

Advances in metalworking

Since 1950 many completely new ways of working with metal have been brought from the research laboratory to the production plant. The aircraft industry pioneered chemical milling, in which sheets of any size are etched away in baths of acid or other corrosive chemicals. Portions of sheet can be protected by a surface mask which prevents them from being attacked. The rest can be eaten away in a controlled way, with no scratches or machining marks on the surface. This is important because even the smallest scratch

or imperfection can induce metal fatigue.

Electrochemical machining (ECM) [8] is a variation in which the liquid bath is not corrosive but an electrolyte (carrier of electric current). The workpiece is connected in an electric circuit and then eaten away by a shaped electrode, rather like electroplating in reverse. Another electrical method is spark erosion [8], in which even the hardest parts are gradually shaped by millions of sparks. Yet another, and totally different, electrical method is electromagnetic forming when massive currents are suddenly switched through magnetic coils which slam the workpiece against shaped dies.

In contrast there are other new methods of extreme delicacy used for shaping on a microscopic scale. Ultrasonic machining [8] is a form of grinding, useful for extremely hard material or non-metals that must be finely shaped. Electron-beam machining uses a concentrated beam of electrons to melt away parts not wanted. Laser machining [8] does the same with an intense beam of light. Such methods can be used to shape electronic circuits that would easily fit on a pin's head.

KEY

In forging, cast ingots or billets of metal are heated red hot and stamped roughly to shape in powerful hydraulic presses.

4 Milling machines cut metal with a rotary cutter [1]. The universal mill [A] has a table which can be moved in all 3 planes [2]. A workpiece is clamped to the table or to the dividing head. Gears [3] allow the workpiece to be rotated during or between strokes. As well as milling surfaces and grooves, the machine can plane a flat workpiece [B]. A cutter can be angled to mill threads. [C] shows the dividing head being used to make a helical cut on a cylindrical workpiece.

6 Turret lathes are for repetitive work where large numbers of the same component have to be made from bar or tube fed through the chuck [1]. Six tools can be used on the turret [2], and a front tool post [3] on the cross slide [4] of the saddle [5] can carry four more. A rear tool post can also carry tools.

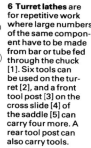

7 Drilling machines of many different kinds are used in workshops. The multi-spindle model shown is a time-saver in mass production. It can be installed as part of a machining line for a particular component, and tooled to drill all the holes in that component in one operation.

5 Grinding is used mainly as a finishing operation giving a smooth surface. The grinding wheel consists of hard particles, usually alumina or silicon carbide bonded together with a resin. Three types of grinding machine are shown: the surface [A], universal [B] and centreless grinders [C]. [A] smoothes plane surfaces, [B] is restricted to internal and external cylinders and [C] to external cylinders or long bars. Sections 1, 2, 3 are soft, medium and hard grinding wheels.

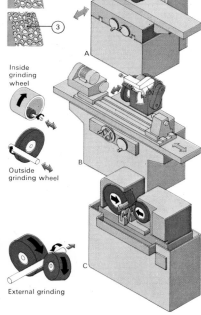

Inside grinding wheel

Outside grinding wheel

External grinding

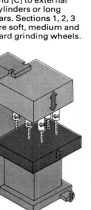

8 New techniques in metalworking include laser machining [A] in which an intense beam of light [1] cuts hard material [2] such as diamond and tool steel. The image of the work and that of the pattern are combined on a closed-circuit TV display. In spark erosion [B], a high-voltage spark etches away small pieces of a hard material. In electrochemical machining [C], corrosion of metal in salt solution (brine) is speeded up using

a high voltage. The hole is cut the same shape as the electrode. Ultrasonic machining [D] uses high-frequency sounds beyond the range of human hearing to vibrate a cutting tool. An abrasive powder then cuts through the workpiece, which could therefore be a non-metal. A cavity is produced which is a mirror image of the tool. Typical shapes produced by these various methods are shown in [3, 4] and [5].

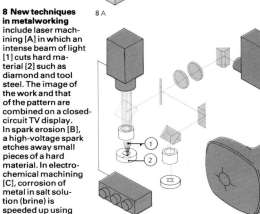

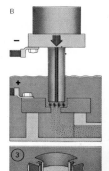

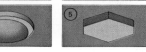

Working with ceramics

Ceramics – originally pottery, glass and enamel – are among the oldest products of human ingenuity. Yet so great is their versatility that, even today, new uses are being found for this class of materials [Key], from high-voltage insulators to radar and computer components and nuclear fuels.

The story of ceramics

During the last Ice Age, hunters made clay images of animals and hardened them near fires. Clay is basically a mixture of aluminium and silicon oxides, together with various impurities. When heated moderately it loses its chemically-bound water and forms a porous, hard material suitable for making hearths, images and, if pre-shaped, pots. To make a non-porous product, higher temperatures are necessary so that a portion of the material melts, or fuses, thus filling up the tiny holes in it.

If silica is fused and then allowed to cool slowly the result is glass rather than pottery. Glazing – the production of a glassy surface on a solid object – dates back to 4000 BC but the first all-glass vessels did not appear until

1500 BC , and it was nearly another 1,500 years before the technique of blowing glass vessels was developed.

The production of ceramics and glassware was a highly mechanized industry, even in the nineteenth century. But in the second part of that century developments in technology put the craft on a more scientific footing. The chemical structures of ceramics contain no free electrons so they are non-conductors of electricity. This property became important when large-scale electricity generation began. The development of satisfactory insulators, for example, required much more knowledge of materials than did the manufacture of teacups. Consequently, scientific studies of raw material composition and firing methods had to be undertaken.

Chemical composition

Chemically, ceramics and glass are composed largely of compounds of oxygen with other elements. Glass, which is composed largely of silica (SiO_2), is a special case – most oxides do not form glasses but many ceramics contain a glassy component. Porcelain, for example, is

made from a mixture of clay, sand and an alkaline flux. Such fluxes, often common minerals such as felspars, lower the temperature at which silica fuses. A fragment of porcelain viewed through a microscope can be seen to have several different kinds of particles bound together by a glass matrix.

Many of the particles in a ceramic are tiny crystals in which the atoms are arranged in simple geometric patterns. In a glass there are no such geometric regularities and the atoms are oriented at random. Physically, glass has a crystalline structure more like that of a liquid than of a solid.

If glass is not cooled properly (annealed) from the molten state [5], tiny crystals do appear and the result is a brittle, semi-opaque material. Some crystal formation can occur with slow decomposition and some very old glass, as in Roman bottles, is partly opaque [3] due to the chemical action that has taken place over the hundreds of years that the glass has been buried. Recently, techniques have been devised to encourage controlled crystal formation in glass by reheating it under specific conditions [8]. The

CONNECTIONS

See also
46 Materials for building
52 Small technology in the home
36 Tailor-made materials

1 Glass was made in antiquity by [A] fusing raw material held by a glass thread on a sand core to form a bottle, or [B] by shaping it in a mould. A wine glass was made [C] by blowing molten glass in a mould, adding a stem, shaping the foot and trimming. Press moulding [D] is a more modern technique. A decorative paperweight [E] was made by fusing coloured glass rods and surrounding them with molten glass. Glass-working tools [F] include a blowpipe, rod, tongs, shears and rolling plate. Glass was originally melted in a crucible [G]; later bell [H] and cone [I] furnaces were used.

2 Stained-glass windows are made from many small pieces of coloured glass joined by lead strips. An artist first makes a full-scale drawing of the design. From this cartoon [B], a cutline [C] is traced on to linen. Glass is cut to the exact shape of the cutline [D] and the pieces are leaded up and soldered [E, F]. [A] Tools used are: [1] cutting knife; [2] stopping knife; [3] lathekin; [4] nails; [5] muller for mixing paint; [6] glass cutter; [7] palette knife; [8] grozing pliers; [9] tray lifter; [10] soldering iron; [11] badger-hair brush; [12] leads; [13] needle point; [14] sharpened wood; [15] and [16] hogshair brushes.

3 Long ago, glass was coloured by the addition of small amounts of metal to the melt. Gold, for example, can be used to make glass red or blue. The colour depends on the size of the individual particles dispersed in the melt. A remarkable example of Roman glass coloured with gold and silver is the Lycurgus cup in the British Museum. If seen by reflected light it appears an opaque green; by transmitted light, translucent purple-red.

4 In enamelling, a coat of glass is fused to a metal surface. Silica, red lead and potash are melted to make a flux, which can then be coloured by the addition of metal oxides; addition of tin and lead oxides gives an opaque enamel. In *champlevé* enamelling [A], pieces of metal are cut away, leaving metal lines between them to form the outline of a design. Pulverized enamel is laid in the troughs and then fused. Afterwards, the enamel is filed and polished. The other techniques of enamelling shown here – *cloisonné* [B], *plique à jour* [C], *basse-taille* [D] and enamel painting [E] – are all variants of *champlevé* enamelling.

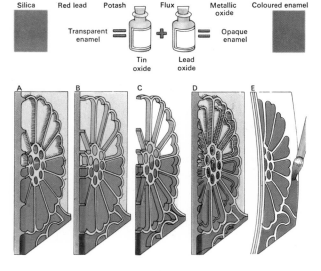

Silica + Red lead + Potash = Flux + Metallic oxide = Coloured enamel

Transparent enamel = Tin oxide + Lead oxide = Opaque enamel

result is a tough heat-proof and flame-proof glass ceramic, which is now widely used in kitchen ovenware.

In firing some ceramics it is not necessary to form much glassy material in order to hold it together. Where the solid has formed solely as a result of sintering (aggregating as a result of applied heat) with no fusing, however, it is usually porous like brick. This is because the particles of material come together more closely but, in most cases, do not drive out all the minute air holes.

It is because of their very high melting points that the raw components of many ceramics have to be sintered rather than fused. Such high melting points mean that ceramics act as good refractories – lining materials for crucibles and furnaces.

Modern uses of ceramics

High-density ceramics can now be prepared as "whiskers", which may be used to strengthen other materials – or even to strengthen more conventional ceramics. Thus ceramics can play an important part in modern engineering, in some cases replacing

metals – as in engine parts that must operate at ultra-high temperatures.

Not all ceramics are oxides, and some of the newer ones are compounds of different elements with carbon or nitrogen [7]. Carbide-tipped drill bits exemplify the everyday uses of new ceramics.

Electricity and magnetism are usually associated with metals but in the developed countries the electrical and magnetic properties of ceramics are widely used. Solar batteries, which convert sunlight directly into electricity, rely on the use of modern ceramic substances.

If some iron-containing ceramics are cooled in a particular way the materials that result are capable of converting mechanical into electrical energy (and vice versa) because of the alignment of electrical dipoles in the material. Much of today's sound transmission and recording is based on such "ferroelectrics". Similarly, "ferrites" are ceramics in which the magnetic rather than the electrical dipoles have been aligned, and they. form an essential part of computers, radar equipment and small electric motors.

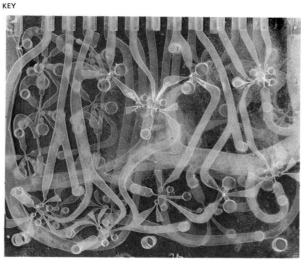

Glass, one of the oldest materials known to man, can be used for some of the most up-to-date tech- nology. These fluid circuits have been etched in photo- sensitive glass. A spe- cial formula means that on treatment with radiation the glass crystallizes, making it easy to etch with acid.

5 The float glass process was develop- ed by Pilkingtons Ltd in 1959 in the UK and is now used throughout the world for the production of flat glass of the type used, for example, in windows. The ingredients are mixed in a hopper [1] and then melted in an oil-fired furnace [2]; the molten glass [3] then passes on to a float bath of molten tin [4] in a special non-oxidizing atmos- phere. The glass spreads out over the molten metal surface to form a uniform, flat sheet. As the glass passes through the bath it is gradually cooled so that it emerges with a firm surface that is not deformed by the rollers that take it into an annealing lehr [5] where it undergoes further cooling be- fore it passes to a computer-controlled cutting and stack- ing operation [6].

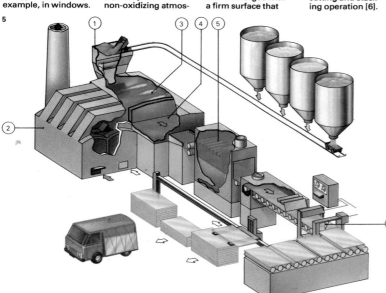

6 Pottery is generally defined as a porous ceramic material made from clay by firing until it is hard. The firing temp- erature is compar- atively low so that the individual part- icles in the clay do not melt and fuse together, as can be seen in this micro- scopic section of a piece of pottery. To make it water- proof a glass-like glaze has to be fired on to the surface. Non-porous ceramics such as porcelain, which do not necess- arily have to be glazed, are fired at higher temperatures so that some part- icles melt and fuse together. Often silica is added to give a glossy texture.

7 Oxides or carbides of fissionable ele- ments such as uran- ium are sintered to form ceramic pellets which are packed in- to metal containers and used as fuel elements in nuclear reactors.

8 Glass ceramic is made by heat-treat- ing preformed glass so that it devitri- fies. By carefully controlling the crystal formation, glass ceramics that combine mechanical strength with good heat-proof proper- ties are produced.

Materials for building

One of man's basic needs is shelter from the natural elements. Primitive man met this need by dwelling in caves and then his descendants began to construct buildings. Key factors in the design of these shelters were the climate and the materials available. For example, in hot, dry countries, houses with tiny windows and thick walls of mud were built to keep out the heat and sunlight; in rainy climates people normally built sloping roofs of grass and rushes, so that the rain ran off the house without penetrating the interior; in earthquake areas, houses were built of light materials – in Japan some internal walls are still made of paper.

Building with timber

Originally an ancient building material, timber [2] is now being applied to major structures, although its widest use is still in house building. Improved design techniques and treatments now overcome many weaknesses. The strength of timber is different along and across the grain, so it is made more uniform by using adhesives to bond multiple layers with the grain running in various directions. Durability is improved by better methods of preservation and the timber is treated so that it does not burn readily. Devices for joining timber structures have also been improved, notably metal plate connectors with multiple projecting teeth, which spread the load over a wide area.

Timber is now used to make fully prefabricated houses [6], which are factory finished and ready for erection on a firm base. More extensively it is used for joists, rafters, window frames, doors, floors and studding [3] for internal non-load-bearing walls which are to be faced with plasterboard. Timber also has appeal for conservationists because it is renewable – we can plant more trees – and its manufacture into a usable form involves no pollution of the environment.

Building with stone and brick

Natural stone is still used for building, although it has taken on a comparatively minor role in industrialized countries, being confined mainly to facings and other decorative finishes. Stone is obtained from quarries by blasting or splitting with wedges. If, like limestone, it is layered, it must be used in such a position that its layers lie at right-angles to the direction of pressure.

Fired bricks are much used for houses in developed countries. Their size and type are governed by national standards and there is a move to set up international standards. The size and proportions are broadly chosen so that a bricklayer can hold a brick using only one hand and the brick can bond with others lying both parallel and at right-angles to the wall face. Despite standardizations, there is still a wide range of bricks to choose from in terms of appearance, strength and durability.

Most bricks are made from clay or shale and are "burnt" in a kiln. However, some are made of silica sand and lime and these are known as calcium silicate bricks. Building blocks are a little larger than bricks and are generally made of concrete.

Building with concrete

Most people think of concrete as a modern material, but its history begins with the Romans, who used it to build their aqueducts and amphitheatres. Another common

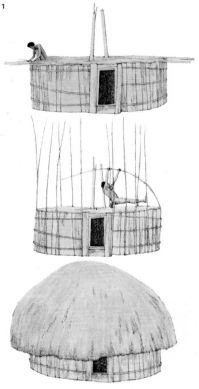

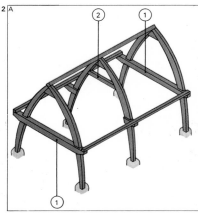

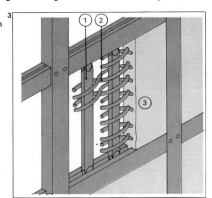

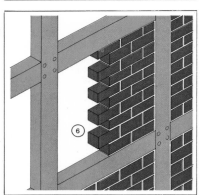

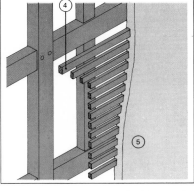

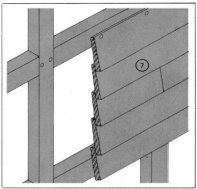

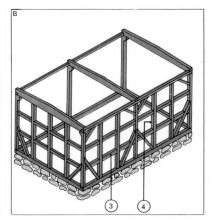

1 A complex house belonging to a woman of the Dani tribe, who live in the forests of New Guinea, illustrates the use of traditional materials. It is made with tools of stone, wood, bamboo and bone. It has a lower room and a sleeping loft; the lower room has wooden walls and a tall doorway. The dome is made with saplings, which form the frame of the whole house, bent over and attached to the centre posts. Then the house is covered with long grass thatch made by the men. The sleeping area in the roof has a reed floor, but it is provided with a hearth on a mud base to keep people warm at night. Grass is trimmed and prepared for use with a stone adze, a primitive tool rather like an axe with an arched blade mounted at right-angles to the handle. Vines are used to lash the building together.

2 Timber was the main structural material used to build small houses in the forested areas of Europe from the Stone Age to the 1700s. Cruck constructions [A] were formed by arched timbers which supported the roof and walls. Horizontal beams [1, 2] formed trusses to take the weight. The box frame [B] had added rails [3] and studs (vertical timbers) [4]. These helped to stiffen the structure and provided attachment points for cladding.

3 In frame walls the space between the timber frames can be filled by timber staves [1] surrounded by wattle [2] and covered by daub and plaster [3]; or timber laths [4] covered by plaster [5]; or with brick nogging [6]. The frame can also be covered with weatherboard or clapboard [7]. In Europe, timber was soon eclipsed as the principal material for frame structures. In the early 19th century, the development of rolled iron and steel girders enabled builders to construct larger buildings. At about the same time glass was being made into large sheets and by the close of the century reinforced concrete was in use. Recent innovation and improvements in timber technology – notably better methods of connecting beams by using metal plates and improved methods of lamination – have reopened the possibilities of its use for making frames.

misconception is that concrete is used only in the construction of large buildings (apart from foundations). Concrete is a mixture of water, stone and sand with a binder (usually Portland cement). Today an industry has grown up that precasts concrete into forms suitable for house walls and roofs, as well as for applications in bigger buildings. When a set of moulds is made for this purpose, the process is called "battery casting". A recent extension of this idea is a system of battery casting concrete panels for rapid building of dwellings in developing countries. They can be set up on site without the need for factories to process them. A "package" is supplied consisting of a vertical battery unit for casting, a transporter vehicle, a tower crane and accessories that enable a sub-contractor to construct medium-sized blocks of apartments at a rate of one per day. The basic design is standard, but even so the system gives great architectural and planning flexibility using only a few units. These units can be assembled in various ways, according to the needs and styles of the individual community or country.

Other composite materials are gaining importance in building homes. In essence they are well established; for example, the old wattle (interlaced rods and twigs) and daub of mud or clay. A modern form used in house building is glass-reinforced polyester resin. Panels of this material are themselves composites; they are applied to timber framing and act with the frame to give so-called "racking resistance", which prevents distortion of the frame. Similarly internal walls may be composites of timber frames or metal frames fitted with honeycombs of metal strips, in both cases faced with plasterboard.

So far, few plastics have been used as structural materials, although they are widely used for a variety of accessory roles in buildings. These include panels, gutters and their supports, drainpipes and decorative ceilings. Water can be stored in plastic tanks and led through plastic pipes. Foamed plastics can be injected into cavity walls to improve thermal insulation. Some designers have built experimental all-plastic houses, but these have not yet been commercially successful because they are too expensive.

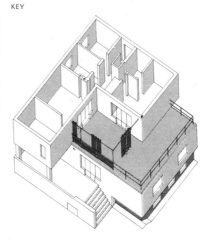

Standardization in building makes use of basic units that can be assembled in various ways in order to provide flexibility in design.

The idea was pioneered by Walter Gropius (1883–1969), the German architect who founded the Bauhaus school of architecture in 1919.

His own house was one of four residences for the teaching staff, sited among the trees of the Bauhaus school in Dessau, Germany. These buildings illustrate Gropius's theory of standardization. The houses are built of similar slag concrete blocks, with the walls as the main supporting elements. His own two-storey house is basically L-shaped with terraces at different levels. The external appearance of the building was as important to Gropius as the functional planning of the interior. Today unit construction, using prefabricated sections, permits the rapid building of homes by relatively unskilled workers in developing countries.

4 Thatching [A] is a roofing technique that involves covering rafters [1] with straw layers [2] vertically fixed to horizontal battens [3]. There is no crossing or weaving, though hazel rods may be used to secure the thatch on a straw roof. Chief materials used are straw and reeds. Stone tiles [4] could also be used, with round tops and a peghole [5] for securing them to battens [B]. Regional and economic conditions have influenced building, particularly since earlier times when transport was difficult, slow and expensive. As it became more efficient, manufactured goods were applied over a wider area. Where slate was available, it was used; if not, then clay tiles were made. With the new rail systems, slate came into use in the countryside and other materials, usually reserved for one area, become readily available almost everywhere.

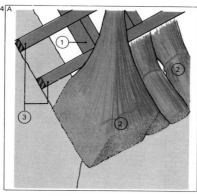

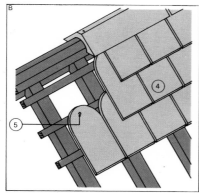

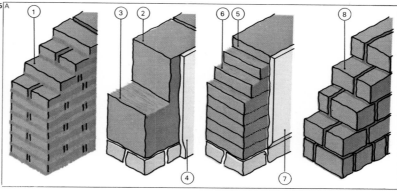

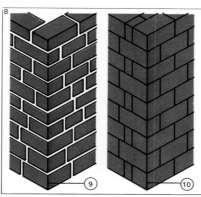

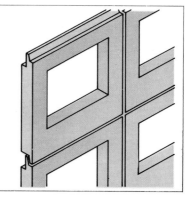

6 Complete home unit systems are now made of timber (with steel corner columns) in factories and delivered to site fully finished and ready for assembly [A]. A concrete base is prepared in advance with points for connections to mains services [B]. The house is ready for occupation within days. Where road regulations limit the width of the load, the home unit may be transported on its side, then turned upright for assembly to the architect's design.

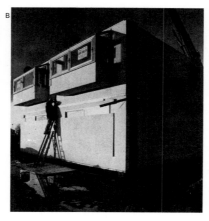

5 Walls may be built of a wide variety of materials [A] including a mixture of sods of turf [1] and thick clay mixed with pebbles [2] linked together with straw [3] and then faced with a plaster coating [4]. Alternatively there may be thin clay layers [5] linked with straw [6] and faced with a plaster coating [7] or a wall built of clay "lumps" [8] like bricks but unburnt. They are generally sun dried, like adobe.

True burnt brick walls [B] may resemble the early irregular brick [9] or be like the standardized brick laid in what is known as English bond [10]. Walls may also be made of precast concrete sections [C]. Other materials used for walls include stone in a wide variety of presentations using regular or unusually shaped stones, sometimes backed by brick and faced with plaster.

Making large buildings

The highest building constructed by 1975 was Sears Roebuck Tower in Chicago, USA. It is 443m (1,454ft) tall with 110 storeys. It houses 16,500 people and has 103 elevators and 18 escalators. Structural engineers have calculated that buildings up to 3km (1.75 miles) tall are technically feasible and such structures may be built in the future.

Why have large buildings?

The chief reason for constructing large buildings has been to make the best use of the limited (and expensive) land area available in the world's major cities. New York City, largely confined to an island, could not spread outwards so it spread upwards. A secondary justification for large buildings for commercial purposes is that the whole of a large company's staff – perhaps a thousand or more people – can be housed together, with obvious gains in efficiency.

Large buildings for homes, such as tower blocks of apartments, also make the best use of land and simplify the installation of services such as electricity and heating. But they can also create social problems – living many metres above street level can lead to lack of contact with the surrounding community. In terms of people, a large tower block is the size of a village but it generally lacks a village's amenities. Many skyscraper hotels, on the other hand, are built complete with shops, restaurants and recreation facilities.

Whatever its size, a building must shelter people from the elements and have some system for controlling the internal climate. It must provide spaces for specific uses and such services as escalators, elevators and stairways for its users. As structures, buildings must be able to carry not only their own loads, which vary according to the heights and materials used, but also the weights of the people and things within them.

Building materials

Most larger buildings are structures of steel, reinforced concrete and prestressed concrete. But a timber home-unit system can be built up to ten storeys high and masonry (blocks or bricks) can take up to 18 storeys built in cellular form – flat slabs arranged in "box-like" fashion giving mutual support.

Aluminium alloys are seldom used on large buildings except for roofing over wide spaces without supporting columns. High-tensile cables may be used in low-rise structures of a novel skeletal design, particularly where a large open space is needed, as in the Olympic tent at Munich, aeroplane hangars or sports halls. Timber tends to be used for skeletal structures such as post and beam constructions and common trusses (the supports for roofs and bridges). So it is more likely to be found in single-storey buildings – although timber structures can be built up to greater heights and spans for schools, hangars, exhibition halls and for lightly-loaded floors, footbridges and roofs. Joints are made by mechanical fixings such as multiple nails, staples and bolts, or with adhesives. Treatments for preserving wood against biological attack and for retarding fire, as well as guidance on timber designs, have all improved considerably in recent years.

Masonry is brittle, strong in compression but weak in tension. Its use is therefore largely limited to columns, walls and arches. Designs have to ensure that lateral forces are

1 Staging can be eliminated by carrying out work close to the ground and then lifting the work up to its final position. One method of doing this is the "lift slab" system. In this, which is applied to the type of structure based on columns and flat slabs, the upright columns [1] are first cast at only a part of their height or sometimes at their full height. Using moveable moulds, the roof and floor slabs are then cast in sequence as a pile or stack into their plan positions around the upright columns at ground level, one on top of the other. The slabs are separated by special linkages which are later used for connecting them to the columns. The stack of slabs are lifted, one at a time, by means of hydraulic jacks [4], until each reaches its allotted level. Each floor [2] is then secured into its final position, using special kinds of anchors [3]. The slabs are lifted by jacks at the very tops of the columns, and for this reason great care is needed to ensure that the columns themselves are upright and stable. Sometimes the method is modified and whole units of slabs with cast-in beams, or shell roofs, are jacked up from below. The system still demands formwork for casting the units, but it means that all this type of work can be completely carried out at ground level. This results in more economical formwork, virtually eliminating the need for raised supports. When each pair of floors has been secured, the outside walls [5] are built between them using brickwork or precast concrete slabs, leaving openings for window frames and doors to be fitted at a later stage.

resisted and that structures cannot buckle or fail by horizontal sliding.

Steel [3] is used in skeletal structures, which have a self-supporting framework clad with other materials. Or it may be used in "surface-active structures", where the frame, the rails supporting the covering sheets and the sheets themselves – particularly when they are of steel – are regarded as acting as a single unit. Ultra-high-rise buildings [Key], with heights of more than 100m (328ft), are rare outside the United States and demand more sophisticated frame techniques.

Today reinforced concrete [6] is widely used for building frames. The concrete may be cast on site or precast, depending on the ease of using formwork (the wooden mould into which concrete is poured) on location and the number of times the process has to be repeated [5]. Other factors are the nearness of the factory to the site, the availability of cranes and the speed with which the building has to be constructed.

Concrete may also be compressed by tensioning its reinforcements and it is then known as "prestressed". This treatment ena-

bles embedded rods or high-tensile wires to carry an amount of direct or bending tension without the concrete itself going into tension, making it effectively much stronger.

Reinforced concrete can be used for skeletal structures, often based on precast units and used mainly for industrial and shed-type buildings. Concrete may also be used in a shell form of construction, where the surface is an intrinsic part of the structure. Exciting and original shapes of shell have been designed, notably in South America and Spain. Probably the most famous such structure, however, is the group of concrete shells of the Sydney Opera House in Australia.

Pneumatic buildings

Large "tents", kept up by inflating them with air under slight pressure, are called pneumatic structures. They are made of strong fabrics and, depending on the degree of curvature and span, may be reinforced with membrane ribs and cables. So far these structures have been used mainly for temporary buildings, sometimes for protecting permanent ones while they are being built.

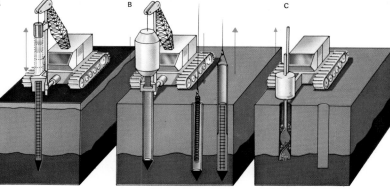

The Empire State Building was the tallest building in the world until 1972 and remained the third tallest after completion of the Sears Roebuck Tower in Chicago and the World Trade Centre in New York City. The latter is 412m (1,350ft) tall, excluding the 68m (223ft) television mast that surmounts it; it has 102 storeys. The main Empire State Building was opened in 1931. It is a steel structure and has been reported to have swayed a total of 75.4mm (2.97in) during a gale of 164km (100 miles) per hour in 1936. Four power beacons shine from the top of the building. The television mast was added in 1950–51.

2 The lift slab system is shown in section.

1 Reinforced concrete floors assembled at ground level
2 Reinforced concrete column
3 Steel rods
4 Hydraulic jack
5 Floors fitted at appropriate levels

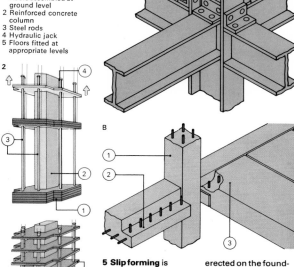

3 Steel beams are riveted together [A] to form a rigid framework. The horizontal beams are fixed to vertical columns. The frame construction using reinforced concrete [B] shows the vertical pillar [1] with its reinforcing rods, the main beam [2], again showing reinforcing rods, and the slab floor [3], which is partly cut away here to show how it is attached to the main beams.

4 Three modern pile-driving methods are: driven piles [A], where a prefabricated pile is driven into hard stratum providing a firm base; driven and cast piles [B] where the vibrator drives a steel tube into the ground: it is reinforced by a steel grid and withdrawn after concrete is cast into it; and bored and cast piles [C] where a hole is lined and a concrete mixture is cast into the hole.

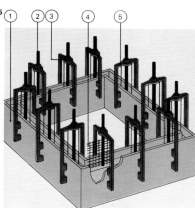

5 Slip forming is the application of continuous casting of concrete to the construction of tall buildings. Formwork in the form of a steel shuttered mould [1] about 1.3m (4.5ft) deep is erected on the foundations. It contains a network of heavy steel wires or rods, which act as reinforcement [4]. Concrete [6] is continuously cast into the mould and the forms are slowly raised by a screw or hydraulic jack [3]. They then effectively "climb" steel pipes or rods [2] previously cast into the foundations. The upward movement of the formwork is transmitted by means of the yokes [5].

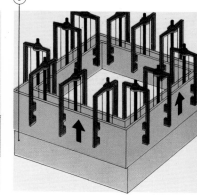
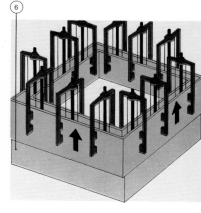

6 A typical reinforced concrete block [A] has steel rod reinforcements with standard hook bends [1]. A load-bearing beam is strong in compression but weak in tension. Without reinforcement, concrete will crack under strain whereas a reinforced beam will not. A concrete block [B] with metal sheathing [2] encloses the tendons [3], which are tensioned [blue arrows]. One end is anchored and the other released so that the tendon compresses the concrete to form a prestressed block. Ordinary cast concrete [C] has a variety of uses: for paths, roads and foundations, with wooden shuttering [4], hardcore base [5], and the shuttering pegs [6].

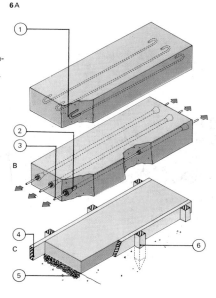

Building with local resources

Over the centuries industrious people with both the will and opportunity to create wealth found ways to make their lives comfortable. The ancient Romans knew nothing of modern technology yet they had a water supply system that provided Rome with nearly 1,000 million litres (220 million gallons) of water each day, homes warmed with hot air flowing in channels under the floor, an efficient sewerage system and thousands of kilometres of excellent roads with fine bridges. These achievements were accomplished by making the maximum use of local resources within the bounds of their own limited technical knowledge.

Overcoming cost barriers

Today the world has developed an enormous fund of sophisticated technology and with its aid men have walked on the moon. Such advanced technology costs a great deal, however, and not all nations have the means of payment. One of the major problems of the Third World is its inability to produce enough to create the wealth needed to buy the technology for increasing production.

The financial problem can be solved with outside aid or through self-help. Each can contribute independently but a combination is often even more fruitful. In Indonesia, Latin America and some African countries, food production is being significantly increased by huge irrigation projects. The engineering know-how is Western, financed by the World Bank and similar organizations, but the works themselves are the product largely of the art of using local resources.

Water, energy and labour

Since the dawn of history, there has been water in the world's rivers running to waste. Today canals and reservoirs are being dug by hand, with animals being used to move the soil. The banks are waterproofed with hand-puddled clay. And even deserts often have vast reserves of water under them. The sun shines daily on millions of square metres of the earth's surface, providing immeasurable quantities of free energy that is never used. Some of the underdeveloped countries have enormous reserves of labour. The problem is to harness this water, energy and labour –

along with the numerous other unexploited resources such as fertile soil, timber and minerals – and make them productive without having to make investments in contemporary technologies beyond the financial resources of those concerned.

There are several methods of developing local domestic water supplies cheaply [2, 3, 4]. If the water has to be raised to a higher level, a simple pump such as the Humphrey pump [5] can be used. And when there is a plentiful supply, as in a fast-flowing stream, a hydraulic ram can be used. This makes use of the energy of flowing water to pump small volumes to a higher level. Where bamboo is plentiful and water pipes hard to get, the bamboo can be drilled out and used to carry water, as is done in Ethiopia and Indonesia.

The sun's energy can be used to heat water in a solar heat collector consisting of a folded length of piping in a frame under glass. By connecting a storage tank at a higher level the water heated by the sun will rise to the tank and be replaced by cooler water. A solar heater of this kind, fitted with a sufficiently large heat collector, will heat water for

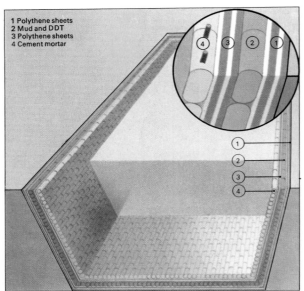

1 Water storage in the Western world is achieved by the construction of huge reservoirs. To convey the water to where it is wanted canals and pipelines are built. A new approach, suited to countries that cannot afford to pay for major civil works, is the provision of a large number of rainwater tanks sited where the water is needed. These "tanks" are dug in the ground and lined with three or more layers of polythene sheet, separated by a mud and DDT mixture to kill termites that bore through the outer skin. The tank shown in the diagram has a wearing surface of cement mortar "sausages" made with polythene tubes.

1 Polythene sheets
2 Mud and DDT
3 Polythene sheets
4 Cement mortar

2 In stored water, before it is filtered, sediment that would quickly block a sand filter must be removed if the water is for human use. The large sedimentation tanks used in water treatment plants in the West are costly. The process can be carried out in a small community supply scheme by means of a simple concrete-lined tank with a sloping bottom. Important features are the inlet baffle [1] (which stops incoming water from "stirring" the tank), the scum board [2] (to stop floating matter from passing out) and the sludge drain at the lowest point of the tank [3].

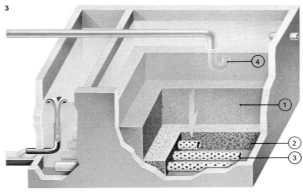

3 The filtration of drinking water for a small community supply scheme can be carried out efficiently by means of a relatively cheap "slow" sand filter in a concrete-lined tank. The sand particles should be between 0.2 and 0.5mm in size and the bed [1] about 1.2m (4ft) deep, supported on a layer of graded gravel [2] surrounding a porous pipe drain [3]. At least 50cm (20in) of clear water must be maintained over the sand and there should be an underwater upward flow inlet [4] that will not disturb the sand. Flow must not exceed 100 litres/m² (18.4 gal/yd²). In such a slow filter a layer of living organisms forms at the top of the sand bed adding biological purification to the mechanical filtration that is achieved by means of the sand.

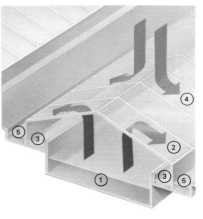

4 The sun's heat can be used to desalinate seawater. The water is led into shallow channels under glass [1]. Water evaporating from the surface condenses on the underside of the glass [2], running down the slope to the edge where it drips into a freshwater channel [3]. Additional fresh water can be collected from rain falling on the outside of the glass [4] and draining into other channels [5].

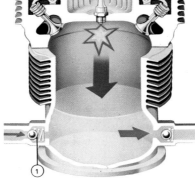

5 The Humphrey pump is an easily made internal-combustion unit in which the water being pumped replaces both the piston and flywheel of the conventional power unit and needs no separate pump. It will operate with an efficiency of up to 20% using almost any gaseous fuel, is cheap to produce and easily maintained. The water inlet valve [1] is automatic. Operating on a four-stroke cycle the "bounce" of the water produces the compression, suction and exhaust strokes.

washing and can even boil it for cooking.

Draught animals walking in a circle can drive simple crop processing machines using a device that works like a ship's capstan coupled by a crown wheel and pinion to a horizontal shaft. Water power can be harnessed to operate small machinery using a simple hardwood turbine of the kind used in Nepal to drive village flour mills.

For small-scale irrigation there are a number of ways of levelling land using ox-drawn graders and scrapers, which can be made by a blacksmith. A metal barrel, cut in half and fitted with a steel cutting edge on one side, can be used to scrape and carry soil, pulled by a team of oxen. Numerous other improved and new farm implements have been designed and found invaluable [8, 9].

Industrial applications
A wide range of village industries can be introduced to create non-agricultural employment. Good soap, for example, can be made from caustic soda and locally available fats. A pottery can be established with a small brick firing kiln to make domestic vessels,

sometimes of original new design. A village foundry for casting aluminium or iron objects can be set up at relatively little cost.

An excellent example of the application of appropriate, as opposed to advanced technology, can be seen in the increased efficiency and productivity being achieved in the coastal fishing industry of Ghana. Most Ghanaian fishermen use 10m (33ft) canoes made from tree trunks. Small trawlers would undoubtedly be more efficient, but a less expensive improvement has been made by fitting outboard motors to the canoes. Between 1961 and 1971 the number of motors used increased from 19 per cent of the number of canoes to 86 per cent.

Another example is found in India where the manufacture of cement is insufficient to meet city needs. In rural districts lime is widely used as an alternative material where the superior strength of cement concrete is not essential. Small-scale lime manufacture is carried on throughout the country. A superior lime mortar, which sets under water, is made by the addition of *surkhi* (a substance similar to the volcanic ash *pozzolana*).

KEY

The march of technology has made men creatures of habit. Someone entering a room at night takes the existence of electricity for granted, forgetting that man lived for many centuries with only oil lamps and candles. The technology of reinforced concrete depends on steel rods or wires to withstand tension while the concrete itself stands immense compression. Yet costly steel is not the only material strong in tension. A concrete stucture requiring some reinforcement against surface cracking can sometimes be adequately strengthened with natural bamboo, a material that is plentiful and cheap in some of the poorer countries.

6 In the developing countries the lack of a reliable, cheap and convenient supply of energy is a considerable problem. Dried cattle dung is a widely used fuel but its value as a fertilizer is lost when it is burnt. By subjecting dung to anaerobic fermentation (without oxygen), methane is produced and the waste slurry has a high nitrogen content which is more readily available as a fertilizer than the natural nitrogen in unprocessed dung. The diagram shows how a cheap 3–4m³ (106–141cu ft) methane generator can be built with provision for continuous recharging with dung slurry, steady production and about 1.5m³ (53cu ft) gas storage.

Slurry hopper

Counterweight

Gas

Water-trap

Slurry

Drain-off tank

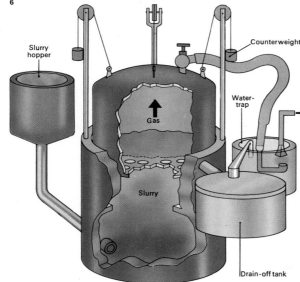

7 A waterwheel is used by the Amish people of Pennsylvania to transfer power over a distance of up to nearly 1km (0.6 mile) by a reciprocating wire power transmission system operating on the principle illustrated. The waterwheel turns a crank which raises and lowers the weighted corner of a triangular frame [left] which pivots on another corner to transfer the motion to the horizontal wire. The moving wire is supported at intervals by chains hanging from pole tops. At the far end of the wire a second frame transfers the motion to a vertical reciprocating pump.

8 Rice cultivation is widespread in poorer countries where it provides the staple diet for the bulk of the population. Because there is usually a surplus of cheap labour in these countries cultivation is traditionally carried out by hand, often by women. In particular the "puddling" of rice fields so that they will retain water as long as possible is normally a tedious hand job. The multi-action puddling tool shown here is pulled by oxen and does the job quickly and economically. Rotating chopper blades first cut the soil transversely. Then discs cut it longitudinally. Finally a row of 16cm (6.3in) long knives cut through the chopped soil like tiny ploughs. This implement was developed in Japan.

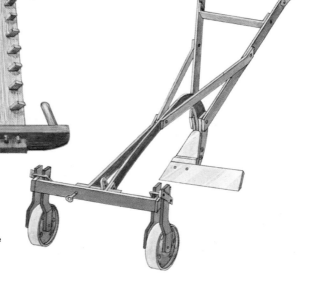

9 The groundnut lifter, drawn by a pair of oxen, is a simple tool developed in Nigeria. The oxen walk each side of a row of groundnuts, which is also straddled by the two "depth" wheels whose height can be easily adjusted. The sharp lifting blade cuts its way horizontally through the soil under the groundnuts, leaving them in loose soil on the surface ready for quick gathering by hand. The tool, assembled with bolts, can be made by any blacksmith. It has proved itself after extensive trials in the six northern states of Nigeria and is now in common use there. It has also been successfully field-tested in southern regions of Zambia.

Small technology in the home

Man has three basic needs – food, clothing and shelter – and it is the last of these that represents a family's most valuable and most durable possession. It is as necessary to an African tribesman as to a European office worker. In the Western world the design and equipping of the home has been developed with the aid of an advanced technology, many of the products of which are beyond the means of people living in less developed countries.

Simple technologies and community needs
To enable people in rural communities to build houses that are more comfortable and will last longer, the designs and construction techniques must be simple. The materials used should come from the kind of technology that can be applied locally, such as the method of reinforcing wattle and daub houses introduced in Zaïre [2] and considered suitable in other countries in Central and East Africa and South America. As a result the people can have better protection from the elements, better cooking [4] and washing facilities, and better sanitation [6].

Some personal needs are also community needs – a reliable and pure water supply, a cheap and effective source of energy (such as methane gas) and access to inexpensive building materials. Rural industries, which provide the community with such essential products as soap, agricultural implements and cooking pots, are another need – and, incidentally, help to create wealth where it is most needed.

Before high-quality, low-cost housing can be built, there must be research and experiment focused on specific local needs. Designs must be appropriate to local style [Key] and to the climate, as well as being flexible, so that small houses can when necessary be enlarged. They must fully exploit local materials and the ability of local labour. Finally, co-operative organizations must be set up to mobilize local resources and to encourage a desire for improvement.

Sanitation and food storage
Sanitation is as much as community as a family responsibility. But it begins in the home, where the introduction of the inexpen-

sive water-seal latrine [7] can be the greatest single method of reducing the incidence of disease.

Methods of storing food are also important for family and community health. In many developing countries the climate is such that a refrigerator would be considered essential by contemporary Western standards. In places where ice is available, an insulated ice-box may provide a partial solution. The box itself is cheap and easy to make, but the ice may well be too costly for most rural families.

In regions where the humidity is not too high an evaporative food cooler may provide a practical alternative. The cooler, which is designed as a cabinet with wire mesh walls and shelves to allow air to circulate, is placed in a shallow pan of water and has a second pan as its top. Jute burlap or sacking "curtains" hang from top to bottom, with their ends in the water. The burlap absorbs water like a wick, thus keeping damp. When the cabinet is placed in a breeze, away from direct sunlight, heat absorbed from the interior is used by the water in evaporating.

1

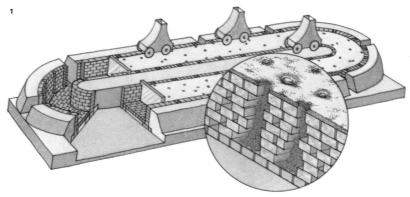

1 In rural areas bricks are made by ancient manual processes. In India, where a rural brick industry provides about half a million jobs, efforts have been made to improve the output of these brickworks by the introduction of an inexpensive kiln designed for simplicity of installation and use. This Indian trench kiln is illustrated here. A typical kiln with a capacity of up to 28,000 bricks each day would be 65m (213ft) long and 25m (82ft) wide overall, the depth of the "trench" being 2.25m (7ft). There is no roof. The top courses of "green" bricks are laid close and covered with ash, leaving only coal feed holes. The kiln is fired section by section, one each day, the 17m (55ft) tall steel chimneys, mounted on wheels, being moved to the section to be fired.

2 This traditional "mud" house, which can rarely survive more than five years of tropical rains, could be made to last three to four times as long if its weak point – the base of its outer walls – was protected. This has been successfully achieved in Zaïre. The method is to dig a shallow foundation 35cm (14in) deep by 35cm wide round the house close to the existing walls. The trench is filled with stones and clay mortar on which is built a low stone wall 1m (39in) high against the house wall. The top is protected with cement, as shown in illustration 3A.

2

3 A   B   C

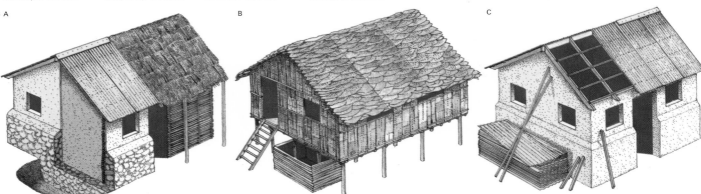

3 Thatch is one of the traditional roofing materials [A], even in regions with a heavy rainfall. Thatch needs support and timber is generally used for this purpose. But sawn timber is expensive and its cost tends to offset the only advantage of thatch – its cheapness. Thatch can also present other problems: it is susceptible to attack by fungi and vermin, and it burns easily, constituting a fire hazard. Commonly used alternatives include sheets of galvanized steel, corrugated asbestos, cement sheets [A] and, in some developing countries, clay tiles [B]. None of these is cheap and corrugated sheets made from asphalt and paper felt [C] provide a good substitute at less cost. The material is made from scrap paper, bagasse (sugar cane waste), jute waste, coconut fibre or rags. These are made into a wet pulp and pressed into sheets which are dried in the sun, trimmed and impregnated in a bath of paving asphalt. After curing for a short time, they are dip-painted to produce any desired finish. The sheets are light and can easily be nailed into position, needing only a ridge "tile" (made of the same material folded) to provide a complete-ly watertight roof. This is an excellent example of how local materials and labour can be used. Timber also has a cheap available alternative in bamboo, which is long-lasting and strong. Until now it has been little used as a building material because local people have not known how to join it effectively. But they now use the results of research that has shown that the application of modern engineering practice can produce bamboo structures with a good strength-to-weight ratio.

As a result the inside of the cabinet is significantly cooler than the surrounding air.

Storage heating
A heating system that has proved successful in Zambia uses an airtight brick enclosure, about 1m (39in) high and 70cm (27in) wide, built against the outside wall of a house. Two holes in the wall lead to the enclosure. The hole at ground level is just large enough to take a lighted coal-pot; the second, 80cm (31in) above it, is filled with a single air-brick. Thick clay over the enclosure reduces heat loss to a minimum.

After a coal-pot is put in, the bricks of the enclosure heat up in about an hour. They hold the heat (in much the same way as does a Western storage heater) long after the fuel has burnt away. In Zambia the charcoal cooking pot is put into the enclosure as soon as the evening meal has been prepared and no more fuel is needed.

Even the electric washing machine has a low-cost equivalent, which has been successfully introduced in Afghanistan. A washing machine works by continually agitating hot water and the laundry immersed in it. Washing by hand has three drawbacks: the skin of the hands becomes sore and unsightly, the temperature of the washing water is limited to that which the hands can bear, and a considerable amount of human effort is wasted bending over and agitating the wash.

The "rural" washing machine is basically a tub in which a vertical plunger, about half the diameter of the tub, is used to agitate the laundry. The plunger has a long wooden handle and is centred in the tub by passing it through a hole in the lid. A long tub can have two plungers – one at each end – operated by a lever pivoted between the two and having a handle at one end. All three disadvantages of hand-washing are avoided: the operator's hands are kept dry, the water can be heated almost to boiling and the plunger makes the best use of muscle power.

These few examples demonstrate that the potential applications of low-cost technology are enormous. Many people could benefit from them today, instead of waiting for their community to "catch up" with sophisticated Western technology.

KEY

In many developing areas of the world houses are small and have no separate kitchens. The wife may squat on a stool by a charcoal stove in the corner of a partly roofed court-yard, pots and basins around her. A replacement "modern" kitchen, as shown here, must reflect her local life-style.

4 Open fires are used for cooking in many village homes in the developing countries. This smokeless clay stove was introduced in India and has proved successful. There is one fire, on a grille over an ash-pit. The heat rises directly to one cooking hole and circulates through channels to the others.

5 This simple solar water heater gives a steady supply of hot water when used in countries that have plenty of warm sunshine. Basically it consists of a flat heater tank [1] angled towards the mid-day sun and a storage tank [2] at a higher level. The tanks are connected by large-bore pipes. Circulation is by gravity.

6 Water-seal privies are more hygenic than the open pits still used in many Third World villages. They improve a latrine by keeping odours in and flies out of the pit. Western porcelain water closets have been successfully copied in concrete in the Philippines at a fraction of the cost. A wooden mould is packed with two parts sand to one cement with enough water to make the mix workable. After 20 minutes the bowl and outlet are dug out with a spoon. Dry cement is sprinkled over the wet interior surface to make it smooth. The mould is finally removed after a period of 48 hours.

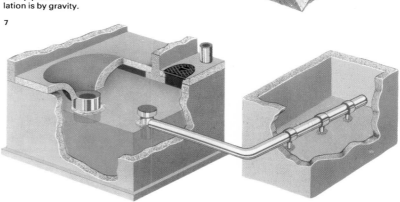

7 Septic tanks are an advance on simple water-seal privies. They consist of a watertight tank in which the waste matter decomposes and a sewer pipe connecting the tank's out-flow to a drain area. If the water privy is situated directly over the tank it need not have its own water seal; in this case the drop pipe should be 10cm (4in) in diameter and dip 10cm below the natural water level of the tank. The pipe is connected direct to the squatting plate with an airtight seal. The decomposition tank can be made of brick or stone faced with rich cement mortar. Alternatively it can be made of concrete or from a 90cm (36in) or 120cm (47in) sewer pipe set vertically, its bottom sealed with concrete. For use by one family it should have a capacity of not less than 1m³ (36 cu ft). Natural decomposition forms a sludge that settles, an effluent that passes into a seepage pit and gas that must be given an outlet.

Rubber and plastics

Rubber and plastics are two much-used modern materials that both consist of complex molecules known as polymers. These polymers have long molecules built up from simpler units attached to each other repeatedly, like links in a chain. The type of chemical reaction by which they are formed is called polymerization. Most rubbers are elastic – that is, they return to their original shape after being stretched or bent. This property, and the fact that rubber was originally a natural product, are generally supposed to distinguish rubbers from man-made plastics. But some plastics, such as Perspex (an acrylic plastic), are springy. Some rubbers and plastics are similar chemically.

Natural rubber: extraction and treatment
Nearly all natural rubber comes from a South American rubber tree (*Hevea brasiliensis*) although several other plants, including some nettles, contain a rubbery liquid sap, or latex. The rubber tree originally grew in Brazil. But in the late 1800s seedlings were raised, first in London's Kew Gardens and then in Malaya, and today most of the world's supplies of natural rubber comes from Malaysia.

Natural rubber is a polymer of isoprene. The latex is obtained from the tree by "tapping" – it seeps out through spiral cuts made in the bark. The liquid latex is coagulated (made solid), dried and exported as sheets of raw rubber [1, 3]. This is a weak, sticky and not very elastic material. Its strength and elasticity are improved by the addition of sulphur in a process known as vulcanization, or curing, of rubber. Vulcanized rubber is harder and springier than raw rubber because atoms of sulphur form cross-links between the long polymer molecules so that these can no longer slide over one another so easily. Rubber containing a great deal of sulphur has many cross-links and so is rigid; this hard material is called ebonite.

The strength and wear-resistance of rubber are also improved by the addition of fillers, which are powdery or fibrous substances mixed into the rubber and include carbon black, silica and cotton flock.

Natural rubber, vulcanized and filled, is a useful elastic material and is widely employed although it is rather expensive to produce. Since the mid-1960s its production has been greatly outstripped by that of synthetic rubbers [2] which have other advantages.

Man-made rubbers and plastic
The first useful synthetic or artificially made rubbers were produced during World War II. Since that time many kinds have been developed. Perhaps the most important, in terms of the quantity produced, is styrene butadiene rubber (SBR). Most of the raw materials for this rubber come from petroleum. SBR is a copolymer made from two kinds of simpler chemicals, styrene and butadiene [5].

Other synthetic rubbers include neoprene and Hypalon [Key], both employed in industry because of their resistance to chemicals. Silicone rubbers are a fairly recent development. They are remarkable polymers because, unlike most others, their long chain-like molecules have "backbones" not of carbon atoms but of silicon atoms. This makes them resistant to extremes of heat and cold and they are widely used as seals for jet engines and aircraft windows.

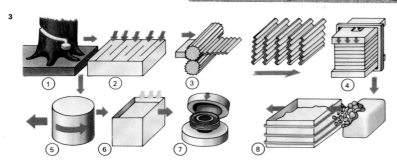

1 Plantation rubber begins life as a watery latex [1] tapped from a spiral cut on a rubber tree (*Hevea brasiliensis*). Foreign matter is filtered out and the latex coagulated into solid rubber with an acid [2]. In coagulation tanks the rubber is separated into slabs by partitions [3]. The slabs are dried, first by rollers [4] then by passing them through a drying tunnel [5]. The raw rubber is then packed for export [6]. Malaysia is the world's largest producer.

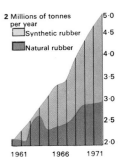

2 Since the mid-1960s the production of petroleum-based synthetic rubbers has far outstripped the production of natural rubber.

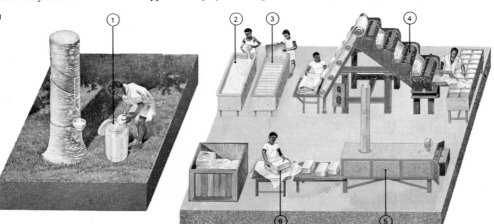

3 At the factory latex from a rubber tree can be acid-coagulated and squeeze-dried [1–4] or concentrated by centrifuge [5]. Rubber is melted [8] or vulcanized [6] (hardened) with sulphur after which it can be shaped in moulds [7].

4 An inflatable boat makes use of rubber's unique properties. Not all rubbers resist stresses equally. Special compounding methods produce rubbers with particular, required properties. The rubber itself is often a mixture of natural and synthetic products. The degree of vulcanization and the presence of reinforcing and filling materials are also important in determining the final properties of the rubber.

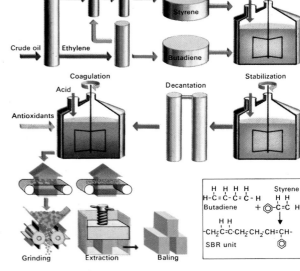

5 Styrene butadiene rubber (SBR) is the major synthetic rubber used widely for tyres, conveyor belts and other commercial and industrial products. It is a copolymer made by polymerizing two chemicals together. These are styrene and butadiene, which originate in crude oil. A catalyst is required, as in most polymerization reactions, to speed up the reaction. In later parts of the process stabilizers and antioxidants are added to prevent the rubber breaking down and to protect it from oxidation by air. It is then dried and baled.

Plastics are now taken so much for granted that it is difficult to imagine what life was like without them [8]. The first man-made polymers to be called plastics were compounds made from a naturally occurring polymer, cellulose. Celluloid (cellulose nitrate) was one such early plastic. It was most famous, or infamous, as the material of highly inflammable films and dolls.

Bakelite, another early plastic, is still used in large quantities for making electrical plugs and connectors. Bakelite was the first of the thermosetting plastics – that is, plastics that set hard on heating and cannot be re-melted without decomposing. More recent thermosetting plastics include urea and melamine plastics, used for plastic "crockery" and decorative laminates and the epoxy resins, most familiar as tubes of household glue. When mixed with a chemical called a catalyst (provided in a separate tube), an epoxy cement polymerizes to become very strong, hard and resistant to chemicals. For these reasons many protective paints and coatings also contain epoxy compounds.

The other great group of plastics are the thermoplastics, so called because they soften on heating. Familiar articles made from them include polyethylene (polythene) bowls and buckets; polyester and nylon fabrics; PTFE non-stick pan liners and PVC clothing.

Like rubbers, plastics become stiffer when their long polymer molecules are cross-linked. Thermosetting plastics are hard and stiff because they contain many cross-links; and most thermoplastics have pliable qualities because they contain few cross-links.

Techniques of moulding

Plastics are generally moulded into the shapes required. Thermosetting plastics are often moulded in pellet form. The pellets are compressed, heated until they flow and allowed to harden to take on the shape of the mould. Thermoplastics are frequently pressure-moulded [7]. Pipes, rods and sheets of plastics and rubbers are generally formed by extrusion in which the material is forced through a hole like toothpaste being squeezed from a tube. The extruder is often equipped with a screw for forcing the materials through a shaped hole or die [6].

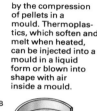

KEY

○ Hydrogen
● Carbon
● Oxygen
● Sulphur
● Chlorine

Polythene

Hypalon

Plastics and rubbers are chemically similar in their structures – they can even be made from the same starting material, called a monomer. This diagram shows [A] the polymerization of ethylene [1] to form the plastic polyethylene (polythene). The same monomer will react with sulphur dioxide [2] and chlorine [3] to produce the synthetic rubber called Hypalon. Natural rubber is composed of similar long-chain polymer molecules. They extend when the rubber is stretched and contract rapidly back to length when the tension is released, giving both kinds of rubber their elasticity.

6 Making plastics is typified by the manufacture of polystyrene. The chemical to be polymerized is styrene, obtained from the petroleum industry. This is first part-polymerized [A], with the aid of a catalyst, in stirred tanks [1]. The material then passes from the tank to a large reactor [2] cooled by a water coil to control the heat given off by the polymerization reaction. At the bottom of the reactor the temperature is 200°C (392°F) and nearly all the styrene has been polymerized to a hot, liquid plastic. This is fed by an extruder [3] to a water bath [4] which cools it to a hard solid. The plastic can then be machined into small chips [5] ready for transport. Products are made [B] from the plastic chips [6] by forcing the melted material, with a heated extruder [7], into a die or mould [8]. Polystyrene is used for many household articles.

7 Most plastic objects, such as these chair seats, are moulded into their required shapes. Thermosetting plastics set hard and cannot be re-melted so they are usually moulded by the compression of pellets in a mould. Thermoplastics, which soften and melt when heated, can be injected into a mould in a liquid form or blown into shape with air inside a mould.

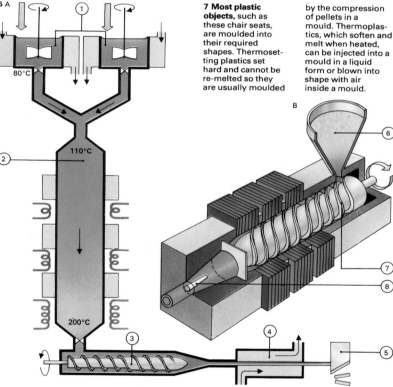

8 The domestic uses of plastic are legion. Utensils range from food containers to washing-up bowls. Even wine is sometimes bottled in plastic. Deprived of plastics, kitchens appear unfurnished.

9 Transparent plastics, such as polyesters and the acrylics, can be used to preserve biological specimens and also find use as materials for ornaments. Clear Perspex is now widely used as a glass substitute for tables.

Fibres for fabrics

The word fabric once described any material made from wool, silk, cotton, or other animal or vegetable fibre. But today the choice of fabrics has been greatly increased by the introduction of man-made fibres. Synthetic fabrics made from them are relatively cheap and have special properties such as high strength and resistance to creasing and rotting. In many modern fabrics a natural fibre such as wool is interwoven with a synthetic fibre such as Terylene, thus combining the advantages of the two.

Animal and vegetable fibres

Cotton is the most widely used of the vegetable fibres. Cotton fibre comes from the seed pod of the cotton plant (*Gossypium* sp) and, like other vegetable fibres, is composed mostly of cellulose. Indeed, the shorter fibres of the cotton plant supply most of the world's pure cellulose and only the longer fibres are used for textiles.

Next in importance is linen, which is made from long fibres in the stalk of the flax plant (*Linum* sp). Jute (*Corchorus* sp), hemp (*Cannabis* sp) and ramie (*Boehmeria nivea*) are other stalk, or bast, fibres. Sisal is a fibre taken from the leaf of the agave plant (*Agave sisalana*). Cotton and linen are the only vegetable fibres widely used for textiles. The others are generally too coarse and are used in the manufacture of sacking, carpet backing and ropes.

Wool is the best known animal fibre. Wool fabrics have low strength but are uniquely comfortable and warm because the soft, springy fibres trap air, which insulates the wearer against cold. The main source of wool is the sheep, but some of the best wools come from goats. Cashmere is a fine fabric made from the fleece of the Kashmir goat living in China, Iran and Mongolia. Mohair is a high-quality cloth made from the wool of the Angora goat, also a native of Asia. Fine, soft wools also come from members of the camel family – the llama, alpaca, guanaco and vicuña [2], all of which live in South America.

Wool, like other kinds of animal hair, is composed mainly of keratin, a fibrous protein also found in skin, nails, feathers and horn. It is one of the few fibres that can be matted together into felt, or fuzz. This mat-ting process adds to its softness. All wools, however, have the major disadvantage of vulnerability to attack by pests such as moths.

Silk, the only other animal fibre of importance, has long been used to make fabrics of delicate lustre and smoothness. Because of its cost, however, it has largely been superseded by synthetic fibres. The home of silk production is the Far East where silkworms [3], fed on mulberry leaves, eventually spin a continuous silk thread up to 800m (0.5 mile) long to make a cocoon.

The methods used in spinning, weaving and knitting natural fibres into fabrics are all very similar to one another [1].

Synthetic fibre production

There are two kinds of synthetic fibres. The first is derived from plant cellulose, and so can be considered as "semi-synthetic" because cellulose is a natural polymer with molecules made up of thousands of glucose sub-units linked together in a chain.

Rayon is a fabric made of fibres of pure cellulose. Plant cellulose is dissolved by various chemical processes into a thick liquid

CONNECTIONS
See also
54 Rubber and plastics
58 Making cloth
244 Colour chemistry

In other volumes
216 The Physical Earth

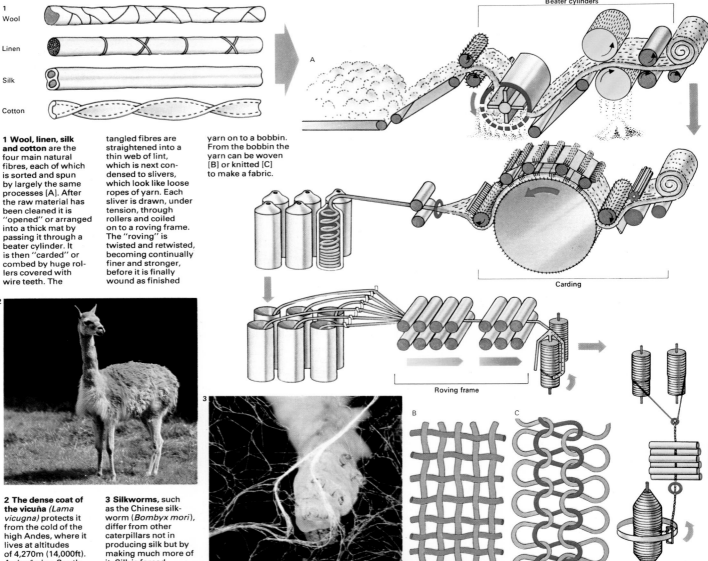

1 Wool, linen, silk and cotton are the four main natural fibres, each of which is sorted and spun by largely the same processes [A]. After the raw material has been cleaned it is "opened" or arranged into a thick mat by passing it through a beater cylinder. It is then "carded" or combed by huge rollers covered with wire teeth. The tangled fibres are straightened into a thin web of lint, which is next condensed to slivers, which look like loose ropes of yarn. Each sliver is drawn, under tension, through rollers and coiled on to a roving frame. The "roving" is twisted and retwisted, becoming continually finer and stronger, before it is finally wound as finished yarn on to a bobbin. From the bobbin the yarn can be woven [B] or knitted [C] to make a fabric.

2 The dense coat of the vicuña (*Lama vicugna*) protects it from the cold of the high Andes, where it lives at altitudes of 4,270m (14,000ft). A vicuña is a South American member of the camel family and its wool is valued for its high quality.

3 Silkworms, such as the Chinese silk-worm (*Bombyx mori*), differ from other caterpillars not in producing silk but by making much more of it. Silk is forced out through a small spinneret between the jaws and hardens on contact with air.

and then regenerated in the form of spun fibres. A chemical solution of cellulose is pumped through a spinneret (named after the silkworm's spinning gland) into a chemical solution, in which it coagulates into fine threads (filaments) [Key]. These filaments are then twisted together to make the rayon yarn. Another cellulose fabric is acetate rayon, the fibres of which are made by treating cellulose with acetic and sulphuric acids. The fibres of cellulose acetate are also made through a spinneret. Rayons dye easily and they also lose their strength when wet, only to regain it when dry.

The second and much larger group contains the purely synthetic fibres – nylon and polyesters known by such trade names as Terylene and Dacron. The materials for these fibres are made entirely from chemicals (often by-products of petroleum).

Spinning plastics into fibres often involves melting and extrusion through a spinneret [4]. For plastics that are sensitive to heating, including the acrylics used to make such fabrics as Acrilan, it may be necessary to dissolve the plastic before spinning the fibre.

Acrylic fibres have a soft, wool-like feel and are often used for making blankets and winter clothing. Synthetic fibres are often superior to natural fibres. They can be stronger, more flexible, and are usually resistant to heat, rot or abrasion.

Dyeing and finishing
Natural fibres, with the exception of silk, are easy to dye but special chemical processes are sometimes necessary to make dyes adhere to synthetics. Unwanted natural colour is removed by bleaching. Natural fibres and rayon can be treated with resins to render them crease-resistant or an elastic plastic to make them waterproof and stainproof.

There are two fibres used in fabrics that are entirely mineral in composition. Asbestos fibres, from the mineral chrysotile, are woven into fire-resistant matting and clothing [5]. Glass fibres are made by melting glass in a tank perforated with tiny holes through which the glass "drips" as filaments which are chilled and snapped into short fibres by air or steam. Fibreglass fabric, hardened with a synthetic resin, is very strong [7].

Spinnerets are finely perforated plates or tubes that make threads from chemical solutions of polymers during the course of manufacturing all man-made fibres.

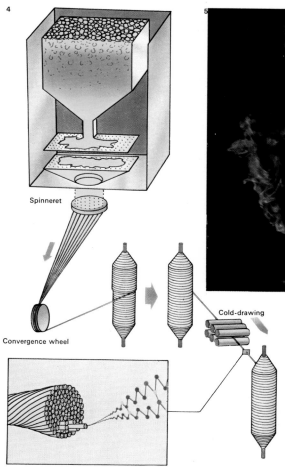

Spinneret

Convergence wheel

Cold-drawing

5 Asbestos fibres are mineral, and so do not burn or char. For this reason clothing woven from asbestos is invaluable for fire-fighting and is often aluminized or "silvered" to reflect heat and keep the wearer cool. Asbestos clothing is also resistant to chemicals and, because asbestos is a poor conductor of electricity, it provides protection against electric shocks.

6 Synthetic fibres are used for many purposes calling for great strength, as in parachute making. Nylon cords and ropes vie with steel in strength and are very much stronger than natural fibres.

7 Motor car bodies and many other products are now made of materials woven from fibreglass. The flexible glass fibres are stronger than steel and resistant to heat, rot, rust and most chemicals.

4 In making nylon two chemical compounds, such as hexamethylenediamine and adipic acid, are melted and combined under pressure in a hopper. The liquid nylon thus formed is filtered through metal gauze or sand and extruded as very fine filaments, or strands, through a spinneret. The filaments are twisted and united into yarn by a convergence wheel, which feeds the yarn on to bobbins. It is then cold-drawn, or stretched, to alter the molecular structure of the yarn filaments and this gives them greater strength and elasticity. This process is carried out by unwinding the yarn from one bobbin, passing it through rollers to stretch it and then winding it on to another bobbin. Nylon is one of the strongest and most elastic of all plastics and is widely used both for domestic and industrial purposes.

Making cloth

Nearly all textiles for clothes, sacks, carpets and other coverings are made by looping or interlacing fibre strands together. The strands can be of natural origin – for example wool and cotton – or they can be synthetics such as nylon and Terylene. Additionally, natural and man-made fibre strands are often woven together, as in the Terylene/cotton mixtures used for shirts and blouses.

Fibres into fabrics

Each fibre strand, or yarn, is composed of many short fibres twisted together by an operation called spinning [1, 2]. When two or more yarns are interlaced in the process of weaving, carried out on a machine called a loom [Key, 3], a length of cloth is made. But when a single continuous length of yarn is looped into a fabric the operation is called knitting. Industrially, knitting is also carried out on machines [4, 5].

Lace fabrics are made by interlacing and twisting yarns together. Felt fabrics are exceptional in that they are not knitted or woven. They are made by pounding hot, wet wool and other fibres together. The soft feel

of "cut velvet" is given by many tufts of severed yarn endings [7] although "figured" velvet, also a woven fabric, is not cut. Its softness is derived from the raised loops of fibres of wool, cotton or synthetics.

The preparation of fibres involves a number of processes. For cotton these include brushing fibres from the seed bolls in a cotton gin; beating the fibres to loosen them; rolling or lapping them flat and scutching (beating) them into fleecy masses. All of these processes are now carried out completely by machines.

Cotton fibres are straightened in a carding machine which also forms them into a loose rope or sliver. The longest fibres only are used for high-quality yarns and are obtained by combing them out on a machine in which the fibres are held by rows of pins. The slivers are then drawn through a series of rubber rollers, each pair of which moves quicker than the previous pair so that the fibres elongate to form rovings. These rovings are then ready for spinning.

Wool needs to be washed in detergent to remove dirt and grease. Fine worsted wool

yarns use only the longer fibres, selected by a combing operation. Retting is the name of a treatment for flax fibres (for linen) in which they are rotted in water to soften them. Silk comes from the cocoon of the silkworm and requires a special treatment with soap or detergent to remove the gum that originally bound the filaments in the cocoon.

Spinning and weaving

The textile industry grew up only in the eighteenth and nineteenth centuries, although spinning and weaving are age-old traditional occupations. The industry progressed as the result of a series of British inventions. The inventors included Richard Arkwright (1732–92) whose spinning frame of 1768 first provided a cotton yarn strong enough to be used as the warp, or lengthwise thread, for machine-operated looms. Two years later, in 1770, James Hargreaves (died 1778) patented the spinning jenny, which spun many threads at once. Other early inventions included the spinning mule of Samuel Crompton (1753–1827) and the power loom of Edmund Cartwright (1743–1823).

CONNECTIONS

See also
56 Fibres for fabrics
244 Colour chemistry
240 Soaps and detergents

In other volumes
66 History and Culture 2

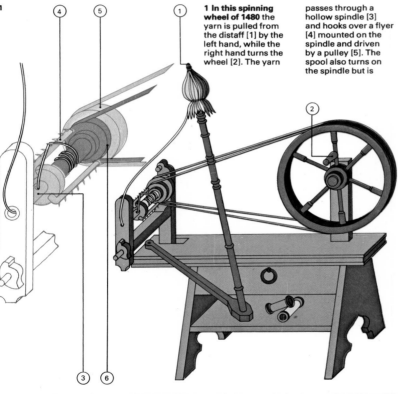

1 **In this spinning wheel of 1480** the yarn is pulled from the distaff [1] by the left hand, while the right hand turns the wheel [2]. The yarn passes through a hollow spindle [3] and hooks over a flyer [4] mounted on the spindle and driven by a pulley [5]. The spool also turns on the spindle but is attached to a smaller pulley [6] which turns faster, so that the flyer twists the yarn at the same time as it is wound on to the collecting spool.

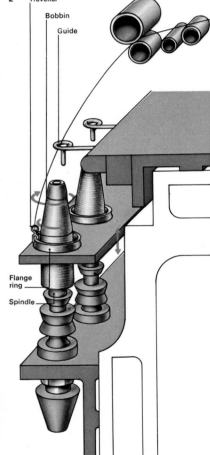

2 **The ring spinning frame**, invented by John Thorpe, is used for cotton. The yarn passes through a series of rollers and a guide, and finally down to the spindle. The flyer is replaced by a small traveller that runs freely on a flange ring which surrounds the spindle. The bobbin is carried on the spindle and rotates very quickly. The traveller is pulled round the flange ring by the yarn and, because of the friction occurring between the traveller and the ring, the traveller lags behind the bobbin so that the yarn is wound on as it is spun. The plate holding the flange ring and traveller rises and falls, distributing the yarn evenly on the bobbin.

3 **A modern textile factory** generally has a number of looms all working at the same time. A machine minder tends one or more machines, joining any broken yarn and supplying the loom with full bobbins. The strength and elasticity of yarn varies with temperature and amount of moisture in the air and so the whole room is air-conditioned to keep breakages to a minimum and stabilize the settings of the fine controls of the high-speed machines.

In spinning, the rovings are passed to rotating spindles carrying bobbins, all mounted on a moving frame of the kind originally invented by Samuel Crompton and called the spinning mule. The frame first moves outwards, pulling out the roving to form a yarn and twisting it. Then it moves back and the yarn is wound evenly on to the bobbins, guided by wires. Worsted and cotton are now usually spun on a ring spinning frame invented in 1828 by John Thorpe in the United States [2].

Linen is spun on a flyer frame that has a hollow inverted U-shaped device, the flyer, mounted on a spindle. Each yarn passes to a bobbin through the inside of this flyer which rotates round the bobbin, twisting the yarn as it does so.

In weaving two yarns are employed: the warp, a set of lengthwise strands held firmly on a loom and alternately pulled apart, and the weft, which is carried sideways through the parted warp strands by a shuttle to which it is attached. The warp strands are then closed together and the new weft strand is pushed back against the previous one by a device called a reed. Factory looms weave quickly because the weft threads are shot through the warp by an extremely fast shuttle. The edges of a cloth are strengthened by doubling the thread or using a stronger warp yarn. This edge is known as a selvedge.

Finishing the fabrics

Fabrics are finished in a number of operations that include bleaching, dyeing and printing. Bleaches such as hydrogen peroxide hypochlorites may be used to whiten a cloth before dyeing. Printing on cloth resembles that on paper: flat or cylindrical printing blocks or silk screen processes can be used.

The lustre of a fabric, particularly cotton, can be improved in a number of ways, including singeing and mercerizing, a process using caustic soda. Wool fabrics are treated against shrinking by chemical treatment of the fibres. Creaseproofing is another finishing process, although it is not necessary for many synthetic fabrics that resist creasing well anyway. Fireproofing, waterproofing and mothproofing are other finishing treatments that are carried out on a large scale.

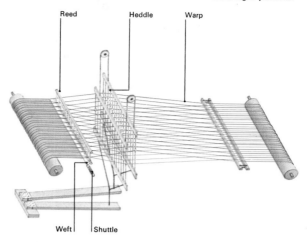

In the first looms, weaving proceeded like darning, the weft being passed over and under the warp. In other early looms every other warp thread was attached to a stick, the heddle, which could be lifted to part the warp allowing passage of the weft thread. The heddle eventually gave way to a device called the shaft, which parts the warp threads in various combinations to allow the weaving of patterns.

Reed   Heddle   Warp

Weft   Shuttle

4 Treadle and pulley operated the first really successful knitting machine which was invented by William Lee in 1589. It could knit at the surprisingly fast rate of 600 stitches every minute.

1 Hooked needles
2 Presser bar
3 Shank of needle
4 Sinkers
5 Handle
6 Hinged arms
7 Bent bar
8 Bent pulley
9 Locker bar
10 Slur
11 Wheel

5 Modern knitting machines resulted in mass-produced hosiery. The power-driven circular knitting frame came into operation in the 1840s and the machine for closing seams 20 years later.

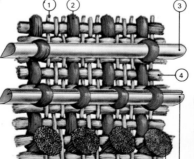

6 Designs are printed on cloth after it has been woven. Flat printing blocks with the design raised in relief or etched below the surface (intaglio) may be used or, as here, the cloth can be printed by means of inked rollers.

7 Velvet weaving originated in China. The pile warps are lifted for the insertion of the wire, which forms a loop in the pile. This wire has a blade at its extremity to sever the loops and leave the tufts as the wire is withdrawn.

1 Foundation warp
2 Pile warp
3 Wire for uncut pile, or terry
4 Foundation weft
5 Grooved wire for cut pile
6 Cut pile tufts

8 The seed heads of the teasel plant form a "brush" of hooked bristles. They are traditionally used for raising a nap on coarse tweed cloth such as Harris tweed and have not yet been bettered by purpose-built machines.

Making paper

Paper is one of the most common everyday materials of industrialized societies and one that man has been using for nearly two thousand years. He does not rely on paper merely to record his noblest conceptions but also uses it to fulfil his most basic needs. Apart from newspapers and books, paper is used for clothing, as containers for food and drink and for decorating houses. As yet, no comparable substitute has been developed.

Most modern paper consists of interlaced fibres derived from wood, although some papers may also contain fibres from rags, other vegetable sources and even synthetic materials. The production principle is as simple as the material itself. Timber is chopped up and, to some extent, purified. The fibres are next treated with chemicals and dispersed into large volumes of water. They are dried out in a thin film and the water is removed, so producing paper.

Industrial papermaking

The industrial manufacture of paper is a far more complex and far-reaching operation. Millions of trees are harvested every year solely to make paper [3]. There has been public concern about the depredations of the paper industry on natural forest areas but responsible paper companies now arrange careful ecologically based programmes to ensure that forests yield regular crops. Screens of standing trees are left round cleared areas to promote natural re-seeding. In many places seedlings are planted to re-establish the forest.

In the forests of North America and Scandinavia the trees are cut and transported to rivers, which are often frozen over in winter. In spring they are then floated downstream in their thousands. In some places the logs are made into huge rafts and then towed to the factory, or trucks are used to convey the logs from the forest to the mill.

After arriving at the pulp mill the logs are stripped of bark and then treated in one of two ways. They may be disintegrated by grinding with huge grindstones and so reduced to what is known as mechanical pulp, or the logs may be "cooked" in large digesters, a process that breaks down the wood chemically. Wood has two principal constituents: cellulose and a complex substance called lignin that holds the cellulose fibres together, thus making the wood rigid. During the "cooking" the lignin is removed.

The product of this second process is known as chemical pulp [5]. The method is more gentle on the wood fibres than the mechanical process and the fibres, because they are less damaged, make stronger paper. Usually, however, the two types of pulp are mixed in proportions that vary with the use for which the paper is intended.

Processing the pulp

If the pulp mill and paper mill are sited close together liquid pulp may be pumped straight from one to the other. If they are distant then the pulp is partly dried and pressed before being dispatched. The dried pulp has to be broken up again at the paper mill in a machine called a Hydrapulper. It is then thoroughly dispersed in a process known as beating or refining. Next come the additives – such as china clay, coated whiting, rosin (probably mixed with alum) – plus other chemicals used as "retention aids" to keep

1 **Papyrus**, the writing material of the ancient world, predated paper by at least 3,500 years. It was made from the papyrus reed (*Cyperus papyrus*), an aquatic plant of the sedge family that still grows in the Nile Delta. It was prepared by laying strips of the reed side by side and then crossing them with other strips. It was then soaked in the water of the Nile, which created an adhesive and stuck the strips together. Finally, the sheet was hammered and left out to dry in the sun. Any surface roughness was removed by polishing with ivory or a smooth shell.

2 **Nicholas-Louis Robert** (1761–1828) patented the first papermaking machine in 1799. It made paper in great lengths using a continuous conveyor belt system and was driven by turning a handle. The prepared mixture of water and pulp was poured into an oval chest then picked up by rotating copper bars and discharged on to the upper surface of an endless wire mesh running on two end rollers. The pulp passed between felt-covered squeezing rollers, removing most of the water so that the web lifted off the wire and could be coiled on a roller. The tension of the wire was adjusted by a screw. The wire was shaken by a cross-bar driven by a wheel. This cross-bar could also be raised or lowered to alter the slope of the wire and thus the rate of water loss.

End roller
Gathering roll
Squeezing rollers
Cross bar
End roller
Shake wheel
Enclosed drum
Oval chest
Vertical bar
Chest support
Screw spring

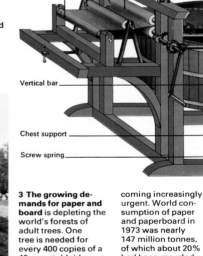

3 **The growing demands for paper and board** is depleting the world's forests of adult trees. One tree is needed for every 400 copies of a 40-page tabloid newspaper. But as forests are being cleared to meet the rising demand, the stock of trees is not being replenished. Even if sufficient land were available to plant new trees, it would still take 20–40 years for them to reach maturity. The need to recycle more waste paper therefore is becoming increasingly urgent. World consumption of paper and paperboard in 1973 was nearly 147 million tonnes, of which about 20% had been recycled. Clearly this proportion will have to increase significantly if there is not to be a world shortfall of pulp and paper, already predicted by the United Nations for the end of this decade. Possible substitutes for wood pulp, such as certain grasses, are also being explored.

4 **Watermarks** have been used for security reasons or as marks of distinction since the latter part of the 13th century, when they were first used in Italy. They are still employed on documents and banknotes as a guarantee of authenticity. The watermark is made by wires bent into a pattern. When this comes into contact with a layer of wet pulp a translucent impression is made, which can be seen when the finished paper is held up in front of a light.

the additives on the fibres when the water is removed. Pigments to colour the paper or to improve its whiteness – titanium dioxide is used – may also be added. Both the beating and the additives influence the appearance and the character of the surface.

The papermaking machine most often used – known as a Fourdrinier machine – has three main sections. At the wet end the slushed pulp of beaten fibres flows on to a moving band of finely woven wire or plastic mesh. There, aided by suction, much of the water drains away, leaving the fibres and most of the additives on the mesh. The wet "web" of paper runs on to the press section where it is carried on felt and passes between rollers that remove more water. The web, by now much firmer, is finally transferred to another felt surface in the drying sections and passes over as many as 60 drying cylinders.

Papermaking machines, some of which are more than 75m (25ft) wide, operate at a rate of more than 900m (3,000ft) per minute. Pulp, which enters the machine with more than 99 per cent moisture, is transformed in a matter of seconds into finished paper with no

more than five to ten per cent moisture. This paper is then wound on to a reel and is sometimes processed even further. It may be remoistened and passed through rollers to impart a polished effect known as calendering, or it may be coated with china clay and latex to produce high-quality art papers.

Paper in the modern age

There has been much discussion in the past concerning the way in which cellulose fibres in paper hold together. Today, this adhesion is largely attributed to so-called "hydrogen bonding". This is a weak link between a hydrogen atom already chemically bonded and a neighbouring hydrogen atom.

Paper is more than just a subject of research in modern science. It is among the components of some of the world's most sophisticated machinery and has some special technical applications – such as forming part of the dielectric in electric capacitors – that are not seen by the general public. It seems that although science produces more and more new synthetic materials every year, it still finds new applications for paper.

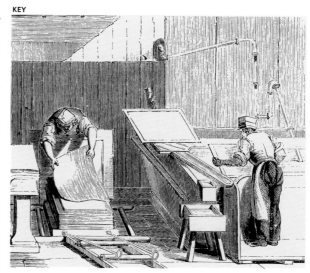

Ways of making paper have changed little in 2,000 years. A suspension of cellul-ose fibres is made by beating the fibres in water, then separating and soaking them. The wet sheet is pressed and heated to remove the water and then further refined.

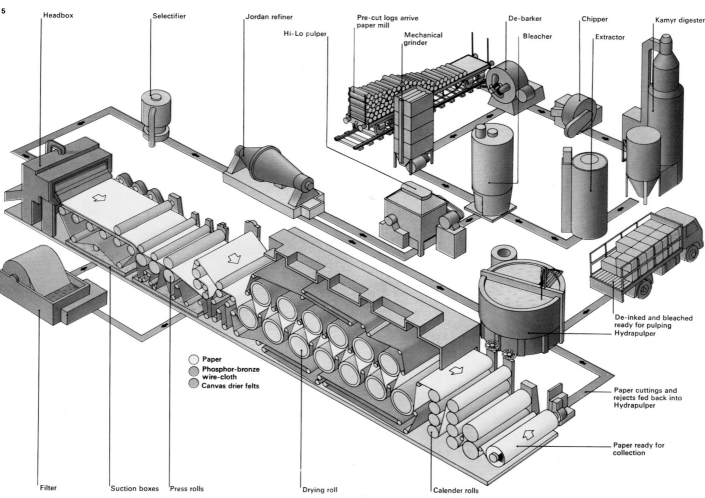

Headbox Selectifier Jordan refiner Hi-Lo pulper Pre-cut logs arrive paper mill Mechanical grinder De-barker Bleacher Chipper Extractor Kamyr digester

Paper
Phosphor-bronze wire-cloth
Canvas drier felts

De-inked and bleached ready for pulping Hydrapulper

Paper cuttings and rejects fed back into Hydrapulper

Paper ready for collection

Filter Suction boxes Press rolls Drying roll Calender rolls

5 The modern paper-making process is shown here in a diagrammatic and simplified flow-chart. Pre-cut logs arrive at the paper mill. They then pass to the de-barker which has cutters that penetrate the bark and force it

off without damaging the timber. The wood may then go to the chipper. This machine has rotating knives that cut wood into pieces about 3mm (0.125in) thick. From there it goes to the Kamyr digester. Here treatment with boiling chemicals

yields a "chemical pulp" from which chemicals are removed at the extractor. Alternatively, de-barked logs may then pass to the mechanical grinder. The two streams meet in the bleacher, pass to the Hi-Lo pulper and from

there to further treatment in the Jordan refiner. De-inked and bleached waste paper is pulped in the Hydra-pulper by means of a spinning multi-vane motor and joins the other pulp at the refiner. All the pulp moves on to

the selectifier, which is a pressurized sieve, and into the headbox. There the pulp is adjusted for consistency and fed at a controlled rate through a sluice gate on to a fine phosphor-bronze wire-cloth which is travelling at high

speed. At this point suction boxes extract most of the water and the paper then forms a web. The drained water is filtered out; both it and the recovered pulp may be recycled. The web is pressed to the required thickness in the press

rolls and then dried over drying rolls. It is given its finish in the calender rolls. The paper cuttings are not wasted but are fed back into the Hydrapulper. Finally the completed paper roll is ready for collection and use.

Basic types of engines

Engines are machines that convert other types of energy into mechanical energy, capable of doing work. The energy is generally heat from the burning of a fuel – oil, petrol, gas or coal – and the work done by the engine can be used in many different ways: to drive other machines, to generate electricity, to pump water or to power vehicles such as cars, locomotives, ships and aircraft.

The rules governing the way in which heat can be converted into work were discovered through experience by early engineers. Later these rules were formalized as the science of thermodynamics. By using them it is possible to calculate how much power an engine will produce, what fraction of the heat will be turned into work and what might be done to improve its performance.

The first engines, invented during the eighteenth century, burned coal to produce steam in a boiler [1]. The steam was then used to drive the engine. An engine of this sort, in which the source of heat is outside the engine, is known as an external combustion engine. An example of this type is the steam turbine as shown in illustration [2].

By far the most successful form of engine, however, is the internal combustion engine. In an engine of this sort fuel is burned and the expansion of gases produced is employed to drive a piston backwards and forwards inside a cylinder and is transmitted via a crankshaft to drive the wheels.

Early gas engines

The internal combustion engine was invented in the second half of the nineteenth century. The first successful design was produced by a German engineer, Nikolaus August Otto (1832–91), who used coal gas as his fuel. Otto's gas engine used the four-stroke cycle first proposed in 1862 by Alphonse Beau de Rochas – and so called because only one stroke in four produces power. The other three are taken up in drawing in the fuel, compressing it and finally, after it is burned, exhausting it.

Since only one stroke in four produces power the output of a single-cylinder four-stroke engine is very uneven and requires a heavy flywheel to smooth it out. Most practical four-stroke engines have more than one

cylinder – usually four or six, but often as many as sixteen. Increasing the number of cylinders produces smoother power because each cylinder is arranged to produce its power in turn.

The internal combustion engine using the four-stroke cycle is the engine found in almost all cars and trucks and many motor cycles. Such engines will run on gas but are normally fuelled with a liquid fuel, usually petrol (gasoline). The fuel is turned into a vapour, mixed with air in the carburettor and ignited inside the cylinders of the engine by the sparking plugs which are timed to produce a spark at exactly the right moment in the four-stroke cycle.

The diesel engine

Many internal combustion engines, including most of those used in modern commercial vehicles, employ a heavier petroleum fuel called diesel oil. This type of engine is named after the German inventor Rudolf Diesel (1858–1913). Instead of a carburettor to mix the fuel with air, a diesel engine has a fuel injector which pumps a measured quantity of

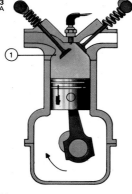
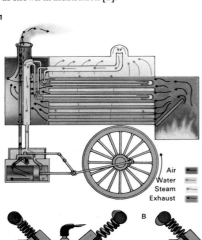

1 In a steam engine, as used in steam locomotives, heat from burning fuel (coal or oil) boils water in a boiler. The pressure of the steam produced then forces a piston backwards and forwards in a cylinder.

2 A steam turbine uses steam from a boiler to drive the rotary blades of a series of turbines, each at a lower pressure than the one before it. Exhaust steam is condensed to water and re-used in the boiler.

Air
Water
Steam
Exhaust

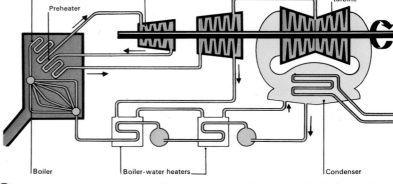

Superheater
Preheater
High-pressure turbine
Low-pressure turbine
Boiler
Boiler-water heaters
Condenser

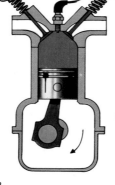
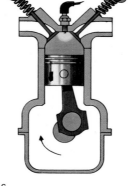
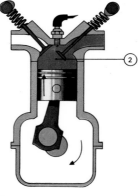

3 In the four-stroke or Otto cycle, used in petrol engines, the expansion of gases forces the piston down in the cylinder. On the induction stroke [A] the descending piston sucks in fuel-air mixture through the inlet valve [1]. On the compression stroke [B] both valves are shut. They remain closed when the spark plug ignites fuel and air during the power stroke [C] and the exhaust valve [2] opens during the exhaust stroke [D].

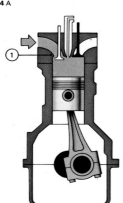
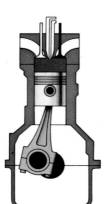
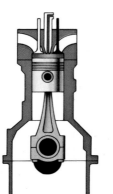
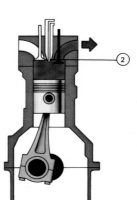

4 In a diesel engine a jet of fuel is sprayed into hot, compressed air at the top of the cylinder. The fuel ignites spontaneously and the expanding gases produced power the piston. On the induction stroke [A] air is drawn into the cylinder through the inlet valve [1]. On the compression stroke [B] both valves are shut and fuel is injected. This burns on the power stroke [C] and the exhaust valve [2] opens during the exhaust stroke [D] to let exhaust gases out of the cylinder.

fuel into the cylinder at precisely the right moment. And instead of sparking plugs diesel engines depend on the compression of the gas in the cylinder to achieve ignition. As the pressure of a gas increases so does its temperature. In a diesel engine the fuel ignites spontaneously at the end of the compression stroke. For this reason a diesel engine is more correctly called a compression-ignition engine.

Converting heat into work

Internal combustion engines do not convert heat into work efficiently. A typical car engine converts no more than a quarter of the energy contained in the fuel into useful work and even a highly developed engine is little better than 35 per cent efficient. The explanation of this disappointing performance lies with the laws of thermodynamics.

The first law says that you cannot get more energy out of an engine as work than you put into it as heat. The second law goes further. It shows that the amount of work produced by an engine is always less than the amount of heat supplied. In other words the

efficiency of any type of engine is always less than 100 per cent.

The efficiency of a heat engine depends on the difference in temperature between the heat used to drive it and the heat rejected as waste. The higher the temperature of the engine's working gas and the lower the temperature of the waste heat the more efficient the engine. Complete 100 per cent efficiency can be achieved only if the waste heat is at absolute zero (−273°C). In practice, of course, this cannot be and thus the efficiency of a heat engine cannot be 100 per cent. Under ideal circumstances a heat engine operating at 1,000°C (1,832°F) – possible for an internal combustion engine – and rejecting waste heat at normal exhaust temperatures (say 150°C [302°F]), has a theoretical efficiency of 67 per cent.

The difficulties of designing a perfect engine mean that all actual engines fall well short of the theoretical efficiency. Friction, vibration and the energy consumed by the engine itself in driving camshafts, fans, and so on, and losses in the transmission reduce the efficiency to much lower levels.

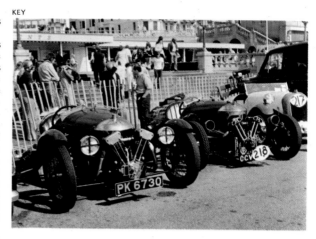

Air-cooled, small light engines were developed in the first quarter of this century, chiefly for powering motor cycles. The British Morgan company used a motor cycle engine as the power unit in this 1927 three-wheeler car. Known as a V-twin, the engine had two cylinders inclined at an angle to each other. It drove a propeller shaft along the floor of the car to a gearbox, with a final chain drive to the single rear wheel. The car had motor cycle-type controls mounted on the steering column.

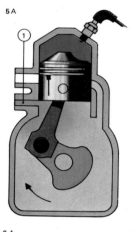
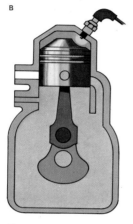
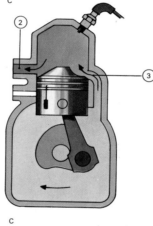

5 In a two-stroke engine the upstroke [A] opens the inlet port [1] and compresses the mixture already in the cylinder. A sparking plug ignites the fuel [B] and after combustion the descending piston [C] first opens the exhaust port [2] and then the transfer port [3], letting in fresh fuel-air mixture. In this way the piston also acts as the engine's valves and oil is added to the fuel to lubricate it. In many two-stroke engines, the cylinder is air-cooled via fins attached to it.

6 The Wankel engine has a rotor in the form of a curved triangle within a chamber. The "points" of the rotor have gas-tight carbon-fibre seals. The four stages of the cycle are the same in principle as the Otto cycle: induction [A], compression [B], combustion [C] and exhaust [D]. All three spaces in the chamber are used at once. There are two sparking plugs and the rotary motion produced can be used as a direct drive, without the need for a crankshaft.

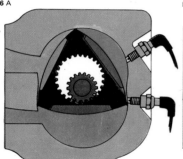
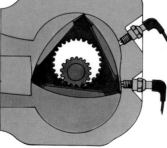
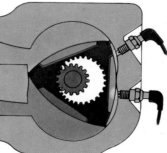
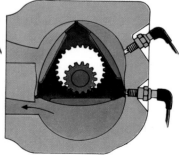

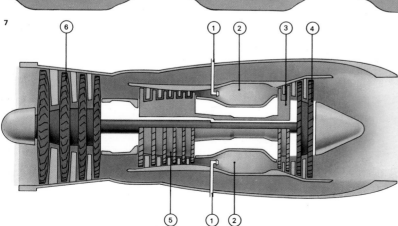
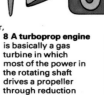

7 A turbofan engine is a type of gas turbine used mainly for powering civil aircraft. Fuel entering the engine [1] mixes with compressed air and burns in the combustion chamber [2]. The expanding gases rotate high-speed [3] and low-speed [4] turbines. These, in turn, drive a compressor [5], which forces air into the combustion chamber, and fans [6], which push air round the combustion chamber and into the tail pipe, providing extra thrust by means of air displacement.

8 A turboprop engine is basically a gas turbine in which most of the power in the rotating shaft drives a propeller through reduction gears. Incoming air is compressed [1], mixed with fuel [2] and burned to rotate the turbine [3]. A typical output is 2,000hp.

Steam engines

The steam engine is generally acknowledged as one of technology's greatest contributions to human progress. In the two centuries following its introduction, commercial life and industry developed at a greater rate than ever before. The steam engine is one of the few inventions that have been almost wholly beneficial, influencing every aspect of life.

The influence of steam power
The steam engine's first impact was on the development of coal mining. Before 1712, deep coal seams were impossible to work because they were flooded with water. Then Thomas Newcomen (1663–1729) developed a steam engine [2] that could be used to pump out water. The engine was extremely simple and, except for a few parts, could be made by local craftsmen. But it wasted fuel because the cylinder had to be cooled after every stroke, and so the engine was used only at mines where fuel was plentiful.

The engine invented by James Watt (1736–1819) in 1769 [5] separated the heated and cooled parts of the work cycle and so avoided the energy losses caused by com-

bining them in the cylinder, reducing fuel consumption by two-thirds. As a result, Watt's engine could be used even where fuel was scarce and costly.

The Watt engine also enabled the development of ironworks, making cast iron readily available and leading to a period of great industrial expansion in Britain. By the mid-nineteenth century, horizontal and vertical types of steam engines had evolved and were developed into powerful mill engines, particularly for textile mills.

The other great impact of steam power was in transport. Steam locomotives, pioneered by Richard Trevithick (1771–1833) [4], led to railway networks offering faster and cheaper carriage of freight than by canal barge. Steamships cut many days off the time needed for an intercontinental sea passage and made vessels independent of the wind and weather.

The next great breakthrough in steam power was the invention in 1884 by Charles Parsons (1854–1931) of the steam turbine. One of its advantages was that its "output" was a rotating shaft instead of the backwards

and forwards motion of earlier steam engines. There was no longer any need for the complicated mechanical linkages of connecting rods, cranks and eccentrics. Within 20 years, 70,000 horsepower Parsons turbines were driving liners across the Atlantic Ocean at up to 45km/h (about 25 knots).

Principles of the steam engine
Any steam engine converts the heat energy stored in steam into usable power. Steam is the vapour produced by heating water until it changes state from a liquid to a gas. If this process takes place in a closed vessel (a boiler), adding more heat raises the temperature and pressure of the steam. Heat from any source – fuel, the sun or a nuclear reactor – can be used. In a conventional steam engine, the pressure of the steam imposes a thrust on a piston connected by a linkage in such a way that its movement is made to rotate a shaft. In a turbine, the steam passes through jets and strikes the blades of a turbine wheel, making it turn.

The process is reversible, and if steam is cooled in a condenser it turns back into

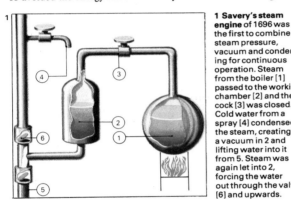

1 Savery's steam engine of 1696 was the first to combine steam pressure, vacuum and condensing for continuous operation. Steam from the boiler [1] passed to the working chamber [2] and the cock [3] was closed. Cold water from a spray [4] condensed the steam, creating a vacuum in 2 and lifting water into it from 5. Steam was again let into 2, forcing the water out through the valve [6] and upwards.

2 Newcomen's steam engine of 1712 was the first to use a piston and linkage to transfer the movement to pumps. Steam from the boiler [1] passed to the cylinder [2], and the weight of the

pump rods pulled the piston [3] to the top. The steam was shut off and cold water [4] sprayed into condense the steam. The vacuum created caused pressure to force the piston down again,

and a link to the crossbeam made it rock and work the main pump. A pipe led air and condensed water out of the cylinder, and a small pump lifted water up to the header tank [5].

3 Small, powerful steam engines were a logical development of Trevithick's locomotive of 1804. Coupled to existing and, later, specially designed machines, they reduced human drudgery, increased production and reduced costs in many

industries. Richard Hoe's steam-powered newspaper press of 1847, shown here, was a rotary machine with the type on cylinders which turned at high speed in contact with a moving band of paper. Steam power also influenced agricul-

ture, where traction engines – the first successful self-propelled road vehicles – could haul and work heavy trailers and threshing machines. Two engines at opposite ends of a field could draw a large many-bladed plough back and forth between them.

4 Richard Trevithick pioneered the use of high-pressure steam and obtained great power from relatively small engines by raising the pressure to 50lb/sq in. He also did away with the con-

denser, thus reducing the weight, and achieved further economies by enclosing the furnace within

the boiler shell, and using the exhaust steam to preheat the boiler feed water. From this, it was but a short step to mount the whole engine on wheels to run on iron rails, and this locomotive

(shown below) proved the feasibility of railways in 1803. The single horizontal cylinder was enclosed within the boiler, and the flywheel drove both axles through a set of gear wheels.

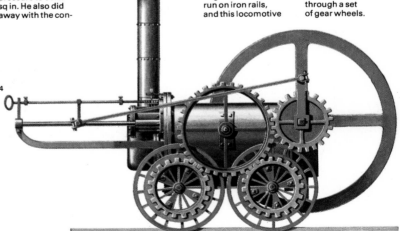

water. If this takes place in a closed vessel, the reduction in volume creates a vacuum which can be used to recover further energy. These two principles were known for nearly a century before Thomas Savery (1650–1715) combined them in a machine for pumping water in 1696 [1].

Condensation of steam during use in the work cycle represents a loss of energy. Steam engines were designed to raise steam to the highest possible temperature, to minimize condensation during the cycle and to expand the steam until it was exhausted at the lowest possible temperature. The idea was to extract the maximum heat energy from the steam. It was achieved by leading the steam on its way to the engine through pipes exposed to hot flue gases from the boiler – a so-called superheater. In a modern power station, superheated steam reaches temperatures up to 600°C (about 1,100°F). Even better efficiency can be obtained by taking the steam after use in part of the cycle and again leading it in pipes through hot boiler gases, in a reheater. In this way, heat from fuel burnt in the boiler is used three times to inject

energy into the steam. Exhaust steam tends to cool the walls of the cylinder, and so it is enclosed by an outer container full of hot steam, called a steam jacket.

Fuel economies

There are other ways of making heat economies. Water fed to the boiler has to be raised to boiling point before evaporation to steam takes place. Any way of heating the water before it reaches the boiler must save fuel. There are two ways of doing this. An economiser consists of a set of pipes exposed to hot gases leaving the boiler, and boiler feed water can be passed through these, heating the water to about 93°C (200°F). But a modern power station, with an evaporation temperature of 370°C (700°F), requires much more preheating. The boiler feed water is heated by steam, drawn from various points along the turbine system, which has already given out power in the main turbines. Called interstage feed-water heating, this system is less costly in energy terms than, say, using the hottest steam straight from the boiler to preheat the water.

KEY

Many early designs for making use of steam engines applied them to road vehicles, generally conceived as steerable mechanical carriages. All designs had to solve the problem of converting the steam engine's reciprocal (side-to-side) motion into a rotary one for driving the vehicle's road wheels.

5 James Watt's engine of 1769 reduced fuel consumption by separating the condenser and the cylinder, but it was still only a pumping engine. The growing industries needed power to drive machines, provided by Watt's double-acting engine of 1784. It included the basic features of modern engines, and was a great advance on the Newcomen engine. A furnace [1] heated the base and sides of a boiler [2]. Steam at a pressure of 7lb/sq in passed into a cylinder [6] which was closed at the top to allow the steam to drive the piston [7] both down and upwards. The effort was transmitted via a parallel motion linkage [22] to the beam [23] which rocked about its centre. Its movement was transmitted to the factory via the connecting rod [24], which had a non-rotating planet wheel [25] fixed rigidly at the end, driving the sun wheel [26] on the flywheel shaft. The sun and planet wheels were kept in mesh by a link behind them. Teeth on the flywheel [27] meshed with those of the pinion [28] at the end of the mill drive shaft. The engine framing was made of timber, and the engine speed was controlled by the governor [11]. Steam was let in and out of the cylinder by two pairs of valves, and the exhaust was condensed in [14] by the cold-water jet [15]. Air and water were removed by a pump [16], and water pumped to the boiler by [18] via the float control [4].

1 Furnace and firedoor
2 Boiler
3 Water
4 Float control
5 Steam pipe
6 Engine cylinder
7 Piston
8 Steam throttle valve
9 Top valve chest
10 Bottom valve chest
11 Governor
12 Governor drive
13 Condensing water tank
14 Condenser
15 Cold-water jet
16 Air pump
17 Feed water tank
18 Feed water pump
19 Feed water pipe
20 Condenser water supply
21 Piston rod
22 Parallel motion
23 Engine beam
24 Connecting rod
25 Planet wheel
26 Sun wheel
27 Toothed flywheel
28 Toothed driving wheel

Power from steam

The power of steam created the Industrial Revolution yet today hardly any steam engines remain. For generating electricity, however, steam (driving a turbine) is still the chief motive force employed.

Modern power stations
A modern thermal power station uses the heat from burning coal or oil, or from a nuclear reactor, to boil water circulating in boiler tubes to produce high-pressure steam [1]. The steam travels along pipes to a steam turbine that consists of a series of propeller-like vanes mounted on a single shaft. Nozzles direct jets of steam towards the vanes, driving the turbine round. A generator coupled to the end of the turbine shaft converts the rotary motion of the shaft into electric power.

Each of the three elements of a power station – the boiler, turbine and generator – has undergone intensive development to produce the most efficient machines possible. As a result the efficiency of electricity generation (the amount of electrical energy output compared with the heat energy input) increased from about five per cent in 1900 to nearly 40 per cent in 1975. In other words, without this increase in efficiency a power station would burn eight times as much fuel to produce the same amount of electricity.

The boiler of a large conventional power station burns powdered coal at a rate of up to 200 tonnes per hour. Rail wagons carry the coal to the plant and dump it in huge coal stores from which it is taken by conveyor to the boiler. It is weighed and pulverized to a powder as fine as flour, then mixed with air and carried by fans along metal ducts to the furnace where it burns fiercely.

Producing the steam
The boiler consists of a tall chimney-like structure lined with vertical pipes carrying water. The heat produced by burning the coal-air mixture boils the water, producing steam. This is first collected in a steam drum, then recirculated through the hottest part of the boiler in another set of tubes which superheat it to even higher temperatures.

From the superheater steam goes directly to the turbines [2]. The steam is taken first to the high-pressure turbine where it passes through a ring of stationary blades. These act as nozzles and direct the blast of steam on to the rotating blades. As it passes through, the steam makes the turbine turn, just as a windmill turns in the breeze. Immediately after passing through the high-pressure turbine the steam is led back to the boiler and reheated. It then passes through the intermediate and low-pressure turbines [3], gradually giving up its energy and generating more rotational power.

Finally the steam, with most of its energy spent, changes back into water in a condenser. This is a large vessel containing cooling pipes carrying cold water from a nearby river or estuary. The cooling water takes the last remaining heat from the steam, producing hot water that passes back to the boiler to be reheated. Condensation creates a vacuum in the condenser, allowing more energy to be extracted from the steam.

The turbine shaft rotates at a speed set by the frequency of the electricity supply. In Britain and many other European countries the speed is 3,000 revolutions per minute (50 revolutions per second), corresponding to an

CONNECTIONS

See also
64 Steam engines
84 Electricity generation and distribution
74 Nuclear power
82 Saving fuel and energy

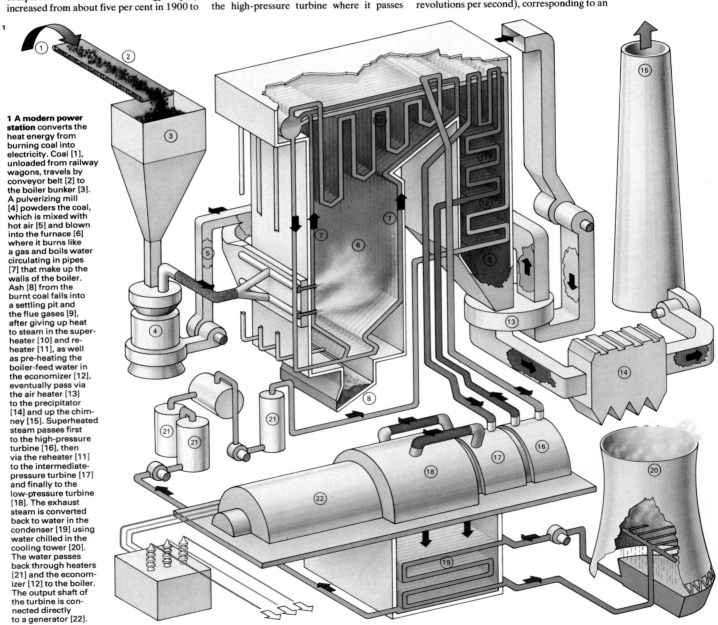

1 A modern power station converts the heat energy from burning coal into electricity. Coal [1], unloaded from railway wagons, travels by conveyor belt [2] to the boiler bunker [3]. A pulverizing mill [4] powders the coal, which is mixed with hot air [5] and blown into the furnace [6] where it burns like a gas and boils water circulating in pipes [7] that make up the walls of the boiler. Ash [8] from the burnt coal falls into a settling pit and the flue gases [9], after giving up heat to steam in the superheater [10] and reheater [11], as well as pre-heating the boiler-feed water in the economizer [12], eventually pass via the air heater [13] to the precipitator [14] and up the chimney [15]. Superheated steam passes first to the high-pressure turbine [16], then via the reheater [11] to the intermediate-pressure turbine [17] and finally to the low-pressure turbine [18]. The exhaust steam is converted back to water in the condenser [19] using water chilled in the cooling tower [20]. The water passes back through heaters [21] and the economizer [12] to the boiler. The output shaft of the turbine is connected directly to a generator [22].

alternating supply at 50 hertz (cycles per second). In the United States the turbine speed is normally 3,600rpm, corresponding to a 60Hz supply.

The electricity generator consists of two electrical windings [2]. One is mounted on the turbine shaft, rotating with it, and is called the rotor. The other is arranged as a shroud around the rotor, fixed to the floor, and is called the stator. The relative motion of rotor and stator generates the electricity.

Generators and efficiency

To get the best performance out of the generator it must be continuously cooled. At one time natural cooling or forced cooling by air fans was used, but since the early 1950s hydrogen gas has been employed because it is much more efficient. The rotor and stator operate in an atmosphere of hydrogen, which removes the heat. In the most recent designs the rotor windings are made of hollow copper tubes with hydrogen circulating through them and the stator windings are cooled individually by tubes carrying hydrogen. With "direct" or "inner" cooling, as this technique is called, the output of the generator can be doubled.

The generator produces electricity at about 25,000 volts. Most domestic supplies operate at only 250 volts but it is much more economical to transmit electricity over long distances at very high voltages. The first step in distribution, therefore, is to step up the voltage to a higher value (275,000 or 400,000 volts in Britain) using a transformer. It can then be fed to the national grid, which links countrywide generating plants.

Linking generating plants helps to operate the entire network in the most economical way. The efficiency of a large generator becomes less when it operates at only part-load. So rather than reduce output to match demand it may be best to turn the generator off altogether and switch in power from another station which is kept going at full power. Electricity authorities compile a "merit order", with the most economical stations at the top of the list, to decide which of the stations to turn off first as demand for power drops. This enables them to keep the cheapest stations in operation all the time.

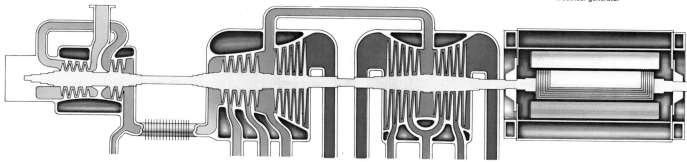

2

High-pressure turbine   Intermediate-pressure turbine   Low-pressure turbine   Electrical generator

2 Turbines and generators are the components that convert the rotary motion produced by steam power into electricity. To obtain the maximum energy from the hot steam, there are several stages of turbines – up to five in a large power station – each using steam at a slightly lower pressure than the pre-vious one. Superheated steam at up to 600°C (1,112°F) gives up much of its energy in a high-pressure turbine. The exhaust steam from this stage is reheated and passed to an intermediate-pressure turbine and then to the low-pressure turbine. The output shaft drives a generator.

3 A low-pressure turbine, shown here diagrammatically, illustrates how steam pressure is made to turn the blades. There are two similar sets, mounted back to back on the same shaft for efficiency. The central blades, where steam pressure is highest, are smaller than those at the ends where pressure is lowest.

4 The turbine hall of a modern power station contains a mass of heavy machinery and insulated pipes carrying steam between the various turbines. Galleries and catwalks halfway up the machines enable engineers to inspect and maintain them. The steam-generating boilers may be sited beneath the floor.

3

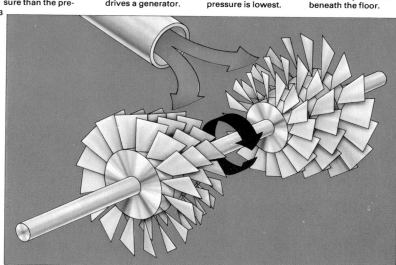

4

Oil and gas engines

In the traditional steam engine, and even in a modern steam turbine, fuel is burned outside the engine to heat water and raise steam to drive the engine. But it is more efficient to burn fuel inside an engine and let the expanding gases produced drive a piston or turbine.

The first such internal combustion engine, running on gas [Key], was built by the German engineer Nikolaus August Otto (1832–91). His engine, demonstrated in Paris in 1867, was large, noisy and not very efficient. But it became the forerunner of 99 per cent of all today's engines.

The four-stroke cycle
Nine years after the first gas engine Otto devised another, based on the four-stroke cycle. The crucial advance in this engine was that the gas was compressed before it was ignited, giving not only a considerable improvement in efficiency but also a marked reduction in fuel consumption.

It takes four strokes of the engine to include one of power, so this system is known as the four-stroke cycle. It is by far the most common type of engine in use today. The four main stages are an induction stroke in which a downward movement of the piston sucks in the fuel-air mixture; a compression stroke in which upward movement of the piston compresses the gas; a power stroke – a second downward piston movement caused by the explosion of the fuel; and an exhaust stroke in which the upward-moving piston forces exhaust gases out of the cylinder.

Many motor cycles and a few small cars use the two-stroke cycle first devised by Dugald Clerk in 1880. In this type of engine the movement of the piston admits the fuel and exhausts the burned gases by uncovering "ports" or holes in the side of the cylinder.

Fuel and exhaust pass in and out of a four-stroke engine using a more sophisticated system of valves, controlled automatically by a camshaft driven direct from the engine's crankshaft. As the engine operates, the valves are successively opened and closed.

The moment of ignition of the fuel must also be accurately controlled. This is done by a distributor, again mechanically connected to the crankshaft, which directs a current of electricity successively to each of the cylinders. This current "fires" a spark in the sparking plugs and the fuel is ignited.

Otto's engines ran on coal gas, a perfectly satisfactory fuel but one that is difficult to store. The gas engine was greatly improved by the use of liquid fuels such as petrol (gasoline) made by refining crude oil. To turn petrol into a combustible vapour it is mixed with air to form a fine mist of droplets that can be drawn into the cylinders. The mixing is carried out in a carburettor.

Unlike steam engines most internal combustion engines do not produce great power at slow speeds. The cylinders are small and each individual ignition stroke produces comparatively little power. To obtain a useful amount of work from such an engine it must be run fast, to put the maximum number of ignition strokes into each second. Motor car engines commonly produce their maximum power at speeds of 5,000 revolutions per minute or more. The upper limit on speed is set by the wear and tear on the engine caused by the oscillating pistons and valve gear. Specially prepared engines, in which great attention has been paid to balance and

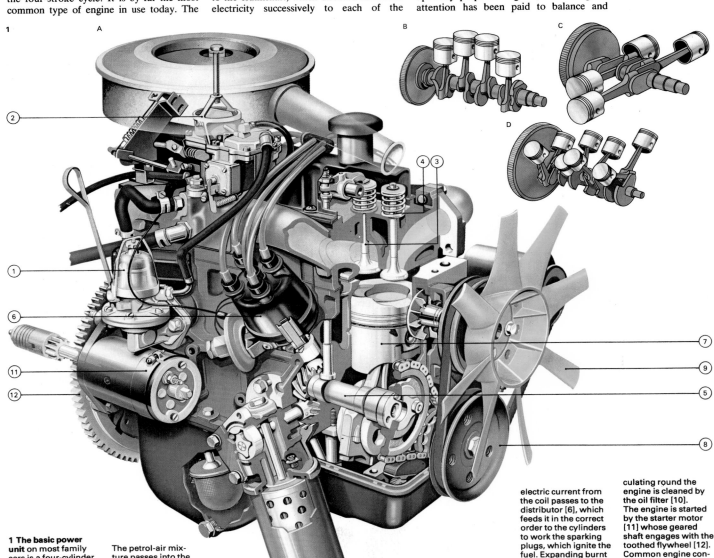

1 The basic power unit on most family cars is a four-cylinder overhead valve petrol engine [A]. The petrol pump [1] draws fuel from the petrol tank and in the carburettor [2] it is vaporized and mixed with air.

The petrol-air mixture passes into the cylinders through inlet valves [3]. The valves are closed by springs [4] and opened by tappets controlled by push rods via the camshaft [5]. High-tension

electric current from the coil passes to the distributor [6], which feeds it in the correct order to the cylinders to work the sparking plugs, which ignite the fuel. Expanding burnt gases depress pistons [7], whose movement rotates the crankshaft. A pulley [8] drives a cooling fan [9] by means of the fan belt. Lubricating oil cir-

culating round the engine is cleaned by the oil filter [10]. The engine is started by the starter motor [11] whose geared shaft engages with the toothed flywheel [12]. Common engine configurations include the in-line four [B], a horizontally opposed or flat four [C] and a V-8 [D], with two sets of four cylinders inclined at an angle.

smoothness, can obtain more power by running up to speeds of 12,000 rpm or more.

The economical diesel engine

The compression-ignition engine, designed by the German Rudolf Diesel (1858–1913) in 1896, dispenses with the carburettor and sparking plugs of the petrol engine. The gas inside the cylinder on the compression stroke is pure air, which is compressed to 1/14 to 1/20 of its initial volume – a much higher compression ratio than is used in petrol engines. At the top of the compression stroke a fine spray of oil fuel is injected into the cylinder. As a gas is compressed its temperature increases, so that the oil spray meets the air charge at a temperature sufficiently high to ignite it spontaneously.

Because of its high compression ratio the compression-ignition or diesel engine is more efficient than a petrol engine. But for the same reason it must be more heavily built, thus offsetting the advantage somewhat. Diesel engines offer economies in fuel consumption at the expense of a loss in performance; they are particularly suited to frequent stop and start duties, and as a result are widely used in taxis, buses and lorries.

The powerful gas turbine

The gas turbine, a completely different kind of engine, was first devised at the beginning of the twentieth century and perfected in the 1930s. It usually has a single shaft carrying a series of propeller-like fans divided into two groups, the compressor and the turbine.

In an operating gas turbine air is drawn in by the compressor fans and its pressure increased. The compressed air is mixed with fuel and ignition takes place, further increasing temperatures and pressures. The burned mixture leaves the engine through the turbine, driving the blades round. Much of the power produced is taken up by the compressor, which is often driven directly by the turbine, but enough is left over to make the gas turbine an exceedingly powerful form of engine. Efficiencies are not high, but the good power-to-weight ratio of a gas turbine makes it suitable for aircraft propulsion. A gas turbine is about three times as powerful as a piston engine of the same weight.

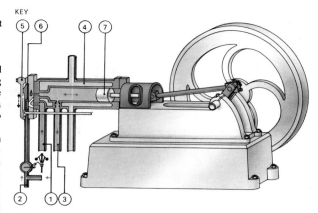

Otto's engine of 1876 was the first successful internal combustion engine. A four-stroke horizontal engine, it used a mixture of gas and air as fuel. The charging stroke drew in air [1] and gas [2] through a slide valve [5] into the cylinder, pulled in by movement of the piston [7]. On the return stroke, the fuel mixture was ignited by a flame carried through a narrow opening in the slide valve from a continuously burning gas jet [6] outside the engine. The expanding products of combustion produced the working stroke. On the fourth and last stroke the exhaust gases were forced out of the engine [3]. A jacket of cold water [4] surrounded the cylinder and kept the engine cool.

2A

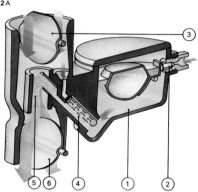

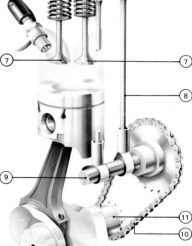

2 In a carburettor [A], petrol enters the float chamber [1] controlled by a needle valve [2]. Part of the air passing the choke valve [3] mixes with the petrol [4]. The mixture passes into the main air stream [5] and past the throttle valve [6]. In an overhead valve engine [B] the valves [7] are worked by push rods [8] moved by cams on the camshaft [9], which is rotated by a chain [10] from the crankshaft [11].

B

3

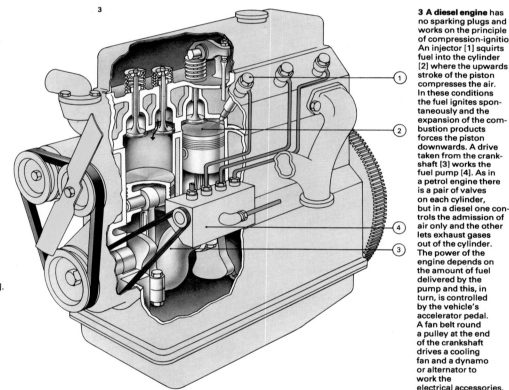

3 A diesel engine has no sparking plugs and works on the principle of compression-ignition. An injector [1] squirts fuel into the cylinder [2] where the upwards stroke of the piston compresses the air. In these conditions the fuel ignites spontaneously and the expansion of the combustion products forces the piston downwards. A drive taken from the crankshaft [3] works the fuel pump [4]. As in a petrol engine there is a pair of valves on each cylinder, but in a diesel one controls the admission of air only and the other lets exhaust gases out of the cylinder. The power of the engine depends on the amount of fuel delivered by the pump and this, in turn, is controlled by the vehicle's accelerator pedal. A fan belt round a pulley at the end of the crankshaft drives a cooling fan and a dynamo or alternator to work the electrical accessories.

4

4 A gas turbine, such as this Rolls-Royce Viper turbojet, uses the hot gases from the burning of fuel to turn sets of turbine blades at the rear of the engine. Other blades, mounted on the same shaft but at the front of the engine, compress the incoming air. Generating up to 4,000 lb thrust, the engine is used in trainer, light attack and business jet aircraft.

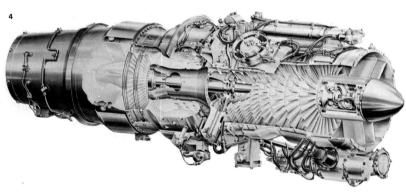

Wind and water power

The windmill [Key] and the waterwheel [1] are two of the oldest forms of power. Waterwheels were in use in Rome in 70 BC to grind corn while the windmill made its first appearance in Persia in AD 644. The modern waterwheel, in the form of the hydroelectric turbine, is more important, but the use of wind as a power source is showing signs of revival.

From waterwheels to turbines

A waterwheel or turbine converts the energy of flowing water into a rotary motion [3]. Early waterwheels used the undershot principle, in which the lower half of the wheel was simply immersed in a flowing stream [1]. They had an efficiency of only 30 per cent. Overshot wheels, in which the flow of water is directed over the top of the wheel, produce efficiencies of 70 to 90 per cent, figures similar to those for modern turbines.

Turbines replaced waterwheels in the second half of the nineteenth century. There are three categories: impulse turbines, reaction turbines and axial-flow turbines [4]. An impulse turbine needs a high head of water pressure. The falling water is directed through a nozzle and the rapidly moving jet produced hits "buckets" on the edge of a wheel. A reaction turbine operates on the same jet principle as a rotating lawn-sprinkler; an axial-flow turbine has a variable-pitch propeller in a large diameter tube.

Hydroelectric schemes and tidal power

Most water turbines are used to exploit the run-off of water from mountain areas which is stored behind a dam. These turbines drive electrical generators. In countries with many mountains such schemes provide cheap, pollution-free power. Hydroelectricity is the fourth largest source of energy in the United States while in Britain it is important only in the north of Scotland.

A considerable capacity of such conventional turbo-power remains untapped: the Fraser River in Canada, for instance, could generate 8,700 megawatts (MW) of electrical power, the Brahmaputra in India 20,000MW. The Yenisey-Angara river system in the USSR, with 11,000MW already installed, has the capacity for generating an additional 53,000MW.

Water turbines can also be used to generate power from the small heads of water produced by the tides [2]. The only full-scale power station of this kind is in the Rance river estuary in the Gulf of St Malo in northern France. The range of rise and fall produced by the tides varies widely from place to place – from as little as 2cm (0.75in) in Tahiti to as much as 15m (50ft) in the Bay of Fundy in eastern Canada. But only places at the upper end of this range make suitable sites for tidal power stations, and even then the economics of such a scheme may not be easy to justify.

Unfortunately the times of the tides do not always coincide with peak electricity demand. A tidal power station might reach its full power in the middle of the night, when electricity demand is lowest. One way of reducing this problem is to divide the tidal basin into two: a high basin that fills between mid-tide and high tide and a low basin that empties between mid-tide and low tide. The setting up of such a scheme allows a continuous difference in level to be maintained.

Another alternative is to use the high basin as a pumped storage system [4]. In

CONNECTIONS

See also
112 Sailing ships
116 Hydrofoils and air-cushion vehicles
198 Building dams
82 Saving fuel and energy

1 The two basic types of waterwheels are called overshot [A] and undershot [B]. In the overshot wheel, a stream of fast-flowing water is directed on top of the wheel, which turns forwards. The paddles of the undershot wheel dip into the mill stream and the water current turns the wheel backwards. In both types of waterwheels the energy of flowing water is converted into rotary power. It can be used to drive various machines, such as pumps or mills for grinding corn. Water power can even be harnessed to drive a dynamo or alternator for generating electricity. Waterwheels are virtually silent and pollution-free.

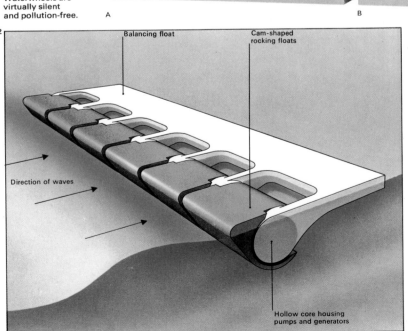

Balancing float

Cam-shaped rocking floats

Direction of waves

Hollow core housing pumps and generators

2 Water waves in the continuous swell of the open sea are a vast and as yet largely untapped source of energy. When a wave passes a certain point no water moves sideways. Instead, large masses of water move up and down. This can be confirmed by watching a cork or other light object floating on a pond. When ripples pass it, it does not move along but only bobs up and down. The assembly of rocking floats shown here is designed to harness wave power and generate electricity. The structure would be about 300m (1,000ft) long – the size of a supertanker. The movement of the floats works pumps and the flowing water from them powers turbines that drive electrical generators.

3 A simple turbine of the 16th century used the power of running water to work a pump for irrigation. Rotation of the turbine [1] turned a wheel [2] with teeth on only half its circumference. The cog wheels [3] turned alternately in opposite directions making the pump wheel [4] oscillate. Self-acting valves allowed the pistons [5] to draw water into one cylinder while pushing it out of the other.

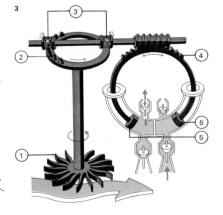

pumped storage electricity generated by conventional plants and not needed by consumers is used to pump water from a low-lying basin into a high-lying one. When demand picks up, the water is allowed to flow down again, generating electricity like an ordinary hydroelectric plant. Such a scheme is not a net generator of electricity but it provides a means of storing large amounts of power.

Harnessing wind power

The use of wind to generate electricity has been less successful. Despite the immense amounts of power theoretically available (Britain could meet most of its electrical energy needs by harnessing just a fraction of its available wind energy) the problem of harnessing it economically has yet to be solved.

The power available to a windmill [Key] is proportional to the cube of the wind velocity and the area swept out by the blades. The ultimate efficiency possible is 59 per cent, but practical machines would be unlikely to do better than 45 per cent. Estimates suggest that the generation of electricity by wind could compete with nuclear power in a

limited number of sites where average wind speeds exceed 32km/h (20mph). But there are not many such sites (30 have been identified in the British Isles) and wind power is thus unlikely to be able to generate more than about one per cent of electricity needs.

For this reason more interest is being shown in exploiting the stored power of the wind in sea waves [2]. Winds blowing over a long stretch of ocean generate powerful waves, and could be used as a source of power. The most promising design, produced by S. H. Salter at Edinburgh University, uses a floating boom that rocks to and fro as the waves pass, operating pumps to compress a fluid which would then be allowed to expand through turbines to generate electricity.

The amounts of wave power available are large. In Britain they add up to 80KW per metre along coasts, or a total in British territorial waters of 120,000MW – three times the peak electricity demand in 1975. The amounts of power available are also greater in winter, when demand is at its highest, but the problems of building and maintaining such booms are formidable.

Windmills have been used for many centuries for harnessing the power of the wind. Originally the power was used for grinding corn – hence the name mill – but later windmills were used to drive pumps, particularly for draining low-lying regions of the Netherlands and eastern England in the Fens and East Anglia. Some early windmills had canvas sails, developed from those used on ships. Then slatted wooden sails were developed, as in this bonnet mill. The sails turned with the bonnet or cap on the top of the mill so that they always faced into the wind. In a post mill the whole structure was pivoted on a central post and turned to face the wind.

4A

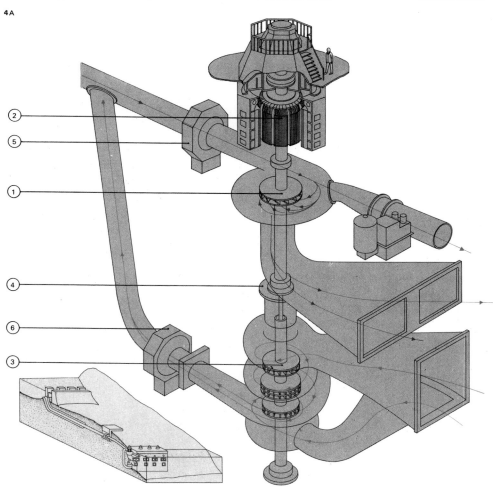

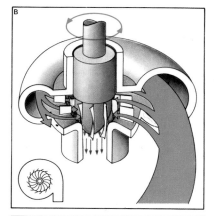

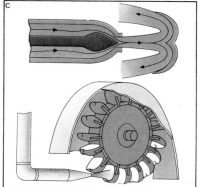

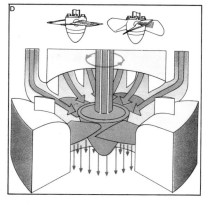

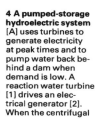

4 A pumped-storage hydroelectric system [A] uses turbines to generate electricity at peak times and to pump water back behind a dam when demand is low. A reaction water turbine [1] drives an electrical generator [2]. When the centrifugal pumps [3] are uncoupled the machine acts as a normal hydroelectric generator. But when the geared coupling [4] is engaged the water turbine drives the pump up to operating speed. Then the generator is connected to the electricity supply whence it acts as a motor. The turbine valve [5] is closed, the pump valve [6] is opened and water is pumped back behind the dam, adding to the volume of stored water available for later hydroelectric generation. Three types of water turbines are in common use. The Francis reaction turbine [B] has adjustable fixed blades that divert the water stream in such a way that it strikes the rotating turbine blades at a tangent. Water flows out of the turbine downwards. In the Pelton wheel or impulse turbine [C] the water passes through a jet and strikes bucket-shaped paddles on the wheel. The direction of the water flow is reversed. The blades on the Kaplan axial-flow turbine [D] resemble a ship's propeller.

Power direct from the sun and earth

Much of the energy man consumes comes indirectly from the sun. Coal, oil and hydro-electricity can all be thought of as forms of solar energy, stored as long-dead plant and animal remains or as water, evaporated from the sea by the sun, which falls as rain and provides water power. Today, however, the search is on for a way to use solar energy directly; coal and oil take millions of years to produce and many of the best hydroelectric sites are already in use.

Collecting solar energy

There are many possible ways of tapping solar energy. It may be intercepted in space by a satellite and beamed to earth as an intense beam. More simply, it can be collected as it reaches the earth's surface by collectors that are basically hot water radiators working in reverse. Such collectors can provide space heating in houses or can be used as the basis for air-conditioning systems. Solar energy can also be converted directly into electricity, at low efficiency, by solar cells similar to those used in satellites. Alternatively, it can be collected by using it to grow plants that are then used directly as fuel, or converted into a liquid fuel by chemical conversion or by the use of micro-organisms. Finally, the heat stored in the sea can be used to provide energy by employing the temperature difference between water at the surface exposed to the sun and cold deep water.

The systems compared

Of these possibilities, only the humble solar collectors are so far used. They consist of panels, usually mounted on the roof of a building and angled so as to obtain the maximum amount of sun. Water is pumped through the panels, picking up heat from the sun as it goes. The pump is controlled by a sensor so that it operates only when the collector is several degrees hotter than the water in the storage tank. Unfortunately, the sun does not shine all the time, particularly in a northern country such as Britain, but despite this it is estimated that a well-designed solar collector system in Britain could halve hot water and space heating costs by half. In warmer countries the advantages are greater and such collectors are widely used in Israel.

Systems for converting solar energy directly into electricity are less attractive. The solar cells used by spacecraft are expensive and typically have conversion efficiencies of only ten per cent or less. In principle electricity could be generated by covering rooftops with solar cells [3] but this would depend on producing cheap cells of the order of only a few pounds per square metre. At present, they cost more than £100 per square metre (£9 per square foot).

An alternative would be to collect solar energy by using transparent pipes carrying a molten mixture of sodium and potassium. The mixture would be heated to well above the boiling-point of water by concentrating the sun's rays on to the pipes with reflectors. The hot metal would then be used in a heat exchanger to raise steam, which could be used to generate electricity. Such a scheme, based in a desert area, would need a lot of land – 50 to 200 square kilometres (19.3 to 77 square miles) for every 1,000 megawatts of electricity – and might be feasible to introduce in the 1980s.

The other possibility is to make use of the

1 A solar furnace harnesses the sun's energy by focusing its heat and light, using large lenses or mirrors. This furnace at Mont-Louis in France gets hot enough – up to 3,000°C (5,430°F) – to melt metals. The heat, which is completely clean and uncontaminated by fuel gases, could be concentrated on a boiler to raise steam for domestic heating or for driving a turbine for generating electricity. Generally a set of flat mirrors is arranged to follow automatically the path of the sun and reflect its rays on to the stationary furnace mirror. The tracking mirror is called a heliostat and may be driven by an electric motor so that it moves through exactly 15 minutes of arc every hour. Alternatively the angle of the heliostat may be controlled by a photocell or thermocouple that detects the absence of light or heat rays.

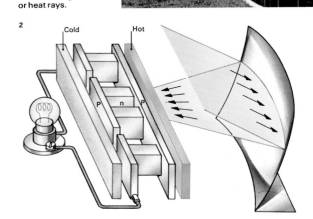

2 In one type of solar cell, heat from the sun is focused on to a surface backed by a semiconductor thermocouple (a series of p-n junctions) which converts the heat into electricity.

3 Solar collectors, seen here on the roof of a house, provide energy for domestic space and water heating. Flat-plate collectors are sheets of metal painted black and they transfer heat to air or water circulating behind them.

difference in temperature between deep ocean water and surface water. The circulation of ocean water is limited, which means that there is often a difference of as much as 15°C (23°F) between deep and surface water. By making use of this difference the vast quantities of solar heat stored in the sea could be used. But the efficiency of operation would be only about five per cent.

Energy from the earth

The eruption of volcanoes is graphic evidence of the energy stored inside the earth. Once thought to be left over from an ancient molten earth, geothermal heat is now believed to be produced continuously by the slow decay of radioactive elements deep inside the earth's core.

In some places on the earth's surface this heat escapes, either as a volcano or as hot springs and geysers [Key]. The first attempt to make use of geothermal heat [5] was at Larderello in Italy where natural steam is now used to generate 390 megawatts of electricity. Other geothermal plants are in operation in Iceland, Japan, New Zealand, the

USSR and the United States. In the United States, geysers 145km (90 miles) northeast of San Francisco generate 302 megawatts.

The number of places where steam or hot brines occur naturally is limited and depends on a supply of water. If a hole is drilled anywhere on the earth's surface it becomes hotter the deeper the drill goes, but the rate at which the temperature increases – the temperature gradient – varies widely. In places with a steep temperature gradient it may be possible to extract heat by drilling holes, fracturing the rocks at the bottom, then passing down water from the surface to be turned into steam and used to generate electricity. Experiments to test the feasibility of this scheme are in progress at Los Alamos Scientific Laboratory in New Mexico [6]. A volume of about 520 cubic kilometres (125 cubic miles) of underground rock a few hundred degrees hotter than the surface rocks contains as much energy as the entire world uses in a whole year. If only a small fraction of this energy can be extracted, geothermal power plants could supply huge amounts of pollution-free power.

Geysers spout steam and superheated water out of the ground. They occur in regions that have active or recently active volcanoes, such as Iceland, Wyoming USA and, as here, New Zealand. It is possible to "tap" the hot water for energy.

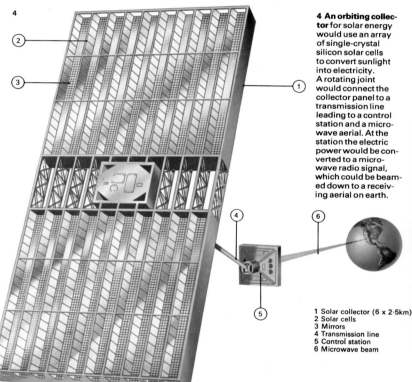

4 An orbiting collector for solar energy would use an array of single-crystal silicon solar cells to convert sunlight into electricity. A rotating joint would connect the collector panel to a transmission line leading to a control station and a microwave aerial. At the station the electric power would be converted to a microwave radio signal, which could be beamed down to a receiving aerial on earth.

1 Solar collector (6 x 2·5km)
2 Solar cells
3 Mirrors
4 Transmission line
5 Control station
6 Microwave beam

5 Geothermal energy (the heat of the earth in deep rocks) can give rise to hot springs when water naturally accumulates underground. A borehole drilled down into the water can be used to get a supply of hot water for industrial or domestic heating. A whole town sited over a geothermal field could obtain most of its heating by this means and experimental systems have been tried in Scandinavia, the USSR and the United States.

6 A man-made hot spring can be used, as in this experiment at Los Alamos, USA, as a source of heat energy. A borehole is drilled several hundred metres into a natural cavity in the earth in which the temperature may be as high as 300°C (572°F). Water pumped down the bore is heated and led up a second bore-hole. At the surface the hot water passes through a heat exchanger, which acts like a car radiator, transferring its heat to air blown over it.

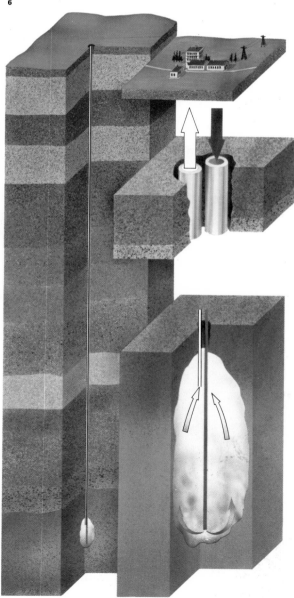

Nuclear power

Nuclear power is the controlled release of the most concentrated source of energy man has yet discovered – the energy of the atomic nucleus. When the nucleus of a heavy atom divides into two, in a process called fission, it releases prodigious amounts of energy – suddenly in an atomic bomb, slowly and controllably in a nuclear reactor, where the energy is harnessed to produce steam and, through turbines, generate electricity.

The only naturally occurring element that will spontaneously undergo fission is uranium. As it is found in nature, uranium ore is a mixture of mainly two isotopes, uranium-235 and uranium-238, and only the first of these will undergo fission spontaneously. (Isotopes are two forms of an element with different masses and other physical properties but identical chemical properties.) In the ore, U-235 makes up no more than 0.7 per cent of the uranium present. The amount of this isotope is increased, in a process called enrichment, to produce a mixture of isotopes that is about 90 per cent U-235.

The fuel is sealed into thin, pencil-shaped containers [5] so that it and the poisonous products of the fission process cannot escape. The fuel elements are supported, usually vertically, so that water or gas can flow between them to take away the heat produced by fission. After flowing past the hot fuel elements and picking up their heat, the coolant is taken away and used to raise steam to produce electricity in turbine generators.

Controlled chain reactions

Most reactors, however, need more than just fuel and a coolant. The fissioning of a U-235 nucleus is triggered off by a neutron, which strikes the nucleus and disturbs it just sufficiently to make it divide into two. In the process of dividing the nucleus produces two or three fresh neutrons, which fly off and strike other U-235 nuclei, creating a chain reaction.

A nuclear reactor [1] must be so contrived that of the neutrons produced by each fission one, and only one, must cause a second fission. Only then will the reactor work steadily and at a constant speed. If on average more than one neutron from each fission causes a second fission, the reactor will accelerate and become a bomb; if less than one, the reactor

will gradually lose power until it stops.

The neutrons produced by each fission travel extremely fast – about 16,000km (10,000 miles) per second – and tend to escape from the reactor altogether before they can cause an additional fission. To make the reactor work they must be slowed down, using a material called a moderator, to increase the chances that they will collide productively with another U-235 nucleus and cause another fission.

Moderators are light atoms which slow the neutrons down by a series of collisions. In commercial designs, three types of moderator have been used – water, graphite and heavy water, which is ordinary water in which the hydrogen atoms are replaced by heavy hydrogen or deuterium atoms.

Safety and refuelling

The reactor is controlled by neutron-absorbing rods which can be moved in and out of the core at will. As the control rods are removed, the number of neutrons absorbed declines, so that more are available for fission and the reaction speeds up [2]. To stop the

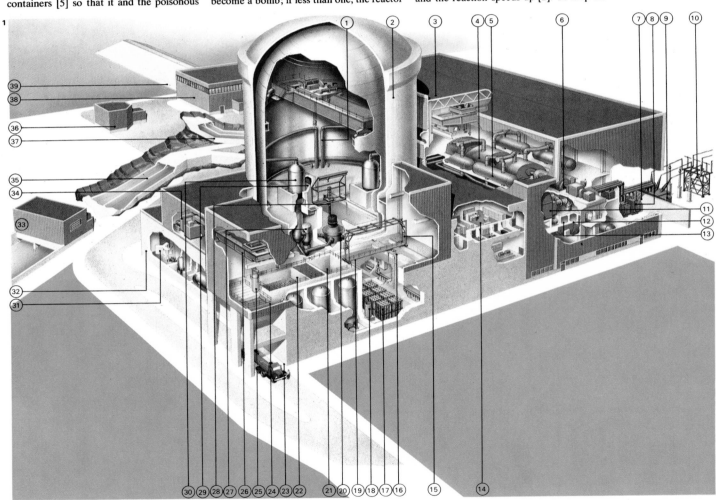

1 Nuclear reactors generate electricity using the heat produced by the controlled fission (splitting) of atoms of uranium or other similar elements. The heat is used in much the same way as in power stations that burn coal or oil to produce high-pressure steam to work turbine generators – only the source of the heat differs. Large volumes of cooling water are needed. Various liquids can be used to "carry" heat from the reactor to the heat exchanger. These include liquid sodium metal or, as here at the Ko-Ri power station near Pusan in South Korea, water under high pressure. Pressurized water reactors are today the most common type worldwide. The various parts of this highly complex installation are: [1] pressurized water pipes; [2] reactor building; [3] turbine building; [4] high-pressure turbine; [5] low-pressure turbine; [6] generator; [7] fire wall; [8] transformers; [9] main transformer; [10] switchgear; [11] cooling water outlet; [12] condenser; [13] cooling water inlet; [14] control room; [15] spent fuel bridge; [16] new fuel lift; [17] spent fuel storage rack; [18] spent fuel drum; [19] drum loading; [20] reactor vessel; [21] decontamination pit; [22] new fuel store; [23] fuel loading bay; [24] fuel handling hatch; [25] spent fuel drum; [26] fuel crane; [27] reactor coolant pump; [28] refuelling gantry; [29] pressurizer; [30] steam generator; [31] drum store; [32] waste drum loading area; [33] cooling water discharge; [34] auxiliary ventilation; [35] cooling water from turbines; [36] seawater (coolant) pump house; [37] cooling water to turbines; [38] cooling water pump house; and [39] seawater inlet tunnel.

reactor in a hurry – a procedure known as scramming it – the control rods are pushed into the core as quickly as possible. The rods "soak up" neutrons, leaving fewer available for taking part in fission, and the reactor slows down.

Surrounding the reactor are concrete and steel walls thick enough to absorb any accidental radiation escape. And, to make sure that the whole system is safe, the reactor must have emergency systems designed to cope with any unexpected failure of the fuel elements or the cooling system.

When the fuel elements are exhausted they are taken out of the reactor and replaced by new ones. The old ones, which still contain some unused U-235, are taken to a reprocessing plant to extract what is left [3]. At this point the nuclear reactor produces a useful bonus. Among the products inside the old fuel elements is a new man-made atom, plutonium-239, created by the neutron bombardment of U-238.

Plutonium-239 (Pu-239), just like U-235, is spontaneously fissionable, so that it can be used to make atomic bombs or to fuel

new nuclear reactors. The first reactors were designed and built as plutonium factories for the production of atomic bombs. Plutonium is, however, extremely poisonous and people have become concerned about the world increase in the amount of this element.

The fast breeder reactor
The latest type of reactor, the fast breeder [6], is also designed to make use of the accidental production of Pu-239. It is fuelled by uranium and around the outside of the core, to catch the escaping neutrons, is a "breeding blanket" in which U-238 is converted into Pu-239. So effective is this arrangement that a fast breeder creates more Pu-239 than the nuclear fuel consumed in the process.

To date, all existing reactors use the fission reaction in which heavy atoms split to produce energy. It has so far proved much more difficult to tame the other powerful nuclear reaction – fusion – that gives the hydrogen bomb its awesome power [7]. In this reaction light atoms heated to two million degrees fuse together to build heavier atoms, producing huge amounts of energy.

KEY

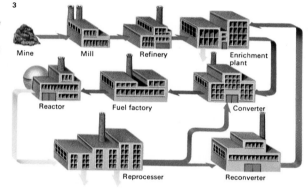

All nuclear power stations have basically the same layout. A reactor [1] heats water and converts it (either directly or indirectly) into steam [2] which then drives a turbine generator [3] to produce electricity. Exhaust steam is converted back to water in a condenser [4], using cold water from a cooling tower [5], and the water is circulated by a pump [6] back through the hot part of the reactor. The cooling tower needs a supply of cold water from a river or the sea.

2 A graphite-moderated reactor also raises steam to power turbine generators. Neutrons, slowed down in a surrounding block of graphite, split atoms of uranium-235. Each of these produces more neutrons, which are again slowed and produce further fissions. Rods of cadmium (blue) are lowered into the "pile" to mop up some of the new neutrons to control the reaction and hence the rate of heat production.

3 Nuclear fuel is used cyclically. From the mine, uranium ore passes to a mill to be converted to uranium oxide. This is refined and, as the fluoride, goes to an enrichment plant before being converted back to the oxide and made into fuel. The enriched fuel is then used in the reactor, from which spent fuel is reprocessed and the recovered uranium reconverted into fluoride and fed back to the enrichment plant.

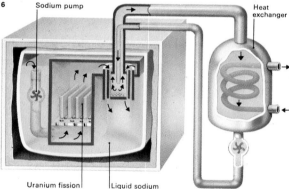

Mine   Mill   Refinery   Enrichment plant

Reactor   Fuel factory   Converter

Reprocesser   Reconverter

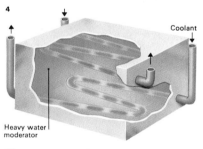

Coolant

Heavy water moderator

4 In a pressure-tube reactor the pressurized coolant (water, heavy water or an organic liquid) passes through the reactor vessel, surrounded by a neutron-absorbing heavy water moderator.

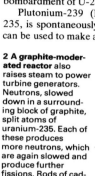

5 A pressure-vessel reactor is in a tank of steel 20–30cm (8–12in) thick and clad with stainless steel.

Sodium pump

Heat exchanger

Uranium fission   Liquid sodium

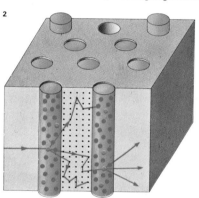

6 In a fast breeder reactor, uranium fuel is used and at the same time converted into the fissile element plutonium. Liquid sodium metal carries heat from the reactor to a heat exchanger, converting water to steam to drive turbine generators. A condenser turns the exhaust steam back to water for recirculation. The plutonium produced is used to enrich the uranium fuel or is used in other reactors.

7 Scientists are trying to harness the fusion reaction that powers a hydrogen bomb. At the very high temperatures the reacting gases, now called a plasma, are confined by magnetic fields.

8 Nuclear reactors for power stations, such as this one at Dounray, Scotland, need a plentiful supply of water for cooling. This water, drawn from the sea or an estuary, is used in condensers to convert exhaust steam to water.

Coal: production and uses

Coal is a hydrocarbon deposit that can be burned to provide heat energy. It is found in layers known as seams, usually below the ground [2], but in some places it appears near the surface immediately under a layer of soil [4]. It is generally hard, opaque and black. It was little used before the sixteenth century, but thereafter it became the basis of a great fuel industry – first in Britain, then elsewhere in Europe and throughout the world.

Origins of coal
Coal is the product of low-lying forest growths from several tens of millions to some 300 million years ago [1]. The residues of the plant debris were submerged in swamps then buried beneath great thicknesses of sandstones and shales. This process was repeated many times. The plant deposits were subjected to pressure, heat and breakdown by micro-organisms; as a result they were transformed into a range of combustible solids.

This process of transformation is often called "coalification" and the degree to which it has taken place determines the "rank" of the coal. Anthracite is the highest rank coal. Rank may vary across a single seam and when this occurs all seams above and below vary in a predictable way. In undisturbed vertical sections the deeper seams are of higher rank than those nearer the surface. This broad generalization, often subject to exceptions, is known as Hilt's law.

To simplify quality control, coal is often specified in terms of its "proximate analysis"; this gives its percentage composition of moisture, volatile matter, fixed carbon and ash as determined by standard methods. The "ultimate analysis" determines the amount of carbon, hydrogen, oxygen, nitrogen and sulphur in coal. In addition, users may wish to know the "heating power" or calorific value of the fuel in units of heat generated per unit of weight burned.

Coals may also be described in terms of their petrographic constituents – that is, considered as rock types. The four physical structures recognized are clarain (smooth, shining constituent), durain (dull, hard, granular), vitrain (shining, black, glassy lustre) and fusain (black, friable, fibrous).

The growth of the coal industry was one of the main spurs to general industrial growth from the seventeenth to the twentieth centuries. Today, coal is not used merely as fuel but is the basis of the manufacture of such diverse commodities as perfumes, nylon stockings, pain-killing drugs and cloth dyes.

History of the coal industry
Until World War II the coal industry was of key economic importance in the industrialized countries. Even in 1950 coal provided 56 per cent of world energy supply; by 1974 this had declined to 29 per cent because of marked increases in both oil and natural gas production. Due to the very large rises in oil prices in 1973, and the estimates of limited reserves of oil and gas, coal production is again being increased in many countries that have conducted a policy of running down the industry since the mid-1950s [Key].

Coal was first scratched from the surface of the earth where it was found by chance. It was later extracted from "drifts" (gently sloping tunnels that followed the seam into the ground) and later still by sinking shafts. Today, coal prospecting is a major scientific

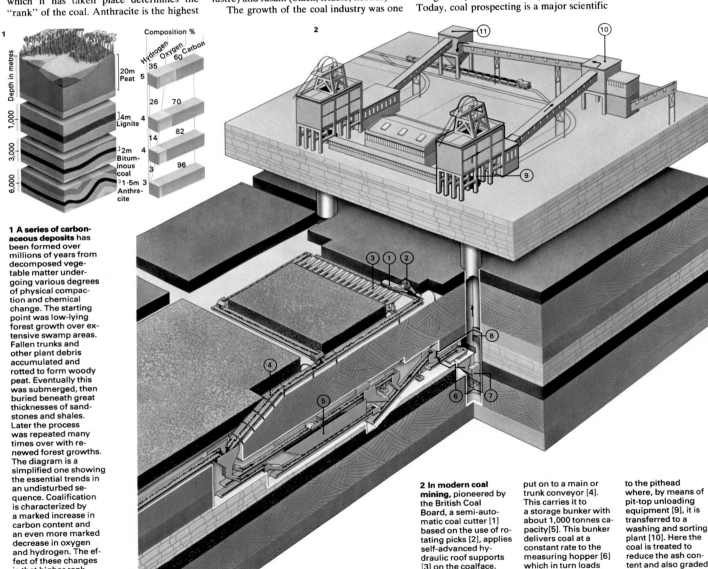

1 A series of carbonaceous deposits has been formed over millions of years from decomposed vegetable matter undergoing various degrees of physical compaction and chemical change. The starting point was low-lying forest growth over extensive swamp areas. Fallen trunks and other plant debris accumulated and rotted to form woody peat. Eventually this was submerged, then buried beneath great thicknesses of sandstones and shales. Later the process was repeated many times over with renewed forest growths. The diagram is a simplified one showing the essential trends in an undisturbed sequence. Coalification is characterized by a marked increase in carbon content and an even more marked decrease in oxygen and hydrogen. The effect of these changes is that higher rank fuels are more chemically condensed and give off less volatile matter when they are tested by standard methods.

2 In modern coal mining, pioneered by the British Coal Board, a semi-automatic coal cutter [1] based on the use of rotating picks [2], applies self-advanced hydraulic roof supports [3] on the coalface. The coal is discharged on to an armoured conveyor which carries the coal to a transfer point where it is put on to a main or trunk conveyor [4]. This carries it to a storage bunker with about 1,000 tonnes capacity [5]. This bunker delivers coal at a constant rate to the measuring hopper [6] which in turn loads the coal skip [7] in the area devoted to pit-bottom loading equipment [8]. In the skip the coal is raised in the shaft to the pithead where, by means of pit-top unloading equipment [9], it is transferred to a washing and sorting plant [10]. Here the coal is treated to reduce the ash content and also graded into size groups. The cleaned and graded coal is then loaded at the train loading tower [11] for delivery.

undertaking employing geophysicists and advanced instruments for investigation before trial borings are made. Sinking a pit is now a major project that takes between five and ten years, demands the investment of many millions of pounds and penetrates into the earth with shafts many hundreds of metres deep.

Coal was originally dug out by pick and shovel and removed to the surface in baskets. Today, in all industrialized countries, coal mining is wholly mechanized and pit outputs may range up to more than a million tonnes a year [2]. The miners operate mechanical cutters and the coal is transferred by conveyor or underground trains. The tendency is towards remotely operated equipment so that men are increasingly withdrawn from the dangerous areas where the coal is actually being won. In one development a coal-cutting machine is automatically steered along the face using a radioactive probe to measure the thickness of the skin of coal left in the roof as it cuts its way along. Coal mining continues to be an occupation demanding special care, but the death and accident rates have been steadily reduced in industrialized countries. The inflammable gas methane may be released from the coal during working and for this reason all equipment used must meet strict requirements to ensure that it cannot cause explosions.

Products from coal

Coal was at one time an important source of gas and chemicals obtained by heating the coal, out of contact with air, in the process called "carbonization". This has sharply declined in importance since the mid-1950s. Nevertheless, widespread research is in progress to make coal once again an important source of chemicals using new processes, such as hydrogenation, that can make it into both gas and liquid fuels. Of all the primary fossil fuels coal has by far the greatest reserves, amounting to probably hundreds of years of use at expected rates; thus it will far outlast gas and oil. The coke produced by de-gassing coal is an essential ingredient in the smelting of iron for making steel. The coal gas produced from coking ovens can be used for heating at the coke-making plant.

KEY
United States
Poland
W. Germany

The five-fold increase in oil prices during 1972–4 caused many countries to reconsider their policy regarding coal production. Some coal producers had seriously reduced coal output, but the USA had generally maintained the level of production while the USSR and other East European producers had continued to develop their coal industries. In East Europe the relationship between prices of oil and coal is the result of political decisions. In the free market economies the state of the coal industry in the postwar period has been determined by the ratio of prices of oil and coal. Today scientists in Western Europe are finding new uses for coal.

3 Pithead equipment, used to transfer the coal to the preparation plant, was once the symbol of a mining community. Today it is concealed in a building designed to blend better with the surrounding landscape.

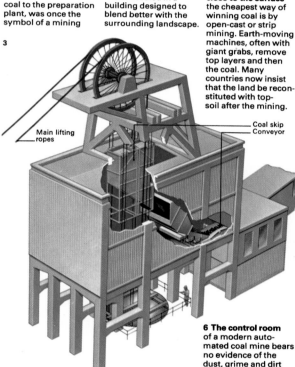

Main lifting ropes
Coal skip
Conveyor

4 When seams are close to the surface the cheapest way of winning coal is by open-cast or strip mining. Earth-moving machines, often with giant grabs, remove top layers and then the coal. Many countries now insist that the land be reconstituted with top-soil after the mining.

5 Coke, shown here leaving a coking oven, is a typical smokeless fuel for industrial use. Coal is heated in ovens, out of contact with the air, until it is white hot. This process removes most of the tarry substances and the fuel produced burns leaving very little ash.

6 The control room of a modern automated coal mine bears no evidence of the dust, grime and dirt traditionally associated with mining. In this one, at Bevercotes Colliery, England, a single engineer monitors the workings of machines both underground and on the surface.

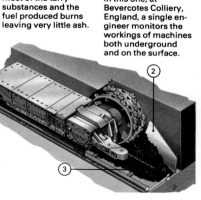

7 The Anderton-shearer [1] is a rotary drum cutter coupled with a diverting plough [2] which loads the coal on to an armoured coal conveyor [3].

Oil and natural gas

The world's first oil well was drilled at Titusville, western Pennsylvania, USA, in August 1859 by Colonel Edwin L. Drake. He found crude oil 21m (69.5ft) below the surface and started an industry that is now the world's biggest. By 1975 there were more than 600,000 oil wells throughout the world, producing a total of more than 55 million barrels of oil a day (a barrel equals 35 imperial gallons or 42 US gallons or 159 litres).

Composition, origins and location

Oil is a complex mixture of hydrocarbons – chemical compounds consisting of carbon and hydrogen. They range in density from light gases such as methane to heavy solids such as asphalt. And the colour ranges from yellow, through green, red and brown, to black. A typical crude oil contains about 85 per cent carbon and 15 per cent hydrogen. Natural gas consists of the least dense fraction of crude oil [7] but it can often occur in the absence of oil.

Oil is thought to be the product of the decay, under special conditions, of single-celled plants and animals that lived hundreds of millions of years ago and settled to form sediments when they died. Some natural gas has a simpler origin; methane can be formed by the bacterial decomposition of vegetable matter, and is found in marshy areas.

After their formation, oil deposits frequently moved until they became trapped in porous rocks at depths ranging from as little as 30m (100ft) to more than 7,600m (25,000ft). For this reason oil deposits are not vast caverns filled with a sea of liquid oil and gas but are more widely dispersed.

The first prospectors drilled for oil in places where it seeped naturally to the surface. Today prospecting has to be more scientific [2] although surface methods can still sometimes be helpful. The appearance of the ground and the presence of the appropriate kinds of sedimentary rocks can suggest likely places to drill. Seismic surveys (measurements of the effects of shock waves in rocks) are widely used. Engineers detonate explosives in a shallow hole and use microphones to detect the depths of the echoes from underground structures. Gravity surveys may also be used; sensitive instruments, some-times carried by aircraft, detect variations in the gravitational field and these may suggest the presence of oil-bearing geological structures. Techniques such as these improve the prospectors' chances of finding oil from about 1 in 30 to nearer 1 in 5.

Methods of drilling

The first drills operated with a pounding action in a dry hole. If oil was found there was nothing to stop it shooting out of the hole as a "gusher". Modern techniques prevent this by using rotary drills [1] immersed in drilling "mud" – specially formulated compounds that fill the hole as it is drilled to create a pressure which prevents gushers.

The drilling bit is driven downwards on the end of a hollow steel tube through which mud is pumped. As the bit drives through the rock the pieces of debris are carried away by the mud and brought to the surface. There, the mud is examined by geologists who check the nature of the stratum through which the drill bit has passed. This should agree with their survey. The mud is screened to remove rubble and then recirculated down the hole.

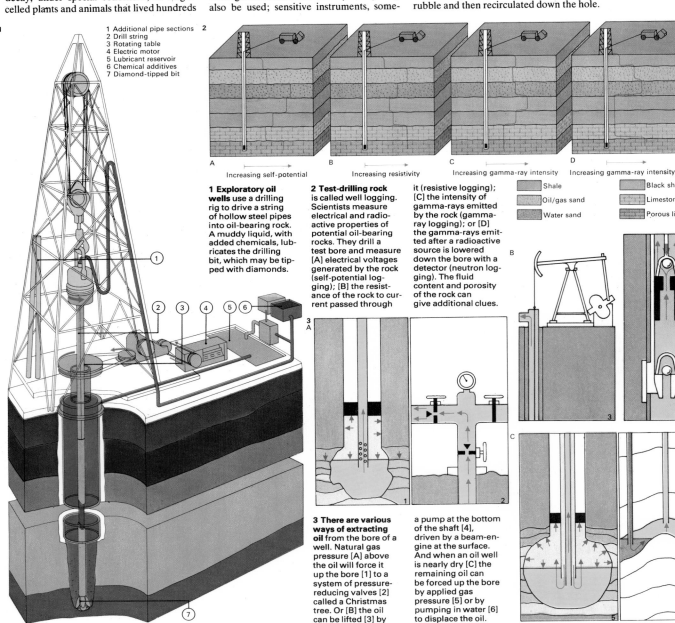

1 Additional pipe sections
2 Drill string
3 Rotating table
4 Electric motor
5 Lubricant reservoir
6 Chemical additives
7 Diamond-tipped bit

1 Exploratory oil wells use a drilling rig to drive a string of hollow steel pipes into oil-bearing rock. A muddy liquid, with added chemicals, lubricates the drilling bit, which may be tipped with diamonds.

2 Test-drilling rock is called well logging. Scientists measure electrical and radioactive properties of potential oil-bearing rocks. They drill a test bore and measure [A] electrical voltages generated by the rock (self-potential logging); [B] the resistance of the rock to current passed through it (resistive logging); [C] the intensity of gamma-rays emitted by the rock (gamma-ray logging); or [D] the gamma-rays emitted after a radioactive source is lowered down the bore with a detector (neutron logging). The fluid content and porosity of the rock can give additional clues.

Increasing self-potential
Increasing resistivity
Increasing gamma-ray intensity
Increasing gamma-ray intensity

A   B   C   D

Shale
Oil/gas sand
Water sand
Black shale
Limestone
Porous limestone

3 There are various ways of extracting oil from the bore of a well. Natural gas pressure [A] above the oil will force it up the bore [1] to a system of pressure-reducing valves [2] called a Christmas tree. Or [B] the oil can be lifted [3] by a pump at the bottom of the shaft [4], driven by a beam-engine at the surface. And when an oil well is nearly dry [C] the remaining oil can be forced up the bore by applied gas pressure [5] or by pumping in water [6] to displace the oil.

Two hazards of drilling are blow-outs of oil or gas and stuck drill pipes. Blow-outs are infrequent, but to prevent them drilling rigs are fitted with safety equipment that can seal off the hole rapidly. Stuck drill bits can sometimes be freed with special equipment. Failing this they are abandoned in the hole, as much as possible of the drill string is recovered, and a new hole is drilled close by.

Once drilled, a well is completed by inserting a 22.9cm (9in) pipe in the bore and pumping concrete round it to seal the sides of the hole. A production string of 7.6cm (3in) tubing is inserted to the right depth. An explosive charge called a "down-hole perforator" is then lowered into the bore and detonated to puncture holes through the casing and concrete to let oil in.

Transportation and storage
Pipelines [6] are generally the most economical way of transporting oil and gas from wells to the processors. Pipes are made of welded steel up to 1.22m (48in) in diameter, covered with asphalted felt for protection and often buried underground. Pumping stations along the pipeline maintain the flow and pressure.

Sea transport is more expensive. The huge tankers used to transport oil from the Middle East are the largest ships afloat, carrying a million barrels of oil or more. Liquefied natural gas can also be carried by sea in even more expensive vessels called liquefied natural gas (LNG) carriers [5].

Oil is usually stored in tanks up to 30m (100ft) across and 9m (30ft) tall. Natural gas can be stored as a liquid either in refrigerated tanks or in ground storage caverns, which are made gas-tight by first freezing the surrounding ground. Then pits up to 40m (130ft) in diameter are dug and the liquefied gas pumped in. Once filled and covered with an insulated lid, the contents themselves keep the reservoir frozen.

An ideal way of storing gas uses depleted oil and gas reservoirs underground and near consumers. This is a technique widely used in the United States. Disused coal mines can also be used, the one at Fontaine L'Evêque, Belgium, can hold 500 million cubic metres (17,650 million cubic feet) of gas and is claimed to be the world's largest gasholder.

Natural gas is often found under the seabed. It is used directly as a fuel, being burned for domestic heating and cooking, or for industrial heating and electricity generation. The gas can also be treated as a raw material for the petrochemical industry and be made into a range of other chemicals.

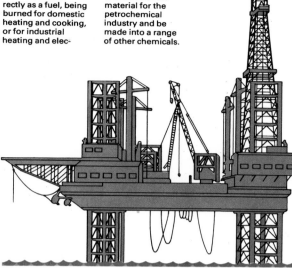

4 A gas drilling rig [1] leaves a string of pipes [2] in the natural gas pocket [3]. These are connected to the mainland by a pipeline [4].

5 Gas is transported by pipelines or in special ships [1]. Before being used it must be treated to remove various chemicals. Water and liquid hydrocarbons are removed in an expansion chamber [2]. Alkali removes sulphur compounds [3], which are purified and stored [4] as a useful by-product. Any remaining liquids are removed [5] and excess gas is liquefied under pressure and stored underground [6]. When gas is required the liquid is evaporated [7], metered and pumped [8] into the national gas grid [9]. Smaller pipes lead to domestic and industrial consumers [10]. Some European countries import gas ready purified and liquefied for storage.

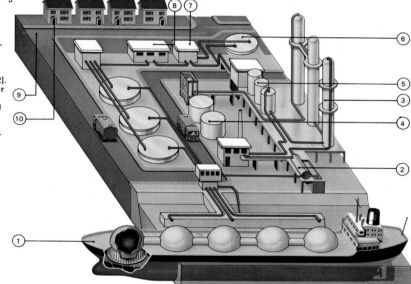

6 Pipelines for oil or gas may be laid underwater [A] or underground [B]. Lengths of pipe from a supply boat [1] are welded together [2], X-rayed for flaws [3] and cased in concrete [4] before being laid. On land a machine [5] cuts a trench and the cast-iron pipes are again welded and X-rayed [6], given a chemical corrosion protection, and laid in the trench.

7 Natural gas is largely composed of methane and higher hydrocarbon gases, with some nitrogen, carbon dioxide and sometimes helium. It varies from locality to locality but the methane generally makes up between 85 and 95 per cent of the gas. It is often economical to extract the helium, which is a light non-flammable gas used for filling balloons. Sulphur compound impurities must be removed.

Methane 85%   Others 2%   Propane 3%   Ethane 10%

Oil refining

Crude oil – often called petroleum, meaning "rock oil" – is the source of a wide variety of chemicals, such as plastics, pharmaceuticals, cosmetics, adhesives, polishes, paints, explosives and pesticides. It is a complex mixture that contains hundreds of different kinds of chemicals called hydrocarbons, because they are composed of hydrogen and carbon. Physically, crude oil is a sticky, inflammable liquid varying in colour from yellow, green, red or brown to black. It may also be fluorescent. Its composition varies considerably from source to source.

The hydrocarbons may be paraffins (noncyclic compounds with the general formula C_nH_{2n+2}), naphthenes (cycloparaffins, largely cyclopentane C_5H_{10} or cyclohexane C_6H_{12} and their derivatives) or aromatics.

How petroleum is refined
Crude oil is processed in a refinery [1]. First the mixture of hydrocarbons is separated into various components (called fractions). Various hydrocarbons boil at different temperatures, and so the mixture can be separated by fractional distillation – heating so that the

fractions boil off, condense and separate at different levels in a long vertical fractionating column [2]. These may then be further refined or chemically changed and sometimes blended back into the straight distilled fractions to improve their qualities.

The eight main fractions, in order of boiling point, are petroleum gases (which pass off from the top of the column), petrol, kerosene, diesel fuel, lubricating oil, fuel oil and wax (which all distil off) and a bituminous residue discharged from the bottom of the column. To achieve the degree of separation needed to yield the great range of petroleum products, the oil is passed through a series of columns. The relative quantities – and properties – of the products are adjusted as required in accordance with the needs of the market. As the number of cars grows, so the demand for petrol increases; similarly the market for kerosene has grown with its use as a fuel for jet engines.

Added flexibility in refinery operation is provided by "cracking" – breaking down larger molecules into smaller ones. In this way, heavy fractions such as those used in

diesel fuel can be made into petrol. For upgrading the quality of the petrol fraction, it may be "reformed"; this is a process in which it is mixed with hydrogen and heated over catalysts. Straight-chain hydrocarbon molecules are rearranged into ring structures, which perform better in car engines. The resulting mixture is again fractionated.

Conversion processes
Small molecules may be reacted to build up larger molecules as a further source of petrol. More processes are designed to remove impurities [3]; these are of growing importance in view of rising concern about atmospheric pollution. Sulphur, for example, is removed by treating the raw material with hydrogen to form hydrogen sulphide, which is then separated; the sulphur forms a valuable by-product.

Chemicals were originally made from surplus refinery gas. This is still one source but it is supplemented by the gases, notably ethylene and propylene, produced by cracking fractions from the oil, particularly the petroleum naphthas. Some chemicals,

CONNECTIONS

See also
78 Oil and natural gas
238 Chemical engineering

In other volumes
138 The Physical Earth

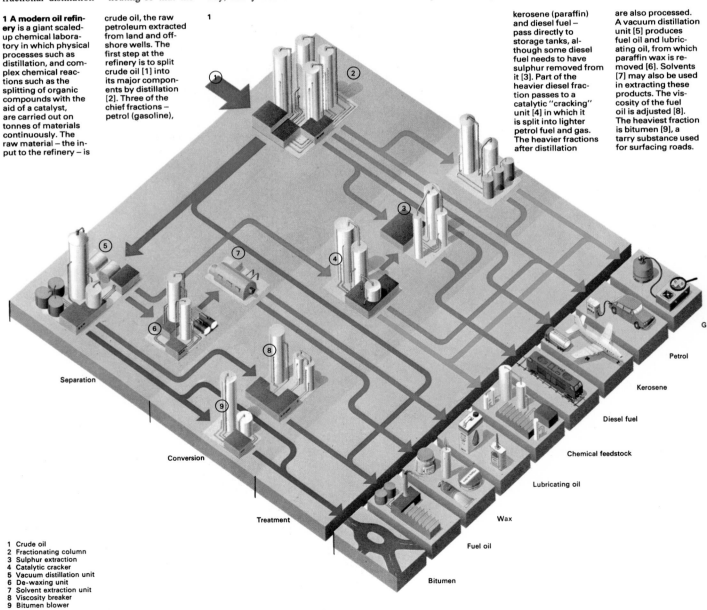

1 A modern oil refinery is a giant scaled-up chemical laboratory in which physical processes such as distillation, and complex chemical reactions such as the splitting of organic compounds with the aid of a catalyst, are carried out on tonnes of materials continuously. The raw material – the input to the refinery – is crude oil, the raw petroleum extracted from land and off-shore wells. The first step at the refinery is to split crude oil [1] into its major components by distillation [2]. Three of the chief fractions – petrol (gasoline),

kerosene (paraffin) and diesel fuel – pass directly to storage tanks, although some diesel fuel needs to have sulphur removed from it [3]. Part of the heavier diesel fraction passes to a catalytic "cracking" unit [4] in which it is split into lighter petrol fuel and gas. The heavier fractions after distillation

are also processed. A vacuum distillation unit [5] produces fuel oil and lubricating oil, from which paraffin wax is removed [6]. Solvents [7] may also be used in extracting these products. The viscosity of the fuel oil is adjusted [8]. The heaviest fraction is bitumen [9], a tarry substance used for surfacing roads.

Separation

Conversion

Treatment

Gas

Petrol

Kerosene

Diesel fuel

Chemical feedstock

Lubricating oil

Wax

Fuel oil

Bitumen

1 Crude oil
2 Fractionating column
3 Sulphur extraction
4 Catalytic cracker
5 Vacuum distillation unit
6 De-waxing unit
7 Solvent extraction unit
8 Viscosity breaker
9 Bitumen blower

such as toluene, are made during the course of the refinery processes. Butylenes are also produced and may be converted to butadiene, the basis of synthetic rubber.

Petroleum derivatives
By far the most important base chemicals are ethylene [6] and propylene. These "building blocks" polymerize directly to form the plastics polyethylene and polypropylene. Ethylene is also converted to such materials as PVC, polystyrene, antifreeze, polyesters and ethyl alcohol and some is used to form synthetic rubbers. Derivatives from polypropylene include solvents, acrylic fibres, polyurethane, foam plastics, nylon and materials called "plasticisers" that are used to give flexibility to paint films and resins that would otherwise be brittle.

Next in importance are the aromatics – benzene, toluene and xylenes. The main source is catalytic reforming but some arise during the special cracking of naphthas after hydrogen treatment of the petrol fraction. This produces more toluene than is needed and the excess is converted to benzene. From this, nylon, polystyrene, synthetic rubbers, resins and detergents are made. Toluene is also a base for making solvents and polyurethane resins. Higher up the series, xylenes are used for conversion to polyester fibres and plasticisers. Acetylene, itself a base for syntheses, is now often made from petroleum sources. Another important inorganic base chemical, after sulphur, is ammonia, for which the hydrogen is supplied from naphtha or natural gas.

Kerosene (which was once known as paraffin oil) is a petroleum derivative that is used for domestic heating, lamps and as a fuel or fuel component for jet engines.

Petroleum provides substantially more than 90 per cent of the world's plastics and resins, synthetic rubbers, fibres (excluding those made from cellulose) and chemical solvents, and about 50 per cent of the world's synthetic detergents. Only a little more than a generation ago, these came largely from vegetable sources, wood and coal. But oil supplies are known to be limited and it seems possible that this picture will be greatly altered in the future.

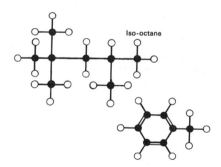

An oil refinery takes delivery of crude oil (petroleum), often directly from an ocean-going tanker, and converts it into petrol, other fuels, oils, waxes and raw materials for making chemicals.

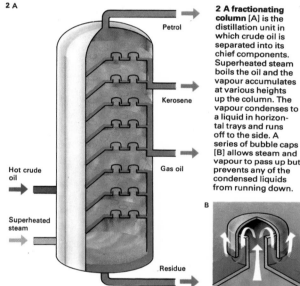

2 A

Petrol

Kerosene

Gas oil

Hot crude oil

Superheated steam

Residue

2 A fractionating column [A] is the distillation unit in which crude oil is separated into its chief components. Superheated steam boils the oil and the vapour accumulates at various heights up the column. The vapour condenses to a liquid in horizontal trays and runs off to the side. A series of bubble caps [B] allows steam and vapour to pass up but prevents any of the condensed liquids from running down.

The most volatile component of all is a gas, similar in composition to natural gas. The next fractions include liquid fuels and solvents such as petrol, kerosene, benzene and domestic fuel oil. Lower boiling heavy oils are used as fuels for marine diesel engines and as lubricating oils. The solid components include paraffin wax (a hydrocarbon known also as paraffin) and the tarry substance bitumen.

3	
Heavy gas oil 300°C	
Light gas oil 200°C	
Kerosene 175°C	
Naphtha 120°C	
Benzine 90°C	
Petrol 30°C	

3 Crude oil has a variable composition, depending on its source. This diagram shows the make-up and boiling points of a typical sample. All the substances named are hydrocarbons – compounds of hydrogen and carbon – although other elements such as sulphur are generally also present as impurities. They have to be removed to limit pollution caused when the fuels are burned, but once extracted constitute valuable by-products.

B

4

n-heptane

Iso-octane

Methyl cyclopentane

Toluene

4 Hydrocarbons from petroleum take the form of molecules that may be linear or long-chain (with a long "backbone" of carbon atoms, with or without a branch) or basically cyclic (with most of the carbon atoms in a ring). The sizes and shapes of molecules determine such properties as boiling point and octane number (fuel rating). Normal heptane, having a straight chain, boils at 98.4°C (209°F), whereas iso-octane, with a branched chain, boils at 99.3°C (210°F); iso-octane is a good fuel but n-heptane has a low rating. Methyl cyclopentane has a five-membered carbon ring and toluene has a six-membered ring.

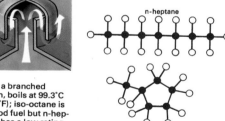

5

	n-pentane
	Iso-pentane
	n-heptane
	n-octane
	Iso-octane
	Methyl cyclopentane
	Toluene

−20   0   20   40   60   80   100   120

Octane rating ⬤ With TEL ⬤

5 The octane rating of an engine fuel – petrol or gasoline – is a measure of its efficiency for modern engines. Hydrocarbons with branched chains or cyclic structures (see illustration 4) are better than straight-chain compounds. All are improved by the addition of the organo-metallic compound tetraethyl lead (TEL), although this can cause pollution.

6

A

Free electron

R   +   →   R

Peroxide radical   Ethylene   Monomer radical

C

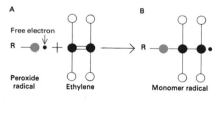

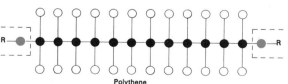

R   —   R

Polythene

6 Simple hydrocarbons such as ethylene can be made from petroleum products. They are valuable for making plastics such as polyethylene (polythene) and detergents. The double bond in ethylene [A] can be "sprung" and reacted with a radical R to give an active compound [B] that can spring other double bonds repeatedly and give a long-chain polymer – a plastic [C]. Such long-chain radicals can also be reacted with acids to form more complex molecules such as detergents.

81

Saving fuel and energy

Historically, the efficiency with which man has used his energy resources has been very low. The steam engine, one of the foundations of the Industrial Revolution, turned only a few per cent of the energy supplied to it into power and the first steam turbine power stations wasted the energy of 95 per cent of the coal they burned.

While fuel – that is, energy – was cheap, this inefficiency was less important. But since the prices of coal and oil have significantly increased many nations have made greater efforts to conserve energy – particularly fossil fuels such as oil [2] and the fuels derived from it. But the inertia of the complex system devoted to energy and the need to spend large amounts of money to make even quite small savings have resulted, as yet, in little significant achievement.

The uses of energy

In a modern economy basic fuel expenditure is dominated by four major uses: domestic space and water heating, industrial and commercial uses, transport and electricity generation [1]. British statistics are fairly typical. The total amount of primary energy used in Britain in 1972 was the equivalent of 332 million tonnes of coal, or rather more than five tonnes of coal for every man, woman and child in the country. The corresponding figures for other European countries are 3.3 coal tonne equivalents for France, 4.5 for West Germany and 2.3 for Italy. When the different sizes of the four countries' economies are taken into account, it shows that Britain uses considerably more energy than any of the others to produce each pound's worth of industrial output – more than twice as much as the equivalent expenditure by France, for example.

Improvement of efficiency

The most convenient form of energy, electricity, is also one of the most inefficient. Steam turbine efficiencies have risen from five per cent early in this century to about 35 per cent in the mid 1970s for the biggest power stations, but this still means that 65 per cent (nearly two-thirds) of the coal or oil burned in power stations is wasted. The generating efficiency of the British electricity network, including losses in transmission, is only 27 per cent. A further small loss takes place when electricity is converted into heat, producing an overall efficiency for electric heating of 22 per cent. This compares with a heating efficiency for natural gas or oil of more than 60 per cent. Electricity, however, can do things that gas and oil cannot: it can drive a whole variety of household machines, power record players and television sets and provide instant, clean and effective lighting.

Various attempts have been made to improve the overall efficiency of electricity generation. One possibility is district heating in which the "waste" heat from power stations is used directly to heat houses or factories. In ideal circumstances overall efficiencies of up to 75 per cent can then be achieved, but this theoretical calculation demands that heat and electricity need keep in step. More practical efficiency figures for district heating are around 45 per cent. Sweden has pioneered the use of combined electricity and heat production. The town of Vasteras, which has a population of about 100,000, is supplied with 600 megawatts of heat and 300

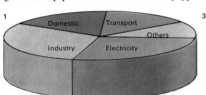

1 In industrialized countries most energy is used in generating electricity. Some waste more than others. Of about 350 million tonne equivalents of coal burned in Britain, for example, 30 per cent is used to produce electricity, 29 per cent by industry, 17 per cent in the home, 15 per cent on transport and 9 per cent on other uses. Energy generation in France is less wasteful.

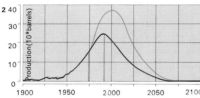

2 Oil reserves will not last for ever. The curves predict the use of oil based on two estimates of total reserves – 2.1 million million barrels and 1.35 million million barrels. Both show an increase in production until the end of the century with total exhaustion of reserves by the year 2100. The areas beneath the curves equal the reserve volume and the barrel quoted is equal to the standard one of 159 litres (35gal).

3 In ordinary homes much energy can be wasted by allowing heat to escape through walls, windows and roof spaces. The secret of conserving heat is insulation and efficient (preferably central) heating. House A has no insulation and is heated by an inefficient open fire burning ordinary coal. House B has lagged cold-water tanks [1] and pipes in its loft, the tiles are insulated with felt or paper [2] and the ceiling joists with glass wool [3]. Tiles hung on the exterior wall [4] provide extra insulation for the bedrooms. The fireplace [5] is blocked and the room heated electrically or by central heating radiators. Heat loss from a room is reduced by draught-excluding strips round the door [6] and by double glazing [7]. A wooden floor [8] to the garage insulates it from the foundations and heat flow through the cavity walls is minimized by plastic foam [9].

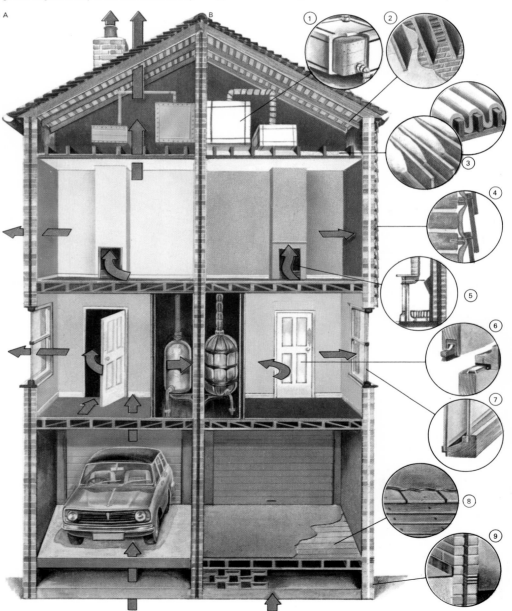

megawatts of electricity, produced from a single power station.

The electricity and transport industries are well aware of the cost of energy and generally make all the economies they can. But a large amount of primary energy – 17 per cent – is used in the home.

Conservation and pollution
Conservation of energy can be achieved in a variety of ways. In the home, draught-proofing, double-glazing, loft insulation and the filling of cavity walls with insulating foam can reduce energy consumption by half [3]. A widespread shift from inefficient private cars to public transport – with one engine trans-porting up to 40 people instead of only three or four – would also achieve savings. Tech-nical improvements in all energy uses, but particularly in industry through the expan-sion of waste recycling programmes, could help further. There is no single recipe, how-ever, to bring these changes about either in the short or long term.

Energy consumption and electricity generation can also create pollution and in some cases the need to reduce consumption may conflict directly with the need to reduce pollution. This is certainly true of the motor car. For reasons of health it would be desir-able to reduce the amount of lead compounds added to petrol [6], and by doing so the lead pollution of the environment would be low-ered. But leaded petrol is more efficient in modern high-performance engines and leaving out the lead would result in greater fuel consumption.

In the same way urban pollution has been reduced by the growth of electricity consumption – but at a cost. The fuel used in power stations to raise steam for turbine generators might instead be burned in houses to heat them. The fuel would be more effi-ciently used, but at a considerable cost in pollution and convenience. And power sta-tions are generally built away from densely populated areas so that the pollution they produce can be dispersed from high chim-neys, thus avoiding the build-up of dangerous concentrations of gases and carbon dioxide in the atmosphere. Corrosive sulphur com-pounds can be "scrubbed" from the gases.

KEY

Efficiency of various fuels

Natural gas is the most efficient fuel for heating. The dia-gram compares the amount of energy pro-duced by burning 1kg (2.2lb) of coal [A], wood [B], coal gas [C], anthracite (the best coal) [D], petrol [E] and natural gas [F]. Each bar repres-ents the number of litres of water that can be boiled using heat from each fuel. Anthracite is as good as coal gas, and both are better than coal. The production of electricity by each one follows the same order of efficiency.

4 The consumption of natural gas, at present supplying about a 20% of the world's energy, has risen by 3.5 times in the last 10 years. The known world re-serves (about 32 million million cubic metres [1,130 million million cu ft]) will last about 20 years at this rate. Even if fresh reserves are discovered natural gas will run out as a fuel supply in about 50 years. The curves compare the consumption of var-ious forms of energy in the USA [A] with total world usage of natural gas [B]. A sig-nificant factor, in the USA and else-where in the world, is the return to coal as a fuel.

4 A USA energy consumption (in million million kilowatt hours)

- Nuclear
- Natural gas
- Oil
- Coal
- Hydroelectricity
- Wood

1850 1900 1950 2000

B World natural gas consumption (in thousand million m³)

5 The impending energy crisis is barely in evidence in a large city, such as Hong Kong, when lit for the night. Thousands of kilo-watts of electrical power are consumed by advertising signs and shop window dis-plays. During the night this electri-city could be "saved" in an energy-storage system. It could then be made avail-able to help meet the peak demand that always occurs in the morning.

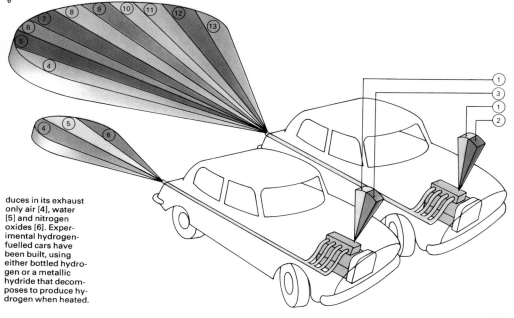

1,500
1,250
1,000
750
500
250

1850 1900 1950 2000

6 A petrol-driven car is inefficient and produces a wide range of pollutants. Running a car on the more efficient hydro-gen as a fuel would almost totally elim-inate pollution. A petrol car uses air [1] and petrol [2] and releases, in its exhaust, air [4], water [5], nitrogen oxides [6], carbon [7], carbon dioxide [8], carbon monoxide [9], lead compounds [10], sulphur dioxide [11], hydrocarbons [12] and aldehydes [13]. The hydrogen car uses air [1] and hydrogen [3] and pro-duces in its exhaust only air [4], water [5] and nitrogen oxides [6]. Experi-mental hydrogen-fuelled cars have been built, using either bottled hydro-gen or a metallic hydride that decom-poses to produce hy-drogen when heated.

Electricity generation and distribution

The electricity that supplies an enormous variety of needs in homes, offices and factories all originates in power stations. These vast buildings contain generators, machines to drive them, transformers [Key] and switching systems. They convert the chemical energy of coal, oil, natural gas or nuclear energy into thermal energy and then into electricity on a large scale to get the best use of the plant, site and engineering needed.

Nevertheless, the generation of electrical power is an inefficient process, although it compares well with other forms of energy use. In power stations burning fossil fuels (coal or oil) nearly two-thirds of the thermal energy released is lost as heat to the atmosphere or surrounding area and only a little more than a third is actually used to produce electricity. There are additional losses as the current is transmitted over the distribution network [3]. When a consumer uses this electrical power he probably turns it back into heat, either intentionally or incidentally during some other process. Imperfect as the system is, it is still the best one available for supplying the power for thousands of dif-ferent uses in homes and factories. At the point of consumption it is clean, convenient and safer than almost any other form of energy supply so far discovered.

Alternative forms of generation

There are some important exceptions to the inefficient consumption of fuel in the generation of electricity. When power is derived from machines driven by water released from behind a dam, or from tidal motion [6], for instance, there are no "fuel" losses. Less common are a few systems that use solar energy [2] or wind power [4]. The capital costs of setting up dams and barrages for hydroelectric schemes or manufacturing cells that convert sunlight into electricity are very high. But the rapid depletion of fossil fuel resources and the increasing pollution from thermal power stations are factors that may in time override purely economic considerations. Where political pressures make access to fossil fuel reserves difficult, the economic case for developing systems that exploit other forms of energy may become even stronger.

For this reason nuclear power stations are being used to generate an increasingly large proportion of power in the industrialized nations. Their overall efficiency is no greater than that of thermal stations [1] but their long-term fuel costs are so small that nuclear stations are economical to operate.

Whether the source of power is oil, coal, nuclear, wind or moving water, electricity is produced by the same kind of machines – turbines and generators. In themselves, generators are highly efficient, converting mechanical to electrical energy with losses no greater than about two per cent.

The supply system

Most power stations have more than one generator to allow for some degree of reliability in case one machine breaks down. Many stations have four, and at periods of peak demand all of them may be working at maximum output. As demand slackens each may be brought out of circuit in turn and shut down although some may be left partly loaded. If there is a sudden demand for power these machines can be brought quickly up to running speed and into use.

CONNECTIONS

Read first
66 Power from steam

See also
74 Nuclear power

In other volumes
112 Science and The Universe

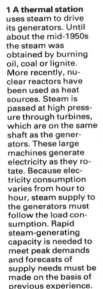

1 A thermal station uses steam to drive its generators. Until about the mid-1950s the steam was obtained by burning oil, coal or lignite. More recently, nuclear reactors have been used as heat sources. Steam is passed at high pressure through turbines, which are on the same shaft as the generators. These large machines generate electricity as they rotate. Because electricity consumption varies from hour to hour, steam supply to the generators must follow the load consumption. Rapid steam-generating capacity is needed to meet peak demands and forecasts of supply needs must be made on the basis of previous experience.

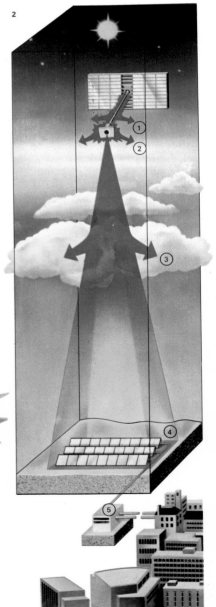

2 A solar panel in semi-stationary orbit is one system proposed for deriving energy from the sun. Clear of the earth's atmosphere, the solar cells would be in the direct rays of the sun, unhindered by dust, water vapour and atmosphere. The electricity (in the form of direct current) generated in the panel would be transmitted along a short conductor [1] to a device for converting the current into microwaves [2]. After a small loss of energy in the atmosphere [3] the microwaves would reach large collecting panels [4] and be passed to a station [5] for conversion from direct to alternating current.

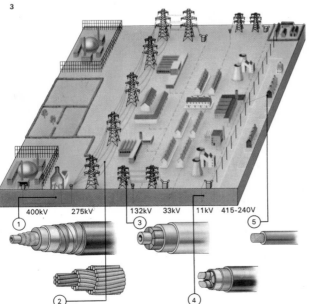

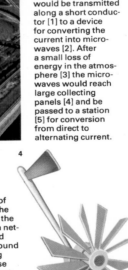

3 Transmission of electricity from the power station to the consumer is by a network of overhead lines or underground cables of varying voltages. Because more energy is lost at low voltages the voltage must be kept as high as possible while still ensuring safety. In a typical network underground cables [1 and 3] use oil, plastic and similar materials for insulation whereas overhead conductors [2] are uncovered, as the air acts as the insulator. A fabric tape cable [4] takes power to a local substation. Then it goes to consumers via lead or aluminium-sheathed cable [5].

4 Wind power was widely used in Europe until the advent of cheap fuels. Windmills need large blades and steady winds, but linked to storage batteries they can often provide cheap power.

400kV 275kV 132kV 33kV 11kV 415-240V

From the generators big, solid conductors in the form of metal bars (called busbars) connect to transformers, where the voltage is increased for transmission. Conductors slung between pylons spread out from the station in all directions. Where necessary or desirable the voltage is lowered by other transformers to a level safe for distribution to consumers. Some of this distribution may take place underground and may eventually terminate at a sub-station where the supply cable is split into a number of feeders, again after appropriate voltage reduction. From there it is taken overhead or underground into industrial or residential areas where further voltage reduction takes place before it is fed into homes and factories. Overhead lines may themselves terminate in sub-stations, thus avoiding an underground section, but almost all consumers in towns and cities are supplied by underground cables.

Typical voltages at the three chief stages are 33,000 volts (33kV), 11kV and 415 volts (or 240 volts single phase). The high-voltage alternating current (AC) in the distribution system is normally three-phase, generally transmitted along four wires – three phase conductors and a neutral, or common, conductor. For domestic use at low voltage, the power is supplied as single-phase AC with two wires ("live" and neutral).

At various points along the line between the power station and the user there are switches, circuit-breakers and similar devices to protect lines and equipment in case of overload or disruption by lightning.

Network links, national and international
The output from each station is linked to that from others by conductors called "interties", running between convenient points, generally on pylons. All the stations in one country are thus electrically connected [5]. This enables some stations to be shut down for maintenance or repair and allows the most efficient stations to run continuously, supplying what is termed "base load". The networks of many European countries are also linked together (England's connection to France is by a cable under the English Channel). This enables some countries to sell or lend power to others.

A transformer (shown here is the core) is used to connect generators to an electricity grid. The transformer's function is to raise the voltage from perhaps 23kV to about 400kV.

5 Control of electricity distribution in England is directed from a room in London. The country is divided into a number of regions, indicated by various colours on the map. Each of these has responsibility for generating and transmitting power and all come under the authority of the national control centre. Because the regions are connected through the network, it is sometimes convenient to send power from one to another, particularly if there are breakdowns of some generators. Regions with more costly fuel supplies prefer to import some power. Nuclear stations are the cheapest to run and attempts are made to operate them 24 hours a day, adding other sources to supplement them when necessary. At the central control room engineers must assess the most economic way of using regional resources, bearing in mind such factors as weather, fuel costs, plant efficiency and cable capacity. They are in constant touch with regional headquarters and can use computer information. The wall map is only a general guide. A computer-controlled display on the engineer's desk shows in more detail the position of all the transmission lines, the state of the equipment and the current-carrying capacity on any part of the national network.

6 Power from ocean tides is being harnessed by a pioneering station at La Rance in France. The experience gained here may be used throughout the world. Special turbines have been developed that can be driven by water flowing in either of two directions. This enables them to function when the tide is either ebbing or flowing. When the tide comes in, water is directed through large tunnels housing the turbines, causing them to rotate. When the tide starts to ebb, the water is held behind a dam until the tide is on the turn and is then released again, driving the turbines. These machines also act as pumps so that when the demand for electricity is low they can increase the amount of stored water over and above that gained naturally, allowing extra power to be generated by the station when the demand increases. The Rance scheme has 24 generating bulbs lying in horizontal tunnels.

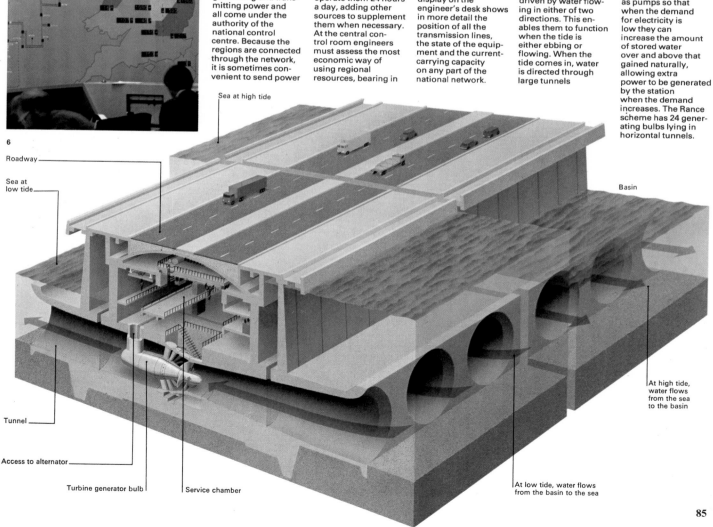

Sea at high tide
Roadway
Sea at low tide
Basin
Tunnel
Access to alternator
Turbine generator bulb
Service chamber
At high tide, water flows from the sea to the basin
At low tide, water flows from the basin to the sea

Levers and wedges

Today man is surrounded by a vast range of machines from clocks, washing machines and other domestic appliances to computers, hovercraft and space rockets. All machines are in some sense "labour-saving" devices; and for once the popular definition matches the scientific one. To a scientist a machine is any device that provides a mechanical advantage – that is, allows a limited amount of effort to do useful work in lifting or moving a load. The mechanical advantage of a machine is the load divided by the effort. In this sense, the simplest machines are levers, wedges and screws – used in their thousands in many complex machines.

Magnifying an effort

Levers have hundreds of uses – a crowbar, an oar, a screwdriver, scissors, a see-saw and a wheelbarrow all make use of the various classes of levers [1]. Their effect is to "magnify" an effort to make it easier to move a load. Each makes use of a pivot, called a fulcrum, and the sizes of the load and effort and their distances from the fulcrum determine the lever's mechanical advantage.

How can a four-year-old child lift a man weighing 75kg (165lb)? One simple way is to sit them both on a see-saw. If the man sits fairly close to the pivot, the weight of the child at the far end of the other side will be sufficient to lift him. The child will move down farther than the man moves up, but this lack of movement is the price that has to be paid for a small effort to lift a large load. The load must move, however, and it is no wonder that the Greek mathematician Archimedes (c. 287–212 BC) is reported to have said "Give me a firm place on which to stand and I will move the earth".

A stationary see-saw – and any lever in which the load and effort balance each other – is said to be in equilibrium. In such cases, the load multiplied by its horizontal distance to the fulcrum equals the effort times its horizontal distance to the fulcrum. If the child on the see-saw in the above example weighs 12.5kg (27lb) and sits 3m (10ft) from the pivot, he would exactly balance the 75kg (165lb) man sitting 0.5m (18in) from the pivot on the other arm of the see-saw ($12.5 \times 3 = 75 \times 0.5 = 37.5$).

The mechanical advantage (the load divided by the effort) is 75 divided by 12.5, equal to 6 in this example. It can also be calculated by dividing the effort's distance to the fulcrum by the load's distance; in this case, 3 divided by 0.5, again equal to 6.

The load multiplied by the distance is called the moment of the load (which is a force). At equilibrium, the moment on the load side equals that on the effort side. If one moment is larger than the other, there is a turning force and that side of the see-saw will move downwards.

Using ramps and wedges

When the ancient Egyptians were making the pyramids or Bronze Age men were building Stonehenge, they were faced with the task of raising huge blocks of stone through several metres. They knew it is much easier to push a heavy object up a slope than to lift it directly, so they probably constructed long ramps of earth up which they dragged the stones (probably on tree-trunk rollers). A physicist calls such a ramp an inclined plane.

To lift a 10-tonne block vertically

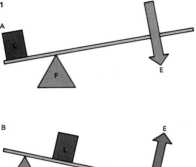

1 All levers belong to one of three classes, depending on the relationship between the effort E, load L and fulcrum F – the pivot. In the see-saw arrangement of the first class [A] the fulcrum is between the load and the effort. In the second class of levers [B] an upward effort raises a load placed between effort and fulcrum. A wheelbarrow uses this principle. In the third class [C] the effort acts between the fulcrum and the load. Many hydraulically operated machines use leverage of this class and some complex machines – such as printing presses – have all the classes of levers somewhere in their mechanism, as have the limbs of the human body.

2 Tall structures such as street lamps can be reached for cleaning and maintenance using a "cherry picker" with hydraulically operated levers, often mounted on a lorry. Similar vehicles are used by fire brigades to provide a high vantage point for hoses or for rescuing people trapped in tall buildings. Hydraulic linkages can work such hinged joints in much the same way as muscles bend a man's arm at the elbow; both are examples of the third class of levers. Because the effort is applied so close to the fulcrum (pivot), a large effort is required to move the load at the end of its long arm. This is why engine-powered hydraulics have to be used.

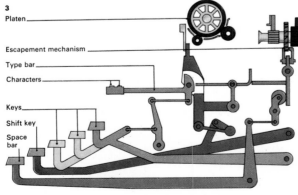

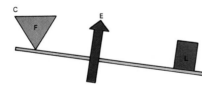

Platen
Escapement mechanism
Type bar
Characters
Keys
Shift key
Space bar

3 A typewriter's keys are operated through a series of linkages acting as levers. As a key is tapped, the levers move a type bar and make a character print on paper wrapped round the platen. Pressing the space bar releases an escapement mechanism to advance the carriage without a character being typed. The levers controlled by the shift key raise the whole lever system so that, on tapping a type key, the lower character on the type bar makes contact with the typewriter ribbon and prints on the paper. Other levers are used to move the carriage sideways and work the tabulator controls.

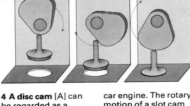

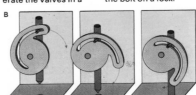

4 A disc cam [A] can be regarded as a lever of variable length that changes rotary motion into an up-and-down or side-to-side reciprocating motion. Disc cams are commonly used to operate the valves in a car engine. The rotary motion of a slot cam [B] drives a vertical arm up or down – or a horizontal one sideways. Using such a cam, a twisting movement can produce linear motion to work the bolt on a lock.

5 A screw can be pictured as an inclined plane wrapped round a cylinder. The mechanical advantage of an inclined plane – it is easier to push a load up a ramp than to lift it vertically – can be realized by rotating the screw, often to exert considerable force. The distance between the threads is called the pitch of the screw and is the distance the screw advances every time it makes one revolution.

requires an effort of 10 tonnes. But, neglecting friction, the effort needed to push or drag it up a slope of 1 in 20 is only about half a tonne. To raise it one metre the load must be moved about 20 metres so that, as with levers, a large movement of the effort is needed for a small load movement to gain significant mechanical advantage. The ratio of the ramp's height to the length of the slope determines how much effort is saved. Modern technology can achieve some space-saving with long ramps by coiling them [8].

Lifting a heavy block slightly (perhaps to pass a rope round it) can be achieved by driving a wedge underneath it. A wedge is like two inclined planes back to back. But instead of moving a load up the plane, the plane is pushed past the load to move it. Driving a wedge into a crack, for example, exerts a tremendous force [Key]. A hatchet or an axe uses this principle, as do chisels, ploughs and pneumatic drills.

Screws – wound-up wedges
A screw thread can be pictured as an inclined plane wrapped round a cylinder [5]. This shape is called a helix and its geometry was studied in about 200 BC by the Greek mathematician Apollonius of Perga (3rd century BC). Archimedes invented screw-cutting machinery and used the screw as the key principle in his famous "screw" pump for raising water.

Just as a wedge can be driven into an object by hammering, the helical wedge of a screw can be driven in by turning. The turning movement requires leverage – a screwdriver or spanner – and the simple-looking screw is in use, a "machine" that combines the lever and the wedge.

The distance between a screw's threads is called the pitch and is a measure of the slope of the corresponding inclined plane. In one complete turn a screw moves through a distance equal to its pitch. The length of the lever turning the screw, divided by the pitch, gives its mechanical advantage. The wedge-shaped section of a tapering wood screw reveals another application of the wedge as it forces its way into the timber. Self-tapping screws cut a helical thread as they wind their way into metal.

A wedge driven into a crack in a log or a block of stone can split it apart. In this quarry in Malta, sandstone is partially sawn into blocks, then wedges are used to split the blocks apart.

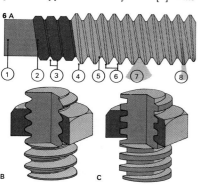
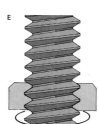
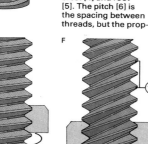
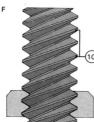

6 Screw threads on bolts generally conform to a few major types and there is a set of specialized terms to describe them. Engineers use this "thread terminology" to define kinds and parts of screws [A], including the two major diameters, the root diameter [1] and the pitch diameter [2], the thickness of the thread [3], the crest [4] and root [5]. The pitch [6] is the spacing between threads, but the properties of the screw depend also on the thread angle [7] and the helix angle [8]. The two main kinds of screw are round-sectioned [B] and square-sectioned [C]. In a single-start thread [D] the lead [9] and the pitch are the same. After one revolution [E] a nut moves along a distance equal to the pitch. In a double-start thread [F] the lead [10] is twice the pitch, and a nut moves through twice the pitch [G].

7 A screw with a fine thread can control precise movements – the distance it moves is equal to the fineness of the pitch. A micrometer, for example, makes use of a fine screw thread to measure small dimensions with a high degree of accuracy. But in other applications a coarse pitch ensures a positive action. In this gate valve, turning the wheeled handle opens it quickly. It normally operates either fully open or fully closed.

9 Levers of the first class – like a seesaw – can be paired, as in scissors or pliers. The same idea can be extended to form a "set of scissors", as in lazy tongs. The principle can, with modern hydraulics, make a powerful machine for raising or lowering loads, such as cargo and luggage at an airport. The same scissors principle is used for some car jacks, which have a screw thread for pushing the lower arms together.

8 An inclined plane is a practical solution to the problem of moving a heavy load up or down through a significant height. In a multi-storey car park the vehicles have to descend to street level from a great height. A long ramp – an inclined plane – would provide such a facility but it would take up a lot of room. Winding the ramp round and round, like the exit from this car park, saves space. The geometry of the ramp is like a screw thread.

Pulleys and gears

Pulleys and gears are wheels arranged to transmit motion and are among the earliest machines invented by man. A gearing system of wooden pins was used to drive the mechanisms in medieval flour mills, windmills and mines. Pulleys were probably known by the ninth century BC and Archimedes demonstrated the efficiency of a compound pulley [Key].

Pulley action and design
In a pulley the wheel is used in conjunction with either a rope, chain or belt. On a gear the rim of the wheel is cut with teeth or a worm thread to mesh with similar projections from another gear. Both pulleys and gears can be used to transmit rotary motion between two or more shafts. If the shafts are close together, as in a clock or a car engine, then gears are generally employed. If the shafts are farther apart, pulleys are more often used. Gear-wheels can also be used to change the direction of rotation by as much as 90 degrees. Wheels of various diameters – with both pulleys and gears – produce a change in the speed of rotation.

Belt-driven pulleys are common driving arrangements for factory and agricultural machines. A circular (endless) belt transmits motion between one pulley on the shaft of a motor and another on a machine shaft, for example that of a lathe. By giving the pulley on the motor or driving shaft a different diameter from the pulley on the machine or driven shaft, the speed of rotation of the driven shaft can be varied. Several pulleys of different sizes are often fitted together on the driving shaft to provide a range of speeds for the driven shaft. This is known as a stepped pulley system. The rim of a pulley may be broad and flat to take a wide belt. Alternatively, the rim face may be grooved to accommodate narrower belts that are V-shaped or circular in cross-section, which prevents them slipping from the pulley.

Cranes and hoists are the other common applications of pulleys. Here the motion is usually transmitted to provide a mechanical advantage. The force or effort employed by the person hoisting is magnified by an arrangement of two or more pulleys so that heavier weights can be lifted than would

otherwise be possible. Hoists with two or more pulleys are widely used in industry for lifting components or packages and transferring them from one place to another [4].

How gear-wheels work
Most gear-wheels have teeth with slightly curved surfaces. These are the surfaces that come into contact with those of another gear. The action of one gear tooth on another generally results in a combined rolling and slipping motion of the curved faces, so that friction and the risk of jamming are very much lower than they would be if the contact faces were flat. Even so, a gear-wheel must be made so that its teeth fit more or less loosely into the spaces between the teeth of its mating gear. The looseness is called backlash.

Pairs of gear-wheels are generally chosen to change the speed, and often also the direction, of rotational movement [5]. If one gear is much larger than the other they are known as the gear (large) and pinion (small). When the gear drives the pinion an increase of rotational speed is obtained and vice versa. The amount of change of speed is directly propor-

CONNECTIONS

See also
96 Machines for lifting
100 Moving heavy loads
92 Machines for
   measuring time

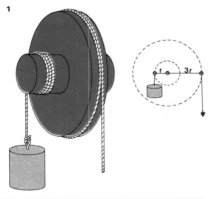

1 Mechanical advantage is the ratio of force exerted by a machine to the force exerted on it. In this wheel-and-axle pulley the radius of the wheel is three times that of the axle. Theoretically, then, the mechanical advantage is three, that is, a downwards force on the rope should result in an upwards force on the weight three times greater. In practice, however, friction always lowers mechanical advantage.

2 A simple arrangement of two pulleys has one fixed pulley [1]; the other [2], to which a weight [3] is attached, is free to move. If someone hauls on the free end of the rope the movable pulley and weight are pulled upwards by twice the hauling force (less the frictional resistance) because they are held by two lengths of the rope. But the weight is raised through only half the distance hauled on the rope.

3 This factory hoist is hooked to a support [1] and has two fixed pulleys of different sizes [2] together with a movable pulley [4] to which a load [5] is attached. An endless chain [3] passes around the whole system of pulleys. A pull on one side of the chain loop extending over the larger fixed pulley exerts a much greater force to lift the load. A pull on the other side lowers the load.

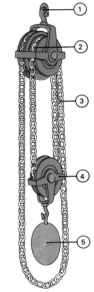

5 In gears the mechanical advantage is related to the number of teeth. The teeth of spur gears [A] are cut parallel to the axis of rotation, while those of helical gears [B] are "twisted" to form part of a helix and often cut double to avoid thrusts that result in wear. The teeth of bevel gears [D] are longer than those of spur gears, giving a greater area of contact that permits the transmission of much greater thrusts. This advantage applies more particularly to bevel gears with spiral teeth [C]. In a worm gear [E] the worm has a single spiral thread and turns a spirally toothed gear at right-angles to its own axis.

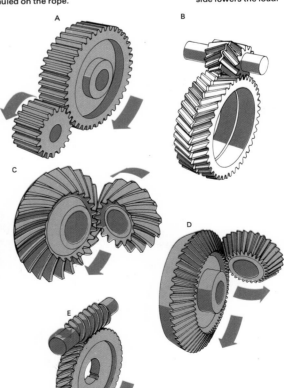

4 Pulley hoists are sometimes used in the motor car industry to lift bulky or heavy components. Hoists are usually suspended from mobile overhead cranes or gantries, allowing the transfer of parts from one place to another.

tional to the numbers of teeth on the gear-wheels. A gear of 100 teeth driving a pinion of 20 teeth, for example, increases the speed of rotation five times; the same 20-tooth pinion driving a gear of 40 teeth halves the speed of rotation.

Motion from a smaller driving gear to a larger driven gear, apart from reducing speed, obviously confers a mechanical advantage. Such an advantage was obtained in the old-fashioned clothes mangle: a small effort in turning the handle was sufficient to turn the rollers against considerable resistance from the squeezed clothing.

When shafts to be rotated by gear-wheels are not immediately adjacent one or more idler gears may be placed between the driving and driven gears to couple them together. Idlers have the same number of teeth as either the driving or the driven gears and so cause no change in the speed of rotation.

Changing the direction of motion

Rack and pinion gears convert rotary motion into linear motion. The pinion is an ordinary circular gear-wheel that meshes with the rack which is a "gear" with its teeth set in a row. This kind of gear system is used, for example, in the focusing mechanism of a microscope or old camera, in which the focusing adjustment turns pinions that move the lens. Cars with rack-and-pinion steering convert the rotary movement of the steering wheel into sideways movement to steer the wheels.

The teeth of gear-wheels may be set parallel to the gear axis, as already described, or spirally as in helical and worm gears [5]. A worm gear is used to drive a shaft at right-angles to its own. Bevel gears [5, 7] also transmit motion through an angle and can have parallel or spiral teeth. A gear system with a central "sun" gear meshing with several "planetary" gears is often fitted to bicycle hubs [8]. This is one example of a gearbox, which consists of a number of inter-meshing gears together with a device for selecting gear combinations, or ratios. Another example is the gearbox of a motor car. Early models had gear systems almost as simple as those of bicycles [9]. Those of today, however, are much more complicated and often have automatic gear selection.

B

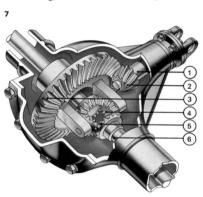

The block in a simple pulley [B] comprises a wheel or sheave that runs inside a housing, the whole block being hung from a hook. The wheel is grooved to accommodate a rope, belt or chain. A single, fixed pulley of this kind confers no mechanical advantage for lifting a load, although a person hauling up a load is able to add his own weight to the pulling force exerted by his arms. However, systems of two or more pulleys can give a considerable mechanical advantage, as shown by the compound pulley [A] said to have been used by Archimedes to move a ship single-handed.

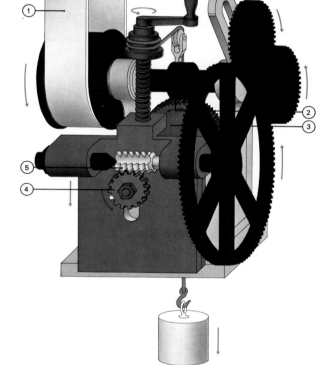

6 **Joseph Whitworth's** gear-cutting machine of 1835 contains a belt and pulley [1] driving a worm gear [2] which, by engaging a cogwheel [3], turns the gear that is being machined [4]. The same drive shaft turns a cogwheel meshed with a second cog and a large wheel that turns the milling cutter [5]. The cutter is mounted in a block that is lowered by means of the screw and counterweight until the gear is fully cut.

7

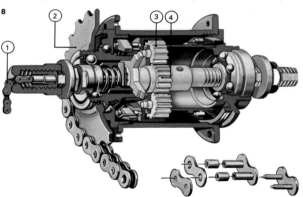

7 **The differential of a motor car** transmits the rotary movement produced by the engine through right-angles to the half-shafts [6] driving the wheels. The pinion [2] of the propeller shaft [1] rotates the crown wheel [3] turning the pinion [4] of the bevel gears [5]. The differential gears let the wheels turn at different speeds when the car turns a corner and the outer wheel rotates faster than the inner one.

8

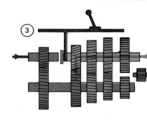

8 **This bicycle hub gearbox** has a central sun gear [4] surrounded by planet gears [3], a typical arrangement that is known as epicyclic gearing. The thrust given by the rider is transmitted to the hub through a chain and sprocket [2] connected to the hub by a cable-operated clutch [1]. The illustration shows the medium (direct drive) gear selected. The components of the chain drive are also shown.

9 **A gearbox permits a vehicle to move** at different speeds while the engine revolutions remain more or less steady. This is done by altering the ratio of input to output gears. More power is provided by a high ratio (low gears) allowing the vehicle to climb hills easily. All gear-wheels except those needed for reverse are always in mesh. When a gear-wheel is engaged it is locked (manually, in this gear box) on to the output shaft, so transmitting power.

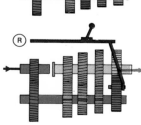

1 First gear	R Reverse gear
2 Second gear	F Input shaft
3 Third gear	L Output shaft
4 Fourth gear	I Idler wheel

Machines for weighing and measuring

Methods of measuring mass, time and distance are among the oldest skills of man. But instruments to gauge temperature, pressure, position and speed have been developed only over the past four centuries. Modern life requires even more complex measurements and depends on the accuracy and consistency of a whole range of machines for the functioning of industry, transport, medical care and meteorology. At one end of the scale are everyday instruments such as the micrometer [Key] which in engineering can measure diameters and thicknesses to an accuracy of 0.000254mm (0.0001in). At the other extreme are specialized machines that set absolute standards – atomic clocks, for example, are synchronized to keep time with the vibrations of caesium atoms pulsing exactly 9,192,631,770 times a second.

Weight, time and temperature
The modern chemical balance [5] is based on one of the oldest principles of all – the idea that the unknown weight of a given mass can be found by balancing it against a known weight. By suspending two pans on cords from a beam the Egyptians used balances to weigh grain and gold against stone weights at least 7,000 years ago. By 1350 BC they were able to achieve an accuracy of 99 per cent. The Romans added an important refinement when they fixed a triangular section to the underside of the beam, thus making balances more sensitive to lighter weights.

Measurement of time with sundials, hour-glasses and clocks is also ancient. The Chinese developed a water clock, or clepsydra, as early as 1000 BC. By the fourteenth century there were mechanical clocks moved by weights with a system of gears connecting these to an escapement wheel – a device to release the energy to the hour hand. Galileo (1564–1642) is credited with introducing a pendulum [2] to control the escapement, achieving a regularity that led to the development of the accurate timepieces of today.

In 1593 Galileo also played an important part in the development of the thermometer with his gas thermometer (air trapped below water). A more accurate alcohol thermometer was invented in 1641, and in 1714 Gabriel Fahrenheit (1686–1736) developed the mercury thermometer and the temperature scale named after him. Thirty years later the Swede Anders Celsius (1701–44) constructed the centigrade thermometer, so called because on its scale the boiling-point of water is 100°, its freezing-point 0°.

The clinical thermometer [8] is designed to measure human body temperature accurately over the range 35° to 45°C (95° to 113°F). Many industrial processes require measurements of temperatures that are much higher or lower than body heat [9]. For high temperatures, an optical pyrometer is used to compare the colour of a hot object with an electrically heated wire filament.

Establishing position
Successful navigation, at sea or in the air, depends on the ability to fix a position in relation to some known point. Instruments to measure the angle of the sun, moon and stars above the horizon have been gradually refined since the invention of the astrolabe, from possibly the third century BC. The positions of celestial bodies could be measured but it was difficult to use accurately on the

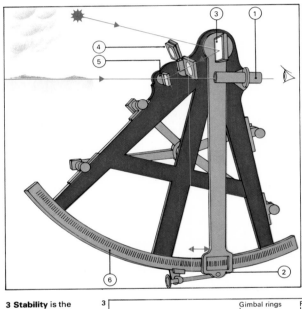

1 **The sextant** is still the basic tool of navigation and is extremely simple to operate. It is held so that the horizon is visible through the telescope [1]. A movable arm [2] carries a mirror [3] and the arm is moved so that an image of the sun reflected from this mirror, and from another half-silvered mirror [5], is aligned with the horizon. On a vernier scale [6] – a scale that measures sub-divisions of the main scale – the angular distance between the sun and the horizon is then read. A piece of dark glass [4] reduces the intensity of the sun's image. The sextant is also used to measure angles in astronomy.

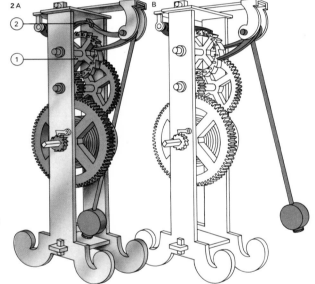

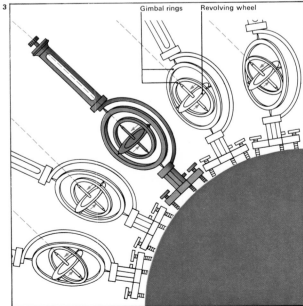

3 **Stability** is the vital contribution made to the science of measurement by the gyro. The instrument depends on a rapidly spinning wheel with a heavy rim which is suspended with a minimum of friction in a system of gimbal rings that allow it to rotate on its axis in any plane. If momentum is maintained at a given speed (by an electric motor for instance), the axis of the wheel maintains the position it took up when first spun. As the earth revolves, the axis continues to point to a particular position in space although the gimbals change their angle relative to it. Early applications were gun-sighting at sea and torpedo steering.

Gimbal rings   Revolving wheel

2 **As a device to regulate speed** a pendulum operates on the simple principle that the longer the pendulum the longer its period – the time taken for one complete swing from side to side. Galileo applied this principle to regulate the escapement wheel of a clock. His wheel [1] had 12 projecting pins lined up with notches round the circumference. As the pendulum swung inwards [A] it lifted a constraining pallet [2] and pushed the pins, allowing the wheel to rotate but only until, with the reversal of the pendulum swing [B], it was once again constrained by the pallet.

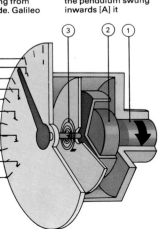

4
50mph
80km/h
40mph
60km/h
30mph
40km/h
20mph
20km/h
10mph
0km/h
0mph

4 **The speed of a vehicle** is read on a speedometer connected by a flexible cable to gears in the transmission system. According to the speed, the core of the cable rotates a magnet [1] which pulls a drum [2] mounted around it. A pointer on the speedometer dial moves with this drum but is stabilized, if the speed is constant, by a hairspring [3] which balances the force of the magnet and holds the pointer stationary.

heaving deck of a ship. In 1730 John Hadley (1682–1744) invented a reflecting quadrant that brought the horizon and the observed object in line by mirrors. It soon developed into the sextant [1]. The sextant, so called because it usually has a scale of 60° (one-sixth of a circle), enabled mariners to measure angles at any inclination and to fix their positions much more accurately.

Measurement of absolute movement relative to the stars was significantly improved after a French physicist, Jean Foucault (1819–68), built a gyroscope [3] to show that the earth revolved on its axis. The principle of the gyro is that the axis of a spinning wheel suspended in gimbal rings holds its original position in space regardless of gravity or magnetic force. By fitting gyros in cases with circular rings, marked by degrees, aids have been developed for the automatic steering of machines ranging from space vehicles to oil drills.

Pressure, speed and radiation
In 1643 Evangelista Torricelli (1608–47), an Italian mathematician and physicist, found that the pressure of air at the surface of the earth was equal to that of a 76cm (30in) column of mercury. At higher altitudes the pressure falls. Working on this principle, Torricelli devised the earlier practical form of the barometer. Measurement of changes in atmospheric pressure were soon being used to gauge the height of mountains as well as climatic conditions. In addition to the barometer and various adaptations of it, a wide range of instruments is now available to measure the pressure of liquids and gases. A common example is the Bourdon tube gauge [7] patented by a French watchmaker in 1850. More sophisticated instruments are based on similar principles.

Accurate measurement of vehicle speed was achieved only in the 1920s with the development of the magnetic speedometer [4]. Today this instrument is usually linked with an hodometer which measures distance travelled. The nuclear age has brought still newer measurement needs and devices. The film badge [6] provides people who work in radioactive environments with a vital means of monitoring the amount of radiation.

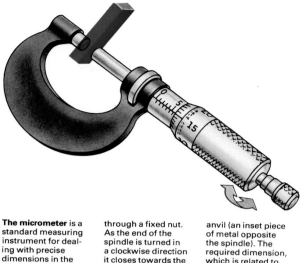

The micrometer is a standard measuring instrument for dealing with precise dimensions in the engineering industry. It consists basically of a screw or spindle that can be screwed through a fixed nut. As the end of the spindle is turned in a clockwise direction it closes towards the workpiece being measured, which is lightly held between the spindle and the anvil (an inset piece of metal opposite the spindle). The required dimension, which is related to the number of turns made by the screw, can be read off a graduated scale.

6 A film badge worn by those likely to be exposed to radiation consists of a film [7] in a plastic holder of known absorption properties [1]. A window [2] lets through all types of radiation. As neutrons [3] do not themselves affect film they are slowed by a lead filter [5] and absorbed by a cadmium filter [6] which emits a gamma ray for each neutron, blackening the film. Gamma rays [4] themselves penetrate all filters. X-rays [9] penetrate the plastic filter, whereas beta particles and other types of radiation [10] blacken the film through the window. The developed film [8] shows blackening.

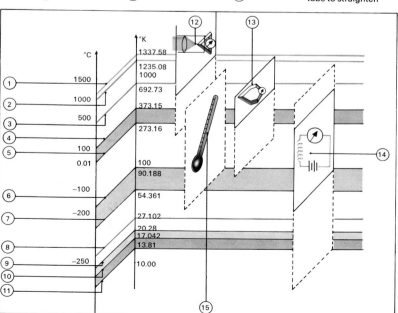

5 Measurement of mass is most accurately carried out by balancing. A high-sensitivity balance used by chemists has screw-weights [1] for fine adjustment. When correctly adjusted, a long, vertical needle in the centre of the unit rests exactly over a central zero on a scale at the base of the balance column. To achieve finer measurements than are possible by placing known weights in one pan a "rider weight" is sometimes used. This small weight [2] is designed to slide along a direct-reading scale marked along the top of the balance arm.

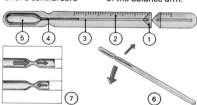
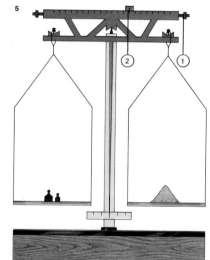

8 A clinical thermometer is an ordinary mercury-glass thermometer with a particularly fine capillary [3]. A constriction [4] allows mercury to flow easily from the bulb [5] but, by surface tension, prevents flow back [7]. A temperature reading can thus be maintained on the scale [2] until the mercury is forced back by shaking [6]. For easy reading the stem is lens-shaped, as in cross-section [1], to visually magnify the mercury when the thermometer is held at the correct angle.

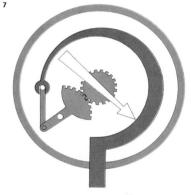

7 Pressure is measured in a Bourdon gauge by allowing a liquid or gas to flow into a curved flattened tube sealed at one end. Higher pressures produce a tendency for the tube to straighten out. The resulting small movement at the sealed end of the tube is amplified by a system of levers fixed to an indicator that moves over a scale to show the pressure applied.

9 Temperatures are defined in the International Practical Temperature Scale in degrees absolute (above 0°K or −273°C). The "fixed points" are melting-points of gold [1], silver [2] and zinc [3]; the boiling-point of water [4] and its "triple point" [5] at which steam, water and ice are in equilibrium; boiling [6] and triple [7] points of oxygen; boiling-point of neon [8]; triple [11] and boiling points of hydrogen at atmospheric [9] and at 25mm of mercury [10] pressures. Standard instruments for measuring temperatures within certain ranges are the pyrometer [12], platinum-rhodium thermocouple [13], electrical resistance of platinum wire [14] and liquid-in-glass thermometer [15].

Machines for measuring time

The earliest mechanical clocks containing movable parts were built about 700 years ago. But the first instrument to measure daily time was made more than 3,000 years ago. This was probably the Egyptian shadow clock, dating from about 1450 BC. Like a sundial, it measured daily time by the movement of a shadow thrown across markers.

The first kinds of clocks

The shadow clock was soon followed by the water clock or clepsydra [1] and the sandglass or hourglass, in which time is measured by the change in level of flowing water or sand.

These remained the only methods for measuring daily time until the Anglo-Saxons began to use candles marked at regular intervals [2]. In medieval times, instruments were made with dials marked in hours. These included the sundial and star dials such as the nocturnal [3].

All familiar clocks and watches work by the regular recurrence of some mechanical movement. The first mechanical clocks, of the thirteenth and fourteenth centuries [5], were driven by falling weights which moved gear-wheels. For the clock to run for more than a few seconds, the power from the falling weights must be released slowly. To do this, one of the gears (the escape wheel) is regularly "held" and released. The mechanism that achieves the controlled release of power in a clock or watch is called the escapement.

Early clocks used the verge escapement. Two projections or pallets, formed on the balance axis, engaged with and disengaged from the teeth of the escape wheel, causing the balance to oscillate regularly. The motion of the escape wheel was transmitted to a single hand on the clock face.

As driving mechanisms, falling weights have the obvious disadvantage of not being easily portable. As early as the mid-fifteenth century, compact and portable clocks driven by springs were developed.

Early spring-driven clocks were inaccurate. A minute hand had appeared on the faces of some of these clocks, but a hand telling the seconds remained almost unknown until the arrival of the pendulum.

In 1657 the Dutch scientist Christiaan Huygens (1629–95), influenced by a suggestion from the Italian scientist Galileo (1564–1642), specified the conditions for a perfectly swinging pendulum and applied it to a clock. From this time, more accurate clocks were made with pendulums.

Problems with pendulums and escapements

Problems of accuracy remained. Pendulums are affected by changes in temperature, which cause them to expand or contract and so change length. In about 1715 George Graham invented the first of many pendulums compensated for temperature changes. Three hundred years ago, however, the principal shortcoming of clock mechanisms was the verge escapement, which interfered with pendulum action. In 1673 a new type, the anchor escapement [4], was invented. It allowed a heavy pendulum to swing in a small arc with such a gain in accuracy that it is still fitted in some clocks.

Another oscillator is the balance and balance spring, or hairspring [6]. One end of the spiral balance spring is fixed and the other end is attached to the axis of the balance. The

CONNECTIONS

See also
86 Levers and wedges
88 Pulleys and gears

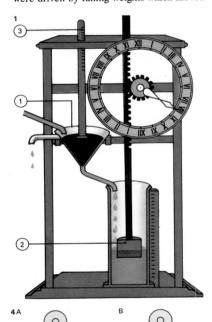

1 This water clock or clepsydra is based on an Egyptian clock made in the third century BC. Water is supplied to the funnel [1] and passes to the cylinder in which the float [2] rises. This is connected to a rack-and-pinion that actuates the hour hand. The rate of water flow is regulated by the graduated stopper [3] and the water is kept at a constant level by means of an overflow tube.

2 The oil clock was a 16th-century development of the candle clock first known among the Anglo-Saxons. Both have a scale recording the dropping level in hours as the oil or wax burns away.

3 To measure time at night, a nocturnal was used. It worked simply: the North Star was sighted through a central hole and the pointer was rotated towards the two "pointer" stars of the Plough.

5 Henry de Wyck was commissioned in 1370 to build a clock for Charles V's palace in Paris. It is a good example of early mechanical striking clocks operated by falling weights. As they fall the weights set gear trains [1] in motion. The crown wheel [2] actuates the pallets of the verge escapement [3] so causing oscillations of the balance [4]. Two inertial weights [5], suspended from the bar balance, can be adjusted to control the rate of the bar's oscillation. The gear train actuates the single dial hand. A second train of gears leading to the striking device is set in motion by a lever [6] actuated by a pin on the hour-hand wheel.

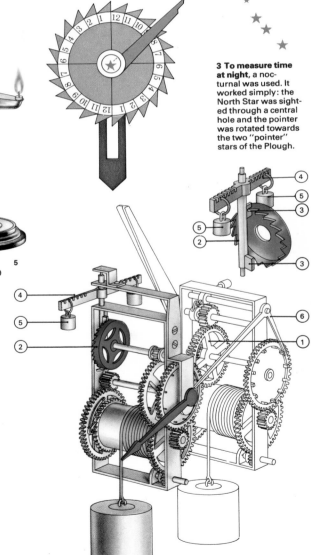

4 The accuracy of a mechanical clock depends above all on the escapement mechanism which releases the energy of a spring or weight regularly in small "bursts" to the time-keeping part of the clock. The anchor escapement of a pendulum-clock [A] has an anchor that swings about its centre and is connected to the pendulum. A main spring (not shown) moves the escape wheel clockwise. A tooth of the wheel pushes one pallet of the anchor [B] until the other pallet checks another tooth [C], the curve of the pallet forcing the wheel slightly backwards. After this recoil the pallet receives a push until the first pallet checks the wheel. In this way, the to-and-fro movement maintains the pendulum's oscillations.

balance spring alternately winds and unwinds as the balance swings.

The balance spring was also introduced by Huygens, in 1675, and he later incorporated it in a watch that he intended to be used for determining longitude at sea. But balance springs, like pendulums, were adversely affected by temperature variations. It was not until 1753 that an effective compensation was made for a watch by John Harrison (1693–1776) whose chronometer of 1759, made in response to a government competition, erred by only five seconds in a sea voyage lasting six weeks. Accuracy had also been improved by the introduction of jewels used as bearings. They reduce friction in the bearings and are extremely hard wearing. Sapphires and rubies were, and still are, the jewels generally used.

Watches were developed and refined with various escapements until the mid-nineteenth century when the lever escapement was almost universally adopted. In this mechanism, invented by Thomas Mudge in about 1755 but neglected for half a century, the pallets are attached to a lever that is detached from the balance for most of its swing. This arrangement, together with the lever's robustness, promotes a high degree of accuracy in timekeeping.

Modern and electric clocks

Other kinds of clocks, including those employing an electric motor to wind a spring or weight, now rival purely mechanical clocks. A mains electric clock has a "synchronous" motor that keeps in step with the frequency of the alternating current supply. Electric pendulum clocks use electromagnets to keep a pendulum swinging accurately.

Oscillators include piezoelectric crystals such as quartz. The crystal vibrates and continues vibrating when the correct alternating voltage is applied accross it. Compact electronic circuits reduce such high-frequency oscillations to only a few per second which then operate gears driving hands. Such clocks can be accurate to a tenth of a second per year. Even more accurate are atomic clocks that employ oscillating energy changes within atoms. Clocks of this kind are now used as international standards of time [9].

Shadow clocks such as sundials were used for centuries for telling the time. They have to be accurately set up with the 12 noon position pointing due north (on a sundial) or on a wall clock, vertically downwards.

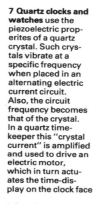

6 This 17th-century watch [B] has its balance spring [1] attached to the staff of a balance [2] which swings first one way, then the other, under the tension of the spring [A]. The regulator [3] shortens or lengthens the spring to alter its tension, which in turn controls the swings and so the accuracy of the watch.

7 Quartz clocks and watches use the piezoelectric properties of a quartz crystal. Such crystals vibrate at a specific frequency when placed in an alternating electric current circuit. Also, the circuit frequency becomes that of the crystal. In a quartz time-keeper this "crystal current" is amplified and used to drive an electric motor, which in turn actuates the time-display on the clock face.

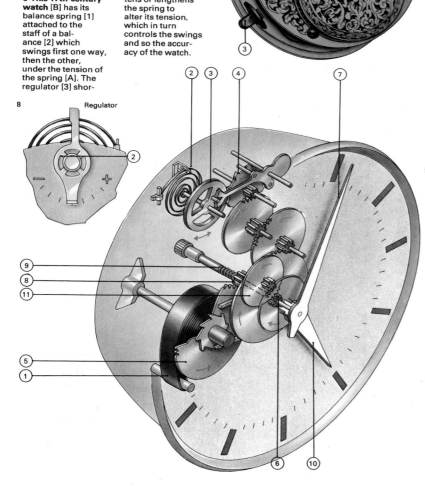

8 A typical mechanical clock of today is operated by energy stored in a mainspring [1]. This energy is released in small controlled "bursts" in an escapement mechanism comprising a spring [2] and balance [3] together with an escape wheel [4]. The mainspring drives a great wheel [5] which in turn rotates the centre arbor [6] and minute hand [7] via the centre pinion [8] and friction spring [9]. The hour hand [10] is turned at one-twelfth the speed of the minute hand by motion work gears [11] which are driven by a cannon pinion in small controlled increments.

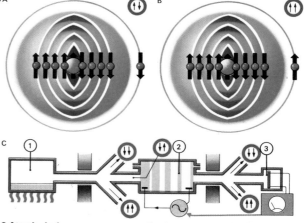

9 Atomic clocks use the frequency of vibration of atoms (about 10,000 million per sec) to regulate a quartz crystal clock. Caesium atoms [A] are normally unmagnetized but radiation can magnetize them [B]. A caesium clock [C] has a boiler [1] giving a supply of atoms whose magnetic axes are lined up by a magnetic field. In the chamber [2] they encounter an oscillating field and the magnetic axes "flip" over so that they are deflected by a second magnetic field towards a detector [3]. Signals from here control a quartz crystal clock.

The factory and assembly line

In modern industrial society nearly all consumer goods are made in factories which are complex organizations of machines, processes, materials, people and products. Underlying the organization of a factory is the principle of the division of labour, formulated by Adam Smith (1723-90) in his book *The Wealth of Nations* (1776), in which he described methods used to make pins. The manufacture of pins was divided into several different tasks, each carried out by a different person. It has been found that this type of organization, in which the work is divided up into a series of separate operations, is essential for large-scale production.

The factory system

A craftsman generally carries out all the stages of manufacture of an article himself. But even he relies on apprentices or other help; few woodcarvers make their own tools, for example, although they certainly sharpen them. Some processes are now being envisaged in which a modern master-craftsman, using computerized systems, will design and control the manufacture of goods by mass production, and in doing so restore the unity of craft techniques.

In Europe the factory system became firmly established for the making of cloth in the seventeenth century. The process was subdivided into carding, spinning and weaving, sometimes with many machines in one factory. Even so, factory production accounted for only a small proportion of the total output of cloth.

A decisive step came with the design and manufacture of machinery with standardized interchangeable parts. In 1803 Marc Isambard Brunel designed machinery with interchangeable parts for making pulley-blocks for the Portsmouth naval dockyard [1]. The machinery was manufactured by Henry Maudsley and there were 45 machines of no less than 22 different kinds. By 1807 the machinery was capable of supplying the entire pulley-block requirements of the Royal Navy and by 1808 there was a yearly production rate of 130,000 blocks.

The first automatic assembly line [4] was begun when the Olds Motor Works in Detroit, USA, was destroyed by fire in 1901.

The factory was rebuilt so that a car could be wheeled from worker to worker. This idea was later employed by Henry Ford who started up the first moving assembly line using interchangeable parts and a moving conveyor belt to transport the vehicle around the factory floor. (This meant that cars of the same model could be repaired or assembled from stock parts.) By 1914 Ford was turning out the Model T at a rate of one every hour and a half. This time-saving cut production costs, and the price of a Model T had fallen from $850 to $400 by 1916.

Modern assembly lines

On modern assembly lines complex products are assembled at speed. Goods are produced in greater quantities and at a lower cost. Production lines even manufacture complex machined parts, such as cylinder blocks for automobile engines, as well as large finished items [5]. Some production lines have equipment that automatically transfers parts from one machine to the next.

The assembly line is undoubtedly efficient but the work is not only extremely

1 The machinery that was set up in the Royal Dockyard at Portsmouth was the first instance of the use of mass production of interchangeable parts with machine tools. The machines were so well designed and built that several of them were still in use in the 1950s. The total saving in the first year was £24,000 of which the inventor, Marc Isambard Brunel (1769–1849), received £17,000.

2 The 1873 Colt revolver is still being made today. Colonel Samuel Colt (1814–62) was the first mass-producer of firearms. His revolvers had single-action mechanisms and therefore had to be cocked before firing. They were first supplied in .44-inch calibre to the US Army after the Mexican War (1846–8).

4 Assembly line techniques are still used in the automobile industry although they are now being superseded by modified systems that are both easier on the worker and less regimented in character.

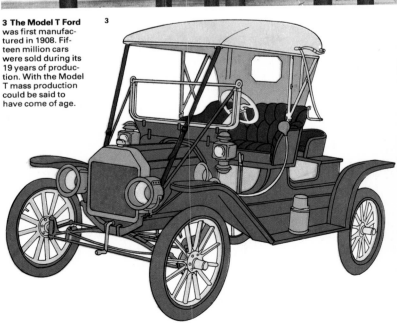

3 The Model T Ford was first manufactured in 1908. Fifteen million cars were sold during its 19 years of production. With the Model T mass production could be said to have come of age.

boring for the workers but has to be performed under tremendous pressure. The whole process of mass production is however, beginning to change. Work is being reorganized to make each person's job more interesting and automated factories under computer control are planned.

The attempt to improve job satisfaction was pioneered in Sweden. Instead of a "one man, one machine" line, departments are divided into a series of teams. Each team receives the same pay, with each person in it performing a variety of tasks. Instead of working to orders, each team devises its own methods. The new system has succeeded. Using it, the Saab-Scania car division at Sodertalje opened a new factory for engine assembly in 1972 and Volvo opened a plant for the assembly of complete cars in 1974.

Other new methods
New methods of organization and manufacture are being developed in most industries, with mass production reserved for the production of identical parts in large quantities. "One off" parts and prototypes remain

expensive. Between these two extremes there are batch-produced items made at moderate expense. This is a surprisingly important area of production and it has been estimated that more than 50 per cent by value of the goods manufactured in the United States are made in batches of 50 or less.

The manufacture of computer-controlled parts [6–8] depends on numerically-controlled machine-tool stations, which can be instructed to carry out a range of machining tasks. There are also versatile handling systems that transfer manufactured parts from one station to another and positioh them according to instructions. The type of design must be made to suit the computerized control system. Japan is the most advanced country in computer-controlled parts manufacture, but the most ambitious system is at Karl-Marx-Stadt in the German Democratic Republic. The system is housed in an air-conditioned building the size of two football fields. Work is transferred round the factory floor on pallets supported on air cushions, like miniature hovercraft, and propelled by means of linear induction motors.

The automatic tool-changing carousel provides a versatile machine tool suited to numerical control.

A computer selects the tool and it is automatically brought into position in a few seconds. This

system is expensive compared with standard tools and its use has not always proved to be justified.

5 Products made on assembly lines, such as these motor car engines, pass from one group of workers to another. Each worker repeats a specific job on a part-built engine as it reaches him [A, B]. Finally the completed product passes to inspectors, who carry out a series of examinations [C] to test its quality. The advantages of "one man, one job" can, however, be offset because of the boring, repetitive nature of the work.

7 A computer is the heart of any fully automatic manufacturing process. It can collect and sort production data, control the ordering of raw materials and stock and, most important of all, instruct automatic machine tools to fabricate various components and other parts.

8 Computer-managed parts manufacturing uses computer technology to lower the cost of machining a small number of parts. Because the computer stores all the information about the part it is able to organize the manufacture of that unit in a "batch" that can consist, if necessary, of only one unit.

6 Computer-controlled parts manufacture uses a versatile handling system. Depending on the complexity of the part, it can be directed to just two or three machines, bypassing the others. The computer is also able to direct the machining of at least 16 different kinds of parts. The engineering drawings of a part to be made resemble a map, and the various angles and dimensions can be coded as instructions to the computer's memory and then later to the machine that makes the part.

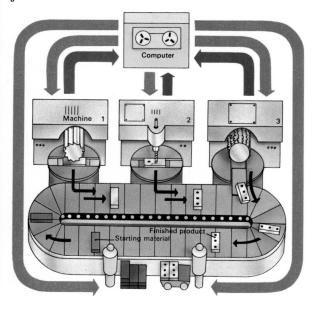

Machines for lifting

The strongest athletes can lift to just above the ground weights of up to 425kg (935lb), but few ordinary people can lift more than 50–60kg (110–132lb). Early man soon developed machines for lifting large stones and tree trunks. A simple device is a single pulley wheel arranged as a hoist. But a rope round one pulley merely changes the direction of the pull and friction in the pulley's bearing in fact makes this simple machine less efficient than a straight pull. If the rope is wound round a wheel or cylinder to form a windlass [Key], a mechanical advantage is gained, by means of which a man can easily lift more than his own weight. A small windlass can be driven by a hand crank – first used in the ninth century – and many early cranes and hoists used this principle. Horses, oxen or other animals could also be harnessed to do the pulling.

Screws and pulleys

Many key inventions of antiquity, such as the screw and the pulley, cannot be credited to any one man. The Greeks were probably using screws by about 400 BC and by the time of Archimedes (who died in 212 BC) the screw certainly had various applications. Archimedes himself invented a type of pump consisting of a long helix in an upward-sloping tube; by turning a handle at the upper end the operator could "screw" water from the lower end of a spiral to its upper end until it flowed out of the top. This kind of pump was used for irrigating the Nile Valley. In Roman and medieval times, screw presses were used for crushing olives and grapes.

In the thirteenth century the French monk Villard de Honnecourt followed Archimedes and made a machine that used a screw for lifting instead of pressing downwards. Today known as the screw jack [2], the device has many applications from lifting a car to change a wheel to jacking up whole buildings while a storey is slipped underneath forming a new ground floor.

Pulleys were also known to the ancient Greeks who used them to lower a statue of a god on to the stage as the climax of a religious drama. By the time of Christ Roman engineers were designing and making multiple pulley blocks for lifting heavy loads. A 200-tonne Egyptian granite obelisk, similar to the so-called Cleopatra's Needles now standing in London and New York, was erected in ancient Rome using many pulley blocks and teams of slaves to provide the muscle power. Today's compact hoists [3] use exactly the same principles.

Hoists and cranes

Machines for lifting can also be made using gears to obtain mechanical advantage. With only horse power, sixteenth-century miners hauled loads of ore and other minerals [1]. Later hoists using steam engines – and even modern ones with electric motors – use similar principles.

Early cranes were merely rope-and-pulley hoists rigged using two or three wooden "legs" straddling the object to be lifted. Power was provided through a windlass or, for heavy loads, by a treadmill. The building projects and dock installations of the Middle Ages depended on such machines for lifting huge blocks of stone.

Modern cranes are of two main types known as bridge cranes and jib cranes. Both

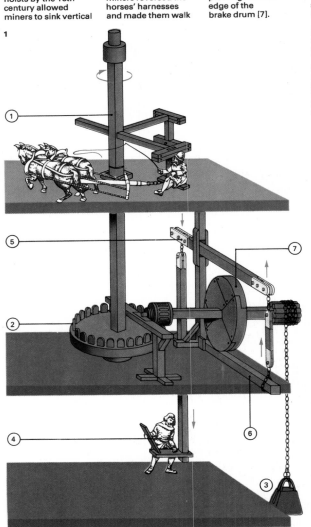

1 In the earliest days of mining tubs full of ore were dragged along horizontal or gently sloping tunnels driven into a hillside. The development of hoists by the 16th century allowed miners to sink vertical shafts. The horse-powered capstan [1] turns a cogged wheel [2] to winch up a leather bucket [3] containing the ore. To lower a load the miners reversed the horses' harnesses and made them walk in the other direction. The brakeman [4] working below stopped the hoist by pulling down the beam [5] to make the timber baulk [6] press against the edge of the brake drum [7].

2 The winding lever [1] of a screw jack is turned once to raise the jack through a distance equal to the pitch of the screw's thread. With a fine pitch a small effort can lift a very heavy load.

3 A compact hoist can be made using two pulley blocks. This one has a mechanical advantage of eight, although 8m (26ft) of rope have to be pulled through to raise the load only 1m (39in).

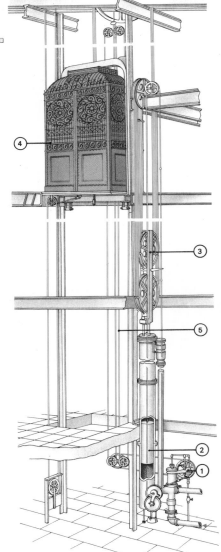

4 An hydraulic lift of the 19th century also uses pulleys. A pump [1] pressurizes water to move the plunger [2]. This movement is transmitted to the lower pair of pulleys [3] to raise or lower the lift cage [4]. In this way, the relatively small movement of the plunger is made to produce a large movement of the cage. The height and speed of the cage are controlled from inside it by ropes [5]; these are connected to the water pump.

use a kind of windlass with steel wire rope wrapped round a powered drum. A bridge crane has a box-girder-like beam (called a gantry) running on long elevated tracks at each of its ends. The gantry can move backwards and forwards along the tracks. The hoisting system is carried in a trolley, which can move from side to side along the gantry beam. Bridge cranes are commonly set up above a working area to handle such loads as tree trunks and steel beams.

A jib crane has a long boom called a jib that can swing horizontally, to move the load sideways. Many such cranes can also "luff" to control the reach of the crane by angling the jib more or less to the horizontal.

Lifting people

Skyscrapers and high-rise blocks of flats would have been impossible without a lift (called an elevator in the United States) interconnecting the dozens of floors and giving access to the ground. In 1857 the American inventor Elisha Otis (1811–61) installed a steam-powered lift in a New York department store.

Early lifts used the screw jack principle, soon to be replaced in the 1870s by lifts using hydraulic pressure. Water, oil or other fluid is pumped to provide pressure against a piston, which in turn raises the load. Many buildings use a combination of hydraulics and pulleys [4] which allows the lift to be made longer. In the twentieth century, buildings were made even taller, especially in the United States. The ultra-high-rise buildings of today need electric-powered passenger lifts that travel at more than 400m (1,312ft) per minute.

In underground railway stations and large stores, there is a more or less continuous flow of people between various levels. Here the people-lifting problem is solved by using escalators, which are continuously moving staircases based on the conveyor belt principle with an endless belt of steps. The original patents of 1891 were obtained and improved by the American Otis Elevator company which, together with Westinghouse Electric Elevators, developed the modern escalator in the 1930s. An escalator 1.25m (4ft) wide moving at 27.5m (90ft) per minute can carry about 8,000 people an hour.

The effort of lifting a man is reduced by the mechanical advantage provided by the windlass on a mine's winding gear. If the shaft has a radius of 30cm (12in) and the windlass 240cm (94in) the mechanical advantage is eight – by exerting 10kg (22lb) of effort an 80kg (176lb) man can be lifted from the mine.

5 A tower crane is used in constructing high-rise buildings. Anchored to the ground or to the building itself, the crane is extended upwards as the work proceeds. Standard lattice sections form the tower which supports a horizontal jib whose weight, and that of the load, is counterbalanced by a block of concrete. The hoist is in a trolley that can travel along the boom and is controlled from the driver's cabin.

6 Each step on the endless belt forming an escalator has two pairs of wheels. The upper pair [1] run on an outer rail and the lower ones [2] run on an inner rail. On the sloping part of the "staircase", the rails are in line. But at the top and the bottom they separate to make the steps line up to create flat sections for people to step on and off. Even when stopped, an escalator can be used.

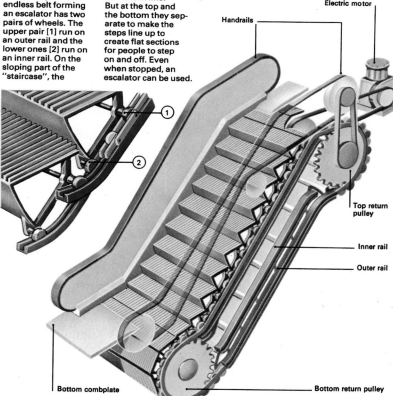

Electric motor

Handrails

Top return pulley

Inner rail

Outer rail

Bottom combplate

Bottom return pulley

7 A fork-lift truck is used for moving, stacking and unstacking goods carried on wooden pallets – platforms that can be scooped up on the truck's forks. A heavy weight at the rear of the truck counterbalances the load. Most trucks are driven by electric motors, although some have diesel or low-pressure gas engines. The same power unit drives the hydraulic mechanism or chains that raise and lower the forks, which may tilt backwards to make the load safer when being moved. Some trucks have telescopic masts which extend upwards to allow the forks to stack loads up to 5m (16ft) above the ground. Carrying capacities vary from five tonnes with small machines to 50 tonnes or more.

Earth-moving machines

Building Iron Age forts, digging canals, making railways and constructing modern motorways have all required the shifting of hundreds of tonnes of soil. As a result, from the Iron Age to the present day men have devised various machines for moving earth.

One of the earliest earth-moving machines was the wheelbarrow, developed in China before 118 AD. The Chinese version had a large wheel 1m (39in) or more across with the load carried above and at the sides of the wheel. The early European wheelbarrow, similar in style to that used today, had a fairly small wheel and the load was carried between it and the handles. Using only wheelbarrows, picks and shovels to assist them, the navvies (short for "navigators") built the whole system of European canals and many of the early railways [Key].

Modern earth-moving machines

Wheelbarrows are still used on small-scale building projects, but today's civil engineers can choose from many specialized earth-moving machines. The digger or excavator was one of the first of these and a mechanical digger was an early nineteenth-century application of the steam engine.

Today there is a wide range of excavator designs, each suited to a particular task. The dragline excavator, for example, has crawler tracks for moving over uneven or soft ground [3]. The digging bucket is suspended from a jib and after scooping up its load is winched back to be dumped. The size of the "bite" taken by the bucket has to suit the material being excavated. So for soft earth and for moving existing stockpiles of earthy minerals, such as crushed ore, a light bucket is used. A medium-weight bucket is employed for general digging duties, but for deep digging, or in rocky terrain, heavy buckets are essential to give enough penetration and prevent undue wear. In all these operations, the digging action occurs as the bucket is dragged back along the ground after being dropped from the end of the jib.

For size and capacity the largest bucket-wheel excavators [2] are among the world's greatest engineering achievements. They belong to the largest of self-propelled land machines and can be used for rapid excavating or for shifting vast quantities of loose material such as crushed ores or coal. They can move up to 10,000 cubic metres (354,000 cubic feet) of material in an hour.

Dredgers are merely floating excavators used for keeping docks, harbours and river channels free from mud and silt. They can also be used for "mining" underwater, to scoop up sand and gravel and other minerals. They have boat-like hulls, which may be passive and thus have to be towed to the site of operation, or fully powered and equipped with the necessary machinery to travel in the open sea. They have diesel engines that drive the machinery directly or power a generator to supply electric motors.

There are three main types of dredgers: bucket dredgers [6], grab and dipper dredgers and suction dredgers [7]. Bucket dredgers have an endless chain of buckets, on the conveyor belt system, that scoop up material from the bottom. A grab and dipper dredger has either a mechanical shovel (the dipper) pivoted at the end of a boom or a "clamshell" (the grab) for excavating materials in bulk. Most grab and dipper dredgers have a

CONNECTIONS

See also
182 Road building
196 Canal construction
188 Tunnel engineering
194 Harbours and docks
138 Small technology and transport

1 A hydraulic shovel has rams hinged at the base of the bucket. By skilfully controlling these rams, the driver is able to excavate material without having to move the machine forwards. For speed of operation a machine may have a crawler hull so that it can swing round to dump into a truck parked behind it in a maximum of 15 seconds.

2 A digging-wheel excavator, with a scoop-tipped wheel up to 20m (66ft) across, can shift many cubic metres of soft material on to its internal conveyor belt. It is particularly useful for shifting dumps of powdered minerals such as china clay or coal-dust.

3 A dragline excavator, which is a kind of revolving shovel with a long boom, is ideal for stripping the topsoil, called over-burden, from near-surface deposits of coal and other minerals. With very long booms, the bucket must be light or the machine must have a counterweight.

4 A shovel dozer, designed for digging and loading, can also be used for moving "spoil" over short distances. The crawler-mounted type can have a bucket holding up to 4 cu metres (140 cu ft) of soil. Special attachments allow the machine to lift and remove rocks and tree stumps.

pair of metal legs, called "spuds", which are lowered to the bottom to stabilize the craft. Suction dredgers work like giant vacuum cleaners and use powerful pumps to suck sludge up from the bottom.

Shovels and dozers

The workhorse of land-based earth-moving equipment is the mechanical shovel, often known today as a hydraulic loader [1]. It may have crawler tracks or wheels (with two-wheel or four-wheel drive). The crawler version can turn in very confined spaces – even in its own length – but the wheeled loaders travel much more quickly. In both the usual cycle of operation consists of loading the excavator bucket, travelling to a heap of "spoil" or to a dumper truck, dumping and travelling back to the excavation position.

The dumper trucks, also essential machines in a modern operation, have a capacity of 20 tonnes or more. Even so, several may have to be used to keep pace with a giant excavator.

If the soil or other material does not have to be moved too far it can be pushed to its new site by a bulldozer or carried there by the hybrid machine known as a shovel dozer. An angledozer merely pushes the material to one side. A shovel dozer [4] can use tracks or wheels, depending on the site. It can dig, load and transport "spoil".

Scrapers and graders

For building level, modern roads the specialized earth-movers employed are scrapers and graders. A scraper may be self-propelled or pulled by a tractor [5]. It has a knife-like cutter that planes off a layer of soil into an internal reservoir called a bowl; this can hold up to 40 cubic metres (1,413 cubic feet) of soil. The depth of cut is controlled by hydraulic rams and the machine can transport its load to a nearby site for dumping. The "spoil" is dumped – gradually or in one lot – by moving the rear tailgate.

For precise finishing of the earth road foundation before concreting, a machine called a grader is used. It has an angled blade 2–4m (7–13ft) wide hydraulically controlled and slung between its wheels. Most of these finishing machines are self-propelled.

In the early days of railways, cuttings were largely excavated by hand; picks and shovels were used to dig out the soil and wheelbarrows to cart it away. The 3km (2 mile) Tring Cutting between London and Birmingham was dug in 1838. Horses pulled the loaded barrows up planks laid up the sloping sides of the cutting, but navvies had the dangerous job of guiding the barrows. Inevitably, accidents were frequent.

5 **A scraper** is one of the key machines in modern road-building projects. Self-powered or hauled by a tractor, large ones can carry 100 cu metres (3,500 cu ft) of soil. For extra power, a second diesel engine may be mounted at the rear. It is this massive power that enables the scraper blade under the machine to skim off layers of earth and force them back into the body or "bowl" of the machine. The heights of the scraper and of the tailgate, which is lifted or slid aside to release the load, are controlled by hydraulic rams, also powered by the main engines. All scrapers have huge tyred wheels to cross uneven ground.

6

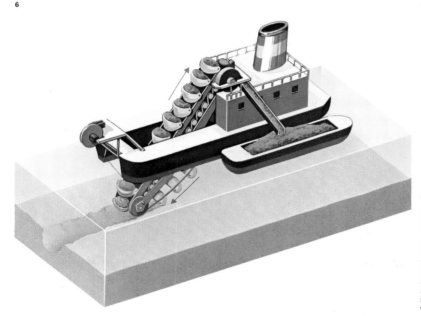

7

6 **A bucket dredger** has an endless chain of buckets that scoop up sludgy material from the bottom of the sea or river. The "spoil" is automatically tipped into a discharge chute and into a barge moored alongside or, when working in a dock, directly into a dumper truck. Most bucket dredgers have no engines and therefore have to be towed into position by tugs, although self-propelled ones are sometimes used for excavating canal banks or other confined areas.

7 **A suction dredger** has powerful pumps that suck up the "spoil" in the form of a watery mud called slurry. Any hard material is broken up by high-pressure water jets or cutters. Most suction dredgers are self-propelled and can move to dump.

Moving heavy loads

Ordinary cranes, used in the construction industry or for loading ships, can lift weights of up to 200 tonnes. But consider the following problems: a prefabricated 1,500-tonne section of a ship (such as the whole superstructure or the front part of the bow) has to be placed in its final position [2]; a 6,000-tonne rocket has to be moved 5km (3 miles) to its launching site [3]; a 7,000-tonne section of a stadium has to be placed in a new position [4]. Each of these involves moving a heavy load, and each has been solved.

What are heavy loads?

The ability to move heavy loads is increasingly important to the engineering industries because the cost-saving of building assemblies on a specific site before moving them to their final places is now accepted. But prefabricated structures are becoming larger and heavier. As new load-moving techniques have been developed, other industries have assessed their usefulness and have had to adopt them.

The word "heavy" is arbitrary, but for these purposes it includes loads ranging from hundreds of tonnes to tens of thousands of tonnes. Moving heavy loads has presented engineers with problems for thousands of years. Many suggestions have been put forward as to how stone was moved in the building of the pyramids and Stonehenge. Certainly a method using tree trunks as rollers would have been known at that period of history and animal or human power could have provided the motive force.

Man started with the lever and soon discovered the arrangements of the moving force, the load and the fulcrum (pivot) that would be most useful in particular applications. Archimedes [Key] is reputed to have claimed, "Give me a firm place on which to stand and I will move the earth". He realized that to use a long lever to gain a mechanical advantage would mean that a small movement of the load could be obtained with a large movement of the applied force.

The problems involved

Moving heavy loads has always involved two different problems: how to reduce the friction underneath the load and how to provide sufficient force to overcome the friction remaining once the load is moving. To reduce friction, rolling logs were used and later wheels of various types. Grease was also applied to ease the movement of the load, particularly in the shipbuilding industry. More recently various "slippery" plastic coatings, such as polytetrafluorethylene (PTFE), have been used, as well as air and water cushions that operate like hovercraft.

There are two kinds of friction involved in moving anything. Static or stationary friction has to be overcome to start something moving and dynamic or moving friction opposes its continued movement. The coefficient of friction between two materials is defined as the ratio of the force required to move the load to the weight of the load. Static and dynamic coefficients have wide ranges. These maximum values drop between the traditional slippery slope of steel on greased steel (used for ship launching) from 0.25 and 0.17, down to 0.10 and 0.05 for steel on PTFE plastic. (They have already been halved.) The values fall to 0.01 for air-bearing systems.

1 **The idea of reducing friction** below a heavy load was known to prehistoric man. Examples include the large stones that were used in the building of Stonehenge [A] on Salisbury Plain in England, probably moved with logs underneath the stones acting as rollers [B]. As the load moved forwards, the logs would be removed from the back and brought round to the front. Large diameter logs would not fall into small ruts and long logs would reduce the ground loading. The motive force could be direct man or animal power, but levers could have been used to roll the logs. Very large loads can be moved in this way.

2 **The principles of multiple pulley blocks** and their use for gaining mechanical advantage has been known for centuries. If the blocks are threaded with a single rope, then a count of the number of lines effectively supporting the load gives the gain. In [A] a gain of six is shown. Nearly every crane system uses this technique. A disadvantage of the technique is that a great length of hauling rope must be winched in to achieve only a small movement of the load. Extremely large portal cranes [B] can lift 15,000 tonnes and are now used in shipyards throughout the world.

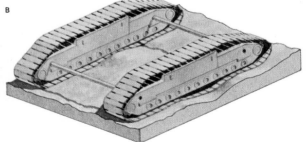

3 **The caterpillar crawler** mechanically transfers linked plates from the back to the front for the load to roll over them [B]. By increasing link width, ground loading is reduced. The motive force is also supplied through the wheels and track. This technique was chosen to transport the 6,000-tonne American Saturn moon rockets [A].

The use of friction-reducing systems can introduce further difficulties. Loads of thousands of tonnes, once they have started moving, also have to be stopped. For this reason, when coefficients of friction are low, suitable braking systems must be incorporated. After a ship is launched, strong chains and cables are needed to stop it.

Any conventional system can be used for motive power, providing that it can overcome the frictional forces that remain. When large cranes such as portal cranes [2] are used to move a load from one point to another, they are merely lifting the load to reduce the coefficient of friction between it and the ground. This means that stresses are then put into the ground at chosen points, which have previously been strengthened.

Strengthening the surface

A third difficulty concerns the amount of preparation required for the surface on which the load is to be moved. This involves a calculation of the maximum permissible ground loading. For many materials this quantity is known – for example, compacted

gravel will withstand a ground loading of 33 tonnes per square metre and wet sand 5.5 tonnes per square metre. At greater loadings than these the gravel or sand "collapses" and the load sinks in. The loading is a pressure – a weight on a given area – and for a given load increasing the area of contact obviously lowers the pressure. This is why snow-shoes support a man on loose snow whereas in ordinary boots he would sink.

In most heavy load-moving techniques there must be a way of spreading point loads. But even this is not a complete answer to ground-loading limitations. A hovercraft or air-cushion system that can carry loads over water has been known slowly to bury itself in dry sand as the sand is blown out from underneath the load.

It rapidly becomes clear that there is no universal way of moving a variety of loads over various surfaces for different applications. The answers to the problems illustrated here are merely a few modern solutions – a portal crane to lift a ship structure, caterpillar crawlers to move the Saturn rocket and air-cushion units to move the stadium stands.

KEY

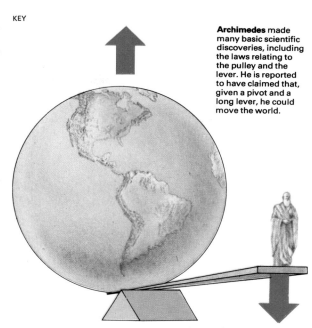

Archimedes made many basic scientific discoveries, including the laws relating to the pulley and the lever. He is reported to have claimed that, given a pivot and a long lever, he could move the world.

4 The air-cushion technique uses high air pressure underneath a load to lift the load slightly so that air can escape [A]. This continually escaping air acts as a bearing for the load and very low coefficients of friction can be obtained. The loading is spread over the area of the cushion. The Ohau Stadium in Hawaii was designed so that the stands could be moved from a rectangular shape [B] for football to a diamond shape [C] for baseball. The air-cushion system was chosen because it reduces friction. Problems on the smooth surface have arisen because of wind loadings and slopes that could cause the loads to run away. Movable friction grippers are used for braking.

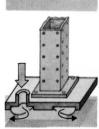

5 Hydraulic jacks can move extremely large loads, such as this 4,000-tonne ship section [B], if they have reaction points. Hydranautic gripper jacks provide these by using hydraulic forces to give grip as well as push [A]. The sequence shows the grippers being pulled up behind the load and repeatedly locked into position.

6 The walking beam uses two "footprints" which take the load alternately [A]. The area of the footprint can be designed to suit any ground loading. By choosing appropriate geometry, the walking beam can be made to move in any direction and also to rotate. Four walking beams carry and rotate this oil platform module [B].

101

Electronic devices

Electronics is the science that deals with electric currents moving in components such as valves, transistors, diodes, cathode ray tubes and many others. When various components are assembled to make circuits to perform specific tasks the resulting apparatus is an electronic device.

Our world is increasingly dependent upon electronic devices of all kinds – in industry, commerce and the home. The devices themselves are legion, yet the types of components from which they are constructed are relatively few. Today, electronic circuit designers prefer to use solid-state components such as integrated circuits and transistors because they are smaller, cheaper and more reliable than valves (known also as vacuum tubes). High-power applications, for which there are no solid-state equivalents, are exceptions. A transistor is a separate electronic component which, with other transistors, capacitors and resistors, can be wired to form a circuit. In an integrated circuit all such components are formed in a single "chip" of semiconductor material only a few millimetres across.

From the countless applications of elec-

tronics the eight devices described here demonstrate the possibilities of electronic devices, which are continually being invented throughout the industrialized world.

Electronics and light

Optoelectronics is the name given to the combination of optical and electronic techniques and hardware. A typical application is a method of counting on mass-production lines. In this, a beam of light is repeatedly interrupted by the products moving along a conveyor belt. The light pulses so produced are focused on to a photoelectric cell, to be counted electronically.

Another labour-saving application of optoelectronics is the remote control of a television set [1]. Basically the system is a torch-like hand-held device. Merely by switching on, directing the light beam at specially fitted "light-control" positions on the TV set and turning the rotary control, it is possible to adjust the volume, contrast and channel received. With such a device a viewer can control his television set without moving from his chair.

About 20 per cent of modern cameras now feature optoelectronic circuits, mostly to control shutter speed [3]. These techniques enable even the most inexperienced photographer to be sure of getting the exposure time right, whatever the light conditions. Some cameras now incorporate automatic systems that prevent the photographer from making an exposure if there is not enough light. Usually, the light sensor is a cadmium sulphide (CdS) photoconductive cell. Its spectral response (the range of light to which it is sensitive) matches that of the human eye and modern photographic emulsions, and so this cell is ideal for automatic exposure and shutter speed control in cameras.

Electronic digital clocks

Unlike mechanical digital clocks, the all-electronic variety is silent. It is designed round several integrated circuits and uses the mains alternating electricity supply to get the necessary pulses of current. For greater accuracy a crystal-controlled oscillator can be used. The pulses are eventually used to trigger a series of special cold-cathode gas-

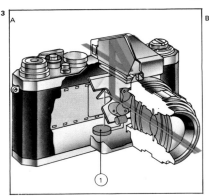

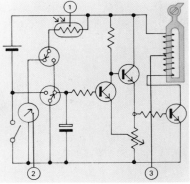

1 In an optical remote controller, a multivator [1] produces pulses [2] using a rotary control [3]. The pulses are amplified [4] and modulate light to produce a set of saw-tooth pulses [5]. These are received by a photo-transistor [6] in the TV, amplified [7] and used to work a trigger to re-form the pulse shapes [8]. This signal can then be used to control the channel selection or the volume of the television receiver.

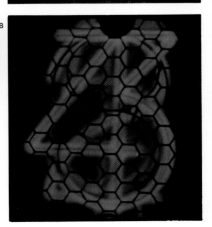

2 A metal locator works by subtracting two supersonic frequencies to produce an audio frequency in headphones. One frequency is produced by an oscillator including the search coil and a capacitor. The other is produced by internal circuitry. When the search coil is not near a metal object a note is heard. As soon as metal is located, the note changes because the coil's inductance has changed.

3 An automatic camera shutter makes the correct exposure for a perfect photograph. The light path diagram [A] shows how some of the incident light is reflected on to a photo-sensitive cell [1]. This is part of the electronic shutter-control circuit [B]. Initially, the cell output powers an exposure meter [2], but is automatically switched into the timing circuit immediately before the exposure. The shutter [C] has an electromagnetic release mechanism [3].

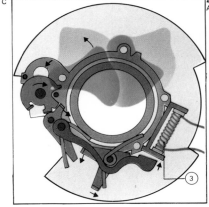

4 A modern digital counting device [A] indicates numerals using several gas-filled valves, known as "Nixie" tubes [B]. These have ten cathodes, shaped as the numbers 0 to 9, one behind the other. Each cathode is a fine wire

and the only one visible is the one that glows under the control of the counter circuitry. The glow itself is caused by an electrical discharge in the gas, when there is voltage between the cathode and a common anode.

filled valves called "Nixie" tubes. Each tube has a series of cathodes shaped to form the numbers 0 to 9 and an anode. Circuitry within the clock causes the appropriate numbers to glow, displaying the time in digital form on a 12- or 24-hour basis. The division between hours and minutes (and between minutes and seconds, if seconds are included) can be either a neon lamp or simply shown by appropriate spacing of the tubes into groups. Similar types of displays are used on electronic counters, such as those used on geiger counters and eletrostatic photocopying machines [4].

The latest method of time indication uses liquid crystal displays. In these, a liquid crystal film is sandwiched between two parallel glass plates and there is a light source in front of or behind the display. When an electric field is applied to the plates the liquid becomes milky and different numbers can be formed. This type of display is used in some pocket calculators and digital watches.

There are two ways of finding old coins, or something more valuable, buried in the ground. The traditional method is to dig. But the modern way is to use a metal locator [2]. It gives a clear audible signal when a metal object down to the size and weight of a coin is hidden just below the surface. Larger objects can be located at greater depths.

Detecting thieves

Robbery is becoming an increasing menace to modern industry and commerce, so the sophistication of anti-theft devices must keep pace. Electronics now play a dominant role in this field, with experts continuously devising new ways of preventing or detecting criminal acts. A jeweller's shop [5], for example, may have ten or more electronic devices to thwart thieves. In addition to round-the-clock visual observation using closed-circuit television cameras, other anti-theft devices detect changes in air pressure, capacitance or vibration. An output from any one of these can sound an alarm at a nearby police station or at a private security organization. The radio signals to police cars – even transmitted images of signatures, fingerprints and pictures of suspects – all make use of modern electronic devices.

Audio units are a fast-growing part of the domestic electronics industry. This stereo music player incorporates a radio tuner, a cassette recorder and a record turntable. To make it into a complete stereo system it is necessary to connect only a pair of loudspeaker enclosures. When dealing with a quadraphonic system, four speakers would be needed at equal distance from the listener. When an audio system is capable of reproducing sound very close to the original programme material, then the term high fidelity, (or "hi-fi") is used to describe it, rather than plain "audio".

5 Electronic anti-theft devices operate by electrodes [1] attached to glass, detecting capacitance changes as the glass shatters. Changes in surrounding light are detected by a phototransistor [2]. A safe and a wire beneath it form a capacitor [3] which is altered by intruders, activating the alarm. Contact devices [4] make or break circuits. A TV camera [5] is linked to a security display. An electromagnetic detector [6] senses vibrations. A fan maintains low pressure; as a door opens a pressure-sensitive diaphragm [7] detects pressure increase. An ultra-violet or infra-red beam is reflected by mirrors on to a photoelectric cell [8]; the current generated changes when the beam is interrupted. The alarm is relayed to outside security guards [9], who investigate. An auto-dialler [10] alerts a police station. The message relayed by radio [11] is received by a patrol car [12].

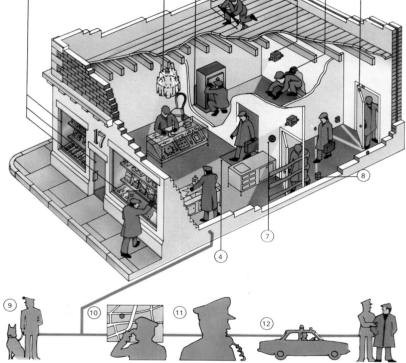

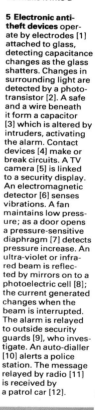

6 A modern closed-circuit TV camera makes good use of miniaturized electronic components and circuits. Its zoom lens [A] is worked electrically by small motors. It uses integrated circuits and other components mounted on printed circuit boards, which are located along the sides of the camera case for easy access and maintenance. As a result, the whole camera [B] is light, compact and portable.

7 An intercom system has an interesting feature – the use of a single unit as both loudspeaker and microphone. A loudspeaker in shape and construction is designed to serve as a microphone as well. In the home, intercoms are often used as baby alarms while the parents, or babysitter, watch television. In large houses, or even small ones where there are several levels, an intercom system can solve a number of communication problems.

Automatic control

In 1788 James Watt (1736–1819) applied his fly-ball governor to maintain the speed of a steam engine – it was probably the first mechanism specifically designed to act as a controller [Key]. Eighty years later James Clerk-Maxwell (1831–79) supplied the mathematical theory of the governor and laid the foundation for the modern science of automatic control. Today machines from household appliances to supersonic airliners are controlled automatically. The ultimate development would be a humanoid robot – an electronic and mechanical copy of a human being. But so complex are the workings of the human brain, and so subtle the ways in which it and various other organs control the living body, that even a primitive mechanical man is still only a dream.

Open and closed loops
There are the two basic categories of automatic control systems, the open and closed loops[1]. An open-loop system is one that does not involve any feedback. Feedback is the routing of information from the output end of a device back to the chief

operating part to control its workings. There are dozens of domestic appliances with open-loop controls. These include automatic dishwashers, record changers and washing machines. They follow a pre-set "program" and perform a series of operations in an ordered sequence within a limited range.

In closed-loop automatic control systems there is negative feedback to bring conditions back to a stable state. An electric thermostat, for example, can detect the temperature produced by an electric heating appliance and, when it reaches a pre-set upper value, cut off the current. Conversely, positive feedback leads to instability or oscillation, as when a pop group's microphones pick up the output from the loudspeakers and produce a screaming sound.

The human body has some remarkable examples of biological feedback systems. To maintain conditions for the various life processes under the most favourable conditions the body follows the principle known as homeostasis. This principle ensures that the status quo is maintained in the organism – for example, the temperature of the human body

[5] is maintained within limits, despite variations in the local environment.

It was not until the twentieth century that feedback systems could be applied over the whole of industry. With the invention of the triode and other valves it became possible to amplify the tiny electronic signals provided by various types of transducers (devices that convert a physical property, such as temperature, into an electric current).

Transducers and servomechanisms
Engineers are now able to design electronic circuits capable of detecting extremely small changes in the signals from transducers in domestic, industrial and medical equipment. These include changes in light intensity (automatic cameras), temperature (refrigerators) [2], liquid flow (kidney machines), gas pressure (heart-lung machines), thickness (steel and paper mills) and colour (paint production). By amplifying and processing the error signals provided by transducers, and feeding back an amplified signal to earlier stages in the process, automatic control of the system is achieved.

CONNECTIONS

See also
94 The factory and assembly line
108 What computers can do
152 How an aeroplane works
156 Space vehicles

1 **With open-loop control** [A] there is no feedback from the system. If, in this hot water system, the valve is set to provide a flow of water at a particular temperature, but incoming water temperature or gas pressure varies later, then output temperature will vary. In a closed-loop system [B] a link between the output temperature and the cold supply is established, providing feedback control and giving water at a steady temperature.

2 **A refrigerant gas** is compressed [1] and changed to liquid [2]. The valve [3] maintains pressure behind it but as soon as the gas passes the valve, pressure drops. The refrigerant is well above its boiling point at the new pressure and so

some vaporizes, cooling the interior to below freezing and eventually causing the thermostat [4] to switch off the compressor. Rising warm air later causes the thermostat to switch the compressor on again, so cooling the refrigerator.

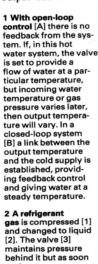

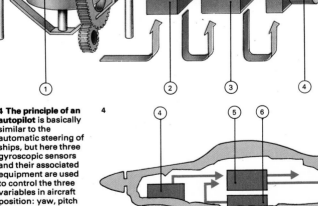

3 **A servo** [2] coupled to a ship's gyro [1], can produce error signals when the vessel deviates from a preset bearing. These signals are passed to a servo amplifier [3] which activates a servo motor [4] driving a rotary/linear converter [5]. This returns the rudder [6] to correct the deviation. As it turns electromechanical feedback [7] returns the gyro and its servo to its original position thereby reducing the signal to zero.

4 **The principle of an autopilot** is basically similar to the automatic steering of ships, but here three gyroscopic sensors and their associated equipment are used to control the three variables in aircraft position: yaw, pitch and bank. Due to additional complexity of the overall system an airborne computer is used to activate servomotors, which bring about the correction. A radio or radar link to the computer allows control from the ground for automatic landings.

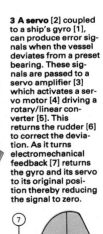

1 Ailerons control "bank"
2 Elevators control "pitch"
3 Rudder controls "yaw"
4 Gyroscopic altitude sensors
5 Computer
6 Radio or radar receiver
7 Servo motor for rudder

A servomechanism, like all feedback systems, is concerned with automatic control. But the term is usually reserved to describe equipment in which control is maintained over the relative alignment of two remote shafts. Typical examples are radar-controlled guns or missile launchers, automatic steering of ships [3], autopilots [4] and numerically controlled machine tools.

By using servos it is possible to detect small changes in the rotation of shafts. Like other types of electrical transducers servos provide error signals and these signals can be amplified and fed back to a servomotor to control shaft alignment. Much power amplification is needed to convert a tiny error signal into a voltage sufficient to drive a correcting device. The radar signal reflected from a target might be measured in microwatts, whereas the power needed to move a gun turret can be measured in kilowatts.

Cybernetics and robots

Unlike other scientific subjects, such as physics or zoology, the science of cybernetics has no well-defined area of activity – it is concerned with control and communication generally. Cybernetics includes in its scope both engineering and biological feedback systems and is derived from various aspects of mathematics, physiology, electrical and computer engineering and psychology.

One practical application of cybernetics enables disabled people to control their own movements. One of the best-known workers in this new and socially useful area of activity is the British scientist Meredith Thring (1915–), who has designed and built many prototype machines. These include a machine to carry a seated adult up a flight of stairs and a mobile seat for paralytics [6].

Robots have fascinated people since the dawn of engineering. But the full possibilities of robot design have had to wait for the comparatively recent introduction of tiny integrated circuits – like other machines, notably the computers, the efficiency of the robot is only as good as its human designer or programmer. Now many more machines and domestic appliances can be made to behave much more like robots and yet retain their familiar shapes and dimensions [7].

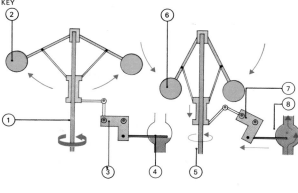

KEY

Designed by James Watt in 1788, this original fly-ball governor controlled the speed of a steam engine. The shaft [1] is driven by the steam engine, causing the revolving weights [2] to move outwards under the action of centrifugal force.

As they move the control linkage [3] gradually closes the steam valve [4] between the boiler and the engine until a point of equilibrium is reached. If the engine speeds up for any reason the weights move farther out, throttling back

the steam and slowing down the engine [5]. As the weights move downwards and inwards [6] the control linkage [7] gradually opens the valve again [8]. Overall, the result is a working engine whose speed is largely independent of its load.

5 Heat changes in the local environment arrive at the body's hypothalamus, an area in the brain. Biological feedback signals trigger the appropriate response to maintain steady body temperature. If it becomes hot, sweating begins, more blood is brought to the skin surface and muscle activity is reduced. If the body becomes too cold less blood flows to the skin surface, sweating is suppressed and shivering starts.

6 By using this mobile seat, designed and developed by Meredith Thring, paralytics can move around. The controls are devised to enable them to operate the seat despite their physical condition.

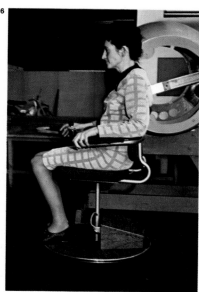

7 A robot lawn-mower is driven by a battery-powered motor [1]. The unit has a sensing coil [2] that locates a cable below the surface of the lawn, in a similar way to a metal locator. A steering motor [3]

is controlled by electronic circuitry so that the lawn-mower is always directly above the cable. By placing the cable correctly the whole lawn can be cut and the mower returned to its starting-point.

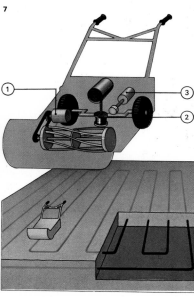

8 There are three common ways of operating automatic doors: by actuating a pressure pad on the surface in front of the door [1]; by cutting a light beam located in a wall near the door [2]; or by actuating a wall-mounted manual pressure pad like a switch [3]. There are various drive systems that power automatic doors; some are all-electric, some are electromechanical, others are pneumatic (air driven). Various door-opening actions are available: automatic slide, with or without manual swing-out side panels for use when the auto-doors are locked or turned off; single or double swing doors; and slide and swing combinations. Special provisions must be made to comply with fire regulations.

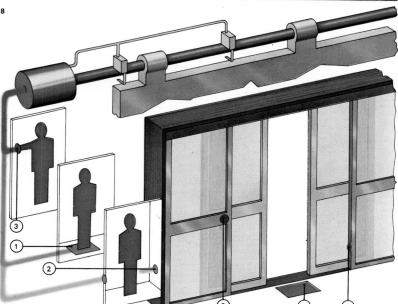

9 An electronic tortoise can even "feed" itself, when its batteries run down, by plugging itself into the socket of a battery charger and replenishing its cells. Features of this electronic animal include

a sensor for locating the battery charger light beam [1]; pins to connect with the charger socket [2]; obstacle detector [3]; driving wheels [4]; steering wheels [5]; and steering sensors (photocells).

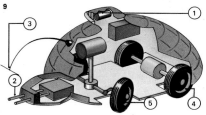

How computers work

Most people regard the computer as an electronic marvel, yet the principle on which it works is relatively simple. The heart of the computer is an arithmetic and logical unit (which adds, subtracts, multiplies, divides and compares numbers at high speed by electronic means) and memory unit, in which many thousands of numbers can be electronically stored and recalled on command.

Programming the computer
The use of the computer is based on the technique known as programming – the conversion of the problem the computer is to solve, or the tasks it is to perform, into the simple steps the computer can carry out. A programmer defines precisely what has to be done in each succeeding step, so that the computer can carry out what is, by this time, a series of simple operations. The value of the computer over the human being lies in its ability to work without error and at immense speed; it can carry out hundreds of thousands of calculations every second, storing intermediate results in its memory and recalling them instantly when required. The various instructions for the stages in the program are stored in numerical form in the computer memory for instant access as required.

To program, for example, the multiplication of 683 by 67 (an unwieldy sum for an amateur human mathematician, although much too simple to be worth feeding into a computer), the programmer would first reduce the number 67 to simpler components, probably as powers of ten, thus: $67 = (6 \times 10^1) + (7 \times 10^0)$. So 683×67 becomes $(6 \times 10^2) [(6 \times 10^1) + (7 \times 10^0)] + (8 \times 10^1) [(6 \times 10^1) + (7 \times 10^0)] + (3 \times 10^0) [(6 \times 10^1) + (7 \times 10^0)]$. Doing a multiplication problem in this way means remembering the method and answers to each step. This is why the computer needs a memory. Also future operations can be made to depend on the result of the calculation so far – that is, the computer makes decisions.

The binary system
Ordinary, everyday calculations are carried out using the decimal system, with the numbers 1 to 9 and 0. A computer can be designed to do the same but electronic engineers found that computers could be designed more simply using the binary system, which employs only two numbers – 0 and 1. This system is simple because an electric current can be switched on and off, using "off" for 0 and "on" for 1. Because the binary system uses only these two numbers it contains many more digits than its decimal equivalent [3]. This makes calculations in binary extremely unwieldy for a human being but a computer "writes" numbers and calculates so fast that the length of the numbers is of no importance.

The binary system also makes the memory much simpler to design. Most modern computers have a magnetic core memory [4]. Each core is a tiny magnetizable ferrite ring, usually about a millimetre (0.04in) in diameter, which can be magnetized by passing an electric current through wires inside it. The polarity of magnetism depends on the direction of the current, one direction standing for 0, the other for 1. A computer memory has thousands of these tiny cores and, provided groups of them have a specific "address", numbers can be stored

1 **Charles Babbage** (1791–1871), an English mathematician, was the first man to realize that if a machine could be built to remember numbers and recognize simple arithmetic situations it could then be programmed to carry out complex calculations automatically. His ingenious "Difference Engine", part of which is shown here, was able to compute complicated tables, although it lacked a memory. A later invention of his, resoundingly named the "Analytic Engine", would have been a true computer but was never completed because the mechanical engineers of Babbage's time were unable to meet the specifications of his design.

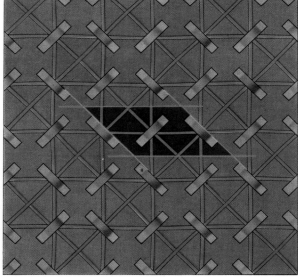

2 **The electronic desk calculator** is built on the lines of the modern computer. The numerical and function keys comprise its input device, the display and print-out facility giving its output. Like the computer it contains an arithmetic unit and an electronic memory but the latter is fairly small, thus limiting the scope of the instrument to simple calculations. Calculator memories vary widely.

4A Write | Read | B | Write

3 **The binary system** is the simple secret behind the computer. The paper tape [right] shows one method of recording binary numbers (the centre row of holes is for tape transport). Binary works on the same principle as the decimal system, but in ascending orders of 2 rather than 10. Decimal units are counted up to 9 and then carried to the "tens" columns where they are recorded as 1, but binary units are counted only up to 1. To add another 1 (total, 2) the 1 is carried to the "twos" column, indicating a 2. Thus 10 in binary means one "2" plus no "1s" – total, 2. The third column in binary (signifying "hundreds" in decimal calculation) is the "fours" column, thus $101 = 4 + 0 + 1 = 5$.

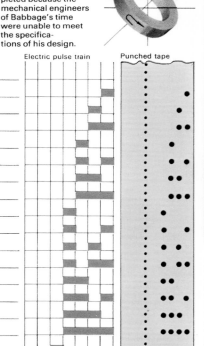

3	Decimal	Binary	Electric pulse train	Punched tape
	0	0		
	1	1		
	2	10		
	3	11		
	4	100		
	5	101		
	6	110		
	7	111		
	8	1000		
	9	1001		
	10	1010		
	11	1011		
	12	1100		
	13	1101		
	14	1110		
	15	1111		
	16	10000		

4 **A computer memory** consists of ferrite cores threaded on address and read-out wires. Each core can be "written on" by passing simultaneous currents along the vertical and horizontal wires that intersect it [A]. This magnetizes the core in one direction indicating a "1". Changing current direction reverses the polarity, thus indicating an "0". The third wire is the "read" wire. In fact the unit cell [B] of such a memory contains two cores (one such pair is in the dark area) which differ only as do mirror-images. A typical array of cores might number 10,000, all identical, and all threaded on a 100 × 100 matrix, each core being defined by its co-ordinates. A computer memory may contain many arrays.

as collections of individual digits and recalled when required.

The magnetic core memory is widely used because it is extremely fast and reliable, but it is only one of several kinds. A solid-state memory is even faster. The computer's own memory is often augmented by a slower backing store, of which magnetic tape and magnetic disc systems are the most widely used. These make possible the storage of a virtually unlimited quantity of information.

The processing unit

Because a computer cannot read more than one number at a time the basis of computer operation is an electronic "clock" producing an endless series of identical pulses – up to millions every second. In one type of computer, these pulses are switched on or off, one after another, to indicate the successive components of a binary number. So the binary number 100110 (which in decimal is expressed as 38) is read as "no pulse, pulse, pulse, no pulse, no pulse, pulse", the unit digit being read first, as in all arithmetic calculations. Each number in a computer cir-

cuit is thus a set of pulses, and to avoid confusion each set must have the same total number. If the full set, in this case, were 16 pulses, the six-digit example would be completed by ten more "no pulses". As each number is defined, it is stored at once in the memory cores where it stays until the next number (or series of numbers) is also stored and ready for use [Key].

Once these numbers have been read in (automatically, controlled by the program), the computer can then be made to juggle with them in accordance with a program already stored in its memory. If two numbers in store are to be added a code number in the calculation program (previously stored in another part of the memory) switches the computer circuits so that the two numbers in store are fed together, pulse by pulse, into an adder circuit. The adder combines the incoming pulses and produces an output train of pulses that defines the added number.

Despite their great flexibility of operation and application, all computers depend for their efficiency on the ability of the programmer to reduce problems to simple terms.

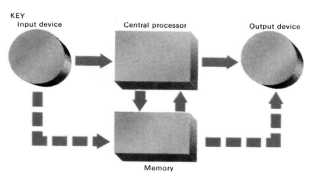

KEY
Input device   Central processor   Output device

Memory

The heart of any computer installation is a processing unit (essentially a high-speed calculator), which operates in conjunction with a memory unit. Data is fed into the computer by means of an input device which converts information and instructions into trains of electronic pulses rep-

resenting numbers. This data is often fed in by means of punched cards or paper tape on which the position of the holes forms a code that the computer can "read". Operators prepare the cards on punching machines, paper tape usually being punched on a special form of typewriter. When

more information is to be stored than can be accommodated in the computer's own memory this can be transferred to magnetic tape or disc. An output device converts the electronic pulses back into information that may be printed on paper, displayed on a screen or communicated via other media.

5 A modern computer installation consists of a number of interconnected machines. Data can be fed into the system by various input devices such as a keyboard (resembling a typewriter), a punched-card

reader, a punched-tape reader and a magnetic-tape input. An operator can even use a light pen to "draw" designs on the face of a cathode ray tube. All the data passes to the central proces-

sor and is stored in its memory or, if necessary, recorded in various back-up stores such as magnetic tape stores or ones using magnetic discs or magnetic drums. The operation of the computer is

controlled by a program, which also has to be fed in and stored. The central processor consults the program as necessary to carry out its various tasks. Information produced by the computer is

presented on an output device, such as a card punch, paper tape punch or a visual display. Some displays have a cathode ray screen and some, called alpha-numeric displays, present infor-

mation electronically as letters or numbers "written" on the screen. Printed information, known as "hard copy", is provided by some form of print-out device. The most common is a line printer, which

generates a whole line of typed copy at a time. Output information can also be provided on magnetic tape, possibly for feeding into another computer or for controlling a typesetting machine.

5

Magnetic tape stores   Tape control unit   Central processor   Control console   Disc control unit   Magnetic disc stores

Card punch   Card reader   Visual display unit   Line printer

What computers can do

The essence of a computer is an arithmetic unit that performs calculations and an electronic "memory" that stores numbers. The computer solves every problem as a series of extremely simple steps, using the memory to store both the details of the program to be followed and the results of each intermediate step while the program is being carried out. In practical terms there are two kinds of problems: those in which the power of the computer to carry out complex calculations at high speed is paramount and those in which its capacity to store immense quantities of information is more important. To solve either kind there must be a means of communication between man and machine.

Information is commonly fed into a computer using a device that resembles a typewriter, which produces a typed record of the fed-in data as a check. Its main purpose is to produce electronic signals – a binary number for each letter and numeral – that the computer can "read". Alternatively it produces either a punched paper tape or a series of punched cards. These are subsequently fed into a "reader" that generates the required

electronic binary numbers. By means of the teletypewriter the operator can feed data and instructions into the computer using a predetermined "code" that the computer is programmed to understand.

There are other means of communicating with a computer. These include the use of magnetic printed characters [5]; an optical "reader" that scans ordinary printed type, passing the information to the computer which is programmed to "recognize" characters and numerals and convert them into its own language of binary numbers; and a "light pen", with which the operator can trace over a diagram on the screen of a cathode ray tube, with the computer programmed to process the diagram as required [1].

Output devices in common use

There are three principal output devices in common use. The line printer is a high-speed typewriter operated directly by the computer. Instead of typing one character at a time it prints a whole line simultaneously. A modern line printer can print 10–15 lines a second, each with 100 or more characters.

The plotter is an automatic computer-controlled drawing instrument. An ink-filled pen moves across the paper under the control of an appropriate computer program, to record in graphical form the output of the calculations [6].

The visual display is a form of cathode ray tube (similar to that in a television set) on whose screen the computer may produce either "written" messages or drawings. This type of output is used where the operator needs continuously updated information, but where a permanent record is not required. The display has numerous special applications such as in airport control, engineering design [1] and in teaching machines [7].

The computer's ability to calculate at high speed and store enormous quantities of information concisely and precisely is being applied to ever-widening fields of activity. In weather forecasting the main problem is to relate continually changing readings of temperature, humidity and barometric pressure from numerous observation points and to build up a chart. If carried out manually, the work can be completed only with

1 A computer can be programmed to simulate perspective. The girl traces the outline of an object with a light pen. The computer retains the spatial relationships of the object's shape in its memory.

2 Various forms of accounting are the most common uses for computers. Banking, stock market transactions, wages control and tax deductions can all be done rapidly and accurately by computer.

3 Moon at day 0   Moon at day 1   Moon at day 2   Moon at day 3   Moon at day 4

Uncorrected course

Corrected course

Day 3
Day 2
Day 1

Computer calculates rocket firing sequence to correct course

3 One of the problems of space navigation is the need to make extremely complex calculations rapidly at short intervals. When a rocket travels to the Moon its precise course depends on the Earth's spin and orbit at launch, the

rocket's changing speed and the Moon's orbit. The rocket has to be taken out of Earth orbit and into Moon orbit at precise moments and small errors must be corrected instantly. Only the computer can calculate fast enough.

4 When an airline booking is made from a computer terminal it is not only the seat availability that is checked for each section of a long-distance flight. The computer can also be program-

med to offer alternative flights that have seat vacancies. It first makes a provisional booking, then calculates the cost, asks for payment, makes the appropriate entry in the airline's accounts and

prints the ticket for the flight with all relevant details. As the flow chart shows there are a number of alternative paths in the program depending on the fulfilment of certain conditions.

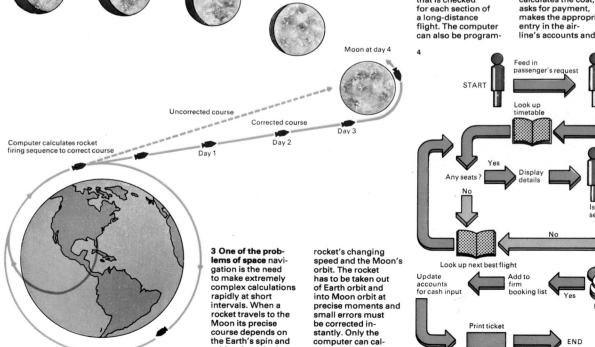

START
Feed in passenger's request
Look up timetable
Any seats? — Yes → Display details — Yes → Add name to provisional booking list
No
Is this seat OK?
Look up next best flight
No
Ask for payment
Update accounts for cash input ← Add to firm booking list ← Yes — Paid? — No → Cancel provisional booking
Print ticket
END

difficulty before new factors have developed to affect the forecast. A computer, suitably programmed, can accept new data as fast as it is received and can produce an up-to-date weather forecast chart in a matter of minutes.

Computers used in engineering

The mathematics involved in the design of a modern long-span steel bridge is so complicated that it takes a large team of engineers working with slide rules and tables several months to calculate the stresses and quantities necessary in the evaluation of a design. Today a computer, fed with data by two or three engineers, can complete the calculations in minutes – not only saving time but releasing experienced men for other work. A consulting engineer's computer can carry out the complex mathematics of reinforced and pre-stressed concrete design; it can also be programmed to specify the steel required for reinforcement and to print out the details for immediate use by the construction engineer.

In anti-missile defence, radar detects a missile a few minutes before its arrival. To destroy it a rocket must be fired, before the missile is visible, so that the rocket will intercept the missile. To calculate the trajectory of the missile from continuous radar data, and fire the anti-missile rocket with the correct direction and elevation within a few seconds, is possible only by using the computer. The computer's ability to carry out complicated mathematics at high speed is used in much the same way in space navigation [3].

Applications of "memory" computers

"Memory" computers also have numerous applications. These include the maintenance of details of licensed vehicles and drivers; world-wide airline booking [4]; recording of personal bank accounts on magnetic tape, including automatic print-out of monthly statements and provision of cash-drawing facilities [Key]; accounts of gas and electricity users and telephone subscribers, including automatic preparation of quarterly invoices; and in factory stores accounting, where the smooth running of a production line depends upon the availability of raw materials and every component part in the quantities required at the moment needed.

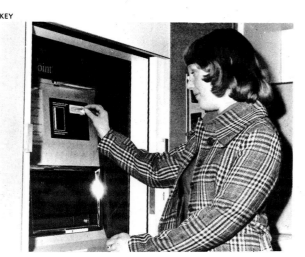

The bank cashpoint, widely used to save customers' and bank clerks' time, typifies the use of computers in everyday life. The customer inserts her personal card in the machine then presses buttons to "read in" her cashpoint number and the sum of money she wishes to withdraw. A central computer checks her balance before the machine dispenses bank-notes.

5 The oddly shaped numerals used on cheques are printed in magnetic ink and can be read by a computer reader as well as by humans. The numerals seen here are designed seven units wide and nine units high. The figures in line A below each magnetic numeral show the number of squares filled with magnetic ink in each vertical column. The electronic reader produces a pulse where four or more squares are filled and no pulse where three or less are occupied. Each magnetic numeral thus generates the binary numbers shown in line B. This is the computer's own language.

| A | 7 2 2 2 2 2 7 | 0 0 0 5 9 4 4 | 0 0 0 6 3 3 6 | 0 0 3 3 9 9 4 | 0 7 7 1 1 4 4 |
| B | 1 0 0 0 0 0 1 | 0 0 0 1 1 1 1 | 0 0 0 1 0 0 1 | 0 0 0 0 0 1 1 | 0 1 1 0 0 1 1 |

| 0 0 6 3 3 3 6 | 0 9 3 3 4 2 4 | 0 0 3 1 6 2 4 | 4 9 3 3 3 9 4 | 0 4 2 2 2 6 9 |
| 0 0 1 0 0 0 1 | 0 1 0 0 1 0 1 | 0 0 0 0 1 0 1 | 1 1 0 0 0 1 1 | 0 1 0 0 0 1 1 |

6 The "contours" of equal light intensity in a lecture hall are seen here. They were drawn by an automatic plotter controlled by a computer that was programmed to calculate these contours from data defining the position and power of the lamps that the architect proposed to install. The "map" shows clearly where more light is needed and enables the engineer to correct the lighting plan accordingly.

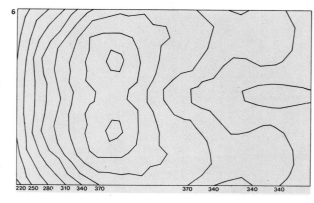

220 250 280 310 340 370   370   340   340   340

7 Teaching machines are designed to involve the student, who must respond to each step in the lesson before it proceeds. In early days machines told the student only whether his answer to a question was right or wrong. Later machines worked on programs that gave the student an explanation when a wrong answer was given, before proceeding with the lesson. The most recent machines go further. They assess the student's performance and adapt the lesson to his potential, relying on a mass of information and an ability to follow a complex branching program.

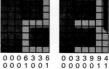

8 PAGE   1   AIMBR2/TXT
A
1.   99999 021   2880   20 15   0
2.   99999 01 315:57:13   44123   76 AIMBR2 1AIMBR2
3.   152
4.   00   12]
5.   11   13] The essence of a computer is an arithmetic
6.   12   21] unit that performs calculations and an electronic
7.   13   221 "memory" that stores numbers. The
8.   21   231 computer solves every problem as a series of
9.   22   331 extremely simple steps, using the memory to
10.   23   321 store both the details of the program to be
11.   31   331 followed and the results of each intermediate
12.   32   411 step while the program is being carried out.
13.   33   421 In practical terms there are two kinds of
14.   41   431 problems: those in which the power of the
15.   42   511 computer to carry out complex calculations
16.   43   521 at high speed is paramount and those in
17.   51   531 which its capacity to store immense quantities
18.   52   611 of information is more important. To
19.   53   621 solve either kind there must be a means of
20.   61   631 communication between man and machine. Do
21.   62   71 Information is commonly fed into a computer
22.   63   721 using a device that resembles a typewriter,
23.   71   731 which produces a typed record of the
24.   72   811 fed-in data as a check. Its main purpose is to
25.   73   821 produce electronic signals - a binary number
26.   81   831 for each letter and numeral - that the computer
27.   82   911 can "read". Alternatively it produces
28.   83   921 either a punched paper tape or a series of
29.   91   931 punched cards. These are subsequently fed
30.   92 1011 into a "reader" that generates the required
31.   92 1021 electronic binary numbers. By means of the
32.   101 1031 teletypewriter the operator can feed data and
33.   102 1111 instructions into the computer using a
34.   103 1121 predetermined "code" that the computer is
35.   111 1131 programmed to understand.
36.   112 1211   There are other means of communicating
37.   113 1221 with a computer. These include the use of
38.   121 1231 magnetic printed characters [5]; an optical
39.   122 1311 "reader" that scans ordinary printed type.
40.   123 1321 passing the information to the computer
41.   131 1331 which is programmed to "recognize" characters
42.   132 1411 and numerals and convert them into its
43.   133 1421 own language of binary numbers; and a "light
44.   141 1431 pen", with which the operator can trace over
45.   142 1511 a diagram on the screen of a cathode ray tube,
46.   143 1521 with the computer programmed to process
47.   151 1531 the diagram as required [1].
48.   152999991 ***

The essence of a computer is an arithmetic unit that performs calculations and an electronic "memory" that stores numbers. The computer solves every problem as a series of extremely simple steps, using the memory to store both the details of the program to be followed and the results of each intermediate step while the program is being carried out. In practical terms there are two kinds of problems: those in which the power of the computer to carry out complex calculations at high speed is paramount and those in which its capacity to store immense quantities of information is more important. To solve either kind there must be a means of communication between man and machine.

Information is commonly fed into a computer using a device that resembles a typewriter, which produces a typed record of the fed-in data as a check. Its main purpose is to produce electronic signals – a binary number for each letter and numeral – that the computer can "read". Alternatively it produces either a punched paper tape or a series of punched cards. These are subsequently fed into a "reader" that generates the required electronic binary numbers. By means of the teletypewriter the operator can feed data and instructions into the computer using a predetermined "code" that the computer is programmed to understand.

There are other means of communicating with a computer. These include the use of magnetic printed characters [5]; an optical "reader" that scans ordinary printed type, passing the information to the computer which is programmed to "recognize" characters and numerals and convert them into its own language of binary numbers; and a "light pen", with which the operator can trace over a diagram on the screen of a cathode ray tube, with the computer programmed to process the diagram as required [1].

8 When a book is to be printed the type is first set in a predetermined format that defines the width and length of each column. While the modern typesetting machine aids the operator in maintaining column width and inserts space automatically between words, it is the operator who must ensure that the rules of style are maintained. These rules lay down the system of paragraphing, punctuation, capitalization, the use of italic and bold typefaces, and hyphenation. Computer typesetting eliminates most of the intellectual effort required. The operator has only to type the copy accurately, the machine organizing the setting according to its program. It produces a coded proof, which is used by the proof-reader for checking errors in the usual way, corrections being fed into the computer before the type is set. The example shows the coded proof [A] and the finished print [B].

History of transport

Transport is arguably man's oldest technology, predating both house building and agriculture. There is no reason to doubt that prehistoric men and women carried burdens in their hands, on their backs and on their heads, and the use of specially constructed litters, sleds and rollers for transport may also be much older than the generally accepted date of about 10,000 years ago.

Water and rail transport
The oldest artefacts that could be described as transport vehicles were crude boats of prehistoric times; men progressed to the dugout canoe about 20,000 years ago. At the same time members of early civilizations in many parts of the world must have developed ways of building rafts of local materials.

Rafts are still made in more or less the same way, using bundles of reeds, logs, grasses and other material, sometimes with added flotation from skin bags or sea-kelp bladders. From the raft, many peoples progressed to making boats with built-up hulls, sometimes using a framework covered by skins or bark (eucalyptus bark was the

favourite in the countries of Australasia).

About 5,000 years ago canals were being constructed, at first to link close river channels and eventually to carry people and supplies over considerable distances. In much more recent times canals were dug throughout Britain and much of western Europe as primary transport routes, largely because of the inadequacy of roads. From 1770 to 1840 canals were almost unchallenged in these areas as the means for the slow but cheap transport of a rapidly growing volume of manufactured goods.

One great advantage of the canal was its ability to move very heavy loads for a minimal expenditure of energy. Compared with the canal barge, no land vehicle could move such loads until the coming of railways.

The first crude railways were used for commercial rather than private or passenger transport. Metal rails had been used in local works such as mines [9] since medieval times, but by about 1800 their use began to spread. By 1830 it had been realized that a uniform kind of track should be adopted and the two-rail system was introduced with smooth steel

rails at agreed spacings and with flanges on the inner edges of the wheels.

Like a canal, a railway could carry heavy loads with minimal resistance to motion. Competition between the two forms of transport in many regions was intense, railways usually enjoying an advantage because of the ease of extension (branches and spurs could be built to serve almost every town) and because they were unrivalled in hilly areas. At the same time progress was made in the design and construction of roads, which in Europe until 1800 had been inferior to the roads built by the Romans.

Transport by road
In theory roads have always had an advantage over railways because of greater flexibility. They cost less to build, which means that a national road network can serve almost every factory and house. They can carry a wide range of vehicles up and down steep gradients. On the other hand, the resistance to motion is much greater than on a railway because the resistance of wheeled vehicles is dependent on the intensity of loading where

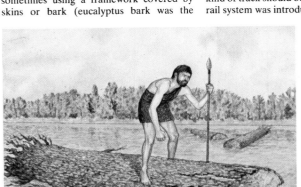

1 Tree trunks were probably man's earliest vehicles. More than 20,000 years ago he began to use wood and other materials to make rafts and the precursors of the later kayaks, coracles and canoes.

2 Egyptian ships date from c. 2500 BC. The broad spoon-shaped hull was made from acacia, a tree yielding only short, irregular timbers. The ship was steered by two oars on each side of the stern.

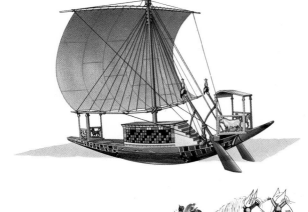

3 The travois was possibly the earliest land vehicle. Used by the Plains Indians of North America it comprised two poles joined to a man, a dog or a horse and dragged along the ground.

4 The chariot is illustrated on one of the tombs at Thebes (c. 1500 BC). Although the Egyptians never perfected the harness, they used many horse-drawn vehicles with wheels of spoked construction.

5 The first hydrogen balloon carried Jacques Charles and M. N. Robert from Paris to a field 40km (25 miles) away in December 1783 in under two hours. It looked like a modern balloon with a "boat" beneath.

6 Henri Giffard's airship could sustain an airspeed of 8km/h (5mph) on its steam-driven propeller. In 1852 Giffard navigated it from Paris to Trappes, making the first powered cross-country flight. Progress was delayed by lack of good engines.

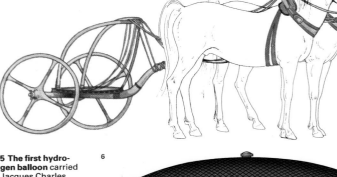

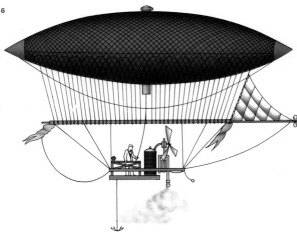

the wheels meet the fixed surface. With a railway this intensity is high and the surface smooth and regular. With a road the intensity is low and until quite recently the surface was irregular. Until late in the nineteenth century roads were marred by ruts, bumps and potholes, which not only slowed vehicles but often caused damage and even serious accidents. In many places it was accepted procedure to take coaches or goods carts to pieces and use teams of men to carry them over the worst sections of road surface.

Mechanically powered vehicles
All the earliest forms of propulsion had relied on natural forces such as wind and river currents or muscle power. By the early years of the nineteenth century mechanical power was being applied to move vehicles – at first steam engines in railway locomotives and ships. Then mechanically propelled road vehicles gained importance with the advent of the internal combustion engine after 1885. Motor cars multiplied in an amazing way, as did those convenient muscle-driven vehicles, bicycles. A key innovation with both of them

was the introduction of pneumatic tyres. The replacement of the horse and horse-drawn carriage by the rubber-tyred car forced governments to build smooth, all-weather roads both within and between cities.

Today's rail and road systems are more complex and many have automatic traffic-control systems involving computers, electronic communication and display systems, as well as methods for dealing instantly with sudden demands or emergencies.

There have also been several revolutions in marine transport. At one time merchant ships had to be equipped to fight off pirates. By 1800 many kinds of true cargo vessels were built and for 100 years they monopolized inter-continental transport.

The newest form of transport, the aeroplane [11], swiftly supplanted the ship as the vehicle for long passenger journeys, while economies of scale led to a startling growth in the size of oil tankers, bulk carriers and other cargo ships. The pressure on shipping lines to reduce journey time has also been relieved because over short distances hydrofoils [8] and hovercraft travel four times faster.

The Air Cushion Landing System (ACLS) enables aircraft to operate from any kind of surface.

This is a converted Buffalo short-field transport plane that helped to prove the concept. The result

is a vehicle able to land on concrete, ice, snow, crops, sand, water or any level part of the earth.

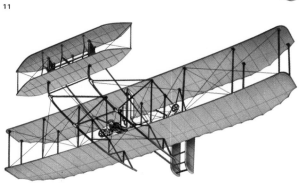

7 This river boat typifies hundreds that served US southern states in the 19th century. Paddle-driven, often with a stern wheel, they had shallow draught and yet could carry heavy loads.

8 The hydrofoil of V. Grunberg ran in 1934. Driven by tandem air propellers, it had two forward floats and a central submerged foil, which was adjusted automatically. It was 30 years before the modern foil.

9 This wooden tramway truck ran on wooden track in a German mine (*c.* 1510) and is probably typical of an early form of railway transport vehicle. By 1670 wagons had flanged iron wheels.

10 This Daimler car (1897) was built by the English Daimler Co, formed initially in 1893 to import cars from the factory in Germany. It had a two-cylinder engine and four forward and reverse speeds.

11 The Wright Flyer III was an improved version of the first aeroplane to make a successful flight. The Wright brothers learned how to control a glider, and then added a light-weight petrol engine.

12 Paul Cornu, the French inventor, made the first heavier-than-air machine to take off vertically in free flight. Tethered to the ground by a rope, his primitive helicopter rose 1.5m (5ft) on 13 November 1907, but lacked stability.

Sailing ships

No one knows when or where man first invented the sail, but it was probably one of the earliest attempts at harnessing a natural force and putting it to work. The earliest evidence of sailing craft comes from Egypt and dates back to the third millennium BC. Ancient Egyptian ships [5] had single square sails spread by two wooden spars, a yard at the top and a boom at the foot; they could sail only downwind but since the wind in the Nile valley is nearly always from the north, this allowed them to sail upstream. Going downstream did not require a sail.

Sailing against the wind

It was some time before man realized that sails could be made that would propel a ship against a wind – not directly into it, but at an angle of less than 90° – and it is only recently that the aerodynamics of such sails has been understood [1]. Another early breakthrough was the invention of the keel, a long plank running from stem to stern, from which the rest of the hull could be built up.

The square-rigged ships of the Middle Ages could not sail much closer to the wind than 90° and even the latest square-riggers cannot sail closer than about 70°. The "fore-and-aft" sail (in which the forward edge is secured to, and pivots around, a mast or stag), such as the Mediterranean lateen and the triangular Bermudan sail of modern yachts, can hold a course as close as 45° to the wind [4].

The fore-and-aft idea appears to have evolved in the Indian Ocean from the Egyptian square sail in the third century AD – leading to the lateen sail of the Arab dhows and to the lug-sail of the Chinese junk. The square sail did, however, survive until the end of commercial sailing ships because it was more efficient than the fore-and-aft sail on long voyages with following winds.

The Romans improved on the Egyptian rig by adding the *artemon* (spritsail) and a triangular topsail [6], but as in the Egyptian ships before and the Viking ships later, steering was done with an oar lashed near the stern. The Chinese, however, had known the axial rudder and the compass since about the first century AD. The former allowed better steering and was more robust than the steering oar; the latter allowed navigation out of sight of land without fear of overcast skies hiding the stars. These discoveries reached Europe only in the later years of the eleventh century and had great consequences in the development of European seafaring.

Development of ship and sail

The multi-masted ship originated in China and the idea was brought to the West at the time of Marco Polo in the thirteenth century. Before this time, vessels such as the single-masted double-ended cog had been used extensively for trade around Europe and the Mediterranean. By the end of the fifteenth century three-masted ships had become a common feature of European waters. The deep, broad carrack [7] was one such common trading vessel and the caravel was also developed at this time. Used mainly by the Portuguese, the caravel was a simpler and lighter vessel. It carried lateen sails (fore-and-aft) but sometimes the foremast was square-rigged.

The galleon [9], which appeared in the mid-sixteenth century, was a cross between the heavy carrack and the slim Venetian

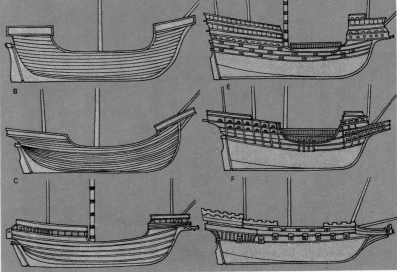

Wind direction
Lateen rig
Square rig

1 Sailing against the wind is possible because the wind acting on a sail creates a lift [L], and a drag [D]. These are equivalent to a driving force [F] and a leeway force [S].

2 "Wearing" is making a downwind turn, a method often used by square-rigged ships, such as carracks and galleons, when sailing into wind. This method loses some ground to windward.

3 "Tacking" is turning into and across the wind. This is a manoeuvre more easily done by fore-and-aft vessels than by square-riggers and loses less ground to windward than wearing.

4 Fore-and-aft rigs are able to point much closer to the wind than square-rigs when tacking in a zigzagging course to windward. After sailing the same number of miles the fore-and-aft vessel is thus much farther ahead.

5 Egyptian ships of *c.* 1300 BC had a square sail and were steered by oars. The hull shape derived from reed boats.

6 The Roman grain ships of the 2nd century AD had a foresail, the *artemon*, and a topsail above the mainsail.

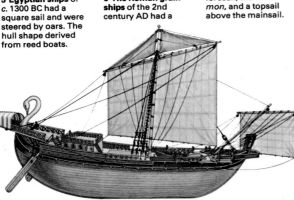

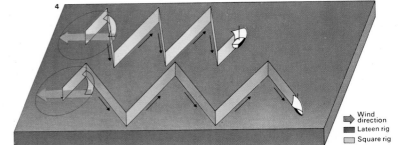

7 The shape of ships' hulls changed considerably between 1400 and 1600. The nef of 1400 [A] was double-ended, with a "pointed" stern, and had "castles" added at both ends. Multi-masted ships (inspired by Chinese junks) appeared with the carrack [B] (1450) and [C] (1465). The flat or *transom* stern appeared with the 1520 great ship [D] whose hull was also pierced for guns. Streamlining and bringing the forecastle inboard produced the galleon [E] (1545) and [F] (1587).

8 Clinker [A] and carvel [B] built hulls were characteristic of North European and Mediterranean ships respectively. From 1520 onwards all large ships adopted the carvel build with the butting of planks end to end.

galley. Galleons had slimmer hull lines [7] than the carracks, a square stern where the carracks were "double-ended" and a forecastle well inboard instead of hanging over the bow.

The transition from the carrack to the galleon was the last big technological "jump" for sailing ships. The difference between a sixteenth-century galleon and a nineteenth-century packet ship, such as the Blackwall frigate, was one of detail although the performance of the latter was far superior. The gradual evolution since the early galleons involved an increase in the size and number of sails, and the introduction of fore-and-aft staysails between the masts and jibs in front of the foremast.

From the mid-eighteenth century onwards, the Western sailing ship branched out in a variety of rigs, from two masts to six or even seven in the instance of a pure fore-and-aft schooner, the *Thomas W. Lawson*.

The last of the sailing ships

The fastest, the most beautiful and the most short-lived of the great sailing ships was the clipper [11]. She was developed in the 1820s in the USA, reached the height of her fame in the 1850s and 1860s and by the end of the century was obsolete.

The clipper was built primarily for speed. She was slim and light, with a limited cargo space, and carried an enormous area of canvas. She fulfilled a variety of needs. The Californian and Australian gold rushes (1849 and 1851), the China tea trade and, from the 1870s, the trade in wool and grain from Australia were all served by clippers. The competition between ships' companies in the race to make fast passages and high profits produced crews of the highest quality.

Such factors as steamships, the opening of the Suez Canal and the transcontinental railways made the clippers obsolete. They were replaced by larger windjammers that were built of steel and where cargo-carrying capacity and labour-saving devices were given precedence over speed. Despite their much greater tonnage (up to 5,800 tonnes) they did not require larger crews than the clippers. Even these ships were eventually killed by steamship competition.

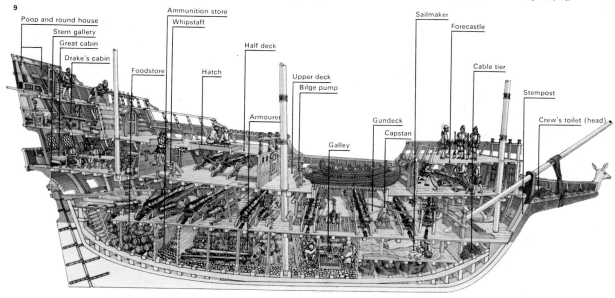

9 Drake's *Golden Hind* was a medium-sized Elizabethan galleon. Her deck was about 28m (90ft) long. The rig consisted of the square bowsprit sail, 2 square sails on the foremast and 2 on the mainmast and a lateen on the mizzen. The hull was characterized by a slim underbody, a projecting head (derived from the galley's ram), a forecastle forming an integral part of the hull and a transom stern with a stern gallery. Steering was done with a whipstaff (the wheel only appeared during the 18th century). There were 2 full-length decks. Such ships were used both for war and commerce.

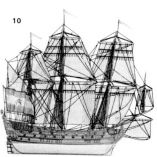

13 The barkentine (19th-20th century) has three masts with a square rig only on the foremast, thus reducing crew size and hence running costs.

10 The Dutch East Indiaman (1720) was larger than the *Golden Hind*, but of the same basic pattern. Two new sails appear: the sprit-sail topsail and the mizzen topsail. These heavily armed and extremely robust merchant ships could be used as men-of-war.

14 The brig (18th-20th century), a common coastal craft, has two masts, both of them square-rigged, and fore-and-aft staysails.

11 The clipper of the mid-19th century was a particularly efficient sailing ship. Its phenomenal speed made trading between Australia, China, USA and Britain much more economical. The name clipper came from the way the ships could "clip off" the miles.

15 The brigantine (18th-20th century) has two masts with a square rig on the foremast and mainsails and staysails on the mainmast.

12 The four-masted steel barks of the late 19th century were built of steel and they carried manufactured goods, grain and nitrates round the world. Their aftermasts had no yards. The bark rig appeared in the late 18th century and steel hulls in the 1870s.

16 The topsail schooner (18th-20th century) is a fore-and-aft rigged vessel with one or more square topsails set on the foremast.

Modern ships

Ships have changed vastly since the first iron-hulled steamer powered by a screw propeller, the *Great Britain* [1], crossed the Atlantic in 1843. Towards the end of the nineteenth century, iron gave way to steel as the standard material for hulls and sail-power declined in usefulness. At first indispensable as an auxiliary to the early crude steam engines, sail was ousted as steam machinery became more reliable. By 1918 there were very few sailing ships trading on main ocean routes.

Types of shipping

Since the introduction of steam there have always been three main divisions of non-naval shipping: passenger-carriers, cargo-carriers (merchant ships) and service craft. The largest passenger-carriers are ocean liners, particularly the North Atlantic record-breakers of the twentieth century. Competition from cheap and speedy air travel has driven big passenger liners out of business. But they survive as cruise liners, carrying holiday-makers to foreign ports instead of travellers on scheduled routes. Many famous liners have ended their careers by becoming cruise liners and now specially designed ships [11] cater for the packaged holiday market. "Short sea" passenger ships and ferries have proved far more resilient in meeting the competition from air travel. Improvements in ship design have enabled road vehicles to drive on and off a ferry [10], allowing faster turnaround and reducing labour costs.

The passenger liner was once the largest type of non-naval ship, but since the 1950s it has been surpassed by the oil tanker [6]. As world consumption of petroleum products has risen, so has demand for tankers. Individual ships have increased in size from an average of about 50,000 tonnes dead-weight in 1955 to 500,000 tonnes in 1976 (deadweight is the carrying capacity of a ship, including cargo, ballast, fuel, water, crew and passengers). The "supertanker" is relatively slow but can be operated by a small crew and it is the most economical way of shipping crude oil. The growth in size is likely to be limited only by safety factors and the risk of pollution from accidental spillage.

The cargo-carrying ship is still the main-stay of world trade, but here too there has been a striking expansion in size. In order to save on operating costs, bulk-carriers handling dry cargoes of iron ore, grain or coal have grown to deadweights of more than 100,000 tonnes. Some are equipped to unload their own cargo, but carriers of grain, salt and coal, for example, still need dockside facilities.

The most important development in cargo-handling is the container ship [5], which loads cargo in prepacked, weathertight containers by means of special dockside cranes. The main advantage is that time in port is reduced and containers are easily handled by an integrated road or rail system. Loading and unloading of diverse cargoes, with its risk of damage and theft, is eliminated. The disadvantage is that container ships can operate only on fixed routes with specially equipped terminals, and in this respect they suffer from the same limitations as the old transatlantic liners.

Smaller vessels

A few general-purpose cargo vessels still "tramp" from port to port seeking any cargo they can find. Or their owners allow the

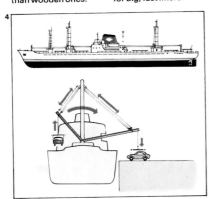

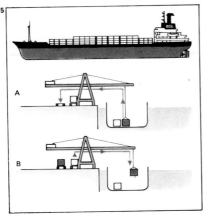

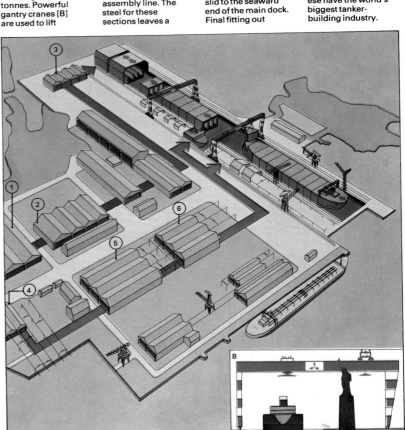

1 The first iron screw steamship, the SS *Great Britain* (1843), was not an immediate commercial success but showed that iron ships could carry more cargo more safely than wooden ones.

2 The *Lusitania* was torpedoed in 1915 by a German U-boat with the loss of 1,198 lives. Introduced in 1907, she helped to establish the Parsons steam turbine as a means of propulsion for big, fast liners.

3 Production line assembly of giant oil tankers is carried out in dry-dock shipyards like the extensive Mitsubishi yards at Koyagi in Japan [A]. Treated steel plate is cut to size [1], assembled in small sections [2], coated [3] and moved to the dockside to be assembled into sections which can weigh up to 600 tonnes. Powerful gantry cranes [B] are used to lift each section into a 990m (3,246ft) dry dock where they are welded together. Steel plate for the stern and bow is handled on another assembly line. The steel for these sections leaves a treatment shop [4], goes to sub-assembly [5] and large assembly shops [6], and is built up in a pocket dock to be added as the ship is slid to the seaward end of the main dock. Final fitting out is done afloat. Yards like the one at Koyagi build the 500,000-tonne super-tankers needed to meet world demands for oil. The Japanese have the world's biggest tanker-building industry.

4 The *Lanka Devi* is a conventional cargo ship of a type still widely used. She can load or unload cargo with her own derricks if cranes are not available on the dockside. Though the *Lanka Devi* has her machinery placed amidships, a more recent trend in cargo vessels is for the bridge superstructure and the propulsion unit to be positioned either at or near the stern.

5 A container ship like the *Encounter Bay* must use special terminals to handle her 1,500 containers, each measuring 36 cubic metres (1,280 cubic feet). Unloading [A] and loading [B] can be carried out simultaneously. Like most container ships, the *Encounter Bay* has a high freeboard and an unobstructed deck, allowing containers to be stacked there. Bigger ships can carry up to 3,000 containers, close-packed because they are a standard size.

whole vessel to be chartered for a single cargo, like a conventional cargo vessel [4]. Now it operates on routes where container ships would be too costly, or it acts as a "feeder" to container routes. The "reefer", or refrigerated cargo liner, also remains important because it handles perishable cargoes such as fruit and meat. Smaller coasters and coastal oil tankers, which trans-ship cargo between ports, act as vital links both as distributors of cargo to lesser ports and as feeders to the major ports and container terminals. Roll-on/roll-off ships are used to carry truck trailers and low-loaders from point to point. Another variant is the LASH, or Lighter Aboard Ship concept, with barges or lighters hoisted aboard a large ship. After being dropped at the mouth of a river, the barges are towed upriver without any transshipment of cargo, as with containers.

Shipping could not function without a host of service craft. Foremost among these is the tug [9], which exists primarily to manoeuvre larger ships in confined waters. Tugs are also used to salvage damaged ships on the high seas and to fight fires, and special

"pusher" tugs handle barges on rivers and canals. Dredgers keep the channels in harbours and estuaries open to deep-sea shipping, and a variety of small craft are employed on pilotage and maintenance of buoys and other navigational aids.

Marine engines
The main propulsion systems of ships are either diesel or steam engines, driving screw propellers. Diesel engines are favoured for their simplicity and economy, and they have steadily driven out the one-time workhorse, the vertical triple-expansion steam engine. Larger ships requiring high speeds rely on steam turbines. The new container ships and bulk-carriers designed for fast running have drawn for the first time on naval experience with high-speed steam propulsion. Gas turbines have been used experimentally, although they have not yet proved sufficiently cheap and reliable. Similarly nuclear power is still uneconomical, and the very small number of nuclear-powered merchant ships is not likely to increase until a cheap and compact reactor plant becomes available.

Terms used on ships include the following: **Ahead** In advance of the bows. **Amidships** Near the middle of the ship's length. **Astern** Behind the ship. **Athwartships** From one side of the ship to the other. **Beam** Greatest breadth of the ship. **Boot-topping** Paint on hull between load line and water-line when ship is empty. **Bows** Foremost part. **Draught** Depth to which ship sinks in the water. **Forward** Towards the bows. **Freeboard** Distance from main deck to waterline. **Lee side** sheltered from the wind. **Port** Left side looking forwards. **Quarter** Direction between stern and beam. **Rake** Slope of funnel, masts or stem. **Sheer** Fore-and-aft curve of a hull rising towards bow and stern. **Starboard** Right-hand side looking forwards. **Topsides** Outer surface of vessel above water-line. **Trim** The way a vessel sits in the water. **Windward** Direction from which wind is blowing.

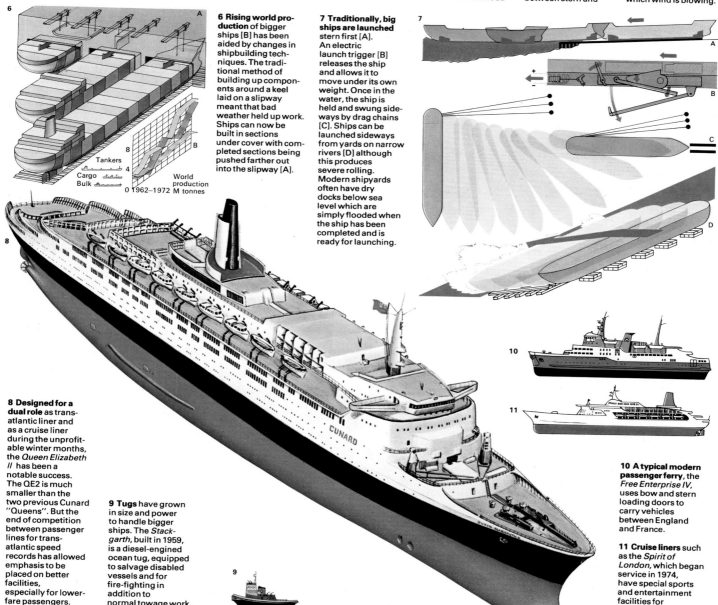

6 **Rising world production** of bigger ships [B] has been aided by changes in shipbuilding techniques. The traditional method of building up components around a keel laid on a slipway meant that bad weather held up work. Ships can now be built in sections under cover with completed sections being pushed farther out into the slipway [A].

Tankers
Cargo
Bulk
World production
0 1962–1972 M tonnes

7 **Traditionally, big ships are launched** stern first [A]. An electric launch trigger [B] releases the ship and allows it to move under its own weight. Once in the water, the ship is held and swung sideways by drag chains [C]. Ships can be launched sideways from yards on narrow rivers [D] although this produces severe rolling. Modern shipyards often have dry docks below sea level which are simply flooded when the ship has been completed and is ready for launching.

8 **Designed for a dual role** as transatlantic liner and as a cruise liner during the unprofitable winter months, the *Queen Elizabeth II* has been a notable success. The QE2 is much smaller than the two previous Cunard "Queens". But the end of competition between passenger lines for transatlantic speed records has allowed emphasis to be placed on better facilities, especially for lower-fare passengers.

9 **Tugs** have grown in size and power to handle bigger ships. The *Stackgarth*, built in 1959, is a diesel-engined ocean tug, equipped to salvage disabled vessels and for fire-fighting in addition to normal towage work.

10 **A typical modern passenger ferry,** the *Free Enterprise IV,* uses bow and stern loading doors to carry vehicles between England and France.

11 **Cruise liners** such as the *Spirit of London,* which began service in 1974, have special sports and entertainment facilities for holiday-makers.

Hydrofoils and air-cushion vehicles

Hydrofoils and air-cushion vehicles [Key] are machines that reflect today's demand for greater speed and flexibility in transport. In common with aeroplanes and helicopters these craft use energy for both upward and forward motion. A hydrofoil's hull is raised out of the water to reduce drag (friction) and thus enable it to travel faster than a conventional ship. An air-cushion vehicle (ACV) lifts itself in order to become independent of the surface over which it travels, which can be water, snow, marsh or sand. Today, each type of vehicle has a different use.

Hydrofoils and their uses

The first hydrofoils were built at the beginning of the twentieth century. Designers calculated that a sea wing (a lifting plane designed for running through water) [2] could be made much smaller than an air wing capable of generating the same lift force. Early experiments used foils stacked in the form of "ladders" [1A] so that, as the craft accelerated and lifted, fewer foils remained in the water.

By 1940 the more efficient surface-piercing foil [1C] had become preferred. In this type of hydrofoil a large foil slopes up on each side of the centre line, looking from the front like a V with the tips emerging from the water. These foils provide built-in stability in banked turns or through waves.

Most modern hydrofoils are of this type, although two other designs are also current. Many Soviet hydrofoils have depth-effect foils [1B], which stabilize automatically an inch or so below the surface of the water. They are ideal for use on shallow draught vessels for inland waterways but are useless in a rough sea. For this application, submerged foils [1D] are the best solution and these are the type used on most high-speed military hydrofoils. Small foils run deep in the water, supporting the vessel on streamlined struts. The angle of the foil can be varied by an autopilot, enabling the craft to run true even in heavy seas.

Most hydrofoils are fairly small, few civil types reaching 150 tonnes and the largest military version being 320 tonnes. As inshore ferries they can travel at high speeds without damaging river banks or disturbing smaller

craft with their wash, and they give a smooth ride across choppy water. Naval hydrofoils provide manoeuvrable platforms for guns, missiles and anti-submarine search and attack systems. At slow speeds, a hydrofoil behaves like a conventional ship and floats with its hull in the water. Control remains precise and the craft can ride out storms.

The story of the hovercraft

A hydrofoil is restricted to water and at low speeds needs deep water to float in. By contrast, a hovercraft – the most common kind of air-cushion vehicle (ACV) – is able to go practically anywhere.

In the nineteenth century engineers such as John Thornycroft (1843–1928) tried to reduce drag by pumping bubbles out of holes in a ship's hull. Later ACV developments sprang from the experimental work in the 1950s of Christopher Cockerell (1910–   ), who realized that some sort of "curtain" to contain the air cushion under the vehicle was essential to provide sufficient lift within the practical limits imposed by engine size.

Modern ACVs ride on a single bubble or

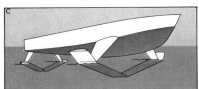
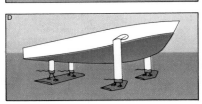

1 Almost all hydrofoils belong to one of four classes. Ladder foils [A] emerge progressively from the water as speed is increased. Depth-effect foils [B] are suited to shallow water without waves. Surface-piercing foils [C] are the most common form for passenger hydrofoils, while submerged foils [D] are preferred for rough seas. The foils are adjustable and their angle is controlled by an autopilot.

2 Foils are sea wings. At speed they generate enough lift to raise the vessel out of the water, thus reducing drag.

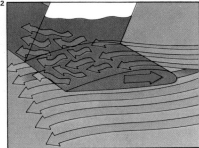

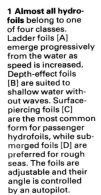

3 Propulsion poses problems because a hydrofoil's screw is lower than the engine. The sloping drive [A] is simplest but the thrust acts at a sharp angle. The Z drive [B] is best for deeply submerged foils, but requires two sets of bevel gears. The V form [C] is general on surface-piercing foil boats. An alternative is to use the engine to drive a pump, which squirts a water jet out at the stern to provide thrust.

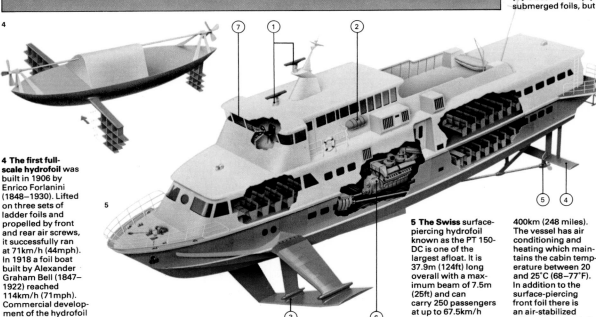

4 The first full-scale hydrofoil was built in 1906 by Enrico Forlanini (1848–1930). Lifted on three sets of ladder foils and propelled by front and rear air screws, it successfully ran at 71km/h (44mph). In 1918 a foil boat built by Alexander Graham Bell (1847–1922) reached 114km/h (71mph). Commercial development of the hydrofoil began in the 1930s.

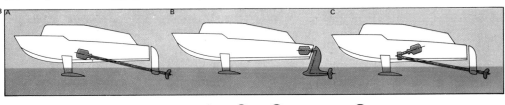

5 The Swiss surface-piercing hydrofoil known as the PT 150-DC is one of the largest afloat. It is 37.9m (124ft) long overall with a maximum beam of 7.5m (25ft) and can carry 250 passengers at up to 67.5km/h (42mph) for 400km (248 miles). The vessel has air conditioning and heating which maintains the cabin temperature between 20 and 25°C (68–77°F). In addition to the surface-piercing front foil there is an air-stabilized submerged rear foil.

1 Radars
2 Liferaft
3 Front foil
4 Rear foil
5 Propeller
6 Two 3,400hp diesel engines
7 Bridge

Twin propellers are sited below the rear foil. There are four passenger saloons, two on the main deck and two on the lower, and passengers can be served with drinks and refreshments in their seats. The Malmo-Copenhagen ferry uses a PT 150DC hydrofoil.

cushion of air blown by fans through slots or jets round the underside of the hull [6]. The main part of the hull serves as a buoyancy tank, enabling a water-borne ACV to float when the lifting power is switched off. Air pumped by the lift fans raises the ACV on a cushion of slightly pressurized air, which leaks away round the edges at the same rate as it is fed in. The propulsion and lift systems may be driven either by one power plant or by separate units.

Early ACVs had flat undersides and daylight was visible beneath them when they were operating, although clearance was often only a few inches. To improve their performance over waves and other obstacles, later ACVs were fitted with flexible skirts [7]. These skirts contain the cushion of air but pass easily over obstructions. This facility makes them truly amphibious and they can operate equally well in or out of water. Unlike hydrofoils they are not limited in size and large passenger-carrying craft and car ferries are being used. Various types of propulsion and steering systems have been tried [8], some combining both functions.

Early optimism about the future of the ACV was not borne out by subsequent development. The craft has not yet shown itself to be completely economical for commercial use, probably because today's vehicles are too small.

Other uses of ACVs

Apart from conventional passenger transport ACVs have proved their value in the military sphere and the principle of the air cushion has been applied successfully to other modes of transport. The ACV system can be used on a train running over a smooth track, the air cushion reducing friction while propulsion is obtained from ducted fans or linear induction motors. As heavy load carriers ACVs are able to transport machinery weighing 500 tonnes over surfaces that could not possibly be traversed by wheeled or tracked vehicles. Air-cushion pallets for industrial use are easy to move and reduce strain on floor surfaces. The ACV principle has also been used in the design of hospital beds to lessen the discomfort of patients suffering from burns and for a "land-anywhere" aeroplane.

Hydrofoils and air-cushion vehicles are lifted and supported by the expenditure of energy – unlike water-borne ships or land vehicles resting on wheels. A hydrofoil [A] lifts itself to reduce drag, so as to run at high speed and travel smoothly across rough water. A hovercraft [B] does the same, but has the ability to traverse almost any kind of surface, including swamps, mud, river rapids and ice. The skirt fitted to amphibious hovercraft helps retain the supporting air cushion and lifts the rigid structure higher above the surface.

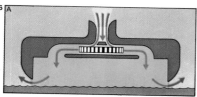
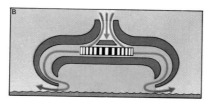
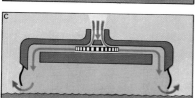
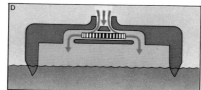

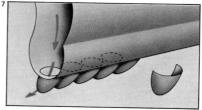

6 There are four basic types of ACV, each with a central lift fan. The simple "plenum chamber" [A] blows in air which escapes beneath the downturned edges. The peripheral jet [B] blows air inwards from a surrounding slit. With a skirt [C] the rigid body is raised above waves and solid obstructions. ACVs designed solely for use over water have side walls [D], saving lift power but adding extra water drag.

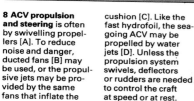

7 The skirt of an amphibious ACV is made of tough, rubberized fabric to withstand heavy wear. Air inflates the inner and outer walls, escaping through inward facing ducts or "fingers". These inflate the main cushion. They are designed for quick and easy replacement, since they are the parts most likely to be damaged.

8 ACV propulsion and steering is often by swivelling propellers [A]. To reduce noise and danger, ducted fans [B] may be used, or the propulsive jets may be provided by the same fans that inflate the cushion [C]. Like the fast hydrofoil, the sea-going ACV may be propelled by water jets [D]. Unless the propulsion system swivels, deflectors or rudders are needed to control the craft at speed or at rest.

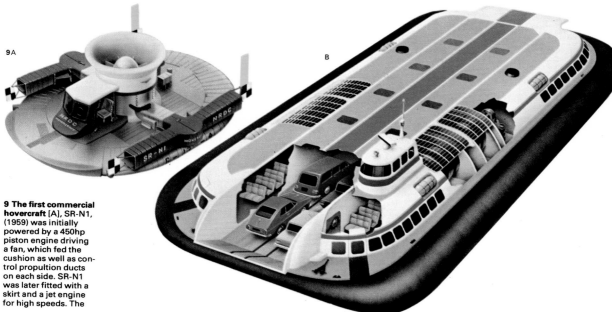

9 The first commercial hovercraft [A], SR-N1, (1959) was initially powered by a 450hp piston engine driving a fan, which fed the cushion as well as control propulsion ducts on each side. SR-N1 was later fitted with a skirt and a jet engine for high speeds. The

VT1 [B] was a large commercial ACV built in 1969 by Vosper Thornycroft. Designed to carry 10 cars and 146 passengers, it was propelled by water screws, yet had a skirt and could run out of the water up a slipway for loading. This design combined advantages of high speed with economy and low noise levels. Subsequent civil and military versions of the VT1 were fitted with a new propulsion system, the water screws being replaced by large ducted air propellers.

Submarines and submersibles

The invention of the submarine is usually credited to an Englishman, William Bourne. In 1578 he described a vessel that could take on and expel water to vary its buoyancy and that used a snorkel-like tube for its air supply. There is no evidence that Bourne's craft ever sailed and the first submarine was probably a leather-covered rowing boat built by a Dutch engineer Cornelius van Drebbel (1572–1634), in about 1620. According to the British chemist Robert Boyle such a boat took King James I (1566–1625) for a ride beneath the River Thames and even had some kind of "liquor" for renewing the air inside. Unfortunately no drawings or detail of this remarkable craft have survived.

Submarines in wartime
The *Turtle* [Key], a one-man submarine invented in 1776 by an American, David Bushnell (*c.* 1742–1824), was the first underwater vessel to be used in war. It pioneered two essential features of the modern submarine – a closed hull and screw propulsion, although the latter was worked by hand. But two further developments were necessary before the submarine could become an effective fighting machine: a good undersea weapon and a source of power.

The Confederate navy used submarines during the American Civil War. Small craft known as Davids – named after the biblical David who vanquished the giant Goliath – were each armed with an explosive charge at the end of a long spar and powered by hand or steam. In 1864, a David, the *Hunley*, rammed the Union ship *Housatonic* off Charleston harbour; both vessels sank and the crew of the *Hunley* were killed. A self-propelled torpedo was also invented.

By the turn of the century the American inventor John P. Holland (1840–1914) was designing submarines powered by petrol engines. Holland's vessels were the first modern submarines and contained all the basic features of those that fought in both world wars. German submarines (U-boats) during World War I were highly effective as raiders of unarmed merchant ships, thereby threatening to cut Britain's vital supply lines. In World War II submarines took a toll of warships as well [6].

The effectiveness of the submarine remained limited by the slowness and short duration of undersea travel when vessels depended on electric motors powered by batteries. The batteries soon became exhausted and submarines had to surface frequently to obtain air for the diesel engines and recharge the batteries (through a dynamo driven by the engines). The diesel engines were also used for surface propulsion. In World War II a snorkel tube was developed to supply air to the diesel engines, which could be run just below the surface. This helped the submarine to escape detection.

How submarines work
Part of a submarine's hull is double-skinned. Water admitted into ballast tanks in the hull lowers the buoyancy of the vessel; diving planes on the sides of the hull control the angle at which the vessel sinks and level it out. Buoyancy is restored by expelling the water from the ballast tanks with compressed air. The central structure – the conning tower or sail – contains a periscope, radio and radar aerials and a snorkel tube. Tracking of an

CONNECTIONS

See also
174 Modern fighting ships
168 Modern artillery
180 Nuclear, chemical and biological warfare
236 Radar and sonar

In other volumes
92 The Physical Earth

1 **Hunter-killer submarines** – this is the USS *Barbel* – were designed to sink other submarines. They represented the ultimate development of undersea warfare until the advent of the nuclear submarine. A nuclear submarine can fire its missiles from beneath the sea, thus remaining invulnerable. The USA and the USSR each possess more than 100 nuclear submarines. Air-conditioned (for the electronic systems as well as the crew), they can travel 644,000km (400,000 miles) without refuelling. The nuclear reactor powers a steam turbine that drives the propellers directly or by means of an electric generator and motors.

2 **Locating submerged submarines** from a ship, or locating surface vessels from a submarine, is achieved by listening for echoes of supersonic sound pulses transmitted through water. The system, originally called ASDIC by the British Royal Navy, is now called sonar. Submarines can be located using sonar buoys, which transmit signals back to a nearby mother ship, or by using a "dunking" sonar suspended from a helicopter and dipped just below the surface of the water.

3 **The torpedo**, invented by Austrian engineer, G. Luppis in 1864 and developed by Robert Whitehead (1823–1905), made the submarine into an effective fighting force. It is fired from a tube inside the submarine. Torpedo attacks on merchant shipping, as shown here, were successful in both world wars, but the torpedo is now primarily an anti-submarine weapon. Accurate positioning is no longer necessary, as modern torpedoes home in on their targets acoustically or are guided by wire. Short-range missiles for use against other vessels have been developed for firing from torpedo tubes.

4 **Rising from the sea** on a test flight, the *Polaris* missile has a range of nearly 4,630km (2,875 miles) and carries three nuclear warheads. The *Poseidon* missile has ten warheads. The missiles are ejected from their launching tubes by compressed gas and the two-stage solid-fuelled rocket motor fires at the surface. They are computer guided and the warheads separate on re-entry.

enemy vessel is accomplished by sonar (underwater) and radar (when surfaced) and communications are by radio. A submarine can receive messages underwater but has to have at least its radio aerial clear of the water in order to transmit signals, although methods of direct transmission are being investigated. Navigation is by means of a computerized inertial system, assisted by sonar and radar in coastal waters or beneath ice floes. The periscope is used for photographing and for checking navigation by sextant.

The nuclear submarine is a true undersea craft. It needs no air for its motors and can make long undersea voyages without refuelling. The first of these vessels, the USS *Nautilus*, was also the world's first nuclear-powered warship of any kind. Launched in 1954, its capabilities drastically changed the military strategy of the Great Powers. Instead of raiding surface vessels, nuclear submarines carry long-range intercontinental missiles armed with nuclear warheads that can be launched underwater. The submarines constitute a strike force that is practically invulnerable because they cannot

easily be located when submerged. They are intended to respond to the first strike against a country and thus deter any nation from making such an attack.

In addition to intercontinental missiles [4] the submarine can carry torpedoes [3] that either home acoustically or are guided by a wire trailed behind the torpedo. Weapons that are part-torpedo and part-missile can also be carried for use against distant vessels.

The role of submersibles

The term submersible may be used to describe any underwater vessel but today it usually indicates a non-military vessel. Small electrically-powered submersibles [7] are used to inspect and carry out maintenance of underwater structures, especially oil and gas drilling heads and pipelines; to check undersea cables and bury them in the mud on the sea floor; to search for minerals on the seabed; and to carry out oceanographic research. They have even been used to search for the Loch Ness monster. The submersibles have bright lights and sensitive manipulators so that they can work on the sea-bed.

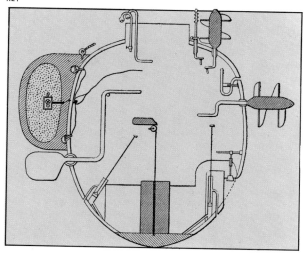

The *Turtle*, an American submarine, was in 1776 the first to operate in a war. The single crew member worked all the controls – lateral and vertical propellors, rudder and pumps – by hand. The packet of explosives [left] was detachable and designed to be fixed to the hull of an enemy vessel.

5 The largest submarine of World War I was the K-class of the British Royal Navy. It was 103m (338ft) long and had a crew of 55. The war proved that submarines were not greatly effective against warships but could be used to sink merchant ships and disrupt supply lines. Germany had no ocean supply lines but the Allies lost 32 ships for every German submarine they succeeded in sinking.

6 The German U-boat *(Unterseeboot)* was responsible for sinking millions of tonnes of Allied shipping during the early stages of World War II. Working in packs and armed with fore and aft torpedo tubes, the U-boats stalked convoys by day and attacked by night. The Allied development of radio direction finding and radar made surfaced submarines vulnerable to aircraft attack.

7 *Star III* is a submersible capable of operating to a depth of 610m (2,000ft). Powered by batteries it is 7.7m (25ft) long and carries a crew of two sealed inside the spherical steel pressure hull amidships. *Star III* is highly manoeuvrable as it has horizontal and vertical motors. It operates from a support ship and has been used to investigate the deep scattering layer of plankton that is so important to ocean ecology. Submersibles have also been used to locate deep-swimming schools of fish to help improve catches. They are capable of delicate undersea work on cables and oil rigs.

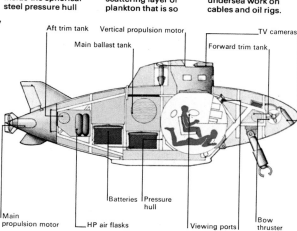

Aft trim tank
Vertical propulsion motor
Main ballast tank
Forward trim tank
TV cameras
Main propulsion motor
HP air flasks
Batteries
Pressure hull
Viewing ports
Bow thruster

8 Modern submersibles are important in oceanographic research. Using them scientists can search the ocean floors for evidence of oil, manganese nodules and other minerals. Similar vessels are used to inspect and maintain underwater structures such as off-shore drilling rigs (for natural gas and oil) and submarine telephone cables. Mechanical "hands", controlled from inside the submersible, are used to collect samples or to hold tools. The vessel can also be equipped with powerful lights, television cameras and underwater cutting torches. At shallower depths, a submersible can act as a base and even living quarters for divers, who leave and enter the craft through an air-lock. Communication to the surface may be by telephone along a cable or a sonar (supersonic sound) beam may be used to transmit suitably coded voice messages or even slow-scan television pictures. All equipment is worked by electricity from storage batteries.

Carts, coaches and carriages

When early man was a hunter he found that he could transport his kill more easily if he dragged it on a crude sled [1] rather than carrying it on his back. Soon he found that lengths of tree trunks used as rollers could move even heavier loads.

A solid wheel fixed to a platform by a simple axle was a logical development of the roller system [2]. The first reported use of a wheel was in Mesopotamia, the land between the rivers Tigris and Euphrates, some 5,000 years ago. Oxen-hauled wheeled transport then developed and its use spread slowly to the Mediterranean, Europe and China. When the Romans turned their attention to the building of wheeled vehicles their fine roads permitted fast travel in horse-drawn chariots – an important factor in the administration of a vast empire.

In the period between the fall of the Roman Empire and the fifteenth century, progress in the development of vehicular transport lapsed. Most travellers were soldiers, pilgrims or pedlars who relied largely on horses or pack animals. Farm carts used for local haulage were drawn by heavy horses

specially bred for the work and in later times used as battle chargers. The few wheeled vehicles of the Middle Ages had no springs [4] and a long journey through Europe could take several uncomfortable months.

Wheeled transport develops

As communication between peoples increased, carriage transport began to develop. The first vehicles used rigid axles until suspensions in the form of flexible wood laths, then leather straps, were introduced. Early sixteenth-century carriages, often extravagantly decorated, were much resented. The public envied those rich enough to afford them, the Church considered private conveyances sinful and the authorities kept a close watch, thinking them ripe for taxing – attitudes very similar to the ones of those who opposed the introduction of the automobile some 400 years later.

By the seventeenth century, the period of mechanical and scientific awakening in Western Europe, some coaches and carriages had metal spring suspensions. Large rear wheels allowed the vehicles to travel at

higher speeds over the poor roads and also provided a more comfortable ride.

During the seventeenth century Britain changed, mainly because of trade, from a farming community to a commercial nation with the need to convey goods and people over long distances. Smaller, lighter vehicles were developed for rapid short trips. But for longer journeys the early stage-coaches could travel little more than 48km (30 miles) a day, stopping periodically at staging posts. A journey from London to Edinburgh – a distance of 675km (420 miles) – took 12 days even by fast coach.

From mail coach to rail travel

The first coaches to carry both mail and passengers travelled from Bath to London in 1783. Mail coaches were so reliable that clocks could be set by their 16km/h (10mph) schedule. Coaching inns, some of them able to cater for up to a hundred coaches a day, provided passengers with meals and accommodation along the main routes.

The nineteenth century saw great changes, including the brief introduction of

CONNECTIONS

See also
110 History of transport
134 Trams and buses
182 Road building
126 History of automobiles

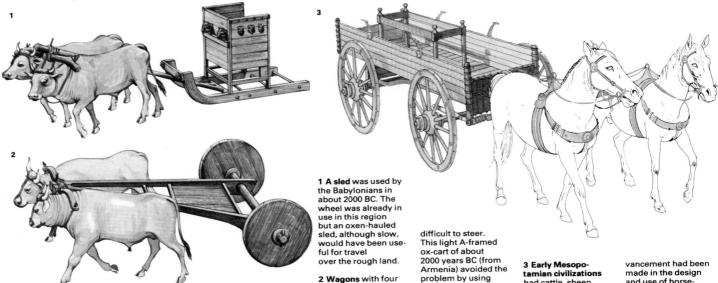

1 A sled was used by the Babylonians in about 2000 BC. The wheel was already in use in this region but an oxen-hauled sled, although slow, would have been useful for travel over the rough land.

2 Wagons with four fixed wheels were

difficult to steer. This light A-framed ox-cart of about 2000 years BC (from Armenia) avoided the problem by using only two wheels.

3 Early Mesopotamian civilizations had cattle, sheep, goats – but no horses. Horses were first found and then tamed on the plains of central Asia. Early Celts had very little knowledge of either wagons or horses but by the first century BC a considerable ad-

vancement had been made in the design and use of horse-drawn transport, as shown by this illustration of a two-horse ceremonial Celtic or Teutonic wagon used on feast days. The picture was reconstructed from fragments found on the western coast of Jutland, Denmark.

4 Horses, mules and pack animals were the main forms of transport in the Middle Ages but this type of "long-wagon" was used to carry womenfolk of rank and wealth in relative comfort.

5 French elegance can be seen in the design of this heavy seventeenth-century funeral coach. Leather strap suspension and the large rear wheels (the front wheels were smaller to allow them to turn on a central steering pivot without fouling the body) gave passengers in this type of coach a reasonably comfortable ride on roads that had improved little for about a thousand years. In Britain Charles II issued a royal

charter to all coachmakers demanding that attention be paid to the problems of transport.

steam coaches. Travelling became more comfortable with the development of elliptical leaf springs made of several thin, flat springs bound together and these are still used for many purposes today. The design of carriages and coaches became more elegant and light following the improvement of roads by the engineer John McAdam (1756–1836) whose type of road surfacing, designed to compact by the weight of passing traffic, greatly facilitated travel.

The Victorians used numerous types of horse-drawn transport for short journeys, ranging from the light gig to the family brake. Wealthy families kept staff to drive or maintain a private coach or drag, a dog cart for sportsmen and gun dogs, a governess cart for the children, a phaeton for rapid journeys, a victoria for park and town use, a brougham for privacy and perhaps a stately landau for formal occasions.

In America the stage-coach enjoyed a longer span of life and helped to open up the West through a vast network of scheduled services operated by coaching companies such as Wells Fargo. The lightweight high-wheeled buggy [7] and canopied phaeton or surrey were extensively used for private transport. Both were distinctively North American vehicles. A buggy seated two people whereas a surrey was essentially a family transporter with two rows of seats.

Cabs, trams and buses

In Europe the hire-cab designed by Joseph Hansom (1803–82) in 1834 (and bearing his name) plied the streets of almost every city and horse-drawn trams and omnibuses provided reasonably inexpensive travel for the masses. The box-like omnibus, French in origin, was first introduced to London in 1829 by George Shillibeer at the time of the demise of the coach. By the 1840s the single-decker bus had developed into the double-decker with seating along the length of the top deck [8]. With further development improvements were added and the double-decker became the design adopted by bus companies when motor-driven transport began to replace horse-drawn vehicles. The last London horse bus in regular service continued to operate until 1914.

The brougham, a type of small, closed carriage for town and winter use, was first made in 1839 for Lord Brougham (1778–1868). It was unique in Britain and led to a revolution in carriage design, although similar vehicles were already in use in Paris. Eventually the brougham became one of the most common town carriages. The original version, known as the single brougham, was drawn by one horse and carried two people. Later models, known as bow-fronted and double broughams, were drawn by two horses and carried up to four people.

6A   B   C

6 The charabanc [A] was useful in large establishments for communal transport. A private omnibus [B], driven by a liveried coachman, was used to carry small groups on excursions. The drag [C] was a private coach based on earlier mail coach designs. The wagonette [D], with seats down each side and a rear entrance, was fashionable for family outings. The dog cart [E], originally designed to carry sportsmen and their dogs, became widely used for everyday transport.

7 European carriage design influenced many early American models. But the buggy had a very definite North American character, although the word "buggy" was originally an English word meaning a hooded gig. The distinctive American buggy, which made its appearance in about 1850, was a light, fast carriage with two or four high wheels and a thin frame supporting the carriage and canopy. It was drawn by one horse and seated two passengers.

8 The word "omnibus" (meaning "for every one") originated in France in about 1825 for a transport service operating in Nantes. Introduced to London in 1829 by Englishman George Shillibeer (1797–1866) the omnibus was an immediate success. The initial service, which ran between Paddington and the Bank for a fare of one shilling, was provided by a horse-drawn vehicle carrying 18 to 22 passengers. A liveried conductor stood guard by the rear entrance door. Soon people adopted the practice of perching on the roof of these single-deckers, leading to the development of the double-decker "knifeboard" bus with no weather protection on the upper deck (fares on top were half price). By the 1880s it had developed into the two-horse "garden seat" bus pictured here, in which forward-facing seats replaced the long bench. Horse-bus services operated in London until 1914.

History of bicycles

Bicycles driven by cranked pedals date from the 1860s and since then the machines have become popular throughout the world, particularly in Britain, France, Italy, the Netherlands and other European countries. Troops riding bicycles were employed by the major powers in World War I and, more recently, by the Vietcong in Vietnam.

The first bicycle
The history of the bicycle begins with non-powered machines developed in France in the late 1700s. In 1791 the Comte de Sivrac built his *célérifère*. It was a wooden machine consisting of two wheels in line joined by a bar that carried a seat. The rider straddled the bar and "walked" the machine along. Similar machines were made by J. Nicéphore Niépce (1765-1833) (also the inventor of an early photographic process) in 1816 and by the German Baron Karl von Drais a year later. Drais' *Laufmaschine* soon became popular in Britain and Germany as the *draisine* or hobby-horse.

In 1839 the Scotsman Kirkpatrick Macmillan produced a "powered" hobby-horse

propelled by pedals at the front which worked backwards and forwards driving connecting rods to turn the rear wheel. Rotating pedal-cranks, driving the front wheel directly, were introduced by the French brothers Pierre and Ernest Michaux in about 1861. They called their machines *Vélocipèdes* and within four years were manufacturing 400 bicycles a year. By 1869 bicycle racing was established on the roads of France.

The "ordinary" or "penny-farthing" bicycle [Key] had a large front wheel driven directly by pedal-cranks and a small rear wheel. Invented in 1871 by the Englishman James Starley (1831–81), it rapidly became the most popular type of bicycle. The size of the large wheel was chosen to suit the length of the rider's legs and varied from 1m (39in) to 1.5m (59in) across.

Chain-driven bicycles
The first chain-driven machine was built in 1874 by H. J. Lawson. Pedal-cranks mounted on the frame turned a large sprocket wheel which drove an endless chain round a smaller sprocket on the rear wheel.

Bicycles again had wheels of roughly equal sizes. The Rover safety bicycle of 1885 was mass produced and within a few years completely replaced the penny-farthing. All these early bicycles had solid rubber tyres mounted on steel-rimmed wheels.

The modern machine
A milestone in the history of the bicycle came in 1888 when John Dunlop (1840–1921) invented the air-filled or pneumatic tyre. The diamond-shaped frame became standard and there were no major changes in bicycle design for the next 70 years. In the 1960s various manufacturers produced small-wheeled bicycles – some of which could be folded to fit into the boot of a car – for town use. Earlier unusual designs included the tandem, a long-framed bicycle for two riders, and the three-wheeled tricycle which became important as the type of machine into which Karl Benz (1844–1929) and Gottlieb Daimler (1834–1900) fitted petrol engines to make the first motor cars in 1885.

A modern bicycle has mudguards, electric lamps powered by batteries or a dynamo and

CONNECTIONS

See also
124 History of motor cycles
138 Small technology and transport

1 The Whippet of 1885 was designed by Lindley and Briggs. It had a pivoted and sprung frame designed to make the handlebars, saddle and pedals independent of the rest of the frame and wheels.

2 The Dursley Pederson bicycle of 1893 was designed by M. Pederson and built at Dursley in England. The frame members were of twin narrow-section tubes, side by side for rigidity, giving a lighter frame.

3 The Raleigh Safety bicycle of 1901 had an all-steel frame joined by a new brazing process using pressed steel sockets brazed to the frame tubes by dipping the joints in molten brass.

4 The Swift ladies' bicycle of 1926 had no cross-bar on the frame, making it easier for a woman wearing a skirt to get on and off. The lightweight frame was an advance over early heavy designs.

5 The Velocino bicycle, made in Italy in the mid-1930s, was an attempt at a compact design that was easy to store and carry. Its inventors also claimed it to be easier to dismount in an emergency.

6 The Moulton bicycle, produced in the UK in 1962, had small wheels, rubber suspension and a low centre of gravity. The frame could be adjusted to "fit" the build of almost any rider and was strong enough to carry heavy loads.

lever or calliper brakes acting on the rims of both wheels [7]. It may have variable gear ratios and a guard to enclose the chain and transmission. Accessories include extra carrying capacity in the form of a rear-mounted rack, a basket in front of the handlebars and a saddlebag or a pair of panniers mounted on each side of the rear wheel.

The frame is generally made of seamless steel tubing brazed or welded together. In the brazing process pre-cut lengths of tubing are fitted over angled sockets and secured in place with molten brass. In welding there are no sockets and the tubes are joined with molten steel to give a joint stronger than the tubes themselves. Using special light alloys manufacturers can make racing bicycles weighing as little as 7kg (15lb).

Between 24 and 40 wire spokes join the wheel hub to the rim. The rim end of each spoke is threaded and a nut or nipple is screwed on to it to keep the spoke in tension. The rim may be made of aluminium alloy, stainless steel or chromium plated stainless steel. Hard rubber brake-blocks, worked by levers (called stirrups) or callipers (which act

like pincers to nip each side of the wheel) press on the rim for braking. An alternative type of brake acts by pressing inside the hub of the rear wheel and is brought into action by back-pedalling.

The rear wheel of a bicycle rotates faster than the large sprocket wheel turned by pedalling. There are generally about 48 teeth on the large sprocket wheel and about 18 on the smaller rear sprocket. This provides a gear ratio of about 2.66 to 1. Variable gears allow different speeds for a constant pedalling effort. There are two main types. In a *dérailleur* gear there are up to six rear sprockets of different sizes plus up to three on the cranks to provide a variety of gear ratios.

An epicyclic gear, developed by Sturmey and Archer, is more complicated. A small cog called the sun wheel inside the rear limb is rotated by the rear sprocket. A ring of teeth line the inside of the hub and the drive is taken to these from the sun wheel via a set of planet-wheel cogs. Three different gear ratios are available. Sturmey-Archer and *dérailleur* types can be combined on one hub to provide the rider with eight gears.

The penny-farthing or ordinary bicycle first appeared in England in the early 1870s, invented by James Starley. The rider pedalled cranks that were mounted at the centre of the large front wheel, which he also turned to steer the machine.

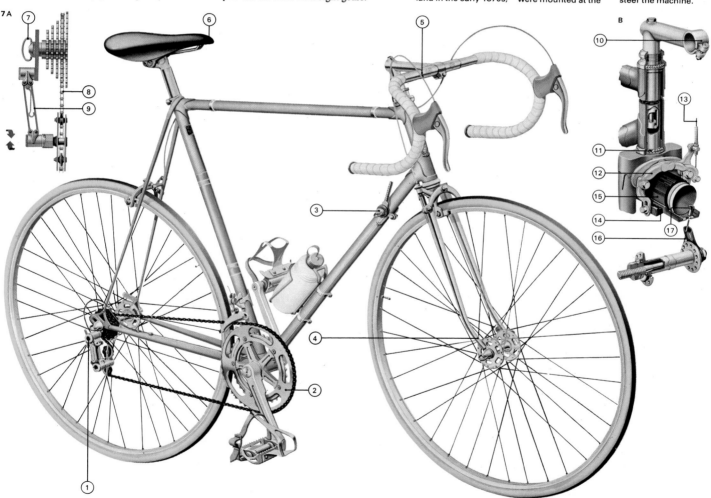

7 A modern racing bicycle has a short wheelbase for manoeuvrability and is made of lightweight alloys to reduce the weight of the machine. It has two sets of *dérailleur* gears, one [1] on the rear hub and the other [2] on the main sprocket, giving it up to 28 different gear ratios. The gears are changed by means of two levers [3] on the frame. The front forks [4] are nearly straight, to reduce the wheelbase, but have enough curvature to act as springs. The handlebars [5] and saddle [6] are both adjustable and the saddle is fixed so that the distance from it to the handlebars is about the same as the length between the rider's elbow and fingertips. The height of the saddle above the lowest point of the pedal is ideally 9% longer than the rider's inside leg length. The *dérailleur* gear [A] on the rear hub has up to six sprockets, of increasing sizes, mounted on a quick-release hub [7]. Small sprockets give high gears and larger ones low gears. The drive chain [8] can be shifted from one sprocket to another by a parallelogram arrangement [9] held in sideways tension by a spring. The stem and fork assembly [B] of a modern bicycle has to bear much of the weight and still pivot freely for steering the machine. Handlebars are fitted to an angled sprocket [10] at the top of the stem. A ballrace [11] provides a friction-free bearing. Most modern machines have brakes consisting of a calliper [12] operated by a bowden cable [13]. When the cable is pulled the calliper nips brake-blocks [14] against the wheel rim. The tyre [15] has a canvas carcass covered by synthetic rubber tread. Each spoke [16] is tensioned by a nipple [17].

History of motor cycles

The motor cycle is an older invention than the motor car. Two Frenchmen, Pierre and Ernest Michaux, built the first motor cycle – a steam powered "boneshaker" in Paris in 1869, sixteen years before Karl Benz (1844–1929) and Gottlieb Daimler (1834–1900) made the first cars. But the advantage of Daimler's petrol engine was soon put to use by motor cycle designers.

Early developments

Other technical innovations soon improved these early machines. The pneumatic tyre of J. B. Dunlop (1840–1921), invented in 1888, helped to absorb some of the bone-jarring shocks from the road surface. The final drive was generally a leather belt, which tended to break or slip in wet weather. The engine was started by pedalling, to turn over the engine, or by "bump starting" in which the rider pushed the machine, running alongside and jumping into the saddle when the engine fired. The Butler spray carburettor of 1889, modified and refined by Wilhelm Maybach (1847–1929) in 1893, was the forerunner of those still used today.

Motor tricycles also date from about 1880. Some were little better than motorized wheelchairs. But the De Dion Bouton of 1898 had a rear-mounted engine, a differential, and was capable of the then staggering speed of 40km/h (25mph).

In Britain the Road Acts of 1861 and 1865 had required that all motor vehicles be preceded by a man carrying a red flag. The repeal of these laws in 1896 removed the restrictions that had cramped British designs for so long. In the same year Colonel Capel Holden patented a motor cycle with a four-cylinder opposed engine. This had a commutator-type distributor powered by a coil and a battery, as on a modern car. External connecting rods drove the rear wheel directly by means of overhung cranks.

Wider applications of motor cycles

The motor cycle movement was also growing in the United States where by 1905 the main manufacturers were Harley Davidson and Indian. Both companies pioneered the use of twist-grips [1] on the handlebars to control the throttle and advance and retard of the ignition timing. The 1.75hp Indian of 1905 had a single-cylinder engine with a steel cylinder machined from a solid casting. Harley Davidson produced their first V-twin cylinder engine in 1909 and have used the same layout in most of their engines ever since. By 1914 the motor cycle speed record has risen to 150.5km/h (93.5mph). In the same year, the first of World War I, the British army began to use motor cycles for dispatch riders and used machines fitted with sidecars that could also carry a machine gun.

By the 1920s nearly all large-engined machines had a chain or shaft as final drive. Overhead valve engines began to appear and some, such as the 1,000cc units in Harley Davidsons and Indians, had four valves to each cylinder. In Germany BMW produced their first motor cycles with a horizontally opposed twin-cylinder engine, an arrangement that has also survived to the present.

As the volume of traffic increased, particularly in the United States, police forces began to use motor cycles for patrol duties. Large four-cylinder machines produced by companies such as Henderson and

1 Indian (1911) became popular by taking 1st, 2nd and 3rd in the Isle of Man TT races. This rarer single-cylinder machine also had twist-grip controls.

2 Brough Superior (1924) was the first production machine generally available with a top speed of more than 160km/h (100mph). This 1930 version, the Black

Alpine, had a JAP 680cc V-twin engine, a heavy-duty four-speed Sturmey Archer gearbox and a bottom link front fork developed by Harley Davidson.

3 Norton International (1932) was so successful it became known as the "Unapproachable Norton". This 490cc version had hair-pin valve springs and others

had rubber-mounted handlebars. Optional extras included plunger-type rear springs straight-through exhaust pipes and a Norton TT-type gearbox.

4 Velocette KTT (1949) resulted from the firm's considerable racing success in the 1930s. The machine had a 348cc overhead camshaft

engine, air-filled "hydraulic" rear shock absorbers and "girder" front forks (later superseded by "tele-draulic" forks), to smooth the ride.

5 Harley Davidson WLA and WLC (1945, produced for the Canadian Government) were adapted from earlier civilian

machines. They had strengthened frames, bearings, gearboxes and clutches. The 750cc V-twin side-valve engine was

rugged and reliable. It was one of the machines for Allied dispatch riders and military police at the end of the war.

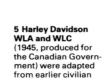

Indian were particularly suited for driving on the long, straight American roads.

Two-stroke engines

A two-stroke petrol engine has fewer moving parts than a four-stroke and is easier to maintain. By 1930 Villiers and other companies were producing a wide range of single-cylinder two-stroke engines. During the late 1920s and the 1930s the motor cycle underwent a social change. It ceased to be a luxury machine and evolved into a relatively cheap and utilitarian form of transport. A pillion passenger could be carried behind the driver and a sidecar gave the "combination" a carrying capacity of up to four (two adults and two children).

By 1937 a Brough Superior fitted with a 1,000cc JAP engine had pushed the world speed record up to nearly 275km/h (170mph). Once again the motor cycle industry was preparing for war. In 1938 BMW produced the R75 model, with a sidecar, for the Germany army. Both sides made collapsible motor cycles for paratroops.

Post-war development was characterized by smaller, higher-revving engines and in Europe thousands of motor scooters were produced. Between 1950 and 1965 manufacturers of dearer "luxury" machines such as the Vincent and Sunbeam (which used rubber-mounted engines) were forced to close down in the face of competition from mass-produced machines from Triumph, BSA, Norton and AMC. A wide range of purpose-built machines became available for scrambles, trials and road-racing.

During the early 1960s the Japanese Honda company began to enter Western markets with their small, 50cc four-stroke machines. Followed by Suzuki and Yamaha producing two-strokes, they soon dominated the market with models ranging from 50cc "monkey bikes" to 750cc four-cylinder machines capable of 210km/h (130mph).

Most motor cycles sold today are economic and comfortable machines. Electric starters and hydraulic disc brakes are becoming standard. The late 1970s may see the introduction of more rotary engine motor cycles, such as those which are already being produced by the DKW company.

The first recorded motor cycle was built in France in 1869, based on an existing Michaux velocipede – a type of "bone-shaker" pedal bicycle. It had a small single-cylinder steam engine. A flexible leather belt linked the engine with a larger one on the rear wheel, thus gearing down the speed of the engine. Within 20 years other inventors constructed steam bicycles and tricycles and in 1886 the German engineer Gottlieb Daimler fitted his air-cooled petrol engine into a wooden bicycle, also of his own design. At about the same time the English inventor Edward Butler (1863–1940) patented his "Petrocycle", a three-wheeler with spoked wheels and a two-cylinder water-cooled engine. The first commercially successful petrol-engined motor cycle was produced in 1893 by Henry and Wilhelm Hildebrand in Munich.

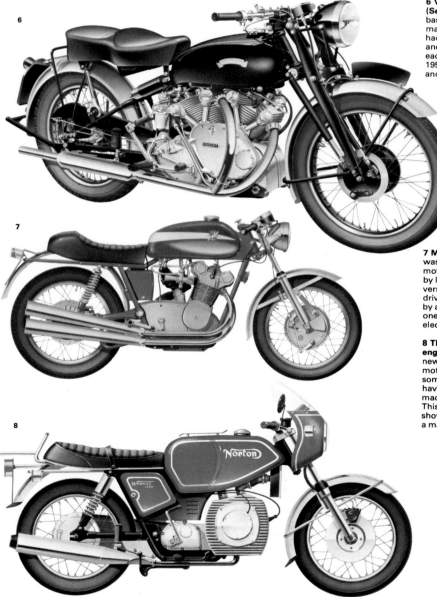

6

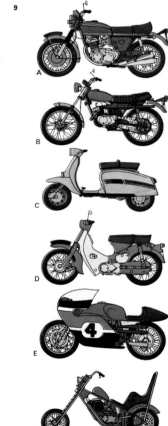

6 Vincent Rapide (Series C, 1950) was based on a 998cc machine of 1937. It had twin carburettors and twin brakes on each axle and in 1955 held the solo and sidecar records.

7 MV Augusta (1950) was a four-cylinder motor cycle designed by Ing Remor. First versions had a shaft drive, later replaced by a chain. It had one of the first electric starters.

8 The rotary Wankel engine may be the new power unit for motor cycles and some manufacturers have experimental machines using it. This illustration shows how such a machine may look.

9 Types of modern motor cycles include [A] standard 750cc road-going sports, capable of carrying two people at speed; [B] trials bike with high-mounted engine, wide handlebars and knobbly tyres for cross-country riding; [C] Italian-pioneered motor scooter with good weather protection; [D] "step-through" motor cycle with small engine and automatic clutch for use in towns; [E] fully modified road racer with a highly tuned engine in a special frame giving a 280 km/h (175mph) machine suitable for only the most experienced riders; and [F] the "chopper" using a standard engine in a highly modified frame with upswept handlebars and a sit-back saddle.

History of automobiles

The automobile was not invented overnight. It took shape from an accumulation of technical advances that resulted in a light and efficient engine. The accepted "fathers of the modern motor car" are two Germans, Karl Benz (1844–1929) and Gottlieb Daimler (1834–1900), who built their first petrol-fuelled motor vehicles within a few months of each other (1885–6).

More than a hundred years earlier, the first self-propelled road vehicle had rumbled through the streets of Paris at nearly 5km/h (3mph) when Nicolas Cugnot (1725–1804) demonstrated his steam-driven wagon [1].

The first automobile

The German Nikolas Otto (1832–91) made the first four-stroke internal-combustion engine in 1876 and in 1885 Daimler had installed a small four-stroke engine in a cycle frame. He drove his first four-wheeled petrol-driven vehicle round Cannstatt in 1886. In neighbouring Mannheim, Benz had tested his three-wheeler car.

Daimler licensed the French firm of Panhard and Levassor to build his engine.

Levassor placed it at the front of his crude car [2] and it drove the rear road-wheels through a clutch and a gearbox. Thus in 1891 the first car to use modern engineering layout was seen. Within three years of the appearance of the first Panhard France was staging motor races on public roads.

At the turn of the century, petrol, steam and electric power shared almost equal popularity for powering automobiles. Steam was well-tried and reliable and electric vehicles held the land speed record. France had several established motor manufacturers – Panhard, Peugeot, Renault, Daracq, Delahaye and others; in Germany Benz had made the world's first standard production car, the Velo (1894), and the Daimler company was just about to present the Mercedes to the public (1901) [3].

In the United States the automobile would develop along different lines. There the car was seen not as a rich man's toy, but as a new method of communication in a continent in which travel had been restricted by a lack of roads and great distances.

Great Britain, slow to start, had legislated

for the car in 1896 when the road speed limits were raised and soon such companies as Lanchester, Daimler (of Coventry), Wolseley and Napier were producing cars.

Motoring in Britain

Encouraged by the keen interest shown by King Edward VII, motoring in Britain became an accepted method of travel – for the rich. Some British manufacturers began to contest French car supremacy and among them the partnership formed in 1904 between Charles Rolls and engineer Henry Royce was one of the most significant [4]. At that time Henry Ford was preparing the motoring world for his Model T, which was introduced in 1908 [5].

By 1910 automobile design had become fairly settled, with a side-valve four- (or six-) cylinder front-mounted engine. Weather protection had been developed, and the electric starter from America (1912) had encouraged women to take to the wheel by removing the physical hardship of the starting handle. Interchangeable parts made to fine limits opened the gates to mass

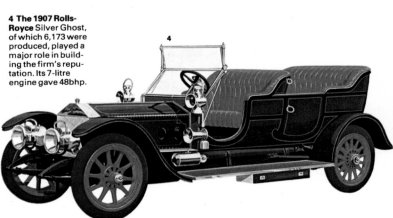

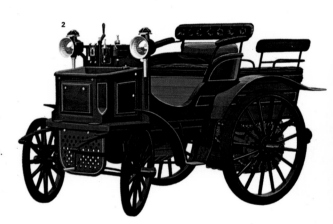

1 The first self-propelled road vehicle, built by Cugnot in 1769, was a tiller-steered, two-cylinder, steam-powered tractor. Its "engine" was on the single front wheel and was designed to pull guns. It involved its inventor in the world's first motor accident when it hit a wall.

2 Panhard and Levassor (1894) was developed from the 1891 design. It had a Daimler engine mounted at the front of the car, with the drive passing through a clutch and gearbox to the rear road-wheels. This French design was to be adopted as the standard modern layout.

3 The Mercedes, built by Daimler, appeared in 1901. Technically advanced for its day, it had a 35bhp, 5.9-litre engine and gate change.

4 The 1907 Rolls-Royce Silver Ghost, of which 6,173 were produced, played a major role in building the firm's reputation. Its 7-litre engine gave 48bhp.

5 Ford's Model T (1908), a simple easy-to-drive car, brought motoring to the world. Nicknamed "flivver", more than 15 million were sold by the end of its run in 1927.

126

production. The Edwardians had laid down the working principles and the following years saw more refinement than innovation. "Balloon" tyres, pressed-steel wheels and four-wheel brakes appeared. Heavy and unstable coach-built saloon bodies encouraged the trend to wood-and-fabric and later to the rigid, welded pressed-steel body.

Cheaper cars

Greater demand by the public in the 1920s brought cheaper cars on to the market from such manufacturers as Morris, Citroën, Opel, Austin and Fiat, although such exotic models as Hispano-Suiza, Maybach, Voisin and Delage still commanded respect – and a deep purse. The economic depression of the late 1920s closed down many companies of both classes, from Clyno to Bentley, and forced the production of even more basic cars.

By the 1930s most cars were being made for the new middle-class "family" driver, uninformed in motoring matters and requiring a near foolproof vehicle. There were some technical milestones, however. In 1934 Citroën produced the Traction Avant [7], the first medium-sized car to have front-wheel drive and independent suspension, and in 1938 the German car that was to become the Volkswagen (people's car) was finalized and tested – the only car to have spanned four decades [8].

The first postwar cars were similar to prewar models, but in 1948 two British cars destined to influence future design appeared – the wide-tracked Morris Minor and the 193km/h (120mph) Jaguar XK120 sports car. In 1955 the hydro-pneumatic suspension system of the Citröen DS 19, a sophisticated successor to Citröen's 1934 car, astonished the motoring world. The end of 1959 saw the introduction of the Morris Mini-Minor/Austin Seven, now universally known as the Mini [9]. It had a transversally mounted engine, front-wheel drive, rubber suspension and short wheelbase.

Since then the automobile has played a more and more important part in modern life, until now its numbers have become a threat to health, to energy resources and to mobility itself – hence the renewed interest in pollution-free electric cars.

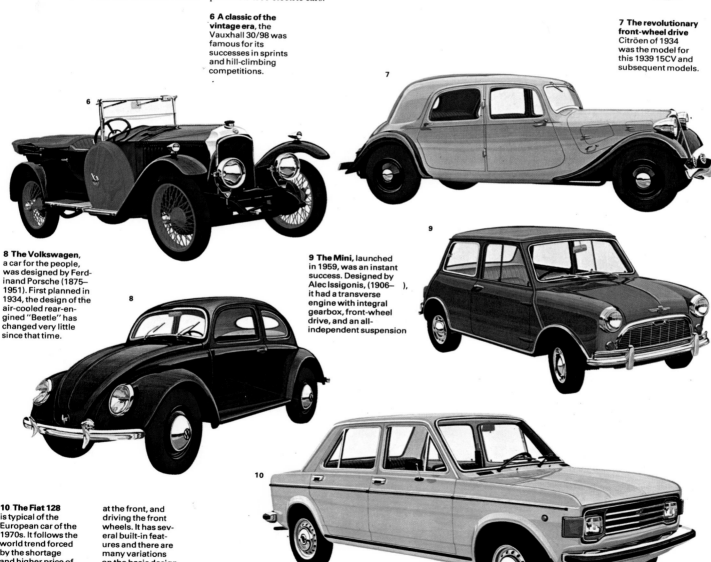

6 A classic of the vintage era, the Vauxhall 30/98 was famous for its successes in sprints and hill-climbing competitions.

7 The revolutionary front-wheel drive Citröen of 1934 was the model for this 1939 15CV and subsequent models.

8 The Volkswagen, a car for the people, was designed by Ferdinand Porsche (1875–1951). First planned in 1934, the design of the air-cooled rear-engined "Beetle" has changed very little since that time.

9 The Mini, launched in 1959, was an instant success. Designed by Alec Issigonis, (1906–), it had a transverse engine with integral gearbox, front-wheel drive, and an all-independent suspension

10 The Fiat 128 is typical of the European car of the 1970s. It follows the world trend forced by the shortage and higher price of fuel and has a small high-revving engine for economy (with an overhead camshaft), mounted transversly at the front, and driving the front wheels. It has several built-in features and there are many variations on the basic design. It is available with two or four doors, as an estate car, and in "Rallye" and sports-coupé versions.

Classic cars

The history of the automobile spans less than 100 years from the spluttering experiments of pioneers such as Benz, Daimler and Panhard to the mass-produced, energy-conserving designs of the 1970s.

Between the two world wars, the fundamental principles of the machine had become well established. Henry Ford (1863–1947) in the United States had proved the economic common sense of mass production and motoring became available to an ever larger section of the population.

Yet during this time there emerged a few really great cars. They were carefully, almost lovingly, built – generally one at a time, to individual customers' orders. Some incorporated radical innovations, although most merely represented the best available combination of design and engineering skills. Their names became synonymous with quality and status. Some, such as Alfa Romeo, Rolls-Royce, Cadillac and Mercedes, survive as names. A whole list of others has gone. But these classic cars were the thoroughbred bloodstock whose descendants form one of today's major industries.

CONNECTIONS

See also
126 History of automobiles
124 History of motor cycles
122 History of bicycles
110 History of transport

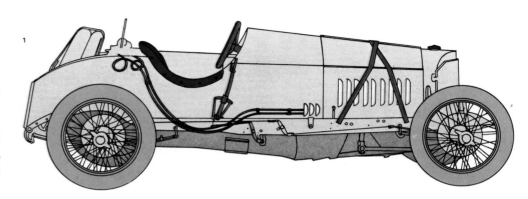

1 Mercedes (Germany, 1914) was designed by Paul Daimler for that year's Grand Prix. Its 4.5-litre engine gave it a maximum speed of 180km/h (112mph). It had two magnetos and three sparking plugs per cylinder. Front-wheel brakes were a post-World War I addition.

2 Hispano-Suiza (France, 1922) had a steel and aluminium engine, based on earlier aero-engines by the Swiss designer Marc Birkigt, of 6.6-litre capacity. It was the first car to have servo-assisted four-wheel brakes, and had a maximum speed of 137km/h (85mph).

3 Isotta-Fraschini A (Italy, 1929) had twin carburettors to feed petrol into its 7.4-litre eight-cylinder engine, giving an output of 120bhp. Its Italian manufacturer pioneered the use of four-wheel brakes.

4 Duesenberg (United States, 1930) was one of America's most expensive cars. Its "straight-eight" 6.9-litre engine had more than 260bhp to give the car a top speed of more than 175km/h (110mph).

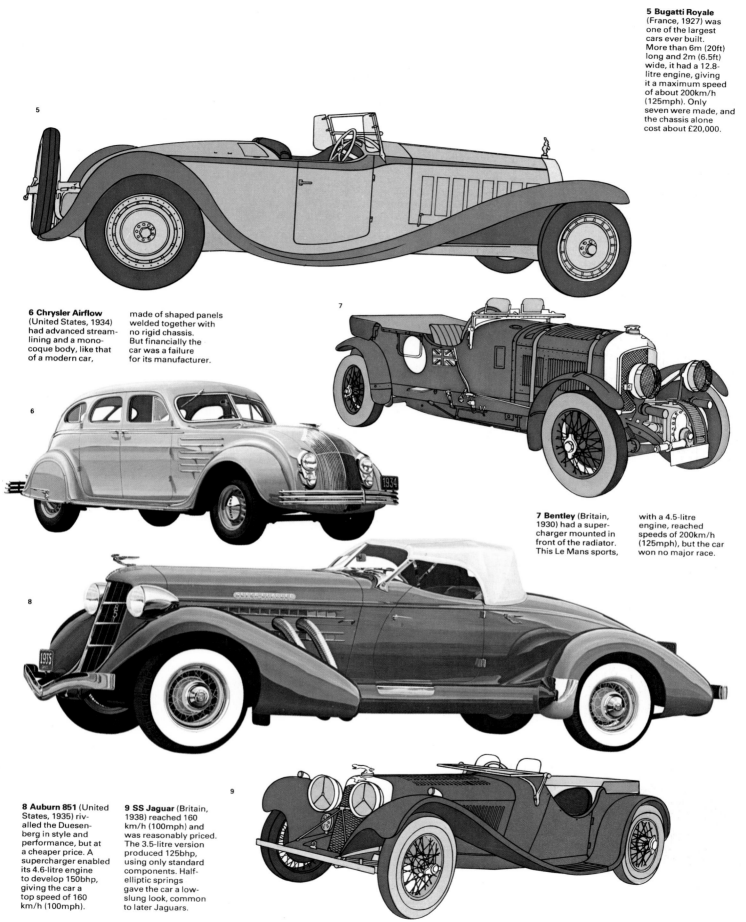

5 Bugatti Royale (France, 1927) was one of the largest cars ever built. More than 6m (20ft) long and 2m (6.5ft) wide, it had a 12.8-litre engine, giving it a maximum speed of about 200km/h (125mph). Only seven were made, and the chassis alone cost about £20,000.

6 Chrysler Airflow (United States, 1934) had advanced streamlining and a monocoque body, like that of a modern car, made of shaped panels welded together with no rigid chassis. But financially the car was a failure for its manufacturer.

7 Bentley (Britain, 1930) had a supercharger mounted in front of the radiator. This Le Mans sports, with a 4.5-litre engine, reached speeds of 200km/h (125mph), but the car won no major race.

8 Auburn 851 (United States, 1935) rivalled the Duesenberg in style and performance, but at a cheaper price. A supercharger enabled its 4.6-litre engine to develop 150bhp, giving the car a top speed of 160 km/h (100mph).

9 SS Jaguar (Britain, 1938) reached 160 km/h (100mph) and was reasonably priced. The 3.5-litre version produced 125bhp, using only standard components. Half-elliptic springs gave the car a low-slung look, common to later Jaguars.

How an automobile works

A typical modern car can be divided into four main component systems: the engine, producer of the power; the transmission, which feeds the power to the road wheels; the electrical system; and the body/chassis, including steering, brakes and suspension [Key]. Wherever the engine is placed – at the rear, driving the rear wheels, in front, driving the front wheels, or even amidships – the working principle is basically the same. In the conventional front-engined, rear-drive car (the construction that is cheapest), the engine feeds rotary power via the clutch, gearbox, propeller shaft and differential to, finally, the back axle and road wheels.

Transmission
By using the clutch [6], the driver is able to connect or disconnect the engine's power from the road wheels, to engage gears, to start smoothly, and to stop the car without stopping the engine. When the driver depresses the clutch pedal while the car is in gear, the drive (power) is disconnected from the gearbox and the rest of the transmission; releasing the pedal reconnects the drive.

Generally located just behind the clutch, the gearbox (either manual or automatic) is designed to vary the ratio of speed between engine and road wheels. The normal petrol engine works best at between 2,000 and 5,000 revolutions a minute (the rate at which the crankshaft turns). To permit this while the car is moving at anything from 15 to 150 km/h (9 to 90 mph), the usual manual gearbox has a selection of four different forward gear ratios, through four pairs of gears [5]. Selecting low (first) gear allows the engine to turn at its working speed while driving the road wheels slowly, resulting in a greater torque or turning effort needed to overcome inertia, heavy loads or a gradient. When the car gathers speed and less effort is needed to power it, successively higher gears are engaged until top may be used.

The propeller shaft, running under the floor along the length of the car, is attached at its forward end to the gearbox and at the rear to the differential. The differential has two functions [7], to "bend" the driving power at right angles and feed it to the rear axle and wheels and, when the car is steered round corners, to allow the outer wheel to travel faster than the inner one.

Electrical system and brakes
In pre-1907 motoring days, the sole function of the battery was to produce the spark for plugs (which ignite the petrol-air mixture in the cylinders), but today the car depends on several electric devices for its operation [4]. These are all powered by the battery, which is re-charged by a dynamo (or alternator), driven by the crankshaft through a belt which also turns the cooling fan. The 6- or 12-volt battery supplies the coil (an induction coil), which produces high-tension electricity for the plugs via a distributor. The battery also provides current for the horn, lights, heater, windscreen wipers, radio and, the heaviest drain of all, the starter motor.

Almost all modern cars, benefiting from racing practice, have disc brakes on the front wheels and drum brakes on the rear. Some cars have discs all round. Metal discs which rotate with the wheels are gripped by stationary pads when the footbrake is applied [8]. As the discs are open to the cooling air, heat

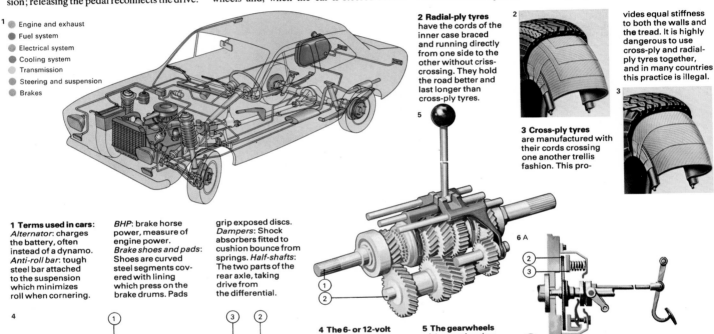

1
- Engine and exhaust
- Fuel system
- Electrical system
- Cooling system
- Transmission
- Steering and suspension
- Brakes

2 Radial-ply tyres have the cords of the inner case braced and running directly from one side to the other without criss-crossing. They hold the road better and last longer than cross-ply tyres.

vides equal stiffness to both the walls and the tread. It is highly dangerous to use cross-ply and radial-ply tyres together, and in many countries this practice is illegal.

3 Cross-ply tyres are manufactured with their cords crossing one another trellis fashion. This pro-

1 Terms used in cars:
Alternator: charges the battery, often instead of a dynamo. *Anti-roll bar*: tough steel bar attached to the suspension which minimizes roll when cornering.

BHP: brake horse power, measure of engine power. *Brake shoes and pads*: Shoes are curved steel segments covered with lining which press on the brake drums. Pads

grip exposed discs. *Dampers*: Shock absorbers fitted to cushion bounce from springs. *Half-shafts*: The two parts of the rear axle, taking drive from the differential.

4 The 6- or 12-volt car battery must, through the coil, deliver 10,000 volts at up to 300 times a second to the plugs, and must also provide current for starting, heating, lighting and electrical accessories. The diagram shows only the starting, ignition and recharging electrical systems.

1 Battery
2 Ignition key
3 Electromagnetic relay, activated by the key when is starting, connecting the battery to the starter motor
4 Starter motor
5 Dynamo or alternator, driven by the engine to recharge the battery
6 Control box
7 Ignition coil
8 Primary coil
9 Secondary coil
10 Distributor
11 Contact breaker
12 Rotor arm
13 Spark plug

5 The gearwheels of the gearbox (except reverse) are always in mesh. Those on the output shaft [1] revolve around it and those on the layshaft [2] are fixed. When a gear is selected, the appropriate gearwheel is locked to the output shaft. In first gear, the widest ratio is used for low-speed driving; second and third gears use progressively narrower ratios, and top gear is obtained by coupling the input shaft directly to the output shaft. Overdrive is a separate and higher top gear fitted to some cars to reduce wear and tear and petrol consumption. It may be engaged automatically or by the driver.

6 The clutch is basically made up of three plates: the flywheel [1], which is fixed to the engine shaft and rotates with it; the driven plate [2], which is connected to the gearbox shaft; and the pressure plate [3], which clamps the

driven plate to the flywheel when the clutch is engaged by releasing the clutch pedal [B]. Disengaging the clutch by depressing the pedal [A] separates the plates so that the flywheel and driven plates rotate independently.

130

is quickly dissipated, avoiding brake-fade, the bogey of overworked and overheated drum brakes. All four brakes are operated by the brake pedal via hydraulic lines. The parking (hand) brake operates on the rear wheels only, usually by a mechanical linkage.

Suspension and construction

Suspension is designed to give the passengers a comfortable, smooth ride, and to protect the body and parts of the car by reducing the shocks from the uneven surface of the road [10]. However, springs alone would give a bouncing ride, and shock absorbers are fitted to "damp" down the oscillation that the springs themselves produce. Traditional elliptical or semi-elliptical springs have in many cars been replaced by helical or coil springs, torsion bar springs (in which a twisting action is used as springing), gas-and-fluid (combined springs and shock absorbers) or rubber springs, or several types combined.

The front wheels of a car are each mounted on separate short axles, so that when the steering wheel is turned and the movement passed to them, each wheel turns on its own axis (the inner one describing a slightly tighter arc than the outer). Rack-and-pinion steering [9], the most popular of several systems, has a pinion on the end of the steering column that engages a transverse toothed rack. The rack, connected at its ends to track-rods attached to each road wheel, is moved right or left by the action of the steering wheel, steering the wheels in the required direction. Power steering makes this easier.

Until the 1930s, the traditional way of building a car was by making a rigid chassis (the wheels, machinery and frame). Everything else was bolted on to the chassis. Now many manufacturers use the body itself as the frame. When welded together, the pressed-steel body panels form a rigid "box", each unit contributing to the strength of the structure. Using unit-construction (monocoque) methods, cars can be made more cheaply, and are considerably lighter than earlier models built on a separate chassis. A number of small manufacturers produce cars using light alloy or fibreglass bodies, which need a separate chassis, often a tubular "space-frame" on which to build the car.

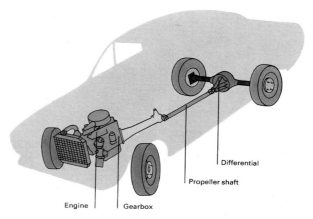

A typical modern car has the engine mounted at the front driving the rear wheels through a gearbox and propeller shaft. The engine, suspension and steering system are all fixed in to the main body of the car, which is constructed as a welded rigid "box" from separate curved body panels. Many manufacturers design cars which can be adapted for right- or left-hand steering, for worldwide sales.

Differential
Propeller shaft
Engine
Gearbox

7 The differential allows one of the half-shafts and its road wheel to rotate more slowly than the other when the car is turning, although both are still being driven, thus improving the cornering and reducing tyre wear. The two half-shafts take the drive from the differential to the road wheels. The diagram shows that when the driver is turning the steering wheel, the rear inner road wheel describes a tighter, shorter arc than the rear outer wheel. Turning a corner would result in tyre scrub and loss of handling qualities without the differential. A pinion on the end of the propeller shaft turns the crown wheel in the differential, which rotates four bevel pinions, allowing the half-shafts to be driven at different speeds.

8 When the brake pedal is depressed, a piston in the master cylinder forces fluid along hydraulic pipes to slave cylinders on each wheel, pushing shoes or pads into contact with drums or discs. (Brake-shoe pads are curved steel platforms covered with tough fibrous shoes which act on the inside of the brake drums. Pads act on exposed discs holding them in a vice-like grip.)

1 Brake pedal
2 Master cylinder
3 Hydraulic pipe
4 Brake shoe and lining
5 Brake drum
6 Slave cylinder
7 Drum brakes on
8 Drum brakes off
9 Disc brakes on
10 Disc brakes off
11 Brake pad
12 Disc

9 Two main types of steering systems are commonly used. Rack-and-pinion steering has a toothed pinion [1] at the end of the steering column [2], which engages with a transverse rack [3] moving it right or left as necessary. Track rods [4] at each end transmit the movement to the wheels. The steering box system (not shown) has a box which houses a worm reduction gear. The gear drives a drop arm, and, via a transverse link, a slave arm. The power-assisted system is a modern refinement of steering, which facilitates driving larger cars by using power steering worked by hydraulic pressure.

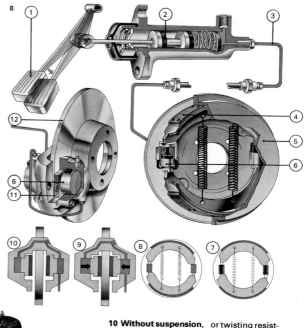

10 Without suspension, every irregularity of the road surface would be transmitted to the occupants of the car. Springing avoids this problem, but to avoid over-springiness, damping must be introduced. [A] shows a rear suspension layout with leaf springs [1] mounted on the axle. Front-wheel suspension [B] incorporates an anti-roll bar [2]. This is a steel bar attached to the suspension to minimize roll by its torsion or twisting resistance when a car corners rapidly. It is not the bar fitted to some cars to prevent the occupants being crushed if the car turns over. Coil springs [3] absorb road shocks. Telescopic shock absorbers [C] are often called dampers and are fitted to the chassis and suspension to cushion bounce from springs. Oil is forced through the constricting valves and slows down the recoil movement.

Cars and society

The internal-combustion engined automobile has been in common use for less than 70 years. At first a toy, then a mode of transport for the rich, and later part of the pattern of living, it was designed as man's mechanical servant. Together with the lorry, it revolutionized the world's trade and social life by its speed and mobility.

The introduction of cheap, mass-produced cars, pioneered by Henry Ford's Model T of 1908, brought personal transport to ordinary people. They could go almost anywhere and the new-found freedom created the beginnings of domestic tourist industries. More expensive cars became status symbols, often representing the wealth or importance of their owners.

The threat to society
Now it is a question of the car's survival or demise. Today, with 220 million vehicles on the world's highways, many people think that the answer to the rapidly increasing problem of pollution and fuel shortages can at last clearly be seen. They consider that sometime in the future society must forget the

automobile as we know it – a five-passenger, four-wheeled vehicle up to 5m (16ft) long and 2.5m (8ft) wide using the appallingly inefficient internal combustion engine, pouring toxic wastes into the air, damaging people's ears and minds with its noise, congesting cities, creating fuel shortages and beginning to take away the very freedom of movement for which it was developed. There may have to be a radical change in size, power unit and people's approach to their personal transport.

What type of engine?
All combustion, from that in a bonfire to a car engine, produces undesirable by-products – carbon monoxide, various unburned hydrocarbons and, from cars, nitric oxide, lead salts, iron oxide and soot in exhaust smoke. New regulations governing the toxic content of exhaust fumes have done much to reduce air pollution. But they cannot (even with the most stringent emission curbs such as those laid down for the future in the United States in 1975, which demand reduction of around 95 per cent) hope to eliminate

automobile-related pollutants entirely. Similarly, the most vigilant noise-abatement organizations cannot hope effectively to damp the roar of heavy, under-powered vehicles working under stress. And even the most ambitious planning of city centres can at best clear only a fraction of shopping space – at the cost of adding to the numbers of vehicles on the roads elsewhere.

In the short term, current and planned regulations will help in certain areas. But automobile designers have long been investigating the day-after-tomorrow, particularly with regard to engines, fuels and overall size reduction, in addition to various public transport systems such as monorail, hover and "bullet" trains.

Modern steam road vehicles have been tested for more than 25 years; their basic problems are weight and water supply. Research continues on low-emission systems such as the stratified-charge engine [1], gas turbine, Stirling (hot-air) engine and hybrid-electric [Key] (in which such an engine powers a generator to charge batteries used for cruising). Other possible power units

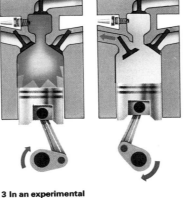

1 The stratified-charge engine is in effect a conventional petrol engine with a modified cylinder head and induction system. In an ordinary engine, the petrol-air mixture is of similar density in all parts of the combustion chamber. In the stratified-charge unit, it is richer near the plug and weaker elsewhere. The rich mixture near the plug ignites readily and the weaker mixture burns more completely.

2 Diesel-fuel exhaust emission from a correctly adjusted engine pollutes the air much less than does a petrol-fuelled unit. Of all the toxic fumes discharged by any type of internal combustion engine, invisible and odourless carbon monoxide is the most damaging. A petrol engine produces thirty times as much as a diesel engine. Can A represents petrol engine emission, B diesel.

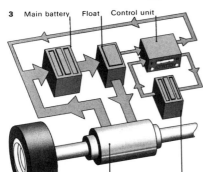

3 Main battery   Float   Control unit

Electric motor   Accessory battery

3 In an experimental urban car electric-drive system, power for accessories is taken from either an accessory battery or through the float from the main source.

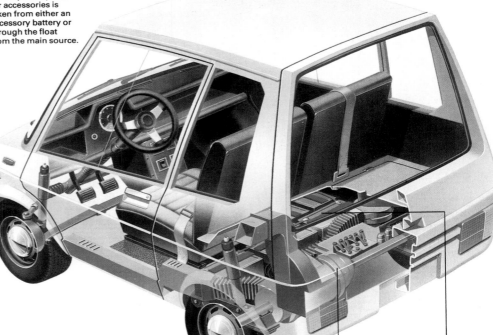

Electric motor   Batteries

4 Short-range personal transport is provided by this experimental urban electric car. A direct-current electric motor mounted on the rear axle is driven by a special 84-volt lead-acid battery, which gives the car a greater range than conventional batteries. A built-in charger can be plugged into a household socket and recharges the battery in 7 hrs. This General Motors car has a range of 93km (58 miles) at 40km/h (25mph).

are pure electric (one of the great goals, because electricity supplies can be made almost unlimited, but still handicapped by heavy batteries and the need for frequent recharging), and fuel cell power, which would convert conventional fuel energy directly into electric power without burning it. This would be one of the most significant developments of our time, but is not likely to be practicable for some years. More than 250 designs of small electrically powered "city" or urban cars have been produced; if the cars were made commercially they could help congested cities and relieve parking problems (three can be parked in one normal bay), although light and heavy traffic would probably have to be segregated.

The future of the car engine
Current research has already produced technical advances aimed at improved economy and ecology, although most of them would, at best, provide only temporary relief. The car industry has produced the catalytic converter [6], a type of de-polluting silencer for use with low-lead petrol. Fuel economy has been improved by electronic ignition, by steel-ply radial tyres which have less rolling resistance, by lower rear-axle ratios, scaled-down engine capacities and, mainly in the United States, the reintroduction of overdrive.

There are three immediate goals. First, future engine design (internal combustion engines will probably remain for at least the next 15 years) must aim at conserving fuel and reducing exhaust pollutants. Second, a way of controlling traffic density in cities, and its flow elsewhere, must be found. Third, and most urgently, greater safety must be built into structural design and additional equipment provided (such as collapsible steering columns, rigid "boxes" and deflecting properties), not only to combat the injuries likely to be received in a collision but also to prevent accidents (improved tyres, brakes, lights, suspension and visibility).

Long-term aims must be to find an alternative power system now that oil supplies – even offshore supplies – are unpredictable. And a fundamental reappraisal in society's attitude to all road transport, its appearance and its function, must be made.

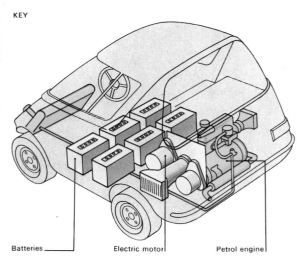

Batteries   Electric motor   Petrol engine

The hybrid city car was developed as a possible answer to pollution in towns. For city and suburban use the car runs on its electric motor, but on out-of-town journeys the small internal combustion engine is started. The electric unit is used on its own only where there must be no pollution and the petrol engine charges the batteries at other times.

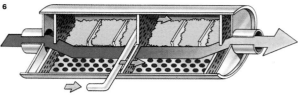

5 The gas-turbine car engine was developed originally by the British Rover Company, which tested the first gas-turbine car in March 1950. It is quiet, powerful, has low maintenance costs and runs on low-grade, lead-free fuel. But it is expensive to manufacture.

1 Air in
2 Radial compressor
3 Compressor turbine
4 Fuel in
5 Power turbine
6 Exhaust
7 Power out

6 Catalytic converters, built into silencers, have been used as a partial solution to present exhaust pollution. Nitrogen oxides are reduced to ammonia at the first catalyst bed and hydrocarbons and carbon monoxide are converted to carbon dioxide and water at the second. The most suitable catalyst is platinum but this is extremely expensive. The system has proved to be efficient enough for some manufacturers to re-calibrate their engines for higher peak performance, using catalytic converters to "clean up" the extra pollution in the exhaust gases.

7 The ideal safety car [1] has a rigid compartment protecting passengers restrained by seat belts and head rests. The car body is designed to absorb impact and, in a collision, the engine deflects downwards, and the steering column collapses. The injuries sustained by unrestrained passengers [2] depend on the speed of the collision (here 97km/h [60mph]), but the location and type of injuries are similar.

Moment of impact

Full impact

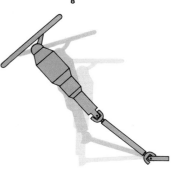

8 A collapsible steering column is hinged mid-way along at a universal joint. With an ordinary rigid steering column, a head-on collision can make the wheel break off and impale the driver on the column. The collapsible column hinges on impact, absorbs some of the force and swings away from the driver's chest. The lower arc of the wheel should be padded or flat to lessen injury.

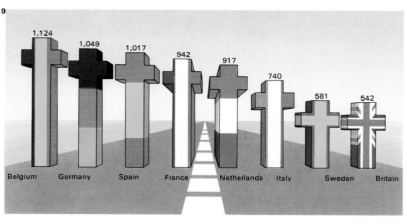

Belgium 1,124   Germany 1,049   Spain 1,017   France 942   Netherlands 917   Italy 740   Sweden 581   Britain 542

9 Deaths in road accidents are quoted here as the number per million vehicles in various European countries during 1971. The USA had 493 per million, and Japan about twice as many as Belgium.

10 A puncture at speed is extremely dangerous – the tyre collapses. A recent development uses a ring of containers [A] that burst when the tyre deflates [B], releasing fluid that seals the hole [C] and vaporizes to re-inflate the tyre [D].

Trams and buses

Industrialization – the transformation of scattered cottage industries into centralized factories and offices – demands good commuter transport. In the early stages of industrialization people moved in from rural areas to live within walking distance of their places of work. But as the developing cities sprawled outwards an efficient system of passenger transport became necessary to feed industry its manpower.

The motor car is inefficient for such a purpose because it takes up far too much road space for the average number of passengers it carries and produces congestion. And it was invented too late to form the basis of a good system. Bicycles are a possible solution, but few people are willing to ride them in all weathers or over long distances. Systems of public transport are essential.

Early passenger transport systems
Cities in Europe and America began to grow large enough to need passenger transport systems in the early 1800s, when the only feasible source of motive power available was the horse. The horse-drawn bus might have

seemed the obvious choice of vehicle, being a logical development of the stagecoach, but roads were in such poor condition in cities that coach-like vehicles gave a very uncomfortable ride. The first extensive transport systems used horse-drawn trams rather than buses because rail-borne vehicles not only carried people smoothly but also allowed a horse to pull twice as many passengers, because of easier rolling. Medieval miners transported minerals by pushing wheeled tubs along primitive rails made of wooden beams. The word tram reflects this origin, being derived from the Low German *traam*, meaning beam, although Americans use the more descriptive word "streetcar".

From tram to trolleybus
Trams and railways developed together. The first city tram network spread through New York in the 1830s, at the time when steam railways began to appear in Britain. Steam soon came to the tramway too and the first steam engine to pull a tram chugged its way around New York in 1837. Europe lagged some way behind – the first horse-drawn

tram network did not open in Britain until 1860 and steam trams did not appear until 1872. Europe pioneered the next and most important development, however, with electric trams [1] in Berlin in 1881.

Coincident with the development of the electric tram came the trolleybus. By the first years of this century commercial systems were operating in Europe and the first English trolleybus routes operated in Leeds and Bradford in 1911. Trolleybuses had twin poles and overhead wires because the rubber wheels insulated the vehicles from the ground. (Trams needed only a single wire for current pick-up, the circuit being completed through the metal wheels and rails.) The first trolleybuses collected current through a trolley slung on the wires, hence their name, but the sprung poles used on trams were soon found to be more efficient.

The trolleybus was a silent and fume-free form of transport capable of fast acceleration and greater manoeuvrability than the tram. Trolleybus networks were built where a city did not want to afford the expense of laying rails for trams and they were often set up in

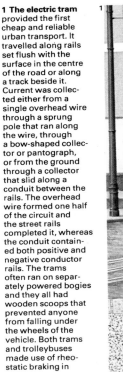

1 The electric tram provided the first cheap and reliable urban transport. It travelled along rails set flush with the surface in the centre of the road or along a track beside it. Current was collected either from a single overhead wire through a sprung pole that ran along the wire, through a bow-shaped collector or pantograph, or from the ground through a collector that slid along a conduit between the rails. The overhead wire formed one half of the circuit and the street rails completed it, whereas the conduit contained both positive and negative conductor rails. The trams often ran on separately powered bogies and they all had wooden scoops that prevented anyone from falling under the wheels of the vehicle. Both trams and trolleybuses made use of rheostatic braking in which the motors act as brakes.

2 San Francisco has the oldest cable tramway, which opened in 1873. Other hilly cities soon introduced cable-hauled trams. In 1884 the first cable tramway in Europe opened on Highgate Hill in London. Other cities noted for cable systems were Melbourne, Kansas City, Edinburgh and Wellington. Cable trams proved unreliable and are no longer in use except in San Francisco and Wellington.

3 An articulated tram makes its way through the streets of Stuttgart in Germany. Modern tramways often couple cars together in this way and thus achieve a more economic method of operation.

4 The first trolleybuses were introduced soon after the turn of the century by the tramway companies. At first current-collecting trolleys ran on top of a pair of overhead wires, but later ones contacted the conductors below them.

the suburbs to feed the inner tram routes.

Both trams and trolleybuses had to contend with buses and found the competition more and more intense. Buses have a long history, the first horse-drawn vehicle being run in Paris by Blaise Pascal in 1662.

Buses versus trams

The name bus – from the Latin *omnibus*, meaning "for all" – appeared shortly before the introduction of the first horse-drawn bus into Britain in 1829. Steam buses soon followed, but the first motorbus – an elegant petrol-driven coach built in Germany by Benz – did not begin service until 1895, by which time trams were well established.

Compared with trams the early buses were small, noisy and smelly and the solid tyres gave the passengers a bone-shaking ride. But buses soon improved and tramways began to decline. In the aftermath of World War II many disappeared as city councils decided to forgo the costs of re-equipping their tramways.

As motor traffic increased trams often impeded cars and severe congestion occurred, exacerbated by dewirements and power failures. Buses became more economic and were more flexible in routing than either trams or trolleybuses, and consequently have now almost entirely superseded them. Trams still operate in several European cities and trolleybuses are to be found in the USSR and Switzerland. In Britain trams still ply the streets only in Blackpool and the Isle of Man and all trolleybuses have disappeared, the last to go being those of Bradford in 1972.

Modern tram designs include articulated vehicles [3] of several interconnected coaches, but flexibility of design has always been a feature more typical of motorbuses. The descendants of the elegant many-doored, open-air touring charabancs of the 1920s are the luxurious air-conditioned, toilet-carrying, reclining-seated, long-distance coaches of today [7]. The old "anymore-fares" city bus is giving way to the pay-as-you-enter, turnstiled bus with few seats and a large standing area. And minibuses and dial-a-bus services now carry passengers on less busy routes.

The first trams were pulled by steam "locomotives", generally with vertical boilers and side panels that totally enclosed the moving parts and wheels.

This tram ran in East London in 1887. Its engine used a steam pressure of 100lb per sq in and it burned about 9kg (20lb) of coal for an hour's working.

Passengers rode in a non-powered trailer, with no weather protection for those on the upper deck. Other trams had compressed air engines.

5 Motorbuses were introduced to the streets of Paris and London at the end of the nineteenth century. Their petrol engines were noisy and smelly, but the new buses soon demonstrated their advantage over the tram – the ability to go anywhere served by a road. This type of English bus, the London General K414, was developed at the end of World War I, and first came into service in 1920.

6 One-man operated buses are a recent development in urban public transport. Passengers either pay the driver or buy tickets from a coin-operated machine. The system saves on bus company staff but can cause delays while passengers are boarding the bus. For short, high-capacity routes in cities some transport companies use "standing-room only" vehicles.

7 The Greyhound Bus network covers the whole of North America, its coaches providing a cheap and reliable means of inter-city transport. Passengers may spend several days and nights aboard.

8 A tourist coach travels through the resort of Sitges in Spain. Such vehicles are often extravagantly arrayed to attract tourists and designed to provide a view of the town rather than a means of rapid urban transport.

Special purpose vehicles

Conventional vehicles such as cars and trucks are designed to run on firm roads, normally with only slight gradients, and carry average loads. A special purpose vehicle has some extra features. It may be able to travel over unusual terrain or to carry a load that is beyond the capacity of ordinary vehicles.

For economy, speedy development and sometimes reliability, specialist vehicles are designed to make the best use of components already available. A standard truck chassis can be fitted with a special body or be used to tow various trailers [7]. Existing components can be built on to a new chassis or into a monocoque hull welded from flat sheets.

Basic design considerations
The first consideration must be the type of load to be carried. This may be two oil explorers and their instruments, 30 tonnes of timber, or a 120mm gun, ammunition and crew. In general, large vehicles are best for bulk loads. Off the road their size makes obstacles relatively easy to surmount – what would be an obstacle to a conventional vehicle becomes less significant. But on the road, many special purpose vehicles are difficult to manoeuvre because of their excessive size.

The terrain over which loads are to be taken is another important consideration in the design of a special purpose vehicle. Soft ground calls for running gear that spreads the load over a large area to lessen the chance that the vehicle will sink into the ground. There must be enough traction to overcome the resistance of loose soil and to cope with slippery gradients. The two main choices are all-wheel drive or caterpillar tracks.

A wheeled vehicle tends to be less expensive and quieter than a tracked vehicle. Extremely large wheels are fitted to small vehicles, making them as good as tracked vehicles on soft ground and giving the added advantage that, with low-pressure tyres, they can float. In some so-called all-terrain vehicles (ATVs) the tyres are so broad that the wheels cannot be pivoted for normal steering. Instead they employ "skid" steering by braking the wheels on one side of the vehicle.

For a given size and weight, caterpillar tracks grip better than wheels and distribute the vehicle's weight more evenly, making them a good choice for cross-country use. Tracked vehicles also have a lower fuel consumption in heavy conditions. They are steered by braking the track on one side.

Suspension for cross-country work
The type of suspension is determined by the speed at which the vehicle has to travel over the roughest ground it is likely to encounter. Trucks and cars designed for smooth roads have only slight suspension resilience. If driven fast across country their suspension is liable to break. For this reason, vehicles adapted with only minimum modification for off-road use have a "hard" suspension giving the driver such an uncomfortable ride that he will not go too fast. Agricultural tractors and their derivatives normally operate so slowly they manage without any suspension at all, although wheeled tractors have resilient tyres, producing bounce. Tracked "crawler" tractors have no resilience; they are slow and are used only for heavy work.

The prime examples of fast cross-country machines are armoured fighting vehicles (AFVs), such as tanks, with tracks and

1 Concrete carriers transport large quantities of concrete that has been mixed away from the site to where it is needed. The drum revolves slowly during the journey to keep the concrete properly mixed. The convenience of off-site mixing justifies the cost of transporting concrete by road.

2 Tracked commercial vehicles are used in snow and on marshy ground. In the Arctic areas of Alaska, Canada and the Soviet Union they are used by the timber industry and by oil and ore prospecting companies. They range from small tractors for hauling sledges, and slightly larger personnel carriers with heated cabins, to huge 40-tonne load platforms.

3 The Jeep was first produced in 1941. It was simple and cheap to manufacture; strong; had good cross-country performance; and yet was small, light and easy to recover if it became bogged down or immobilized. The name "Jeep" is shorthand for General Purpose – GP. The Jeep was adapted for many special tasks. With its windscreen folded flat and a machine gun fitted, the Jeep made a good scout car; with locally produced frames to take stretchers it could be used as an ambulance. Most rivals to the Jeep are complex and heavier.

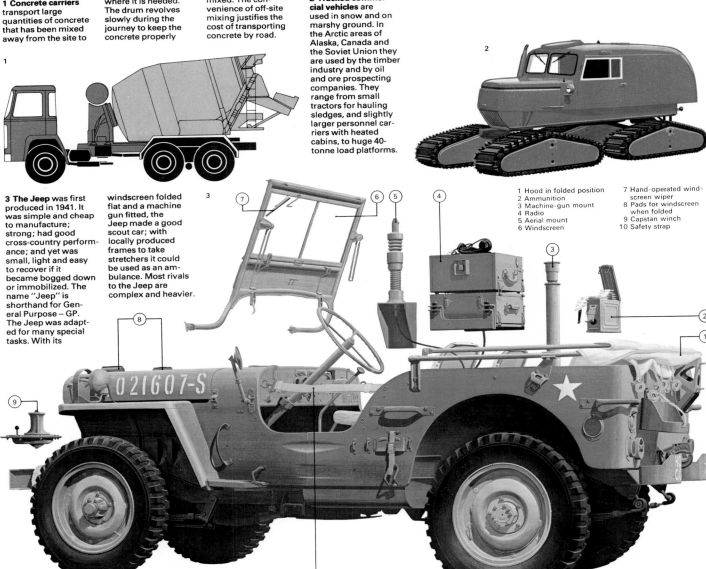

1 Hood in folded position
2 Ammunition
3 Machine-gun mount
4 Radio
5 Aerial mount
6 Windscreen
7 Hand-operated windscreen wiper
8 Pads for windscreen when folded
9 Capstan winch
10 Safety strap

suspensions of high resilience. Vehicles such as earth-movers would benefit from suspensions when hauling "spoil" but at present an increased operating speed would not justify the extra cost that would be incurred.

A small European family car carrying the driver alone has power in relation to the vehicle weight of about 60bhp per tonne, whereas an expensive sports saloon might have 180bhp per tonne. At the other extreme the European limit for trucks is a minimum of 8bhp per tonne (6kW/tonne) and manufacturers have problems achieving this modest power for large trucks at the limit of 38 tonnes laden weight (requiring 304bhp).

For economy of operation, a diesel engine is supreme in commercial vehicles, unless the vehicle is small or high power output is needed for minimum weight, when a petrol engine is a better choice.

Generally the transmission – the linkage that connects the drive from the engine to the wheels or tracks – must suit both road journeys at relatively high speeds and slow, heavy work across country. It must also provide the power for winches and other equipment.

Except in simple vehicles, the transmission is usually as heavy as the engine.

Steering a special purpose vehicle

Short tracked vehicles can be steered by using the transmission to drive the tracks at different speeds. A long, narrow vehicle is built in two parts, with a powered and articulated joint in the middle, and the vehicle steered by "bending" it sideways at the joint. Such vehicles have a good performance on soft soil such as clay or on snow. In the same conditions, a wheeled vehicle should have wheels of large diameter rather than of broad section, but there is seldom enough room to fit them. A sharp tread on the tyres is necessary when the surface is weak. In dry sand fat tyres perform well, but the tread should not bite deep and make the sand grains flow from under the wheels of the vehicle.

Reliability is difficult to achieve in special purpose vehicles. Often their use cannot be simulated accurately in trials and over-insurance in design – making all parts stronger than necessary – must be avoided or the vehicle will be excessively heavy.

A tipper truck justifies a special design, because its body is so short. The loads it carries are massive yet consist of small particles, such as sand. The tipping mechanism allows rapid unloading by the driver working alone, so providing economy and a rapid turn-round time.

4 The Coles Colossus is a 14-wheeled truck designed as a crane carrier. It has hydro-pneumatic suspension and the engine is a Rolls-Royce 300hp turbocharged diesel. Sections can be added to the jib-strut crane in order to convert it into a tall tower crane.

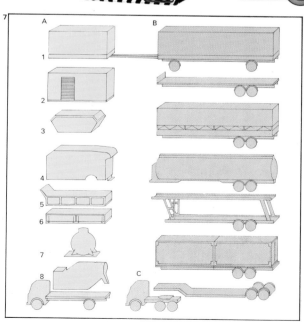

5 The racing car's wedge shape helps to keep it on the track. Its low wind resistance and its wide tyres give it stability on corners.

6 Swamp buggies use large, low-pressure tyres to carry men and materials over inland waters, shifting sand and deep, soft swamp.

7 A basic truck [A] can have various types of vehicle bodies, such as a plain van [1], side-entry van [2], rubble skip [3], dust cart [4], tipper [5], open truck [6], liquid gas carrier [7] or cement carrier [8]. It can also be adapted to tow a trailer [B]. One tractor [C] can be used to haul various kinds of articulated semi-trailers, giving good manoeuvrability and giving the operator full and economic use of the vehicle.

8 This riot vehicle weighs 20 tonnes, has bullet-proof steel and glass for protection, and carries a 15-man squad. It is armed with a water cannon, special sprays and a high-intensity siren. It can assist authorities in controlling crowds and rioters in situations where minimum force is desired.

9 A hovercraft is also a special purpose vehicle although the craft shown is limited to water travel. It uses two gas turbine engines to drive the fans that supply high-pressure air on which it "floats" over deep or shallow water.

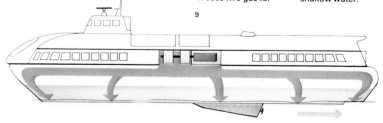

Small technology and transport

Physical communications – basically roads and railways to transport people from place to place – are an essential ingredient of progress. When irrigation canals were built in Indonesia more than half a century ago, the building of roads parallel to the canals did not seem worthwhile. But the lack of such roads led to a decline in standards of inspection and maintenance and as a result the canals steadily deteriorated. Today, to transport additional food for growing populations, the canals are being rebuilt – with service roads, and costing much more than they would have originally.

An adequate transport system is equally necessary for the distribution of food and for the more general trade without which there can be little continuing improvement in the quality of life in any community.

The basic ingredients of physical communication are products of technology. The richer the community, the more sophisticated the technology that develops. Progress is the result of man's mastery of the world he lives in and of his ability to use the resources within his reach. In regions in which modern mechanized road building and modern transport is too costly for the local communities, and where progress is limited by the lack of adequate physical communication, there are two ways in which the problem can be solved. The first is by foreign aid – the system in which the richer nations provide the poorer ones with money, materials or trained personnel. The second is by the use of simpler, less costly technologies. Even where outside aid is available, it may be more productive to use it for a wide range of schemes based on low-cost local technology than on a few highly sophisticated and correspondingly more expensive projects.

Building low-cost roads

The principles of road design are simple, and road building can also be a relatively simple process [1, 2]. Earth, and sometimes rock, must be moved and stones must be quarried, crushed, collected and spread according to a plan. The checking of levels is important because the drainage of the finished road depends on it. But even levelling does not require expensive instruments and techniques. A cheap but accurate method uses a length of transparent plastic tubing with its ends tied along a pair of wooden rods. The rods are placed upright on the ground and the tubing filled with water. If the rods are graduated, say in centimetres, from the bottom up, the difference in ground levels where each is placed can be quickly found by reading off the height of the water-level in each tube, and subtracting one from the other. The tube can be up to 30m (100ft).

The building of bridges

Some roads have to cross a watercourse. It may be possible to construct a shallow ford, but the roadway under the water will be quickly worn away unless it is made of well-cured concrete. In most cases therefore, a bridge is needed.

In the Western world a modern road bridge is generally a structure made of reinforced concrete or, for longer spans, steel. Both materials have advantages where the loads to be carried are frequent and heavy. But this does not mean that they are necessarily the best for a bridge carrying light

CONNECTIONS

See also
182 Road building
190 History of bridges
38 Hand tools
40 Hand-working metals

1 Low-cost methods of road construction vary according to the climate and availability of materials and labour. The basic principles are universal and the following alternatives to more generally accepted methods can often be adapted to suit varying situations. To build an earth road, the first step is to clear trees, shrubs and roots [1]. Trees are cleared to keep the road in sunshine and therefore dry. The topsoil is removed [2] and dumped not closer than 8m (26ft) from the centre of the road. Wide side ditches [3] are dug and soil from them spread to raise the road level between them. The road surface is compacted by rolling [4], ensuring a cross slope (for drainage) of not less than 1 in 20. The ditches must be graded along their length so that excess water can run away [5]. The original topsoil should be relaid on the ditch slopes to encourage the growth of grass [6]. A waterproof surface can be given to a compacted earth road [7] by spreading a layer of 5cm (2in) stones, then brushing and watering in finer grades of crushed stones to fill the spaces. A final rolling produces a dense surface [8]. Bridges can be built using a number of standard prefabricated timber sections supported on an iron beam and fixed together with interlocking metal plates (shown in inset).

traffic in undeveloped areas where cement and steel has to be brought at considerable expense from distant sources of supply, and yet where suitable timber, quarried stone, and bricks may be available locally. Before the general use of iron (from about 1830) and concrete (from about 1890), most bridges were built of masonry or timber, or a combination of both.

The design of bridges is not usually considered a field for standardization. But a recent scheme by the forestry department in Kenya has shown that enormous savings in cost can be achieved by standardization. A British civil engineer working for the department has designed a standard 30m (100ft) timber truss panel made from Kenya cypress. The panels are prefabricated at a central workshop, carried to each site and joined together there to form a bridge with a carriageway running over the top cord. Two panels set parallel to each other, can safely carry a 20-tonne truck. If a route has to be upgraded for heavier vehicles, more panels can quickly be added to an existing bridge.

For longer bridges intermediate timber piers can be built up from the river bed to support two or more spans. A pilot project has been set up that produces an average of one bridge span every three days. Some steel is used for panel couplings and ties, but most of the bridge is built from local materials, using local labour. As a result, the proportion of imported (and hence expensive) materials is kept to a minimum.

Low-cost transport

Modern motor vehicles may be necessary for carrying heavy loads in rural areas, but medium-sized loads can be transported by traditional means provided friction and gradients are kept to a minimum. The systematic construction of roads provides an opportunity for avoiding steep gradients. It also ensures a smooth, hard road surface on which rubber tyres roll with a minimum of friction. Modern wheel bearings eliminate the other main source of friction. By such means [3] an ox-team can be used to carry heavier loads more efficiently. Other means of providing low-cost rural transport depend on the local environment and materials.

A well-designed road does not collect water [A] but allows it to drain off [B]. John McAdam (1756– 1836) first applied the basic principles of a serviceable road – it is the subsoil that supports the traffic and any soil sufficiently compacted and kept dry can support any reasonable weight.

2 Earth- and water-bound roads can be built without costly modern machinery. The 2.4m (8ft) drag-grader [A] has metal edges and is drawn by two oxen. This form of grader was used on the construction of many early US roads. One man can shift a load of soil with a team of oxen and a fresno scraper [B]. It is made from a strong oil drum. The steel-nosed V-drag [C] is used for cutting ditches.

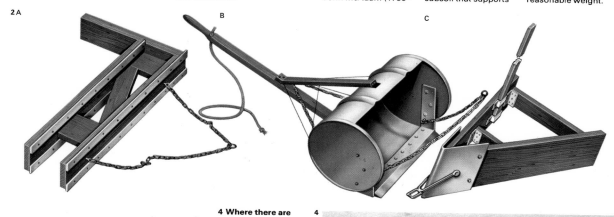

3 The ox-cart is still the most widely used load-carrying vehicle in the rural areas of most of the world's developing countries. Among its advantages is the fact that, unlike a motor vehicle, it is simple to maintain, easy to repair and uses the energy of fodder and not of oil. The ox causes no air pollution and provides manure as a valuable by-product. The main disadvantage of the traditional ox-cart is the energy wastage due to primitive wheel bearings and the friction between solid wheels and the road. By fitting the back-axle, wheel and tyre assembly of scrapped automobiles, the disadvantages of the ox-cart can be overcome at low cost to produce a more efficient vehicle.

4 Where there are water routes the barge provides an economical method of transporting heavy loads. In coastal China an experiment in the mass-production of reinforced concrete sampans has resulted in a cheaper product with a longer life. The six- and ten-tonne sampans are hand-built upside down over a pit, using prefabricated bulkheads over which steel rods and wire mesh are laid before cement platering.

5 The cycle rickshaw is a cheap and convenient form of transport in several South-East Asian cities where there are marked differences in earnings and much unemployment. It often competes directly with the more sophisticated buses and taxis. Opinions may differ about the moral desirability of public transport propelled by human energy, but cyclists such as the one seen here resting are able to work and support their families as well as providing a service.

Locomotives

The spread of railways, which transformed life in the nineteenth century, is linked inextricably with the steam locomotive. To its devotees the steam locomotive was one of the most romantic and beautiful machines ever built. It first appeared in 1804 in a simple version [1] invented by a Cornishman, Richard Trevithick (1771–1833).

Early rail systems

The first steam locomotives to do useful work were ordered and used by coal mines in northeast England in 1813–20. In 1825 a public railway was opened between the English towns of Stockton and Darlington. It had been planned for horse traction, but George Stephenson (1781–1848), a leading builder of colliery locomotives, persuaded the directors to operate a steam locomotive hauling trains heavier than horses could manage. The success of this line led to the much bigger and more important railway between Liverpool and Manchester. It was opened in 1830 after bitter opposition from landlords, coachmen, canal bargees and the large sector of the population that abhorred

any change and considered smoke-spouting locomotives to be engines of Satan. Against spirited competition, Stephenson's *Rocket* [2] was chosen to provide the motive power. It was small and light enough to run on only four wheels without breaking the flimsy iron track. Thanks to steady improvements in manufacturing, it became possible to make boilers stronger, cylinders and pistons more accurate and better fitting, and the whole locomotive capable of developing more power at higher speeds.

For a century steam locomotives provided nearly all the traction for the world's railways. There were no dramatic technical advances but size, power and speed grew constantly. In Europe many rail systems used excellent track, capable of bearing 100-tonne locomotives running at up to 160km/h (100mph). But in the United States in early days, and in most other young, developing countries, track was lighter, and often badly laid by men racing to complete more miles each day. This called for more wheels to spread the load. Engines came to be identified by their wheels, so that 4–6–2 desig-

nated the number of leading, driving and trailing wheels. Speeds were limited and seldom exceeded 80km/h (50mph), apart from one or two short record-setting runs.

The steam locomotive had reached its zenith by the 1930s. European "steamers" were clean, splendidly painted in the livery of their operating companies and, when designed for express passenger haulage, often capable of reaching 160km/h (100mph). American locomotives tended to be more utilitarian. Demands for greater power led to increases in size until they became the biggest land vehicles in history.

Electric locomotives

The first rival to steam came in the form of the direct-current electric motor, adopted in cities (especially in underground railways), to avoid smoke pollution. The first electric train ran at an exhibition in Berlin in 1879. Soon countries such as Switzerland and Norway found that it was cheaper, with the development of hydroelectric power, to generate electricity than to burn brown coal or wood, and their networks became all-electric.

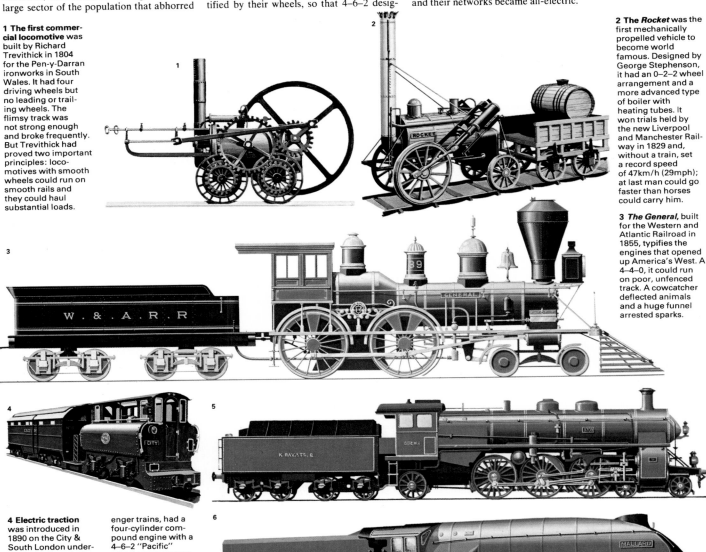

1 The first commercial locomotive was built by Richard Trevithick in 1804 for the Pen-y-Darran ironworks in South Wales. It had four driving wheels but no leading or trailing wheels. The flimsy track was not strong enough and broke frequently. But Trevithick had proved two important principles: locomotives with smooth wheels could run on smooth rails and they could haul substantial loads.

2 The *Rocket* was the first mechanically propelled vehicle to become world famous. Designed by George Stephenson, it had an 0–2–2 wheel arrangement and a more advanced type of boiler with heating tubes. It won trials held by the new Liverpool and Manchester Railway in 1829 and, without a train, set a record speed of 47km/h (29mph); at last man could go faster than horses could carry him.

3 *The General*, built for the Western and Atlantic Railroad in 1855, typifies the engines that opened up America's West. A 4–4–0, it could run on poor, unfenced track. A cowcatcher deflected animals and a huge funnel arrested sparks.

4 Electric traction was introduced in 1890 on the City & South London underground railway in the heart of London.

5 Class 53/6 of the Bavarian State Railway, dating from 1908 and used to pull prestige pass-

enger trains, had a four-cylinder compound engine with a 4–6–2 "Pacific" wheel arrangement.

6 Britain's *Mallard* set the world steam speed record at 203 km/h (126mph) in 1938, pulling a seven-coach train.

Today the electric motor, with its linear form under development, is regarded as the best form of traction for railways but the huge capital costs impede its introduction except on the busiest routes. As long ago as 1955, French Railways demonstrated that electric trains of conventional type [9] could run at more than 320km/h (200mph), but average speeds of public trains have risen only slowly. The Japanese New Tokaido line [10] achieved a sudden jump in speed because the line was laid for high speed. Even so, track and trains need constant maintenance.

Diesel power

About 1920, the first diesel locomotives and railcars came into general use, powered by compression-ignition oil engines developed by the German Rudolph Diesel (1858–1913). Though often noisy, diesel engines pick up speed faster and convert 25 to 45 per cent of their fuel energy into useful haulage, whereas the fuel efficiency of steam traction seldom exceeded eight per cent. Despite greater capital cost, diesel locomotives gradually ousted steam from 1935

onwards, until today steam engines are confined to a shrinking number of railways in Africa and Asia and a few local lines elsewhere. Diesels can be started and stopped easily, burn no fuel when not working, and can run at close to maximum power for hours at a time with no strain on either machines or crew (in contrast to the grimy slavery of the former stoker, or fireman, on the steam footplate). Many diesel locomotives run more than 160,000km (100,000 miles) a year and modern examples are highly reliable, versatile and relatively efficient, as well as being fast [11, 12].

Only in the smallest sizes does the diesel engine drive the wheels through a mechanical gearbox, as on a lorry. Generally the two are linked hydraulically or electrically. Hydraulic transmissions are arrangements of turbines linked by oil under high pressure, and they can transmit smoothly 2,000 horsepower with any ratio between input and output speeds. In the diesel-electric locomotive the engine drives a generator or alternator. This is used to supply current to traction motors similar to those of electric locomotives.

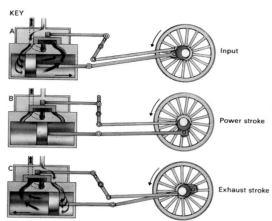

KEY

Input

Power stroke

Exhaust stroke

The double-action engine used to drive a steam locomotive uses superheated steam generated in a heating-tube boiler. Steam entering the cylinder [A] drives the piston back and exhausts steam from the other side of the piston. [B] shows the mid-position, with both inlet valves closed. In [C], steam is led to the other side of the piston and the cycle is repeated. The exhaust steam passes directly into the atmosphere, and this wasted energy, coupled with the design of the boiler, gives this type of steam engine the remarkably low efficiency of only eight per cent.

7

7 The "Big Boy" class of the Union Pacific were the heaviest locomotives built (540 tonnes). With a 4–8–8–4 layout, they hauled heavy freight trains in the Rocky Mountains at up to 120km/h (75mph).

8

9

10

8 The *Beyer-Garratt* (4–6–4 + 4–6–4) of Rhodesia Railways shows how articulated design can fit powerful locomotives to light track. The boiler supplies a pair of engines pivoted at their ends.

9 Class CC 7100 of the SNCF (French Railways), pulling a light train, ran at 331km/h (205mph) in March 1955. It was a standard electric locomotive modified with a high gear ratio.

10 New Tokaido trains began running between Tokyo and Osaka in 1964. With 12,000hp per train, the new electric route of 515km (320 miles) is covered in three hours, but costs are high.

11 Typical of the locomotives of today is a diesel-electric built in Montreal in 1972 for East Africa. Devoid of frills, it spreads its weight on eight axles in two pivoted bogies and can haul big loads.

11

12

12 The High Speed Train, introduced by British Rail in 1973, set a world record for diesel trains when it reached a speed of 230km/h (143mph). Planned for general service from 1976, it has a light but powerful diesel at each end and with 4,500hp can carry passengers at 200km/h (125mph). To increase normal service speed to 250km/h without major alterations to track and signals, British Rail is developing an electrically propelled Advanced Passenger Train. A prototype is planned to run between London and Glasgow in 1978.

Railway transport

Man built railways long before he built steam engines and before he had even mastered the basic technology of engineering with iron and other metals. The earliest railways were constructed of timber and were in operation not later than the fourteenth century. They were built to overcome the severe limitations of other forms of land transport. The provision of a purpose-built track avoided pot-holes and ruts and deterioration due to the weather and simultaneously reduced the friction and rolling resistance of vehicles so that a given force could transport a heavier load.

The first railways

Most railways built before 1825 [1] were what might today be called tramways. They ran over distances of 3km (2 miles) or less and carried a single commodity such as coal or stone (for example between a mine or quarry and a loading wharf for ships). Wagons were pulled by horses or by human beings. Rolling stock was solidly made of wood, reinforced and joined by metal. Wheels were crude, but the smoothness of the track meant that they could be much

smaller than those for carts and coaches running on roads. The track, variously of wood or iron, was built with inner, outer or double flanges to keep the trucks running along it.

At first trucks were pushed individually. Then growing traffic led to two or more being coupled together, by simple iron links or even by ropes, to operate as a train. Usually the horses or human beings had to push only over short sections or on the return journey, the falling gradient from a pithead to a wharf being mainly downhill. Often a horse would ride down in an empty truck, ready to pull the unladen train back to the mine again. There was no signalling or traffic control and the absence of brakes meant that coal-laden trains ran downhill completely out of control.

Improvements and standardization

Not until after 1820 were crude friction brakes fitted, but the really big advance was the invention of powerful air, steam and vacuum brakes. The air brake, introduced by George Westinghouse (1846–1914) in 1869 [6], cut the stopping distance needed on level track by as much as 90 per cent.

By 1820 several of the variable features had been firmly decided upon. Track was no longer of wood but of iron, with high-strength steel in standard sections following at the end of the 1860s. The flange was no longer on the track but on the inner edges of the wheels, which were fixed in pairs to the axles. Although not understood at the time this feature enabled trains to run faster and more smoothly.

Rolling stock had to be designed to fit the loading gauge of the line so that no part projected far enough outwards or upwards to strike a tunnel, bridge or signal. The restriction of the loading gauge meant there were capacity advantages in making vehicles longer, but they had to run round corners without forcing the flanges off the rails. Most early stock had only two axles. By 1845 three-axle goods and passenger stock was common, all axles being held in axle-boxes fixed to the vehicle frame. By 1875 the biggest coaches were fitted with pivoted two-axle bogies, which allowed increased length and reduced the minimum radius of curve that could be traversed.

1 Early trucks ran on wooden rails. Invariably the plain wooden wheels were kept on the track by flanges, grooves or other guideways built into the track. This truck hauled iron ore in the 1500s.

2 Railway gauges throughout the world range in width from 5ft 6in (1.68m) to less than 2ft (60cm). Most European countries, and North America, use the standard gauge of 4ft 8.5in (1.43m).

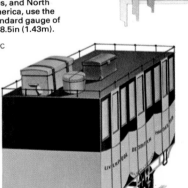

3 Early passenger cars established the idea that there should be three different classes of accommodation for rail travellers, priced at different levels. Third class [A] was simply an open truck, second class [B] had bench seats and first-class travellers were housed in something resembling three opulent horse carriages on a single chassis [C]. All these had to be designed to function together.

4 Underground trains have been carrying the workers of the world's major cities since London's Metropolitan line opened in January 1863. In its first year the line carried 9.5 million passengers. Today the entire London network carries more than two million passengers each day. Trains of some major cities are shown here.

4   New York   Berlin   Montreal   London

By this time trains often comprised 12-bogie passenger cars, or had standard couplers, and the different operating authorities gradually standardized on preferred systems.

In the twentieth century it became necessary to try to standardize rail gauges [2] and to fit all rolling stock with standard braking and control systems, vehicle heating and lighting supplies and, above all, with standardized couplers. Early couplers were merely heavy hooks and links, connected by hand, but by 1925 automatic couplers were beginning to come into use. These resembled strong claws that could snap shut by merely pushing vehicles together and prevented an individual vehicle from overturning if its wheels should come off the track [5].

Railway passenger transport

The early years of the twentieth century saw the growing construction of urban underground or rapid-transit railway systems. This led to a fresh class of rolling stock [4], designed for passenger transport only, often with a small loading gauge for underground lines and propelled by electricity (picked up from one or two extra current-carrying rails). Unlike most earlier trains these were powered by electric motors placed along the train, instead of having a separate locomotive to move the rolling stock.

Such "multiple unit" (m.u.) trains are completely flexible in that they can be made up of any number of small groups of cars and can run equally well in either direction. They are also capable of rapid acceleration and braking because they have a very high ratio of power to weight and powerful brakes. Some have been fitted with rubber tyres (to reduce noise) and almost all have power-driven sliding doors. The latest stock is fitted with automatic control so that if a driver rides with the train he does so only as a passive overseer. Similar technology operates on long-distance passenger trains for surface use. Propulsion is being applied to most or all axles right along the train and water-turbine brakes are fitted to slow trains down from very high speeds. The latest trains have bodies that can tilt smoothly on bends [Key]. Steel and wood have given way to light alloys and fibre-reinforced plastics.

The most advanced rail rolling stock of the late 1970s is that of the British Rail Advanced Passenger Train (APT). It has a dramatically lightened body and a new form of bogie designed to exceed 250km/h (155mph) even round the bends of existing track.

5 Couplings between railway wagons began as simple hooks and chains [A] with buffers to absorb the shock. Automatic couplers and uncouplers [B] were introduced in the USA in 1882 and led to today's buck-eye coupler [C]. Some include connections for the brakes, electrical controls and heating. Instead of buffers American wagons have "draft" gear that uses springs, friction or hydraulics to absorb operating shocks.

6 George Westinghouse's pneumatic brake, patented in 1869, is used more widely by railways than any other type. The brake pipes are kept full of compressed air. When the brake is applied air in the main pipe [1] escapes. Auxiliary reservoirs [2] that still contain compressed air are then automatically connected to the brake cylinders, in which the pistons [3] move outwards and force the brake-shoes on to the wheels.

Stockholm   Paris

7 Modern freight stock is designed to carry particular kinds of commodities. Vehicles have now probably reached the maximum size that can be accepted with today's track but research by British Rail from 1963–8 opened the way for freight trains of the future to run with greater safety at speeds of at least 235km/h (145mph).

A CIE (Ireland) containers on flat car
B SNCF (France) car transporter
C Austrian Federal liquid-gas tank car
D Canadian Pacific box car
E Western Pacific (USA) box car
F New Zealand Railways coal hopper
G Penn Central (USA) open gondola
H South Australian bulk grain hopper
I Finnish State flat car (timber)
J Italian State refrigerated van
K British Rail bulk cement
L Indian Railways hopper

Railways of the future

The railways of the future will probably not involve any spectacularly new principle but will be developments of those we use today. Most rail administrations have built their entire system to a stereotyped model, with two steel rails of a particular type separated by a "gauge" of about 4ft 8.5in (1.43m), 3ft 6in (1.07m), 1 metre (3.28ft) or 5ft 6in (1.68m). Sums of money equivalent to hundreds of millions of pounds were invested in this track and, in today's economic environment of inflation and rapidly rising material and labour costs, it is not easy to see how any major change can ever occur. Only in a place with no existing railway at all can new railway principles be adopted without the financial loss of scrapping an existing system. Yet already much is being done to improve efficiency of "traditional" two-rail systems.

Rewarding trends in railways

One of the most rewarding efforts is the elimination of traffic bottlenecks, level crossings, sharp bends and permanent-way restrictions of all kinds. Another improvement is to construct the track in a different way. Instead of laying rails and ties (sleepers) on a bed of ballast, which constantly needs attention, track could be made up into prefabricated concrete-based sections laid directly on to firm ground [2]. The maintenance cost of such track could be cut to less than one-tenth of that of a traditional track.

Another vital area for improvement is automatic train control. Today railways are introducing electronically based control and communications systems [1]. These involve fixed beacons or cables laid along the track for communication between trains and a control centre with a computer [3]. Trains can be started automatically, accelerated at an exact rate, held to precisely the best speed at all times and automatically guided on to the right track or made to comply with any special limitations. Any emergency can be instantly known throughout the system, the computer changing its program to re-instruct all traffic. Using early forms of such control, city transit systems such as San Francisco's BART and London's Victoria Line have, since their first day, been operating automatically, the driver riding as a passenger to keep an eye on things.

Use of such automatic control, coupled with arrangements for bringing trains to rest in a safe distance, has been allied with advances in the design of rolling stock to allow dramatic increases in speed. One alternative to the traditional railroad is the monorail [Key], which has been in use for most of this century in various forms. Its main advantage is not extra speed but the fact that it is easy to erect on stilts across a city.

With changes to the track, speeds greatly in excess of 250km/h (155mph), beyond the limit for conventional track, will be possible.

Wheels obsolete

The most radical new developments in the final quarter of this century involve the elimination of wheels. High-speed vehicles can run along smooth tracks by air-cushion lift or magnetic levitation [Key]. Although both methods demand the consumption of energy – doing what the wheel does for nothing – the wheel-less train can run much faster than a wheeled one and needs less costly track. By getting rid of any contact bet-

1 Marshalling yards offer a foretaste of the semi-automated railway of the future using the same track as today. Freight cars enter the yard over the hump in the foreground, roll down the slope beyond, are scanned by an electronic eye and switched to the correct track. They are slowed and halted by hydraulic retarders beside the rails which "squeeze" the wheel flanges. The whole process is under computer control.

Hydraulic retarder

Electric eye

Hump

2 Railway track cannot break away from the established form and spacing of rails, but it is continuously being developed to reduce costs, especially the cost of maintenance. The British track in the foreground, laid on prefabricated concrete base sections, needs no ballast and in theory little attention for years. In a large rail system this type of advance would save a larger sum than the total annual bill for the cost of the energy to drive the trains. The possibility of eventually changing to a different form of track is remote.

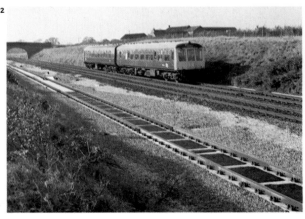

3 Automatic train control – applying the brakes when a train is approaching an obstruction – has been employed on railways in Britain for many years. But this fully automatic system is still in the experimental stage. A set of "wiggly wire" conductors is laid between the running rails of the track and the electric currents in them detected (using induction) by coils suspended from the underside of the locomotive. The presence of a current is shown on equipment in the cab as a digit, say a number 1. If a second wire runs in the opposite direction to the first, its current effectively cancels that of the first wire and the equipment records the number 0. In this way, the pattern of wires can relay a series of coded instructions, in the form of a set of consecutive digits, standing for "reduce speed to 50km/h" or "stop train in 2km", and so on, controlling a train automatically.

ween train and track the need for maintenance would also virtually be eliminated, or so it is hoped. There would be no noise save for the rush of air past the vehicle and the only power used to drive the train at cruising speed would be that needed to overcome drag. The track, basically made of concrete box sections, would be rather straight in comparison with today's metals because at, say 800km/h (500mph) it would be impossible to climb gradients comfortably, to breast summits or to negotiate curves. The TACV (tracked air-cushion vehicle) [4] is a well-developed technology, although no extensive system has yet been built.

From present to future

The magnetic-levitation method (maglev) dates from as recently as 1968 and many systems are operating over short test stretches. The fundamental feature of maglev is the use of the same magnetic field both to lift and to propel the train (probably a single vehicle). Superconducting magnets are used to effect dramatic savings in consumption of current.

For the more distant future there is a wealth of fantastic possibilities. Undoubtedly the most rewarding, most frightening and least likely in the foreseeable future is the gravity tunnel train [6A]. If a tunnel were to be bored from, say, London to New York, it would seem at each end to dive quite steeply down into the earth. A vehicle placed in the tunnel would fall towards its destination. If the tunnel were empty of air the vehicle would reach a speed of many thousands of kilometres per hour at the mid-point, when it would appear to be travelling on level ground. It would then increasingly seem to climb, coming to rest at its destination (with no expenditure of energy, save that of pumping out the air). Man does not yet possess the technology to build such a "railroad", but he could build shorter gravity-vacuum systems that would dive down in curving tunnels between stations only a few miles apart, with atmospheric pressure behind the car and a near vacuum in front. Another "train" of the future might zoom along an air-filled tube by sucking in air in front, compressing it and discharging it as a propellant jet behind [6C].

KEY
A
B
C
D

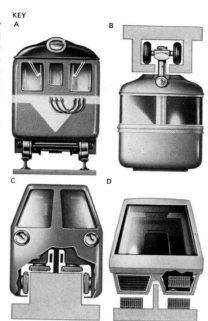

There are many kinds of railways already in use. Almost all the world's public rail track is of the two-rail variety [A], with steel rails spaced side by side. In urban areas the monorail is sometimes found, with the cars riding above the track or hanging from it [B]. Advocates of air cushion trains [C] claim that the track, made from reinforced concrete box sections, is cheaper. One of the latest systems uses magnetic levitation [D] to support the train. Examples of all these have been tested. Various futuristic designs exist only on paper and nobody can foretell with certainty what the railways will look like even 25 years ahead.

4

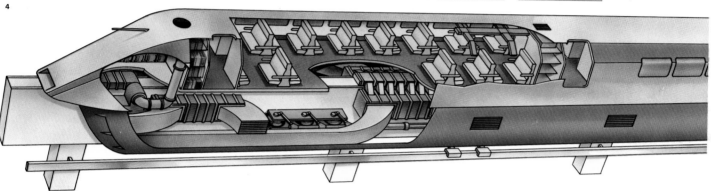

4 The hovertrain, a British tracked air-cushion vehicle (TACV), illustrates the kind of system that one day may supersede the steel two-rail. The track is assembled from cheap concrete boxes carried on short stilts. The only physical contact between vehicle and track is the sliding electric current pick-up that serves the linear motor. Electric blowers form an air-cushion to lift and guide the vehicle.

5

5 A city could be linked to a distant airport using three kinds of rail links. A conventional surface railway (diesel or electric) could serve the city and a junction to other rail routes. An underground railway could serve outlying suburbs and the centre, and a high-speed monorail could form a non-stop rapid link directly to the city. But only an underground can be built with little disruption to existing buildings.

6 Futuristic ideas for railways, not yet even in the experimental stage, include a vacuum train and an airtube system. The vacuum train [A] is "sucked" along by low pressure in front of it; gravity aids acceleration and deceleration because the tunnel slopes down steeply from one station and slopes upwards to the next [B]. The tunnel for the airtube train [C] is filled with air and air cushion pads keep the train centred. Airflow through the train propels it along.

6 A

C

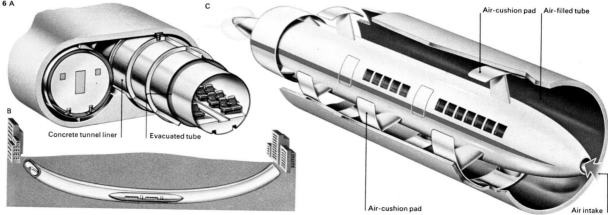

Air-cushion pad   Air-filled tube

Concrete tunnel liner   Evacuated tube

B

Air-cushion pad

Air intake

Balloons, blimps and dirigibles

When early men dreamed of flying they usually imagined "flying machines" that resembled artificial birds. It so happened that, more than a century before the technology existed to make flying possible in this way, a totally different method of flying was developed in France. This suddenly developed method – the use of balloons – is called lighter-than-air flight. The technical term for such aircraft is an aerostat. Aerostats are buoyant in the atmosphere and float at a particular level that depends on their mass, on the surrounding atmosphere and also on the volume of air that they displace.

Balloon construction

The concept of making a balloon from some light material and filling it with a gas having a density less than that of air dates from medieval times. In 1670 Francesco de Lana proposed an aerial ship to be lifted by four large copper spheres from which the air had been pumped out (he did not know that spheres strong enough not to collapse would weigh many times more than the mass of air they displaced). In the next century the bal-

loon came much closer with the discovery of the gas we now call hydrogen, the least dense of all the elements. The British chemist Joseph Black (1728–99), who studied hydrogen, thought of making a hydrogen-filled balloon, but probably did no experiments.

A few years later, in France in 1782, the papermakers Joseph and Etienne Montgolfier (1740–1810 and 1745–99, respectively) had been watching charred fragments spiralling upwards above a bonfire. Why, they wondered, did the fragments rise? Thanks to their skill with paper the Montgolfier brothers were able to build small balloons which, when filled with hot air from a fire, took off and sailed upwards. On 4 June 1783 they publicly flew an 11m (36ft) balloon of linen and paper that climbed to about 1,830m (6,000ft).

The balloon caused a sensation. Jacques Charles (1746–1823) had meanwhile set to work to build a hydrogen balloon, while the Montgolfiers constructed a hot-air balloon large enough to lift a man, carrying a fire beneath it. On 15 October 1783 Pilâtre de Rozier was carried aloft in the tethered bal-

loon, while five weeks later [Key] he and the Marquis d'Arlandes flew the first aerial journey in history, covering 8km (5 miles) in a gentle breeze in 25 minutes. Charles's hydrogen balloon made a manned flight the following week, on 1 December 1783. For the next 100 years lighter-than-air flight dominated man's attempts to fly, with balloons reaching heights of 6km (3.7 miles) and travelling hundreds of kilometres (for example, in 1859 John Wise flew 1,300km [804 miles] from St Louis to Henderson, NY). In 1870, during the Franco-Prussian War, balloons were the only link that existed between Paris and the outside world.

Uses of balloons and airships

It was natural for early aeronauts to wish to devise some means of locomotion to free themselves from the mercy of the wind. Some attempted to use oars and others tried propellers cranked by hand, but it was not until the invention, in 1852, of the steam-driven dirigible (meaning "steerable") by Henri Giffard (1823–1921) that the airship emerged as a vehicle. The earliest airships

CONNECTIONS

See also
148 History of aircraft
154 Helicopters and autogyros
110 History of transport

In other volumes
72 The Physical Earth

1 Ferdinand von Zeppelin (1838–1917) pioneered the rigid airship, which was to take his name. His first vessel, LZ1, flew in 1900 and was followed by a whole series, including this LZ13 of 1912. It was 141.5m (460ft) long and 13.8m (45ft) in diameter and could carry more than six tonnes of passengers and cargo. This airship, named *Hansa*, made nearly 400 flights carrying more than 8,000 pas-

sengers 45,000km (30,000 miles). LZ14 (remumbered L1) was ordered by the German navy; some commercial airships were taken over by the army, which used them in World War I for reconnaissance and bombing.

2 Non-rigid airships all have flexible envelopes that are stabilized by being inflated to a pressure slightly higher than that of the surrounding atmosphere. The load is suspended by a sys-

tem of ropes or wires that distributes the weight around the fabric envelope. The sea patrol blimp illustrated dates from 1913 but closely resembles many used in World War II. Similar ships, filled with

non-inflammable helium are still flying. Some are used for advertising and carry huge arrays of lights that can be made to display slogans or pictures. Others are being developed as freight carriers.

3 Semi-rigid airships are uncommon but the *Norge* was one of a series of Italian airships of the 1920s and achieved fame by making a voyage to the North Pole in May 1926. There was no

rigid structure inside the gas envelope, but a rigid keel ran from bow to stern and served as a structure to which everything could be attached. Cords and cables extended

upwards from the keel to secure the envelope, preserve its shape and enable it to lift the whole ship. Below the keel were braced frames carrying the control car and engine nacelles.

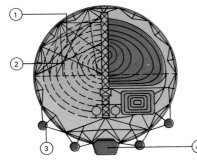

4 Rigid airships have a skeleton framework that contains the lifting gas in a series of bags. The largest ever built were the German LZ129 *Hindenburg* (1936) and LZ130 *Graf Zepplin* (1938), named after the earlier LZ-127 of 1928. Each had 200,000m³ (7 million cu ft) of hydrogen, giving a lift of about 232 tonnes,

contained in gas bags [1]. These were housed inside the aluminium framework [2]

from which were hung the four 1,050hp diesel engines [3] and the payload [4]

comprising 50 passengers and their baggage and 12 tonnes

of freight and mail. The *Hindenburg* exploded in flames in the US in 1937.

| 0 | 30 | 60 | 90 | 120m |

| 0 | 100 | 200 | 300 | 400ft |

5 Relative sizes of the three types of airship show the greater size possible with rigid [A] compared with semi-rigid [B] and non-rigid [C] ships.

were what are today called non-rigids [2]: each had an envelope of flexible fabric from which the load was suspended by cords. In the semi-rigid airship [3] there is a rigid keel and in the rigid type [4] the entire envelope is built round a rigid framework. All were fully developed by the start of World War I in which airships, tethered "kite balloons" and non-rigid "blimps" all played major roles.

The operation of all aerostats depends on balancing their mass against the volume of displaced atmosphere. With gas balloons and airships the normal technique is to vent gas from the envelope to descend and to release sand or water ballast to rise. Airships in cruising flight can also change their direction or height by using aerodynamic tail controls, but these are not effective at low speeds. With hot-air balloons the lift depends on the difference in temperature between the air inside and that outside the envelope.

The future of the airship
By the beginning of World War II the large airship was dead – killed by a series of major disasters. Among the last of these was the destruction by fire in 1937 of the German airship *Hindenburg* [4], with 36 deaths. Barrage balloons and blimps had their uses but after 1945 there remained very few devotees of the sporting hydrogen balloon. Then, in about 1965, the scene was transformed. The modern hot-air balloon gradually became a worldwide best-seller and today many hundreds are sold each year. They use immediately controllable propane burners with which climb and descent can be governed for a whole day if necessary.

The airship is also being recognized once more as a potential cargo carrier. All over the world designers and freight carriers are studying plans of completely new kinds of airships that would make use of modern technology to carry loads of hundreds or thousands of tonnes in safety and at low cost. Such giant cargo airships – the *Skyship* [8] is an example – could become a reality before the end of the century. They may be especially useful in opening up undeveloped regions and in carrying large freight containers and possibly even bulk cargo direct to their final destination.

Man's first balloon flight took place on 21 November 1783 when two men travelled about 8km (5 miles) across Paris. They rose aloft standing on a gallery of wickerwork suspended beneath the painted envelope of the Montgolfier brothers' largest hot-air balloon. Made of paper-lined linen and coated with alum to reduce the fire risk (although not with complete success) the envelope was 15m (49ft) high and the whole balloon weighed about 785kg (1,730lb). The air inside, a volume of about 2,200m³ (77,700 cu ft), was heated by a large mass of burning straw resting on a wire grid in the centre of the gallery.

6 Balloons rise because the gas they contain is less dense than air. Control is limited to upward motion by releasing ballast [A] and downward by releasing gas [B]. A drag rope [C] gives stability and slows the balloon on landing, when the rip panel is opened [D] to let the gas out. The lifting power of 28m³ (about 1,000 cu ft) of gas is shown in [E].

8 *Skyship* is a proposal for a modern cargo airship. This 10m (32ft) model was demonstrated in Britain in 1975. A full-scale *Skyship*, 215m (700ft) across, would cruise at about 160km/h (100mph) and have a crew of 24 with a payload of 400 tonnes.

7 The British R101 airship crashed on a hill at Beauvais, France, in 1930. The ship was on a voyage from England to India when it crashed, killing all but six of the 48 people on board. Burning hydrogen created a fireball hot enough to melt the metal of which it was built. This and the fatal disaster to the *Hindenburg* seven years later sealed the fate of the hydrogen airship.

History of aircraft

Contrary to popular belief, the brothers Orville (1871–1948) and Wilbur Wright (1867–1912) were not the first men to build an aircraft that could fly. Otto Lilienthal (1848–96) in Germany, for example, had already made hundreds of flights in his gliders. The Wright brothers' place in history is assured by the fact that their aeroplane was the first powered, controllable heavier-than-air machine craft to fly, in 1903.

Louis Blériot's pioneering flight

After the Wright brothers one of the important contributors to aeroplane design was Louis Blériot (1872–1936). He introduced a number of new features in his design, including a tractor (pulling) propeller, single wing (monoplane) and rudder and elevator at the rear. His Type XI achieved world acclaim on 25 July 1909 by flying from France to England. As with almost all other flying machines of its day it was of mixed construction; the main spars of the stubby wing and the four longerons of the lengthy fuselage were made of ash and the whole structure braced by numerous wires. Like the Wrights,

Blériot covered both the top and bottom surfaces of his wings, although many other designer-aviators used only a single surface of fabric, stretched over the top of the wing.

By 1912 Deperdussin, helped by research in Scandinavia, had built a racer of monocoque (single-shell) construction, which made for strength, lightness and a completely new streamlined form. He built his fuselage from multiple thin veneers of tulip wood, wrapped to shape and finally glued and covered with doped (lacquered) fabric. Most of the 100,000 aircraft built in World War I used the traditional wire-braced wood structure, but then the monocoque design developed gradually and found new expression in metal. Some early military aircraft, such as the Voisin L series, had all-metal structures. Some were made of steel tubing assembled by welding, riveting or with bolted joints, and others used the new aluminium alloy Duralumin. Whatever the material, the basic method was to make a strong skeleton and cover it with fabric.

A few wartime machines designed by Hugo Junkers (1859–1935) were not only

all-metal in skeleton but were also skinned with metal. In 1919 Junkers flew his F13, the first all-metal monoplane in commercial service. The low-mounted wing was completely unbraced by struts or wires and, like the fuselage, was skinned with Duralumin sheets having fore-and-aft corrugations for rigidity. From this stemmed a family of transports used all over the world, the best known being the Ju 52/3m – the leading European airliner in the 1930s and an aircraft built in large numbers for Hitler's *Luftwaffe*.

The only other family of transports able to rival the Junkers all-metal monoplanes were those from the Dutch Fokker company. These were also monoplanes, but they had deep wooden wings mounted above the fuselage of welded steel tube with a fabric covering. Together these two companies dominated Europe until the mid-1930s.

The Schneider Trophy

Throughout the 1920s, much attention and money was lavished on the Schneider Trophy – an international competition for racing seaplanes. Seaplanes rather than conven-

CONNECTIONS

See also
110 History of transport
152 How an aeroplane works
150 Modern aircraft
154 Helicopters and autogyros
176 Early military aircraft
178 Modern military aircraft

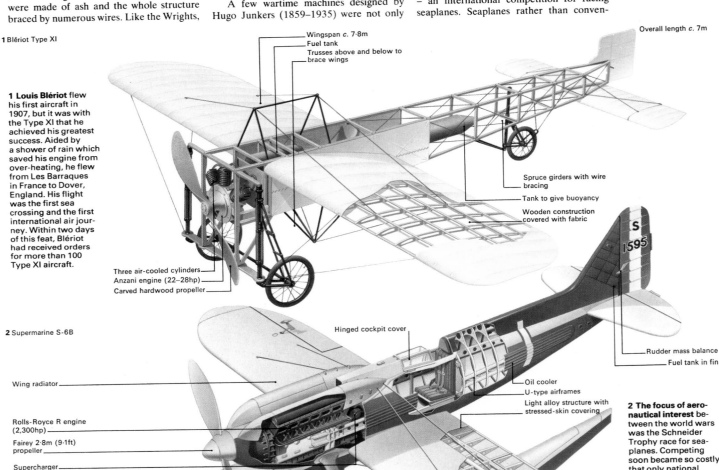

1 Blériot Type XI

Wingspan c. 7·8m
Fuel tank
Trusses above and below to brace wings

Overall length c. 7m

1 Louis Blériot flew his first aircraft in 1907, but it was with the Type XI that he achieved his greatest success. Aided by a shower of rain which saved his engine from over-heating, he flew from Les Barraques in France to Dover, England. His flight was the first sea crossing and the first international air journey. Within two days of this feat, Blériot had received orders for more than 100 Type XI aircraft.

Three air-cooled cylinders
Anzani engine (22–28hp)
Carved hardwood propeller

Spruce girders with wire bracing
Tank to give buoyancy
Wooden construction covered with fabric

2 Supermarine S-6B

Hinged cockpit cover

Rudder mass balance
Fuel tank in fin

Wing radiator

Rolls-Royce R engine (2,300hp)
Fairey 2·8m (9·1ft) propeller
Supercharger

Oil cooler
U-type airframes
Light alloy structure with stressed-skin covering

Radiators on floats

2 The focus of aeronautical interest between the world wars was the Schneider Trophy race for seaplanes. Competing soon became so costly that only national (ie air force) teams were able to take part. The 1931 race was won by the British S-6B, which influenced design of the Spitfire, but the Macchi MC72 would probably have won if it had been ready in time. In 1934 the Macchi improved on the world speed record it had set the previous year with a speed of almost 710km/h (441mph).

Aileron mass balance
Air-speed indicator

Fuel tanks in floats

148

tional aircraft took part because Jacques Schneider thought that future international airline operators would make wide use of waterborne aircraft.

In addition to the public interest it stimulated, the competition had great influence on the mainstream of aircraft design. Also significant was the steady improvement in aviation technology, especially in the United States. The most important factor was the perfection of all-metal stressed-skin construction, in which the light-alloy skin was not just a covering but a crucial load-bearing part of the structure (so permitting a lighter skeleton underneath). Engines were improved and installed in better ways, with cowlings giving reliable cooling and reduced drag. Propellers were no longer fixed blades of wood or metal, but consisted of hub mechanisms carrying blades whose pitch (setting to the airflow) could be varied to suit the different demands of take-off and high-speed flight. Wings were fitted with flaps to give more lift at take-off and both lift and drag for landing. Landing gears were made retractable to reduce drag. Inevitably, air-

craft acquired "systems" worked by electricity, hydraulics, compressed air and other methods, which grew ever more complex.

The Douglas aircraft
One of the first modern airliners was the Boeing 247 of 1933. In the same year Douglas flew the DC-1, only one of which was built, and took orders for the slightly improved DC-2. In 1934 Britain held an air race to Melbourne, Australia; the outright winner for speed was a special racer, carrying no load, but the second and third places were taken by a DC-2 and a Boeing 247. On 17 December 1935, the thirty-second anniversary of the Wrights' first flight, Douglas flew the DC-3. Over the next ten years this was to become the world standard airliner and the standard Allied transport in World War II, with some 11,000 being built in the United States and the Soviet Union. Large numbers are still in use and individual DC-3s have flown as many as 80,000 hours. Previously, few aircraft flew for more than 1,000 hours without either being wrecked or suffering from severe structural fatigue.

The frailty and small size of early aircraft is emphasized by the comparison of their silhouettes with that of a modern Jumbo jet. This growth has, to some extent, been forced and made possible by huge improvements in propulsion and in airfields. Even the Ju 52/3m and the DC-3 had to operate from small, rough grass fields. These planes represented the most efficient compromise between conflicting demands that was possible in the 1930s, just as the Jumbo does today.

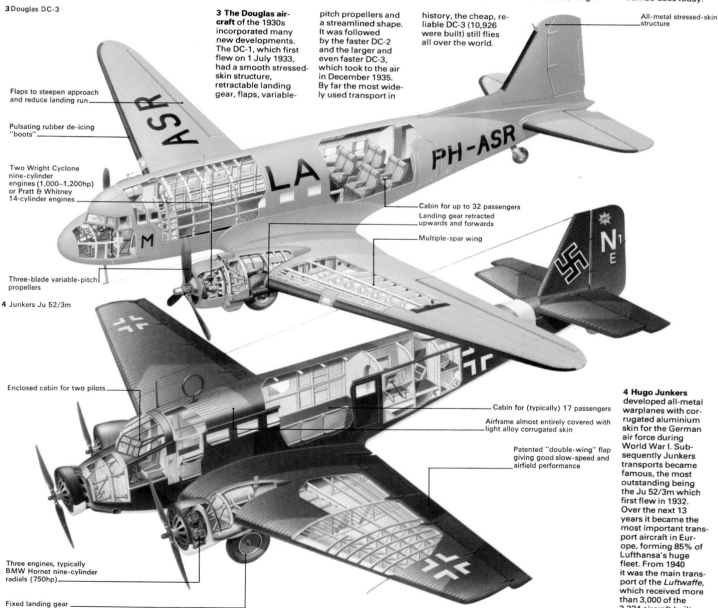

3 Douglas DC-3

3 The Douglas aircraft of the 1930s incorporated many new developments. The DC-1, which first flew on 1 July 1933, had a smooth stressed-skin structure, retractable landing gear, flaps, variable-pitch propellers and a streamlined shape. It was followed by the faster DC-2 and the larger and even faster DC-3, which took to the air in December 1935. By far the most widely used transport in history, the cheap, reliable DC-3 (10,926 were built) still flies all over the world.

All-metal stressed-skin structure

Flaps to steepen approach and reduce landing run

Pulsating rubber de-icing "boots"

Two Wright Cyclone nine-cylinder engines (1,000–1,200hp) or Pratt & Whitney 14-cylinder engines

Three-blade variable-pitch propellers

Cabin for up to 32 passengers

Landing gear retracted upwards and forwards

Multiple-spar wing

4 Junkers Ju 52/3m

Enclosed cabin for two pilots

Three engines, typically BMW Hornet nine-cylinder radials (750hp)

Fixed landing gear

Cabin for (typically) 17 passengers

Airframe almost entirely covered with light alloy corrugated skin

Patented "double-wing" flap giving good slow-speed and airfield performance

4 Hugo Junkers developed all-metal warplanes with corrugated aluminium skin for the German air force during World War I. Subsequently Junkers transports became famous, the most outstanding being the Ju 52/3m which first flew in 1932. Over the next 13 years it became the most important transport aircraft in Europe, forming 85% of Lufthansa's huge fleet. From 1940 it was the main transport of the *Luftwaffe*, which received more than 3,000 of the 3,234 aircraft built.

Modern aircraft

The advent of the gas turbine engine late in World War II revolutionized aircraft design. The light plane – although still a recognizable descendant of earlier models – has been dramatically improved by fitting turbine engines. More advanced machines – airliners, heavy freighters and virtually all military aircraft and helicopters – have been utterly transformed. Of course the basic principles of wing lift, control, lift drag ratio and structural design necessarily remain unchanged. However, development of ever more complex and sophisticated control, navigation and guidance systems has continued to the point where the systems today often account for more than half the cost of an aircraft.

The early pace setters

This strange equation is true even for the class of aircraft bought by wealthy private owners and companies. In the 1930s the Percival Gull and Percival Vega Gull set several world records for long flights, establishing these planes as reliable private-owner and light business aircraft for many years. Built entirely of wood, they were neat low-wing monoplanes powered by air-cooled piston engines of 130–200hp and seated three or four adults in a comfortable enclosed cabin. Such aircraft were among the first to be capable of undertaking long flights to distant parts of the world with some certainty of getting there. However, it is almost impossible to compare them with similar planes today, which are loaded with pilot-aids of all kinds. Intense competition means that modern aircraft, such as the Beechcraft Super King Air 200, have to be improved and up-dated constantly, resulting in a superior and more reliable product.

Instead of being made of wood, this Beechcraft model has a light alloy stressed-skin structure designed for fatigue-free use over perhaps 30 years (a factor never even considered 40 years ago). It is as large as several airliners of 1935, with seating for up to eight passengers. Maximum take-off load is 5,670kg (12,500lb), four times that of the Percival Gull. The two 850hp turboprop engines are almost nine times as powerful. Instead of flying for 1,130km (700 miles) at 225km/h (140mph) at a height of 4,880m (16,000ft) the pressurized King Air 200 cruises at 9,850m (32,300ft), far above most bad weather, for up to 3,300km (2,050 miles) at over 510km/h (320mph). Yet perhaps the biggest contrast is in complexity of manufacture and operation; the number of items of equipment built into the aircraft that contribute directly to its flight – such as pumps, valves, radios, instruments and controls – amounted to 33 in the Gull. The total for the modern machine is 4,408.

The arrival of Comet I

When the British Comet I came into service in 1952, most airlines thought it premature and continued buying piston-engined machines for a while. But new jet aircraft offered the passenger a wholly new experience – flight that was not only faster but also significantly smoother and more comfortable. The early jets were, however, not so impressive to the man on the ground although they stimulated unprecedented and sustained growth and profitability for the airlines; they were outwardly extremely noisy and burned fuel rapidly.

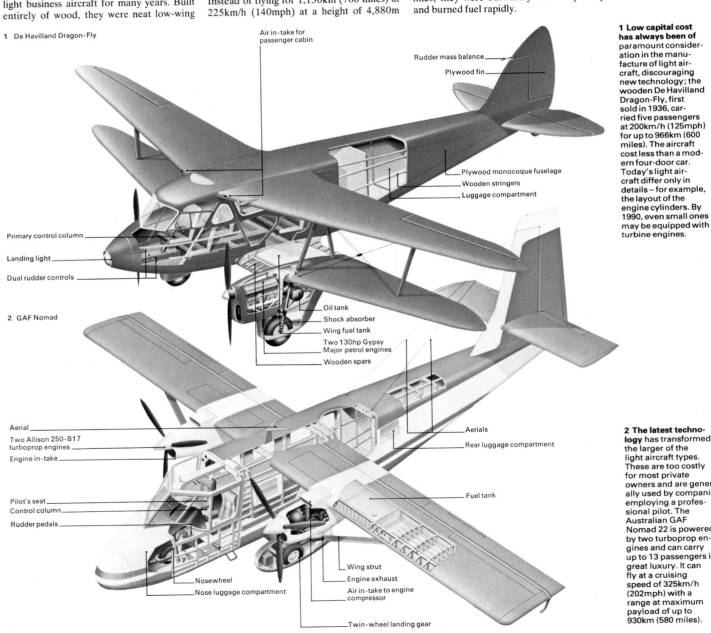

1 De Havilland Dragon-Fly

Air in-take for passenger cabin

Rudder mass balance

Plywood fin

Plywood monocoque fuselage

Wooden stringers

Luggage compartment

Primary control column

Landing light

Dual rudder controls

Oil tank

Shock absorber

Wing fuel tank

Two 130hp Gypsy Major petrol engines

Wooden spars

2 GAF Nomad

Aerial

Two Allison 250-B17 turboprop engines

Engine in-take

Pilot's seat

Control column

Rudder pedals

Aerials

Rear luggage compartment

Fuel tank

Nosewheel

Nose luggage compartment

Wing strut

Engine exhaust

Air in-take to engine compressor

Twin-wheel landing gear

1 Low capital cost has always been of paramount consideration in the manufacture of light aircraft, discouraging new technology; the wooden De Havilland Dragon-Fly, first sold in 1936, carried five passengers at 200km/h (125mph) for up to 966km (600 miles). The aircraft cost less than a modern four-door car. Today's light aircraft differ only in details – for example, the layout of the engine cylinders. By 1990, even small ones may be equipped with turbine engines.

2 The latest technology has transformed the larger of the light aircraft types. These are too costly for most private owners and are generally used by companies employing a professional pilot. The Australian GAF Nomad 22 is powered by two turboprop engines and can carry up to 13 passengers in great luxury. It can fly at a cruising speed of 325km/h (202mph) with a range at maximum payload of up to 930km (580 miles).

Curiously, an answer to these antisocial qualities had been lying unheeded since the early days of Frank Whittle (1907–), inventor of the first British jet engine. The turbofan engine, virtually a cross between a turbojet and a turboprop, could propel airliners just as fast as a turbojet engine, with much less noise and a considerably lower fuel consumption. The large-diameter turbofan was rediscovered during the development of a US Air Force freighter in the mid-1960s. The losing bidder for that contract, Boeing, took four of the giant yet quiet engines and used them to propel the Boeing 747, the first of the huge, wide-body transports, popularly known as the "Jumbo".

Increasing public demand has resulted in a progressive increase in the size of transport aircraft. The DC-3 cabin is 1.7m (5.5ft) wide, and a Constellation of the immediate postwar years had a cabin 3m (10ft) wide at its widest point. The first of the big jets, the Boeing 707 of 1958, had a much longer cabin, which was 3.5m (11.5ft) wide throughout. In 1969 the first Boeing 747 was delivered, with a cabin nearly twice as long and 6.1m (20ft) wide.

The capability of such aircraft is much greater than mere size suggests, because they travel faster and fly far more hours a day than older machines. Several airlines have just one Boeing 747F freighter, but a single 747F can carry more cargo each year than all the world's airliners of 1939. And the turbofan-powered wide-bodied jet is, by comparison, relatively clean and quiet.

Supersonic transport
These "selling points" are not present in either the Concorde or the Tu-144, the first supersonic transports to enter commercial service. It is extremely difficult to make supersonic propulsion systems quiet, although the rapid and steep climb of these aircraft creates a local noise problem only around the airport itself. The move towards wider cabins is reversed in the SST (supersonic transport), because at Mach 2 (twice the speed of sound) aircraft must be relatively slim. On the other hand, journey time is greatly reduced so the SST offers as much extra to the passenger as did the first jets when they were introduced in the 1950s.

KEY
Boeing 747 "Jumbo"

AIR FRANCE

BAC-Aerospatiale Concorde

GAF Nomad 22

De Havilland DH90 Dragon-Fly

These aircraft silhouettes, reproduced to a common scale, illustrate the size of the Boeing 747 and the slenderness of Concorde. Supersonic aircraft must be relatively slender if they are not to be uneconomic, whereas the subsonic 747 can have a much wider cabin. The next generation of long-haul subsonic airliners will almost certainly be even larger. The differences between the two light aircraft are less apparent externally and involve mainly the structural materials.

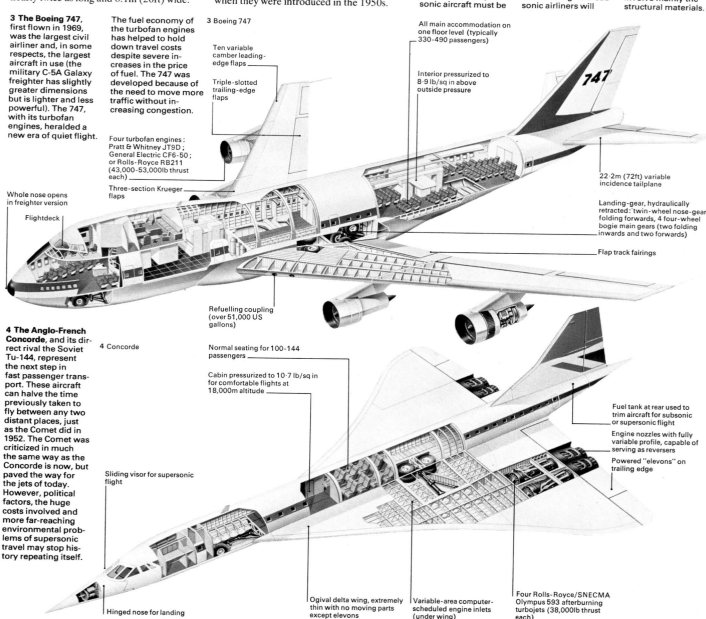

3 The Boeing 747, first flown in 1969, was the largest civil airliner and, in some respects, the largest aircraft in use (the military C-5A Galaxy freighter has slightly greater dimensions but is lighter and less powerful). The 747, with its turbofan engines, heralded a new era of quiet flight.

The fuel economy of the turbofan engines has helped to hold down travel costs despite severe increases in the price of fuel. The 747 was developed because of the need to move more traffic without increasing congestion.

3 Boeing 747

Ten variable camber leading-edge flaps

Triple-slotted trailing-edge flaps

All main accommodation on one floor level (typically 330-490 passengers)

Interior pressurized to 8.9 lb/sq in above outside pressure

747

Four turbofan engines: Pratt & Whitney JT9D; General Electric CF6-50; or Rolls-Royce RB211 (43,000-53,000lb thrust each)

Three-section Krueger flaps

Whole nose opens in freighter version

Flightdeck

22.2m (72ft) variable incidence tailplane

Landing-gear, hydraulically retracted: twin-wheel nose-gear folding forwards, 4 four-wheel bogie main gears (two folding inwards and two forwards)

Flap track fairings

Refuelling coupling (over 51,000 US gallons)

4 The Anglo-French Concorde, and its direct rival the Soviet Tu-144, represent the next step in fast passenger transport. These aircraft can halve the time previously taken to fly between any two distant places, just as the Comet did in 1952. The Comet was criticized in much the same way as the Concorde is now, but paved the way for the jets of today. However, political factors, the huge costs involved and more far-reaching environmental problems of supersonic travel may stop history repeating itself.

4 Concorde

Normal seating for 100-144 passengers

Cabin pressurized to 10.7 lb/sq in for comfortable flights at 18,000m altitude

Sliding visor for supersonic flight

Fuel tank at rear used to trim aircraft for subsonic or supersonic flight

Engine nozzles with fully variable profile, capable of serving as reversers

Powered "elevons" on trailing edge

Hinged nose for landing

Ogival delta wing, extremely thin with no moving parts except elevons

Variable-area computer-scheduled engine inlets (under wing)

Four Rolls-Royce/SNECMA Olympus 593 afterburning turbojets (38,000lb thrust each)

How an aeroplane works

If someone holds a sheet of paper in both hands and raises it in front of his face, tilting the near edge slightly downwards, most of the sheet (beyond the hands) arches down towards the floor. What happens if he blows across the top of the sheet? Oddly enough, it rises until it is stretched out horizontally. One might have expected this by blowing on the underside, but why is the sheet lifted by blowing over the top?

The answer lies in the effect that speed of flow has on the pressure of air. In the case of the paper, and the arched shape of an aeroplane's wing, air flowing over the top has to flow farther than that crossing underneath. The upper air therefore speeds up and as a result, its pressure falls. The higher pressure of the slower air below creates "lift". This effect is responsible for about 80 per cent of the lift on a normal aeroplane wing moving at less than the speed of sound [Key].

Wings and air flow

Modern aeroplane wings are not like sheets of paper but are more-or-less fixed in cross-section. A conventional subsonic section always has a strongly arching top. The underside may be fairly flat in slow aeroplanes, with the thickest part of the wing well forward (towards the "leading edge"). In fast aircraft the wing looks almost symmetrical, so that it would work either way up, and the thickest part is about half-way between the leading edge and trailing edge. In every case most of the lift is generated by reduced pressure as air curves faster over the top. A little lift is added by increased pressure underneath.

Factors affecting lift

For any given wing the lift depends on the angle of attack (the angle at which the wing meets the oncoming air). The greater the angle, the greater the lift – up to a point. Some modern aircraft with short but wide wing shapes, such as Concorde, can go on generating lift to seemingly impossible angles; but ordinary wings soon run into trouble. At an angle of about 16° there is little more lift to be had. By 18° the lift is erratic, the flow finds it hard to remain attached across the top of the wing and, suddenly, it breaks away. Instead of smooth, streamlined

flow, the air billows away in great eddies. Lift is largely lost and any aircraft flown at such an angle of attack stalls [2].

Wing lift also depends on the square of the airspeed; lift at a given angle of attack may be 1,000kg at 100km/h, 4,000kg at 200km/h and 9,000kg at 300km/h. So how does an aeroplane take off? The pilot knows its gross weight, the airfield's height above sea-level and the air temperature (hot or high places have thinner air and give less lift). For each set of conditions he knows how fast the wing must move through the air in order that maximum lift (with a safe margin left to avoid stalling) shall exceed the weight. In all but the most simple aircraft the lift at low speeds can be greatly increased by extending flaps behind and below the trailing edge and, often, by using "Kreuger flaps" along the leading edge that serve to increase the curvature of the wing section. These devices change the apparent cross-section of the wing, making it act far more effectively on the airflow and greatly intensifying the difference in pressure between the underside and upper surface. At the correct speed the pilot

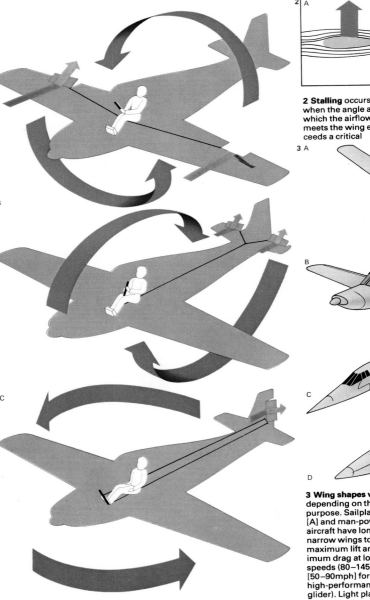

1 Control surfaces behave like miniature wings. These surfaces are: ailerons for roll [A] about the longitudinal axis, giving lateral control; elevators for pitch [B] about the lateral axis for longitudinal control; and a rudder for yaw [C] about the vertical axis for directional control. Each surface is hinged and is controlled from the cockpit. When deflected the surface's main effect is not to push the whole aircraft but to cause it to rotate about one of its axes. Sideways movement of the control column moves the left and right ailerons in opposite directions because, as one wing comes up the other must go down. Fore-and-aft movement of the stick deflects left and right elevators together because both act together in raising or lowering the tail for a dive or climb. In supersonic aircraft there are sometimes no ailerons and the left and right tailplanes can twist in opposite directions for roll control. The rudder is moved by foot pedals and one of the things a pupil pilot must learn is how to "harmonize" the various controls, working the stick and pedals together to just the right degree. In very large or fast aircraft, some or all of the control surfaces are hydraulically powered and an artificial "feel" is fed back to the pilot's controls in the cockpit.

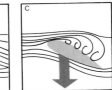
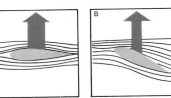

2 Stalling occurs when the angle at which the airflow meets the wing exceeds a critical value. At high speed [A] the angle is small. At lower speeds the nose must be pulled up more and more to maintain height [B]. Suddenly the aircraft stalls and a spin [C] may result.

3 Wing shapes vary depending on their purpose. Sailplanes [A] and man-powered aircraft have long, narrow wings to give maximum lift and minimum drag at low speeds (80–145 km/h [50–90mph] for a high-performance glider). Light planes [B] have simple, thick wings that provide good lift at low speeds. A supersonic airliner [C] has a wing that is optimized for cruising at twice the speed of sound (Mach 2: 2,175 km/h [1,350mph]). It is rather like a Gothic window flying apex-first and is less efficient at take-off and landing. Supersonic combat aircraft [D] often need variable-sweep "swing wings" that can be folded back for operational flying or spread out sideways for cruising or "loitering".

pulls back on the control column [1]. This deflects the horizontal tail (elevators or, often, the whole tailplane) so that it is tilted sharply, with the leading edge pointing downwards. Immediately the airflow pushes the tail down, the whole aircraft "rotates", the wing reaches the angle at which lift exceeds weight and the aircraft climbs away.

Airborne control

As speed and height are gained the pilot "cleans up" the aircraft by retracting the various flaps. The wing can lift the aircraft without them and, as speed continues to increase, the angle of the wing becomes less and less and the aircraft levels out. At full speed the wing skims virtually edge-on to the airflow. But in any sudden manoeuvre, such as a tight turn, the wing has to generate vastly increased lift. In the most violent manoeuvres it is possible to reach stalling angle even at high speed. Normally a stall-warning system alerts the pilot of any approach to stall and this is especially important in thin air at great heights where wing angles are greater under all conditions. The "absolute ceiling"

that the aircraft can reach is, in fact, the altitude at which level flight demands the stalling angle even at maximum speed.

In straight and level cruising flight the control surfaces on the wings form an integral part of the lifting surface. Any movement of the cockpit controls deflects these surfaces or the tail rudder to make the aircraft rotate about one or more of its axes [1]. Roll and pitch commands are often needed; the rudder is used mainly in making a "co-ordinated" turn, with the nose of the aircraft moving left or right along the horizon, the wings tilted to give a lift force acting towards the centre of the turn and the rudder and elevators acting together. When landing the pilot places the aircraft in exactly the right position, aiming along the runway centreline, at such a speed that the wing (with all high-lift devices extended) reaches stalling angle just as the wheels make grazing contact with the runway. Many aircraft extend "spoilers" as they touch down to kill remaining lift. The same spoilers can be used with or instead of ailerons to produce roll motion in high-speed flight or as air brakes on a landing approach.

KEY

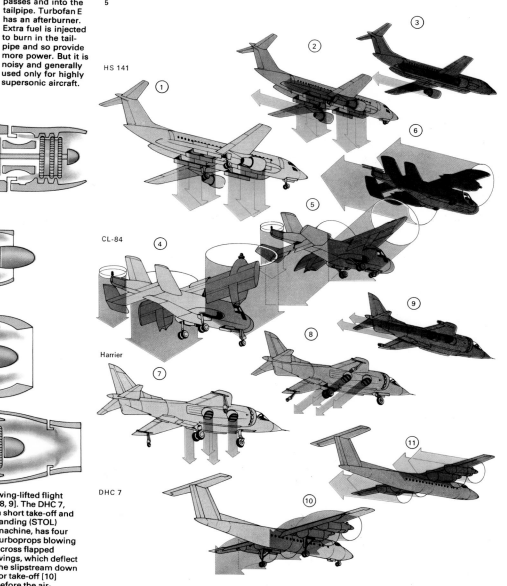

Streamlines (paths of particles of air) show airflow over a wing [1] giving lift [2]. The airflow rises across the leading edge [3], arches across the top of the wing and leaves travel- ling sharply down- wards. When high- lift devices such as leading edge slats and trailing edge flaps are extended, the effect is accen- tuated and the air- flow over the wing almost resembles an upside-down U. De- signers try to achieve the highest possible ratio of lift to drag (aerodynamic resistance) by choosing the cor- rect form of wing.

4 Aircraft propulsion is tailored for the job. The piston en- gine [A] is cheap and efficient for slow aircraft. The turbo- prop [B] also handles a huge airflow but it is better suited to larger aircraft.

Turbofans [C–E] are best for aircraft that fly at high subsonic speeds, such as air- liners. In [C] the fan at the front of the engine displaces air round the main engine. In [D] several fans force air along by-

passes and into the tailpipe. Turbofan E has an afterburner. Extra fuel is injected to burn in the tail- pipe and so provide more power. But it is noisy and generally used only for highly supersonic aircraft.

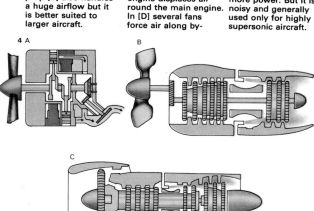

4 A

B

C

D

E

5 Vertical take-off and landing (VTOL) designs have includ- ed the HB 141, with lift jets for vertical take-off [1]; the pro- pulsion turbofans started [2] and the lift engines shut off [3]. It was never built. The CL-84 had two turboprops on a

tilt-wing set at 90° for take-off [4]; the wing slowly rota- ted [5] and then com- pletely straightened [6]. The Harrier has a directional thrust engine blast- ing downwards for take-off [7]; the nozzles then rotate for acceleration into

wing-lifted flight [8, 9]. The DHC 7, a short take-off and landing (STOL) machine, has four turboprops blowing across flapped wings, which deflect the slipstream down for take-off [10] before the air- craft levels out [11].

5

HS 141

CL-84

Harrier

DHC 7

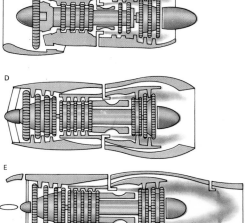

Helicopters and autogyros

The conventional aircraft copes efficiently with its main task, the moving of people and cargo from one point to another. But it is limited because it cannot hover above the ground to lift or set down an object in one precise position and it cannot land on ground too uneven for the building of an airstrip. For difficult terrain, and difficult jobs, a helicopter is needed.

Helicopters in history

All heavier-than-air flying machines stay in the air using the principle known as lift. As an ordinary fixed-wing aeroplane moves quickly through the air the shape of the wings makes the air pressure below them greater than that above. The difference in pressure "lifts" the aircraft and enables it to fly. If the wings can be made to rotate instead of being fixed, lift can be obtained without the machine itself moving forwards. This is the principle on which a hovering machine works.

The principle has been known for a long time. Leonardo da Vinci (1452–1519) sketched a design for a rotating-wing machine [Key] and dubbed it *helix pteron*,

which is Greek for "spiral wing". The name is in use today in slightly changed form in the word "helicopter".

No helicopter could fly until an engine could be made sufficiently light and powerful. With the development of the petrol engine about 1900, the power problem was solved and full-scale helicopters just managed to become airborne. These first rotating-wing machines ran into stability problems and test pilots were reluctant to try them out unless the machines were tethered. The problem of keeping a hovering machine aloft without tilting proved very difficult to solve and yet deliberate tilting proved to be necessary to enable the machine to fly in a particular direction.

Tilting a windmill

In 1923 the Spanish inventor Juan de la Cierva (1896–1936) successfully flew a machine that was a strange hybrid between a helicopter and a fixed-wing aircraft [4]. It had wings and a propeller, but it also had a freewheeling rotor on top. As the machine flew, the motion of the air past the rotor

whirled the blades like a windmill and provided extra lift, enabling it to fly slowly and to take off with a short run. Cierva called it the "Autogiro", from the way the blades automatically gyrated as it flew. This effect is called auto-rotation and it can enable a helicopter to land safely if it loses power.

A most versatile machine

By World War II the helicopter had been perfected, due principally to the work of Igor Sikorsky (1889–1972) [5], an American of Russian origin. One of the main difficulties was torque – as the engine turned the rotor in one direction, it also turned the body of the helicopter in the opposite direction. Torque is a consequence of action and reaction and has been overcome in two main ways. Either a small vertical rotor, fitted to the tail of the helicopter, acts as a propeller to oppose the torque or the helicopter has two horizontal main rotors that spin in opposite directions, thus cancelling out the torque. By adjusting the tilt of the various rotors the helicopter can be held steady or made to turn. Another solution to the problem of torque is to power the

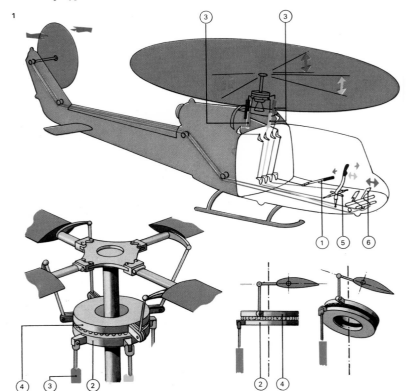

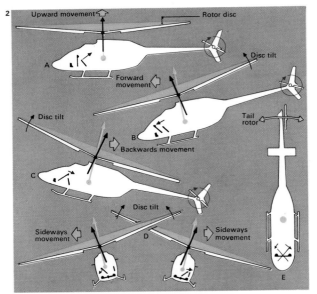

1 A tail-rotor helicopter has a cyclic pitch stick [1] that operates jacks [3]; these tilt the lower swashplate [2]. The upper swashplate [4] also tilts, tilting the main rotor to propel the helicopter. The collective pitch stick [5] raises or lowers the swashplates, thus changing the pitch of the rotor blades and altering the lift of the main rotor. The foot pedals [6] change the pitch of the tail rotor blades, swinging the helicopter round.

3 Rescue is a task at which the helicopter excels. Injured mountaineers, shipwrecked sailors, stranded tourists and flood and earthquake victims often owe their lives to helicopters.

4 The autogyro was designed by Juan de la Cierva in order to make flying safer. The free-wheeling rotor (as well as the wings) provided lift. The propeller produced forward motion. Ironically, Cierva himself was killed in an aeroplane crash.

2 Vertical and hovering flight occurs when the axis of rotation of the rotor is in line with the centre of gravity [A]. Moving the collective pitch stick increases or decreases lift. Pushing the cyclic pitch stick forwards tilts the rotor disc (the space swept out by the blades) forwards and moves the helicopter forwards [B]. Pulling the cyclic stick back [C] makes the helicopter move backwards and pushing it to the right or left enables it to "crab" sideways [D]. The tail rotor is controlled by the rudder bars [E]. This swings the helicopter round in order to allow the pilot to change the direction of flight.

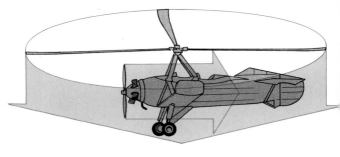

rotor by having a jet engine at the tip of each blade; the blade's motion is the reaction to the jet's thrust and no torque occurs. A small ram jet may be used or exhaust air ducts may be connected to a gas turbine in the body of the helicopter.

Once stable in the air, a helicopter can fly easily in any direction. The rotor produces a downwash of air and the reaction to this downwash forces the machine upwards. If the lifting force equals the weight of the helicopter, then the craft remains stationary in the air. If the lifting force is lessened by slowing the rotor or if the angle at which the blades sweep through the air is changed, the machine descends. If the rotor is tilted slightly as it whirls, part of the downwash of air is directed to one side and the helicopter moves in the opposite direction.

Forward speed, however, is not very great and manufacturers are experimenting with aircraft that have the manoeuvrability of the helicopter and the speed of the fixed-wing machine. Winged aircraft with tilting rotors that can face upwards or forwards are being tried, although the use of jet engines with swivelling exhausts has so far been more successful in aircraft such as the Harrier "jump-jet".

Helicopters are used for passenger transport – for example, as a rapid service between airports and city centres and to link islands without airstrips. But helicopters are expensive to buy and to run and thus have generally failed to compete economically with other forms of passenger transport. They are used mainly in special applications for which no other machine would be suitable. Only helicopters can be used for many kinds of rescue work [3] – to lift people trapped in burning buildings, to rescue sailors from shipwrecks or holiday-makers swept out to sea and to remove people from areas devastated by earthquakes or floods. Helicopters are invaluable for transferring both workers and materials to remote places and are also used as cranes to hoist heavy objects into position on top of buildings – some modern churches have their spires placed in this way. The flexibility of the helicopter also makes it a versatile war machine for carrying men and weapons, and for hunting submarines.

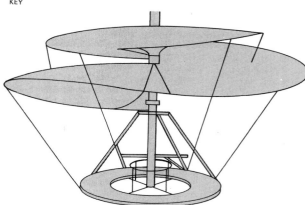

Leonardo da Vinci made this design in 1483 for a rotating-wing aircraft. The spiral wing, which Leonardo suggested should be made of starched linen, would have lifted the machine in much the same way as the rotor of a helicopter. But it is certain that Leonardo's machine never flew. There was no engine in existence that was capable of powering the device and even if there had been it would have spun wildly out of control, for Leonardo did not realize that the engine would also rotate the body in the opposite direction to the wing. This turning effect is known as torque.

5 The VS-300, built by Igor Sikorsky, was the first practical helicopter. It first flew in about 1939; but went through many modifications before being perfected. At one stage it could fly in every direction except forwards. These problems were overcome by 1941 and production models were tested in World War II. Its single main rotor and small tail rotor became the predominant helicopter design.

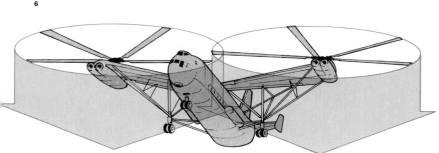

6 A twin-rotor helicopter achieves twice as much lift as a single-rotor machine with a similar rotor. The Soviet Mi-12 is the world's largest helicopter. Built by Mikhail Mil, it set a world record in 1969 by lifting 40 tonnes to a height of more than 2,000m (6,560ft). It has a span of 67m (220ft) across its rotors, which spin in opposite directions. The Mi-12 is a transport aircraft.

7 A tandem helicopter has twin rotors mounted one behind the other. The first tandem helicopter, the "Flying Banana", was built by American engineer Frank Piasecki in 1945. Such helicopters can be made large enough to serve as passenger vehicles or troop transports. Other twin-rotor machines have side-by-side rotors intermeshing like an eggbeater, or coaxial rotors one above the other.

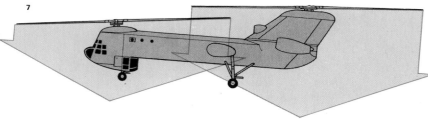

8 A "flying crane" lifts a heavy field gun during army manoeuvres. Such operations are quick and dispense with the need for towing vehicles. Standard helicopters are useful for carrying freight and some interesting tasks have been proposed for the most powerful machines. They could rapidly unload containers from ships offshore and do away with the need for deep harbours. Another idea envisages the carrying of a Sky-lounge – a bus-like vehicle that would gather passengers in a city and then be lifted by helicopter straight to the door of an aircraft at the airport. Flying cranes are particularly useful in wartime for retrieving aircraft that have crashed but are not beyond repair.

Space vehicles

Since the Soviet Sputnik 1 was launched on 4 October 1957 several nations have between them launched hundreds of artificial satellites into orbit round the Earth. The United States and the Soviet Union have sent exploratory "probes" to orbit or soft-land on the Moon, Mars and Venus. Other planets have been studied at close quarters using unmanned space vehicles. The greatest achievement was probably the landing of men on the Moon during the American Apollo programme and there is good reason to suppose that by the end of the twentieth century the whole Solar System will have been explored – but by unmanned probes.

Getting into space

The use of vehicles to explore space dates from the development of rockets powerful enough to launch the vehicles and enable them to "escape" from the Earth's gravity [Key]. This requires a velocity of 11.2km per second (approximately 40,000km/h or 25,000mph), which is technically known as the Earth's escape velocity.

To reach such speeds, multi-stage rockets are employed. These make use of the piggyback principle [1] – the rocket entering orbit is fired at the edge of the atmosphere, having been carried there on top of another rocket that, in turn, may also have been lifted on an even more powerful first-stage rocket. This technique of overcoming gravity was first proposed by the Russian pioneer of rocketry Konstantin Tsiolkovsky (1857–1935). In 1949 an American multi-stage rocket sent a vehicle up to a height of more than 390km (242 miles) above the surface of the Earth.

Rockets and satellites

All rockets are reaction motors. They work using the principle of Newton's third law of motion which, describing the behaviour of any moving object, states that action and reaction are equal and opposite.

In a rocket the "action" is the escape of hot gases roaring out of the tail; the "reaction" to this action forces the body of the rocket in the other direction. The principle can be demonstrated by blowing up a balloon and releasing it; the action of the jet of air escaping from the neck of the balloon is balanced by an equal and opposite reaction that pushes the balloon through the air. For this reason a rocket will work in the airless near vacuum of outer space; it does not, like a jet engine, require a supply of air for the burning of fuel. The presence of air is actually a handicap because it sets up a resistance to the rocket's motion.

The fuel in a firework rocket is a solid propellant explosive such as gunpowder. But solid fuels are too weak and uncontrollable to be used alone in space rockets. Instead two liquids are used – a fuel and an oxidant. When mixed in a combustion chamber they react together to produce hot gases that are expelled from the exhaust and create thrust. The first successful liquid-fuelled rockets were made in the United States in 1926 by Robert Goddard (1882–1945). By the time of Goddard's death German scientists, led by Werner von Braun (1912–   ), had developed the V2. This was a liquid-fuelled rocket that carried a one-tonne explosive warhead and was the direct ancestor of modern space rockets. After World War II, von Braun and his colleagues went to the

1 **Multi-stage rockets** consist of a number of smaller rockets combined to make one big one. At the start of the flight the large, lower stage is used; here it accounts for 83.3 per cent of the propellant but accelerates the rocket to only 33 per cent of its final velocity. When it has used up its fuel, it drops away and the second stage takes over. Only the third stage goes into orbit.

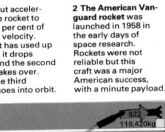

3rd stage
2nd stage
1st stage

2,542m
127,868m
922
118,420kg
4,489m/sec
481,441kg
71,526m
2,972m/sec
2,283,034kg

2 **The American Vanguard rocket** was launched in 1958 in the early days of space research. Rockets were not reliable but this craft was a major American success, with a minute payload.

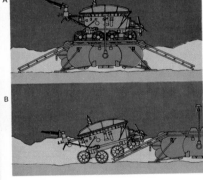

3 **The Soviet "Moon crawler"** Lunokhod 1 [A] was taken to the Moon by a Luna rocket probe; after landing it was sent down a ramp [B] on to the surface. Lunokhod 1 landed upon the grey plain of the Mare Imbrium and crawled along for months, controlled from the USSR, sending back invaluable data. A second Lunokhod operated in the Mare Serenitatis near the landing site of the Apollo 17 module.

4 **Since 1957** hundreds of space probes have been launched. The Soviet Cosmos vehicles [A] are artificial satellites, brought back to Earth after limited flight times. Mariner 9 [B] went into Mars orbit in late 1971. Continuing well into 1972 it sent back detailed photographs of the Martian landscape. The first probe launched by the Soviet Union to Venus [C] was not a success.

United States to continue with their work.

Sputnik 1 was the size of a football and carried little apart from a radio transmitter. Some of today's satellites are the size of a large truck. They have been used in many ways: for mapping [7], communications and scientific research into phenomena impossible to study properly from the ground because of the Earth's atmosphere. Communications satellites have evolved from the early passive type, which consisted of a "silvered" balloon that acted like a mirror to reflect radio signals beamed up to it back down to Earth, to the modern active satellites that amplify received radio signals before re-transmitting them.

Manned satellites have now become relatively common and in 1973 Skylab was placed into orbit as the first true space station. Docking procedures have also been carried out between two spacecraft; in 1975 an American vehicle docked with a Soviet one, the first such meeting between traditional rivals in space. All such manoeuvres require precise information about orbits and velocities. This is provided by radar sets on the craft and on Earth and the necessary complex calculations are carried out using computers suitably programmed for their task.

Probes to the Moon and planets

The first target for unmanned space probes was the Moon. In 1959 the Soviet Luna 3 made a circumlunar voyage and subsequently the whole of the Moon's surface was mapped by automatic probes. Soft landings on the surface were also made [3] and mechanical fingers used to collect dust and rock.

The lunar probes were followed by the first attempts to explore the planets. In 1962 the American Mariner 2 made a fly-by pass of Venus and vehicles have since been sent to Mars, Mercury and Jupiter. By 1975 a Soviet probe had soft-landed on Venus and in the same year a vehicle was on its way to Saturn. The American Vikings made soft landings on Mars in 1976.

Vast amounts of invaluable information have been collected. Telescopes and spectrometers in space vehicles aid Earth-bound astronomers. And infra-red photographs of Earth can reveal new resources.

All space vehicles are put into orbit or sent on their journeys to the Moon, other planets and beyond by powerful rockets.

This rocket launched the American Mars probe Mariner 9 in 1971, carried in the top part of the launcher. It was the first probe to be put into a close path round Mars and transmitted back to Earth thousands of high-quality pictures.

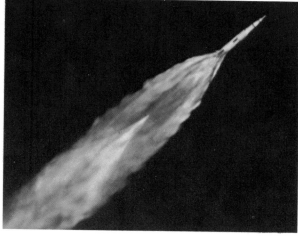

5 Various paths for space probes can be drawn assuming that the vehicles are fired horizontally from the top of a tall tower reaching above the Earth's atmosphere. At low velocity [1] the vehicle soon falls back to the ground. With greater velocity [2] the vehicle travels farther before landing. But with orbital velocity [3] it does not land at all and enters a closed and stable orbit.

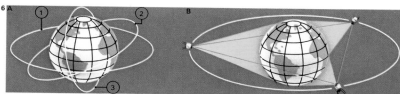

6 Satellites may travel in orbits of various kinds [A]. Some move in the plane of the Equator [1], others have inclined orbits [2] and some use polar orbits [3]. For a "stationary" satellite of the Syncom communications type [B] the period is exactly one day. Its distance from Earth is 35,900km (22,300 miles); it appears stationary and is ideal for television relays.

7 An orbiting satellite is better for photographing the Earth than is an aircraft. One exposure from a space vehicle [A] can cover an area that would need hundreds of photographs from an aircraft [B]. The whole area can be shown with greater accuracy and detail. Also, a vertical space photograph does not have the distortion inherent in aerial mosaics of a wide area. Aerial photographs require lengthy and specialized processing to make them into a mosaic map, whereas this task is greatly reduced using space photographs. Any necessary revisions can therefore be made more easily.

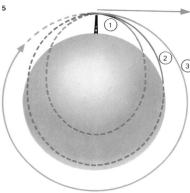

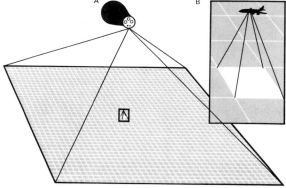

8 A comparison of space and aerial photography clearly shows the far superior structural definition of the former. The Richat craters in Mauritania, West Africa [A], probably of volcanic origin, are well defined in the photograph from the orbiting Apollo 9 vehicle. A mosaic of aerial photographs of the same area is shown in [B]. The Apollo photograph clearly reveals previously unrecorded features including depressions up to 1,500m (nearly 1 mile) across.

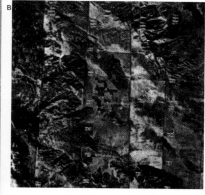

Man in space

Yuri Gagarin (1934–68) of the Soviet air force was the first man in space. In April 1961 – less than four years after the launch of the first artificial satellite, Sputnik 1 – he made a complete circuit of the Earth in Vostok 1, above the bulk of the atmosphere. before landing safely in a prearranged area.

Gagarin and zero gravity
Gagarin's flight was a truly pioneering venture. Nobody at that time had any real idea of how the human body would react to a prolonged period of weightlessness. Yet during his flight Gagarin experienced conditions of zero gravity – something that cannot be simulated on Earth for more than a brief period.

Zero gravity [1] does not mean that the orbiting astronaut has completely escaped the pull of the Earth's gravity. The best way to picture it is to think of a book placed on a piece of card: the book presses on the card and with reference to it the book is "heavy". If both are then dropped the pressure of book on card ceases; the two objects move in the same direction at the same rate.

The same situation occurs when an astronaut is inside his vehicle; the two move at the same rate so that the passenger does not press down upon his craft and "weight" vanishes. (His mass – the quantity of matter in his body – does not change.)

Gagarin found that zero gravity was neither inconvenient nor unpleasant. This has been confirmed by all later space travellers although "walking" in space [4, 5] is extremely exhausting. The first man to venture outside an orbiting vehicle was the Russian cosmonaut Alexei Leonov, and this has since been repeated many times by both Americans and Russians.

Americans in space
The first American in space was Alan Shepard, who made a sub-orbital flight lasting for about 15 minutes in May 1961. In the 1960s manned satellites carrying two or three astronauts were sent up and there were elaborate docking operations in which two independent spacecraft were skilfully manoeuvred together and joined.

One initial difficulty facing international docking was that American and Soviet designs differed because their space programmes had been developed independently of each other. But following the success of the US Skylab space station plans were made for a joint exercise and this was accomplished with the Apollo-Soyuz mission in 1975. Both vehicles had been suitably modified – the Soviet cosmonauts generally breathe ordinary air at normal pressure, for example, whereas the Americans prefer pure oxygen at reduced pressure. For the joint mission a special "adaptation chamber" had to be set up between the two control cabins.

During manned flights many experiments are carried out. The Earth can be closely studied and there have been vast improvements in man's knowledge of the circulation of the atmosphere, which should bring better weather forecasting; plant and mineral resources can be assessed; and nearly all other sciences can benefit.

Predicted space-flight dangers from meteoroids (solid particles moving in space), cosmic radiation and weightlessness have failed to materialize. On the other hand mechanical mishap both on Earth and in

1 **Free fall** is the condition of weightlessness or zero gravity. [A] shows an astronaut training in a flying aircraft. In [1] he is experiencing normal gravity; in [2] the plane is put into a curving dive that simulates the free-fall state for a very brief period; in [3] he takes to a pressure couch to counter the extra g force as the aircraft levels off. [B] shows how, in an orbiting capsule, gravitational pull (mg) is balanced by centrifugal force (mv^2/r) to produce zero gravity (m = mass; r = radius of orbit; v = velocity and g = acceleration due to gravity).

2 **The Gemini programme** followed the first US manned programme (Mercury). Gemini 7, shown here, was able like her sisters to carry two men, could conduct docking procedures and allowed for "spacewalking".

3 **Apollo 15's command and service modules** were photographed by lunar module pilot David Scott. At this time the probe was orbiting the Moon above the Sea of Fertility (Mare Fecundatis).

4 **Astronaut Alfred M. Worden** spacewalked outside Apollo 15 on the return trip; he recovered film equipment that had been used earlier. At the Moon he did not go to the surface with Jim Irwin and David Scott but remained in orbit in the Apollo command module.

5 **Astronaut Edward White** carried out the first US spacewalk in 1965 during the Gemini programme. (The first spacewalk ever was made by the Russian Leonov earlier in 1965.) Astronauts outside a craft do not drift away but remain in the self-same orbit. White was later killed in the fire at Cape Kennedy that destroyed a capsule under test.

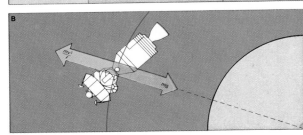

space is a danger and there have been deaths of both Americans and Russians. Manned flights to the Moon have so far been the preserve of the Americans; Russians have concentrated on automatic exploration.

The Apollo programme [6–9], initiated in the early 1960s, reached its climax with the Apollo vehicles of 1968 and 1969. During the Christmas period of 1968 Frank Borman, James Lovell and William Anders orbited the Moon in Apollo 8. In the following year the lunar module was tested close to the Moon's surface. Finally in July 1969 Neil Armstrong and Edwin Aldrin landed on the waterless Sea of Tranquillity. The gap between Earth and Moon had finally been bridged by man.

Inherent problems of Moon missions
The fuel problem is such that it is not yet possible to send a single-stage vehicle to the Moon and back. The initial launching is by step-vehicle; the command and service modules combine then travel to the neighbourhood of the Moon and enter closed orbit. Next, two of the astronauts make the final descent in the lunar module, the only function of which is to shuttle its crew from the main spacecraft to the Moon's surface and back. Nevertheless, the procedure has its dangers. The explorers depend entirely on the ascent engine of their lunar module; if this fails there can be no chance of rescue for the men on the surface.

All the landings made so far have fortunately been successful. The only in-flight failure came when an explosion aboard Apollo 13 on the outward journey put the main propulsion unit out of action. The astronauts were forced to use the motors of the lunar module to pass round the Moon and return safely to Earth.

Apollos 11 and 12, and 14 to 17 have made great progress in lunar study. ALSEPs (Apollo Lunar Surface Experimental Packages) have been set up and are still operating. During the last three journeys the astronauts were able to drive across the surface in lunar rovers or Moon cars. Yet the Apollo system was limited in scope: before men can go to the Moon in large numbers there must be provision for rescue, and this may not be possible for another 15 years at the earliest.

An astronaut's eye-view of Earth taken by an Apollo crewman *en route* for the Moon shows both North Africa and Arabia. There is considerable cloud-cover – but a great part of the Earth was still visible to the astronauts.

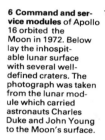

6 Command and service modules of Apollo 16 orbited the Moon in 1972. Below lay the inhospitable lunar surface with several well-defined craters. The photograph was taken from the lunar module which carried astronauts Charles Duke and John Young to the Moon's surface.

7 Lunar rover vehicles (LRVs) considerably extend the area of exploration for astronauts on the Moon. Charles Duke is seen here with the LRV of Apollo 16, near the peak that was soon unofficially named Stone Mountain. In the background the bright rays are from South Ray Crater.

8 The first Moon landing was made from Apollo 11 in July 1969. Edwin Aldrin stands on the lunar surface filmed by Neil Armstrong who was first to descend the ladder from the lunar module. The entire mission – and the "Moon walk" – was shown on TV.

9 Hadley Delta, one of the peaks of the lunar Apennines, forms the background for David Scott and the Apollo 15 LRV. It is farther from Scott than it looks – the distance is more than 30km (19 miles). On the Moon there is no atmospheric scattering so distances can be deceptive and the sky is always black. The US flag does not flutter on its pole since the Moon has no wind – the fabric has to be wired to make it stand out.

10 The ascent motor of the lunar module worked perfectly on each occasion that an Apollo craft left the Moon – this is the view from Apollo 15. Yet this was the weakest link in the entire programme. If for any reason the ascent engine failed there could be no hope of rescue; it is not likely that men will return to the Moon until rescue provision is made. The ALSEPs left on the lunar surface are powered by solar cells and are still functioning.

Early weapons and defence

To provide himself with food and protection, man has always needed weapons. Primitive man probably armed himself with sticks and stones that he found lying around [1]. These hand-held weapons were adequate at close range but were obviously no use for beating off a large predatory beast or bringing down an animal such as a fleet-footed antelope. With an increase in manual dexterity and intelligence, man began to adapt the materials at hand to make tools and weapons.

The first weapons

A curved stick, carved to the correct cross-section, can be thrown with great accuracy and will return if it misses its target. This weapon – the boomerang – is still used today by Australian Aborigines.

Stones with a groove carved round them and tied together by a length of leather or cord produce a bolas, a throwing implement still used in South America. This weapon also has a long history and similarly shaped stones have been found in Stone Age sites in Europe. Such weapons are good enough for hunting game, but man needed better

weapons and the techniques for using them in order to overcome his fellow men.

The spear [5, 6] has been used for thousands of years as both an offensive and a defensive weapon. Originally it was merely a straight stick with a fire-hardened point. With the gradual improvement of weapon-making techniques it was successively tipped with stone, bone, bronze and eventually iron.

Once a spear had been thrown or an opponent had managed to get through the defence, a more convenient weapon was needed for hand-to-hand fighting. Zulu warriors solved this problem with a short-handled, long-bladed stabbing spear called an assegai. Other peoples arrived at a slightly different solution with the sword [9], although there is little evidence of when or where it originated.

At first, swords were made of bronze and subsequently of iron. The Greek armies needed special swords for cutting as well as short weapons for stabbing. The Romans [Key] fought with an short iron sword – the *gladius*. Wielded from behind a barrier of large shields the *gladius* enabled the Roman

legions to rule most of the known world for about 400 years. The slow-moving phalanxes of Rome were eventually overcome by the lightly armed cavalry of the barbarian hordes. With the collapse of the Roman Empire the Roman way of fighting disappeared almost completely.

From their bases in Scandinavia came the "sea-wolves" in their longships. Vikings dealt death and destruction with their long-bladed swords and battleaxes. Resistance to them produced organized armies whose soldiers were armed not only with long-sword and battleaxe but also with the bow. Their discipline beat the loosely organized Vikings.

Bows and arrows

The bow [8] used in Europe was a much heavier weapon than the light recurved bow of the Eastern Steppes (made of laminated horn and wood and small enough to be fired from horseback). Not much used in the west for warfare, it was a hunting bow designed for use on the ground and from cover. Its ultimate form was the English longbow – 1.4m (5ft) of yew with a horsehair bowstring cap-

1 Neanderthal man lived in Europe some 40,000 years ago and represented Palaeolithic (Old Stone Age) culture. The weapons of these people were mostly stones, clubs and spears. The spears were merely straight shafts sharpened to a point and hardened by fire. The technical development of weapons did not occur evenly throughout the world but eventually Neanderthal man's culture was supplanted by more sophisticated cultures using stone-tipped weapons – the Neolithic or New Stone Age peoples. The Australian Aborigines remained at the Palaeolithic stage. Stone weapons were superseded by bronze and eventually by iron.

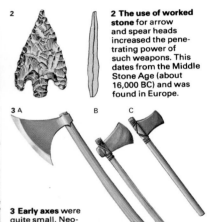

2 The use of worked stone for arrow and spear heads increased the penetrating power of such weapons. This dates from the Middle Stone Age (about 16,000 BC) and was found in Europe.

3 Early axes were quite small. Neolithic [B] and Bronze Age [C] axes measured about 7cm (3in) across the blade, compared with 28cm (11in) for the Viking axe [A] of the 7th century.

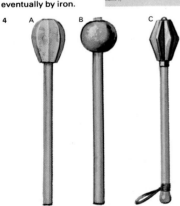

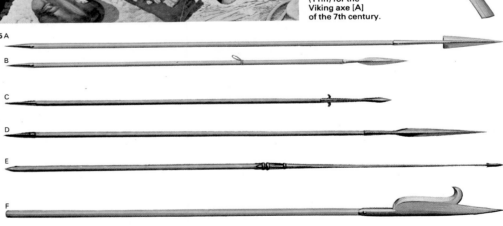

4 The war club or mace grew gradually more formidable. The 2,000-year-old iron mace [B] resembles a 4,000-year-old Egyptian wooden mace [A] but was less likely to break. The fluted head of a 14th-century mace [C] had much more crushing power.

5 The early spear [A] and javelin [B] used in 650-500 BC were little different from the Frankish-Gothic spears [C, D] and Roman javelin [E] of the 5th–7th centuries AD. The bill [F] was a medieval development of the spear.

6 Spear heads changed little in their design, despite the different materials – stone, bronze and iron – from which they were made. Bronze spear heads from Greece [A, B] made in about 700 BC resemble the later

iron ones of the Celts [D], Vikings [E] and Saxons [F]. Broadheaded stabbing spears, such as the Macedonian example [C], often had a collar or flange to prevent the spear from being pushed right through the victim's body.

able of loosing an arrow 1m (39in) long. The Norman troops of William the Conqueror, who in 1066 invaded England, carried bows and each was also armed with a shield and sword and they wore long corselets of mail.

Later arms and armour

Mail became standard equipment, at least for those who could afford it, and gave some protection from arrows. But it did not prevent the wearer from being bruised by blows aimed at him. As armour developed, small plates were introduced into the mail at vulnerable points until eventually a knight was completely encased in steel. The mounts which resembled great lumbering cart-horses, also had their own armour.

On their mighty chargers the knights thundered into battle – although somewhat slowly with the weight they were carrying – equipped with a variety of death-dealing weapons. A 3.7m (12ft) lance of ash, tipped with iron, a shield and a long-sword were standard equipment. Other weapons included a short dagger and often a mace and battleaxe or morning star.

Swords of this period were generally long, double-edged and straight with a reach of about 2m (6ft). They could be used with one or both hands. The age of chivalry came to an end with the success of the English longbow at the battles of Crécy (1346) and Agincourt (1415). The bow's rapid delivery, accuracy and range defeated the knights and demoralized foot soldiers. Opposing the longbows were Genoese crossbows that could be operated by untrained troops and did not require the practice needed to use the longbow. Their arrows could also penetrate armour. But although a more useful weapon in these respects, the crossbow had a short range and a low rate of fire and was therefore no match for the longbow.

Swords became narrow and light, with large guards to envelop and protect the hand. By the eighteenth century they had become very light and fairly short, often with decorated hilts. Horsemen favoured curved swords, which were used until the present century. In modern times foot soldiers carry bayonets instead of swords, which are now used almost exclusively for ceremony.

A large shield and short-sword were standard equipment for the Roman legion-ary. The shield provided a defensive wall behind which the soldier was safe from almost all enemy weapons. In phalanx formation shields could be used to form a roof and walls known as the *testudo* or tortoise. In this way soldiers could attack the walls and gates of fortifications vir-tually with impunity. A short-sword lacked the reach of a longer weapon and was prob-ably used with a stabbing action. But it was also less cum-bersome and less fat-iguing to use than a long-sword. A helmet and breastplate helped to protect the body from blows that passed the shield.

7
A

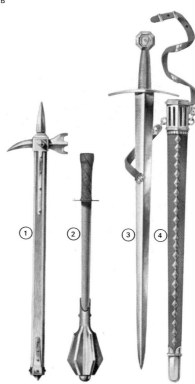

B

7 The medieval knight [A] was a product of the arms race between more effective arm-our and improved arrows. More vulner-able than a man, the horse also had its own armour; but if the knight was unseated, as often happened, he found it hard to remount and use his weapons, and was at the mercy of any nearby foot soldier less heavily encumbered. With the introduc-tion of firearms, against which armour offered little pro-tection, armour declined in use. The lance was the knight's primary weapon, but if this broke he still had a variety of other weapons [B] to use. Most knights carried a war sword [3] with a decorated scabbard and belt [4]. But other popular weapons included a mace [2] and a battleaxe. The war hammer [1] was a direct development of the earlier battleaxe.

① ② ③ ④

8 A

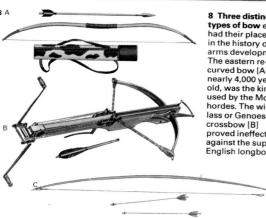

B

C

8 Three distinct types of bow each had their place in the history of arms development. The eastern re-curved bow [A], nearly 4,000 years old, was the kind used by the Mongol hordes. The wind-lass or Genoese crossbow [B] proved ineffectual against the superior English longbow [C].

9 A B C D E F G H I

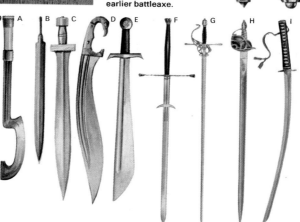

9 Primitive swords include the Egypt-ian sickle sword [A] of 2000 BC and a Swiss sword in bronze [B], 1,000 years later. The Greeks used a double-edged stabbing sword [C] and a single-edged cutting sword [D]. The single-edged falchion [E] dates from medieval times. The straight double-edged sword [F] was a cut-ting weapon. The later rapier [G] was used with only the point. The cavalry sword [H] and Samurai [I] were slashing swords.

161

Development of firearms

Firearms, whatever their age, are similar in principle. They consist of a tube or barrel along which a projectile is propelled by an explosion, a means of ignition and a way of controlling the ignition. The development of firearms is marked by improvements to the firing mechanism or "lock", so called because firearm mechanisms or actions were originally made by locksmiths.

Development of ignition systems

The first firearm was the infantry hand cannon. It was merely a tube with a spike at one end on which it could be supported when fired. It was ignited by a glowing match thrust into a hole (called a vent) in the breech (the closed end of the barrel).

The first spring-powered mechanical ignition system was the matchlock [1] developed in the late fifteenth century. The arquebus was the earliest matchlock musket. Matchlocks remained in use for over two hundred years, establishing many of the accepted characteristics of the longarm. The first matchlock weapon to be fired from the shoulder was a sixteenth century arquebus,

but the early weapons were massive, often needing a support for aiming [2]. Moreover, loading was slow and dangerous. Powder and a ball were rammed down the barrel from the muzzle, with a wad to hold them in place. Then priming powder (fine gunpowder) was placed in the priming pan – all while the mechanism held the glowing slow-match. These difficulties made the weapon unsuitable for mounted men. Almost all matchlocks were smoothbore (with unrifled barrels), and breech-loaders were rare.

The wheel-lock [3] was an enormous advance over the matchlock because it could be loaded and held ready for long periods for immediate use. It was invented in the early sixteenth century (many of the finest surviving examples are German [4]) but was complex, fragile and expensive. Because rich men rode on horseback, the wheel-lock in carbine and pistol [5] form became the weapon of the horseman. Wheel-locks were favoured by German mercenaries (sixteenth century) and English cavalry troops (early seventeenth century).

The chief need was for a weapon lighter

and less cumbersome than the matchlock, yet cheaper and more reliable than the wheel-lock. Both the snaphance and flintlock relied on a flint for ignition. They differed in that the battery or steel (the piece of metal struck by the flint) and pan-cover were one piece on a flintlock [8], two on a snaphance.

The flintlock soon replaced all other firearms. Muskets such as the Brown Bess and the Charleville, and later rifles such as the Ferguson breech-loader, Jaeger rifle and the Kentucky rifle [7] have contributed much to the development of firearms and to the histories of nations.

Many inventors had tried to increase firepower by using multi-barrelled arms, superimposed charges and other means. But none was really practical and only double-barrelled weapons had lasting popularity.

Percussion and the repeating firearm

The Reverend Alexander John Forsyth (1769–1843) began the percussion era in 1805 with his lock that used a highly sensitive explosive called fulminating powder as priming. Like other later devices the lock

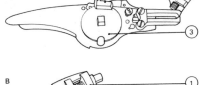

Matchlock musket (England, 1630)

Muzzle support

1 The matchlock mechanism was introduced in the late 15th century. A slow-burning match [1] was secured in a serpentine cock [2] which, when the trigger was pressed, brought the match down on to the powder in the pan [3]. A pan-cover [4] kept the powder dry and retained it in the pan when the arm was not in use. The lock shown here is typical of those used from the early seventeenth century and became

common on infantry firearms. The mechanism was too cumbersome to adapt for a pistol or handgun, but it was used on the petronel, a carbine with a short, curved butt, intended for use by horsemen.

2 Matchlocks of this type were in use in Europe as the common infantry firearm throughout the seventeenth century. Musketeers protected from the enemy by pikemen, performed the intri-

cate, lengthy drill of loading and firing the massive weapons. Weighing as much as 11kg (25lb), these muskets required a forked support to enable soldiers to aim and fire them accurately.

4 Made in southern Germany about 1540, this carbine probably belonged to a rich man. At that time wheel-locks were extremely expensive and usually beautifully decorated. Repairs were also costly.

3 The wheel-lock is shown here externally [A] and internally [B]. Most had to be pre-wound with a key. To ignite the charge the dog [1] holding iron pyrites [2] was lowered into contact with the serrated

edge of a steel wheel [3] projecting through the bottom of the pan [4]. Pressure on the trigger released the wheel, causing a shower of sparks that ignited the priming, which fired the main charge.

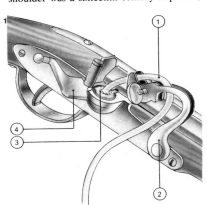

Wheel-lock carbine (Germany, c. 1540)

5 This military pistol of the mid-17th century is similar to the weapons used by the Cromwellian cavalry and other European armies. A trooper with two such pistols had great fire power.

3 A

B

Military wheel-lock pistol (England, c. 1640)

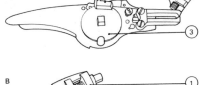

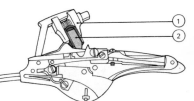

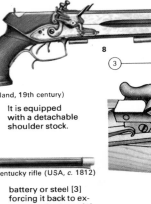

Double-barrelled pistol (England, 19th century)

6 The flintlock in a late form is shown in this officers' pistol.

It is equipped with a detachable shoulder stock.

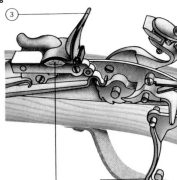

7

Kentucky rifle (USA, c. 1812)

7 The Kentucky rifle was a notably accurate flintlock firearm.

Believed to have evolved from the Jaeger rifle introduced to America by German colonists, it was developed in Pennsylvania and used widely in

the Appalachians. The hinged box cut into the butt carried patches to wrap around the bullet before it was rammed down the muzzle.

8 The flintlock was developed in the early seventeenth century. Pulling the trigger [1] released the cock [2] that held a flint. This struck against the

battery or steel [3] forcing it back to expose the flashpan [4], and creating a shower of sparks that fell into the priming and ignited it.

used the explosive properties of fulminates when struck by a hammer blow as the means of ignition. Of the various percussion ignition systems that followed Forsyth's invention the cap-lock [9] was the most successful.

At first the general design of weapons did not alter, and many flintlocks were converted to cap ignition. Then in 1835–6 Samuel Colt (1814–62) patented his revolving pistol and the repeating firearm had arrived. Even so, Colt's business was not an unqualified success and in 1842 his first venture failed from lack of orders. But, in 1847, Colt was asked by Captain Walker of the US Dragoons to produce a new weapon of 0.44in calibre for military use. This large, six-shot saddle-holster pistol was the "Walker" Colt.

The Walker Colt was followed by other dragoon pistols, the 0.31in calibre pocket models, the 0.36in Navy [11] and Police models, the 0.44in Army, as well as revolving shotguns, muskets and rifles. Muzzle-loading revolvers can, if badly loaded, occasionally "chainfire", when more than one chamber discharges in succession. Because a longarm is held in both hands, this hazard made revolving longarms unpopular – a right-handed shooter, for example, could shoot off his left hand.

All percussion Colts were "single-action" pistols – that is, the shooter had first to cock the hammer with his thumb. Most were open-frame revolvers and lacked the rigidity of solid-frame pistols [12].

Cartridges and the modern firearm

Cartridges had been used for centuries, but not combining the bullet, the charge and primer. The first successful combined cartridge was made in 1812 and later perfected by the German gunsmith Johann Dreyse (1787–1867) for use with his needle-fire rifle in 1837. In America Daniel Wesson (1825–1906) developed an improved rimfire cartridge in 1856 and a similar cartridge was used in the Henry rifle [13B]. In rimfire the primer is sealed in the rim of the cartridge case. Centrefire cartridges (with a central primer) followed, and were used in the 1873 Colt [14] and Winchester. Centrefire is still used for most modern firearms, including machine guns and cannon guns.

KEY
A

B

C

D

E

The flight of a bullet is stabilized by spin, caused by spiral grooves (rifling) in the sides of the barrel. Here eight grooves are shown [A]. The distance between the grooves is about the diameter of the bullet; the raised "lands" cut into the sides of the bullet, causing it to rotate. Ammunition consists of bul-let, charge and primer, today combined in a metal or plastic case [E]. Earlier firearms used separate bullet [B], powder [C] and percussion caps [D]. The cap consists of a small charge of sensitive explosive sealed in a cap of metal foil. Struck by the gun's firing pin or hammer, it ignites the main charge.

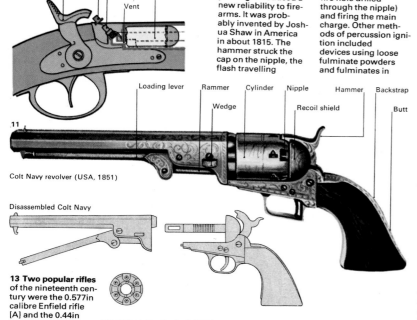

9   Hammer   Cap   Nipple   Charge

Vent

9 Percussion-cap priming introduced a new reliability to firearms. It was probably invented by Joshua Shaw in America in about 1815. The hammer struck the cap on the nipple, the flash travelling through the vent (a fire hole drilled through the nipple) and firing the main charge. Other methods of percussion ignition included devices using loose fulminate powders and fulminates in pills and tapes, but the cap was the most advanced external ignition system. It made possible the future work on repeating firearms, cartridge weapons and, finally, machine guns and other automatic arms.

10

10 Powder flasks were used until the introduction of cartridge weapons. A pistol flask such as this was designed to measure the charge as well as pour it; such flasks were both easy and safe to use.

Loading lever   Rammer   Cylinder   Nipple   Hammer   Backstrap

Wedge   Recoil shield   Butt

11

Colt Navy revolver (USA, 1851)

Disassembled Colt Navy

11 The Colt Navy was not the first of Colt's revolvers, but is probably the most famous percussion revolving pistol. In general appearance it is similar to the larger Dragoon and small Pocket models. The Navy was a "belt" pistol, as distinct from the saddle-holster and pocket pistols, and was open-framed, having no strap over the top of the frame. It had a six-chambered cylinder and was of 0.36in calibre. Each chamber was normally loaded from the muzzle with powder and ball (or conical bullet) rammed home and the nipples primed with percussion caps in readiness for firing.

12

13 Two popular rifles of the nineteenth century were the 0.577in calibre Enfield rifle [A] and the 0.44in Henry [B]. The Enfield was made in England from 1853; it was a single shot muzzle-loader using a paper cartridge [D] which had to be torn open on loading to expose the powder. It was primed with a percussion cap. Several sorts of bullet for muzzle-loaders are shown [C]. The American-made Henry carbine, developed by Tyler Henry, was produced from 1862–6. It was an early magazine cartridge rifle and the forerunner of the Winchester. It had a tubular magazine beneath the barrel holding fifteen rimfire cartridges [E], with a sixteenth in the chamber.

12 An early double-action revolver was this Beaumont-Adams, developed in England in 1856 from the Adam's self-cocking revolver. Trigger pressure cocked and fired it, or alternatively it could be cocked with the thumb before firing.

Beaumont-Adams revolver (England, 1856)

13A Enfield rifle (England, 1853)

B Henry rifle (USA, 1862)

E

C

D

14 Colt cartridge revolver (USA, 1873)

14 The most famous revolver is the Colt "45", which first ap-peared in 1873 and is still produced today. A centre-fire revolver, it was the first "modern" handgun to be manufactured in quantity for both civil and military use.

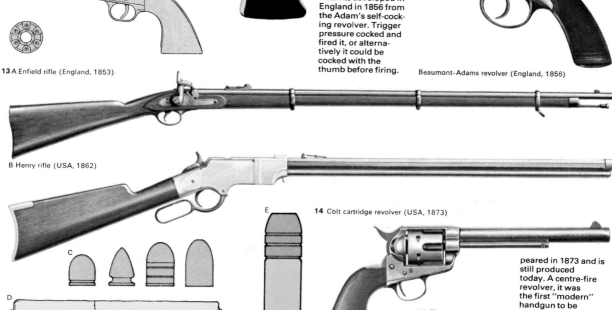

Automatic weapons

An automatic weapon is a gun, rifle or pistol that fires continuously without any external aid such as hand-cranking (or even electric power) for as long as the trigger is pressed. Early designs were hand-cranked in one way or another, but they laid the framework for the first true automatic designs.

Early American designs
The best known early mechanical weapon was the Gatling [1], developed by the American inventor Richard Gatling (1818–1903) and first demonstrated in 1863. This used six barrels rotated by a hand crank and for 50 years the design was sold and copied throughout the world using a wide variety of calibres (barrel sizes). Later true automatic weapons used electric motors or gas pressure taken from the barrel. In the blowback action, part of the explosive force propelling the bullet ejects the spent cartridge and re-cocks the firing mechanism. In a gas-operated weapon hot gases led from the barrel force back a piston that works the ejection and re-cocking mechanisms.

Another American design was the Lowell gun, produced in 1875. This was also a hand-cranked weapon but one that overcame the heating problem affecting all machine guns after 300 or 400 rounds (bullets) have been fired, when the barrel becomes too hot for further use. It had four barrels, one of which was used at a time. When this became too hot a new barrel was rotated into position.

The next development came from another American, Hiram Maxim (1840–1916). He designed an automatic weapon [2] in London, starting by modifying a Winchester rifle. He used a hook fixed to the barrel to lock the bolt in place for firing. The recoil drove both hook and bolt back until the hook was lifted by passing under a bridge. The bolt continued back, driving round a crank to extract and reload the cartridge and was then forced back by a spring. Ammunition was fed into the gun by means of a belt, which could be joined to successive belts to provide continuous feed for long periods of firing.

Satisfied with the success of this design, Maxim simplified it and set up a company to produce the gun with Vickers, the ship-building company. Demonstrations in Europe impressed everybody who saw the gun in action. Vickers eventually took over the company and developed an improved version of the Maxim, which became the standard machine gun for many years.

As designs proliferated three main mechanisms came into use: blowback, recoil (as in the Maxim) and gas-operation. With the recoil method, the breech is locked to the barrel and they move back as one to start the cycle of extraction, ejection, cocking, feeding, chambering, locking and firing again.

World War II
By the start of World War II machine guns were grouped into three main types: the light machine gun such as the Bren [5], which can quickly be set up and fired; the medium machine gun such as the Vickers, capable of sustained fire but of necessity heavier and more robust; and the heavy machine gun used against aircraft and similar targets. The differences between the types are, however, more tactical than technical. The light or sub-machine gun was not popular until it proved

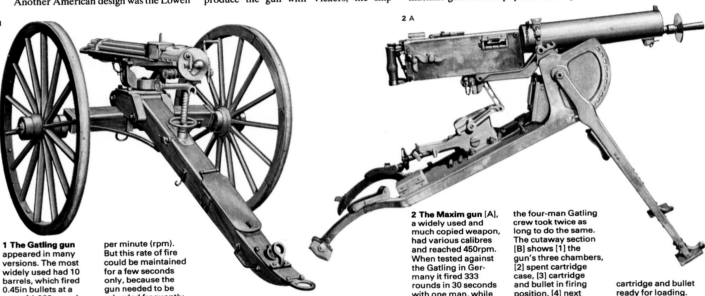

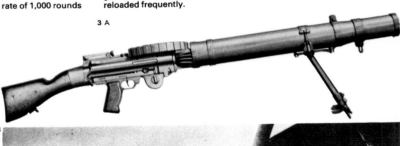

1 The Gatling gun appeared in many versions. The most widely used had 10 barrels, which fired 0.45in bullets at a rate of 1,000 rounds per minute (rpm). But this rate of fire could be maintained for a few seconds only, because the gun needed to be reloaded frequently.

2 The Maxim gun [A], a widely used and much copied weapon, had various calibres and reached 450rpm. When tested against the Gatling in Germany it fired 333 rounds in 30 seconds with one man, while the four-man Gatling crew took twice as long to do the same. The cutaway section [B] shows [1] the gun's three chambers, [2] spent cartridge case, [3] cartridge and bullet in firing position, [4] next cartridge and bullet ready for loading.

3 The Lewis gun [A] was the most successful light machine gun of World War I; more than 100,000 were produced for the Allied armies. Unloaded it weighed 11.8kg (26lb); it was 128cm (50.5in) long and held a 47- or 97-round pan magazine firing 550 rounds per minute. It could be made more quickly and cheaply than any similar gun, could be carried and operated by one man, and used the same .303in ammunition as infantry rifles. Isaac Newton (1858–1931), a US Army colonel, had developed the gun in 1911, but it was first used by the Belgians in 1914 and soon after by the British. It was also widely used on aircraft, being first installed on pusher biplanes. When used in single-seat tractor aircraft the Lewis was mounted on a quadrant on the top wing [B] so that the pilot could pull it down to change the drum. Since aerial gun-aiming was imprecise the Lewis, relatively inaccurate, gave an effective "spray" of bullets.

its worth in the war in the shape of the many excellent German weapons such as the Erma, numerous cheap and simple Russian weapons and the Sten and Thompson [4].

The Americans relied on massed firepower from the Garand or M1 automatic rifle, described by General Patton (1885-1945) as "the best battle implement ever designed". This was a gas-operated rifle of great simplicity and reliability. It was augmented later by a shorter carbine for use by troops (other than front-line infantry), who found a full-size rifle a hindrance. This was the concept behind the Belgian FN automatic rifle [6] after the war.

Automatic pistols and later developments

"Automatic" pistols (strictly speaking, semi-automatic weapons) started to come into their own at the turn of the century [Key]. One of the best known – the 9mm Luger – went into service in 1908. It was officially replaced by the Walther P38 in 1938, but remained in general use although, being a particularly well-made gun, it was dependent on the quality of the ammunition. The advan-

tages of the automatic over the revolver was its more rapid rate of fire and larger magazine. But it was more likely than the simpler revolver to jam in dirty conditions. This is one of the crucial points in the designs of all weapons, particularly automatics. These guns have to be able to operate in extreme conditions without loss of efficiency.

With the coming of fighter planes, then bombers that needed to defend themselves, the machine gun took to the air. The Lewis gun, Spandau and Hotchkiss were used in World War I fighters, with eight 12.7mm (0.5in) guns being mounted in fighters in the later stages of World War II, and up to 13 in the Flying Fortress bomber. Brownings were the most popular choice of Allied forces.

With the coming of fast fighters the target was often in range for only a split second and was generally heavier and stronger. Cannon (with explosive bullets) began to replace machine guns long before the war ended but the ultimate weapon for this purpose came with the development of the rotating-barrel Vulcan [7], which uses a similar principle to the Gatling for its phenomenal rate of fire.

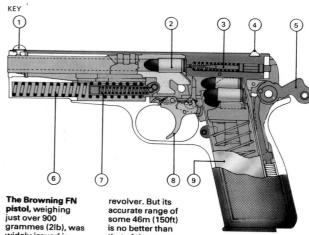

KEY

The Browning FN pistol, weighing just over 900 grammes (2lb), was widely issued in World War II as the standard British automatic pistol. It carried 13 9mm rounds in a magazine in the butt, an advantage over the usual 6-shot revolver. But its accurate range of some 46m (150ft) is no better than that of the revolver. With both weapons, accuracy depends principally on the man behind the gun. Strictly speaking, this and similar weapons are semi-automatic.

1 Foresight
2 9mm cartridge in breech
3 Firing pin
4 Rear sight
5 Hammer
6 Return spring
7 Return spring guide
8 Trigger
9 Magazine

4 Sub-machine guns proved their worth in World War II. The Thompson [A], built by Colt, was a good weapon but expensive. It was replaced by the Sten [B]. More than 2 million of these were produced at a cost of £1.50 each. The German Erma MP 40 [C], like the Sten, was made largely from stamped-out parts; only the barrel and breech were precision made. Many regarded it as the best light machine gun of the war.

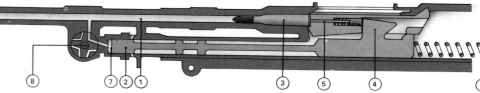

5 The Bren was highly regarded and is still in service today [A]. Built by the Royal Small Arms Enfield factory from a Czech design, the Bren was a good light machine gun and standard equipment in nearly all the Allied armies. The mechanism [B] is gas-operated. Gas from the barrel [1] drives back the piston [2] and the breech block, which ejects the empty case and then, as it is driven forward again by the return spring [6], picks up a new round. The hammer [4] then hits the end of the firing pin [5] and fires the round. The volume of gas in the chamber [7] is controlled by a gas regulator [8].

6 The Belgian FN was adopted as the standard NATO rifle in the 1950s. It can be used as a single-shot weapon or as an automatic and fires the standard 7.62mm round. The operation of the gun is simplified because the left hand works the cocking, feed and safety while the right hand is left to hold and fire the gun. Like the Bren, the FN is gas-operated, but its 20-round magazine is mounted underneath the gun. The gun is self-loading and fully waterproof and is one of the best infantry weapons in the world. It has a maximum rate of fire of 600rpm.

7 The GEC Vulcan was designed for use in fighters in its 20mm form. Later the 7.62 mm MiniGun was developed for helicopter use in Vietnam against ground targets, where the 6,000rpm rate had devastating results. The Vulcan is here mounted on a tracked chassis for anti-aircraft work with a power-operated traverse to track low-flying supersonic aircraft. Although wasteful of ammunition, the Vulcan represents the optimum in current machine guns.

History of artillery

The term "artillery" strictly includes any weapon that projects a missile farther than a man can throw it by hand. But with large missiles, the mechanism used is too heavy to carry, and the definition of artillery is now generally limited to weapons of this type.

Early artillery "engines"
All artillery makes use of stored energy that can be released and expended quickly to propel the missile. Before the invention of gunpowder there were three ways of doing this: the compression and tension of fibres, as in a bow; the torsion (twisting) of sinew or fibre as in the catapult; and – much later – the use of a counterweight, as in the trebuchet.

Artillery using the bow principle was known in Syracuse in 399 BC, at about which time a repeating bow firing bolts was developed by Dionysus of Alexandria. Similar weapons were still being used by the Chinese in the 1890s.

A development of the bow, known to the Romans as the ballista, used a single arm inserted into a band of fibre that was tightened by a winch [2]. It could be used to throw a variety of missiles, including stones, Greek fire (a mixture of pitch, sulphur and naphtha), live or dead prisoners and filth. A ballista had a range of about 500m (1,640ft) and fired a missile weighing up to 150kg (330lb). The only siege engine to be invented in the Middle Ages was the trebuchet [4], which used a heavy counterweight that imparted velocity to the missile.

Development of firearms
Gunpowder became known in the West about the end of the thirteenth century. Guns first appeared in Europe in the first quarter of the fourteenth century and there seems to have been an industry making and exporting guns in Ghent at that time. Early reports of their use in combat include the Siege of Metz (1324), the Battle of Halidan Hill (Scotland) in 1333, and the Battle of Crécy (1346). The first guns fired arrows [5] and later ones shot stones or balls of cast or wrought iron.

Early light guns were breech-loaders, each with a separate barrel or chamber, mounted on a wooden frame or sledge. These guns were individual pieces with little standardization. Some, such as Mons Meg [Key], were huge and became legends. Mons Meg is 4m (13ft) long, its calibre 49.5cm (19.5in) and its weight 5 tonnes. It fired 150kg (330lb) granite balls. A weapon known as the Dardanelles gun, of cast bronze beautifully made in two pieces that screwed together, is at the Tower of London. Its length is 5m (16ft), its calibre 63.5cm (25in), its weight 17 tonnes and the weight of the stone shot 304kg (670lb). Tzar Pouchka, in Moscow, cast in 1586 was 5.4m (18ft) long, its calibre 91.4cm (36in), its weight 38 tonnes and the weight of the stone ball 998kg (2,200lb). The weight of powder used to fire this monster was 90-136kg (200–300lbs). In 1544 Emperor Charles I of Spain (1500–58) limited artillery to seven models, permitting standardization of shot.

The barrels of the first true guns were made from a bundle of wrought-iron rods arranged in a cylinder and welded together. Molten lead was generally poured into the grooves and the whole barrel wrapped in iron hoops. The trunnion, a short stub-axle fitted to the barrel to enable the muzzle to be raised

1 The bow introduced the principle of storing energy for sudden release. It was used in warfare by the Ancient Greeks – classical Greek technology could be very competent when there was a slave shortage. Bows and arrows continued to be used in warfare even after the invention of gun-powder because they were cheap, effective and accurate. The crossbow was developed in the Middle Ages.

2 The most powerful catapults of antiquity used the torsion principle rather than the sprung bow. The stone missiles were of varying weights and the artilleryman had to make allowance for this.

3 This Graeco-Roman ballista of the first century AD used the principle of the bow. The long wooden arm was pulled down (like half a bow) and suddenly released to hurl the missile.

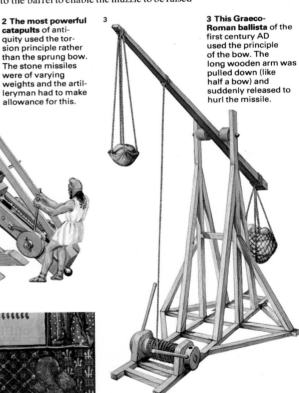

5 The pot-de-fer, the earliest type of gun used in the West, was first employed in the 14th century. It fired a bolt or heavy arrow. Similar Chinese guns probably fired arrows from bamboo tubes.

6 This breech-loading cannon of the late 14th century is from Castle Rising, Norfolk, England. The illustration shows a spare breech and balls. The problem of recoil was a major one at this time.

4 The trebuchet, a medieval siege weapon, was designed to hurl stones and similar missiles for great distances. Also called a mangonel, it worked with a heavy stone counterweight that fell to provide the necessary power.

or lowered, was not invented until the mid-fifteenth century. Artillerymen filled cases with small stones or scrap metal, or loaded them loose in the barrel, and used them as anti-personnel missiles.

Tactics developed slowly. Initially artillery was used to batter down walls or to fire from walls at an approaching army. Turkish siege guns were largely responsible for the fall of Constantinople in 1453. Between 1537 and 1551 the Italian mathematician Niccol Tartaglia (c. 1500–77) put forward the first theory about trajectories. He showed that a flying shot follows a curved path – previously gunners had thought that it went straight for a certain distance and then fell vertically downwards.

In 1626 Gustavus Adolphus of Sweden (1594–1632) tried to devise light field pieces to support his troops. He had his armaments industry make copper barrels bound with iron rings and covered with leather. But these barrels rapidly overheated and therefore could not quickly be reloaded. He also organized his artillery into three roles: field, regimental and siege. This became a fairly standard form of division until 1776 when the French Inspector of Artillery Jean-Baptiste de Gribeauval (1715–89) grouped them into field, siege and coastal defence. By that time guns were carried on light horse-drawn carriages with second carriages, the limbers, holding ammunition and spare parts.

Artillery at sea
Heavy naval guns were of greatest use for broadsides at close quarters, although longer-range light guns could be used for harassment. Ships were rarely destroyed by ship-based guns but heavy damage could be inflicted. Pebbles or almost any pieces of iron could be used on land or sea as anti-personnel missiles; grapeshot was bags of round-lead shot. Chain-shot, two hemispheres of iron keyed together and joined by a chain, was particularly effective against masts and rigging. Shore artillery could be far heavier and more accurate. Iron shot was sometimes heated before firing in order to start fires on ships. After loading the main powder charge, the gunners added a small damp charge followed by a red-hot ball.

7A

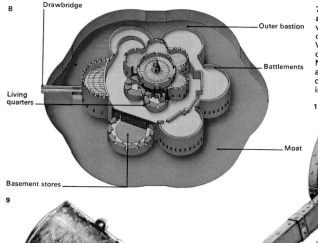

Drawbridge
Outer bastion
Battlements
Living quarters
Moat
Basement stores
9

7 Commanders played a crucial part in developing the technology of artillery. Henry VIII [A] developed the British Navy, founded an arsenal, and took a detailed interest in the development of fortifications [B]. He introduced some of his own ideas in fortifications of the north against the Scots and in Boulogne against the French, and encouraged the science of military engineering.

10

10 A field piece of this type was used by all armies during the War of Spanish Succession (1701–14). It had a chased and elaborately decorated gun barrel of brass or iron [1] and its gun carriage was reinforced with iron binding [2].

8

8 This muzzle-loading Parrott rifle – a 10 pounder with a 3in bore – was used by both sides in the American Civil War (1861–5).

9 Used in sieges with devastating effect, this 13in mortar was deployed by Union forces during the American Civil War.

11 During the Crimean War (1854–6) this British 18-pounder field gun [B] with its ammunition wagon [A] was used. Much of the allied artillery, however, was of a lighter calibre. Many of the field pieces in use were 9-pounders and some were as much as 40 years old.

12

11A

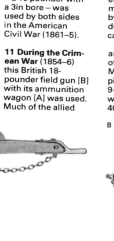

B

12 Gun tools of the Napoleonic Wars included a damp sponge [1] – used to get rid of glowing residues in the bore – a rammer [2], which drove the projectile into position, and a worm [3] for removing any obstruction.

Modern artillery

There was little change in heavy guns for the 500 years from the time of their introduction in the 1300s to the early 1800s. Their barrels were cast in bronze or iron and they were loaded from the muzzle using black powder (gunpowder). Then the nineteenth century advances in metallurgy and chemistry were applied to artillery.

In 1855 William Armstrong (1810–1900) of England designed a three-pounder gun with a barrel made of wrought iron wrapped round an inner tube. The barrel was rifled to give spin to the shell and therefore greater accuracy and it was loaded from the breech. Many such guns were made in various calibres [1], but the design was not completely successful. Better was the sliding breech-block, designed by Krupp of Germany, and the interrupted screw system.

In 1888 black powder was replaced by a new propellant explosive – guncotton or nitro-cellulose. It burnt more slowly, generating more hot gas and therefore more force. Gun sights were also improved.

In an explosive shell fired from a muzzle-loaded gun the fuse was ignited by the hot gases from the propellant. By 1880 steel projectiles were in general use. A soft metal collar, on a steel shell, called a driving band, sealed it from the gases, thus another type of fuse had to be used. Some detonated when they hit a target and others used clockwork or timed powder trains.

A new explosive for the shell, based on picric acid, was introduced in 1886. The shrapnel shell – a case filled with small steel balls – was invented in 1784 and continued in use until 1916. Breech loading and the new explosives gave rise to "fixed ammunition" with a cap detonator, main propellant charge and the projectile all in one metal case. For calibres of more than about 15cm (6in) separate charge and shell continued to be used.

Developments to World War I

A French field gun of 1897 had a device to absorb much of the recoil. As the barrel was forced back by the explosion in it, the movement was arrested by compressing oil in a reservoir; air in a second reservoir was also compressed and its pressure forced the barrel forward again. A spade at the end of the gun carriage helped to keep it steady. This gun, the famous "75" (it had a 75mm calibre), was still in service in World War II. Its flat trajectory, however, limited its use against well-protected installations for which high-trajectory howitzers were sometimes needed.

Britain had a variety of guns: a 13-pounder (3in calibre, about 75mm), 4.7in and 6in guns, and 4.5in, 5in, 6in and 9.2in howitzers. Germany used a 77mm field gun firing a 7kg (about 15lb) shell and howitzers of 42, 105 and 155mm calibres. Nearly all gun carriages were pulled by horses, although there were a few trials using lorries. Some large guns were mounted on railway wagons. The most famous of these was the 21cm (8.3in) "Paris Gun" with a range of 132km (82 miles), successful only in its value as an item of propaganda.

Developments in ammunition included poison gas shells (1915) and white phosphorus smoke shells (1916). The high explosive TNT replaced shrapnel. Trench warfare led to the development of the midget howitzer, or mortar. Mortar calibres increased from 75mm (3in) to 150mm (6in) and even

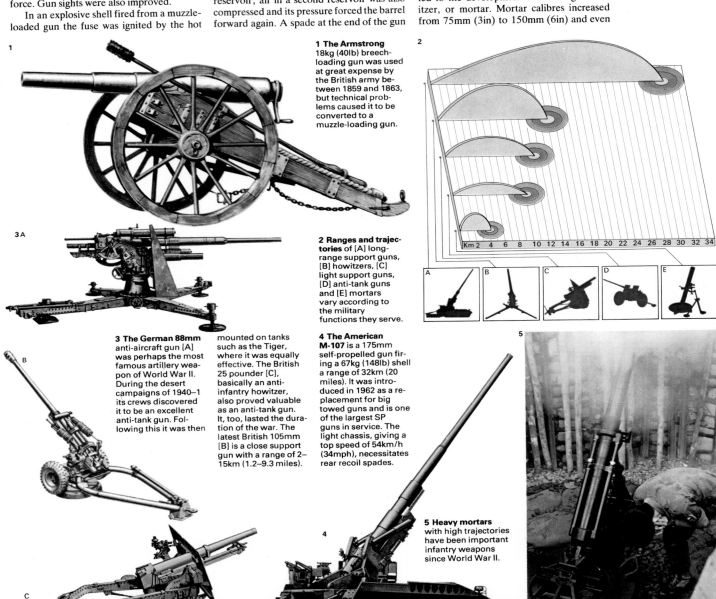

1 The Armstrong 18kg (40lb) breech-loading gun was used at great expense by the British army between 1859 and 1863, but technical problems caused it to be converted to a muzzle-loading gun.

2 Ranges and trajectories of [A] long-range support guns, [B] howitzers, [C] light support guns, [D] anti-tank guns and [E] mortars vary according to the military functions they serve.

3 The German 88mm anti-aircraft gun [A] was perhaps the most famous artillery weapon of World War II. During the desert campaigns of 1940–1 its crews discovered it to be an excellent anti-tank gun. Following this it was then mounted on tanks such as the Tiger, where it was equally effective. The British 25 pounder [C], basically an anti-infantry howitzer, also proved valuable as an anti-tank gun. It, too, lasted the duration of the war. The latest British 105mm [B] is a close support gun with a range of 2–15km (1.2–9.3 miles).

4 The American M-107 is a 175mm self-propelled gun firing a 67kg (148lb) shell a range of 32km (20 miles). It was introduced in 1962 as a replacement for big towed guns and is one of the largest SP guns in service. The light chassis, giving a top speed of 54km/h (34mph), necessitates rear recoil spades.

5 Heavy mortars with high trajectories have been important infantry weapons since World War II.

230mm (9.4in) and large mortars evolved into infantry weapons with recoil systems and accurate sights [5].

Developments to World War II
Light close-support guns had been the only addition to artillery since 1918 when war broke out again in Europe in 1939 and lorries began to replace horses. The British used the 25-pounder gun-howitzer, with a range of 12,370m (13,400yds) [3]. The United States and Germany had 105mm howitzers firing 4.5kg (10lb) shells over a range of 10 to 12km (6.25–7.5 miles).

Improvements in design allowed heavier shells to be fired over greater ranges – and calibres varied from 20mm (0.79in) to 800mm (31.5in). New fuses exploded shells at a pre-set height above the target. Mountings included wheels, half-tracks and full caterpillar tracks. Special defensive guns were developed for anti-tank use (some self-propelled) and for anti-aircraft [3].

Then new weapons appeared – rockets that had been tried and abandoned in the mid-1800s. Germans and Americans used Panzerfaust and bazooka anti-tank rockets. Britain and the USA developed a 3in anti-aircraft rocket and a 5in multiple assembly that could lay 455kg (1,000lb) of high explosive per second on a target for nearly a minute. The Russian equivalent was a 130mm (5.1in) lorry-mounted multiple rocket system called Katyusha.

After World War II
In the closing stages of the war Germany introduced its two "vengeance weapons", the V1 flying bomb and the V2 rocket. The V1 was a small pilotless aircraft using a pulsejet engine with rocket-assisted take-off. It was used to bomb London and nearby targets from launch sites in France and the Netherlands. The V2 [6] was a supersonic liquid-fuelled rocket carrying a warhead of nearly a tonne of high explosive. It served as a model for later American and Russian experiments with rockets – soon to be called missiles [9]. For speed of preparation and firing, modern missiles have solid-fuel propellants and, armed with nuclear warheads, they can be fired from underground silos or submarines.

Russian rockets, on display in Moscow's Red Square, maintain a balance of power with US weapons.

6 The German V2 rocket was one of the most destructive and sophisticated weapons of World War II. Launched mainly from sites in the Netherlands they killed 2,855 people in England.

Built like finned shells, 14m (46ft) high and guided by automatic pilots, each carried nearly a tonne of explosive across a range of 320km (200 miles) at a top speed of 5,794km/h (3,621mph).

Launching table

Control room
Trailer and uplift jack

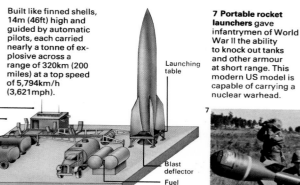

Blast deflector
Fuel containers

6

7 Portable rocket launchers gave infantrymen of World War II the ability to knock out tanks and other armour at short range. This modern US model is capable of carrying a nuclear warhead.

8 Minuteman was the first in a series of US intercontinental ballistic missiles. These three-stage solid-fuelled rockets are always fired from below ground and carry nuclear warheads.

8

9 The command system of a surface-to-air guided missile sometimes has a long-range radar [1] system that detects the target and sends flight-path information to a control computer [2], which

activates a tracking radar [3] to lock on the target. In tactical systems the tracking radar both detects and tracks the target. Once the target has been established as an enemy, the control computer

launches a missile [4] and activates the missile radar [5]. Information from the target and missile radar is then fed back to the control computer which guides the missile through a command radio [6].

7

10 Rockets with considerable range and firepower, such as *Swingfire*, can be launched from the backs of tanks as well as from fixed sites. They can locate and destroy enemy vehicles before

they get within gun range. Another means of achieving speed and mobility in placing rockets for attack or for defence is to launch guided rockets from high-speed naval patrol boats.

10

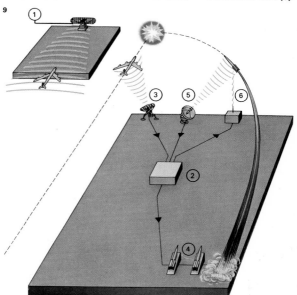

9

Armoured fighting vehicles

Of the many types of armoured fighting vehicles (AFVs), the tank is the most significant. The idea of such a vehicle can be traced back hundreds of years. But it was not until the internal combustion engine and caterpillar tracks were developed at the beginning of this century that it became practicable to devise a war machine that could carry men and weapons protected by armour plate.

Britain took an initial lead in tank development, the early World War I machines being designed by the Admiralty as landships. Then an American, Walter Christie, designed a heavy tank with a more advanced suspension (the T3), which influenced the new Russian tanks. By 1941 the Russians had their T34s [3] – then the best and most numerous tanks in the world and not matched until 1943.

Guns and armour

Between 1939 and 1945 typical gun calibres grew from 37 to 120mm and armour thickness from 30 to 240mm (1.2 to 9.4in). Tank weights rose from 20 to 70 tonnes. The heaviest tank ever to go into production was the limited traverse *Jagdtiger* of 1944, with a 128mm gun, 250mm (9.8in) armour and a weight of 72 tonnes. Such tanks were too expensive and their size and weight hindered their mobility.

In recent years tank design has been aimed at flexibility in use, with sizes and weights less than the highest of World War II. Advances in gunnery have improved the shooting on the move and at moving targets. Night shooting uses infra-red gunsights.

In the 1970s, conventional tanks weigh about 50 tonnes. They have a long-barrelled gun of 105–120mm calibre, which can fire high-velocity armour-piercing shot against tanks and high explosive shells against other targets. It can also fire cannister shot (containing large scatter-gun pellets) against massed infantry. The gun is generally mounted, together with a machine gun for use against infantry, in a turret that provides all-round traverse. The combination of hitting power, mobility, armour and good radio communications allows a tank to make a flexible, swift response and gives it the shock-action needed to penetrate defences.

Most tanks have four-man crews. The commander, gunner and loader/radio operator are in the turret and the driver down in front, in the hull. The engine is mounted behind the turret and at the back is the combined gearbox and steering unit. The tank runs on tracks, with five or six large wheels each side and large springs and shock-absorbers to permit it to travel at speed over rough ground. Modern tanks have steeply sloped armour up to 120mm (4.7in) thick.

Among the most prominent tanks of the 1970s are the Russian T62, a 47-tonne tank with a 115mm smooth-bore gun firing spin-stabilized ammunition, and the German *Leopard 1*, a fast 43-tonne tank with a 105mm gun. The British Chieftain, heavily armoured and with a 120mm rifled gun, is one of the heaviest tanks currently in use. The American M60 A2 is a 52-tonne tank armoured with both a 152mm conventional gun and Shillelagh guided missiles.

Tanks rarely operate effectively alone, needing infantry, artillery, engineers and aircraft in support. But past battles have shown that, boldly used, armour can be decisive

1 The Daimler armoured car first served in 1941 and was still used in the 1960s. It was a reliable reconnaissance vehicle. Despite four-wheel drive, it had only limited cross-county ability. Its 95 hp engine could achieve 80km/h (50mph) and had five forward and reverse gears. If the vehicle had to reverse along a road, the commander could steer it by means of a wheel mounted in the turret. It carried a crew of 3 and a 2-pounder gun.

2 PzKpfw III was Germany's main tank at the beginning of World War II and an important weapon in the *Blitzkreig* attacks on Poland in 1939, France in 1940 and Russia in 1941. Early models had a 37mm gun, but this later version had a 50mm gun. It was outclassed by the Russian T34. A limited traverse version, *Sturmgeschutz III*, with the 75mm gun of the PzKpfw IV, gave excellent service until the very end of World War II.

3 The Russian T34 with a 76mm gun and 45mm (1.8in) of sloped armour was, in 1941, the best tank in the world, and came as a shock to the Germans. Its 500hp diesel engine was a model of reliability with a speed of 51.5km/h (32mph). The Germans were able to advance deep into Russia despite the T34 because Russian tactics and organization, compared with those of the Germans, were poor.

4 The American M4 Sherman tank was the main allied tank of World War II after 1942. It had a five-man crew, 80mm (3.1in) armour protection, and was originally equipped with a short 75mm gun. The later version, shown here, had a larger, 76mm, gun.

because of its great mobility and because of its shock value against infantry.

Light armoured fighting vehicles
Some AFVs, such as the Daimler armoured car [1], do not have the strong hull armour of main battle tanks and are protected only against shell splinters and machine-gun bullets. This keeps their weight down and enables them to travel at high speed.

Reconnaissance vehicles seek out the enemy, patrol open flanks and act as an early warning screen against surprise attack. Some have wheels and some tracks. A tracked AFV tends to be smaller and have better cross-country performance, but is noisier than its wheeled alternative.

Armoured personnel carriers, such as the Saracen [7], take infantry into action. There are wheeled and tracked versions, which have guns to provide cover for their infantry or to engage other similar vehicles.

The artillery have armoured self-propelled guns, such as the Abbot [8]. The engineers have armoured bridge launch vehicles, carrying a folding bridge span on a battle-tank hull. Armoured bulldozers dig fire positions or clear debris. Armoured recovery vehicles rescue tanks for repair, clear routes or assist tanks to cross water. Many of the lighter vehicles are now made of aluminium. They float, or can easily be made to do so with a screen.

The future of tanks
Anti-tank guided weapons can be launched from the air, the ground or from light vehicles and are controlled in flight. But control by the firer is difficult and automatic guidance is expensive. Unless they are mounted in an armoured vehicle, they are vulnerable and too expensive to use against targets other than tanks. Thus as a weapons system they have not yet supplanted tanks, although they do complement them.

The advent of modern weapons has not reduced the value of tanks. Just as an infantryman is vulnerable to every weapon, yet plays a vital role in a battle, so tanks have always been vulnerable to some weapons and to mines. It is probable they will be a part of most armies for years to come.

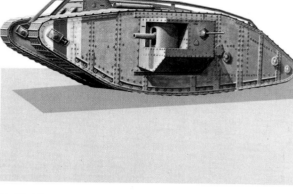

The Mark IV tank was used by the British in World War I. Its shape was largely dictated by conflicting requirements: it had to cross wide trenches and also be transported to the front on narrow railway wagons. Mk IV tanks were first used in the Battle of Cambrai, in France, on 20 November 1917.

5 A minesweeping flail was one of the many special AFV devices developed in World War II, here seen mounted as the "Crab" on the Sherman tank. The flail consisted of chains on a rotating drum, which beat the ground in front of the tank and exploded any mines in its path. Now tanks push ploughs or rollers, or launch sausages of explosives by rockets over a minefield to clear their paths of mines.

6 The Swedish S tank is a novel concept with no turret. This allows a low profile and slight weight yet excellent armour. The gun is aimed by moving the whole tank on its suspension and steering. The crew is three, although one man can operate the tank in an emergency. Ammunition is loaded automatically; the magazine, is at the rear, limiting the fire risk if hit. The gun breech at the back of the chassis reduces protuberance of the barrel, allowing traverse in confined spaces. The engine must be running to traverse and the tank cannot fire while it is moving.

7 The Saracen is one of the armoured personnel carriers that uses wheels rather than tracks, giving a silent run and a long range on roads with little maintenance. Such wheeled APCs are useful for internal security operations. A Saracen has smoke dischargers and a machine gun in its turret. All six wheels are driven, but it has a limited performance off the road. Tracked APCs have a better cross-country performance.

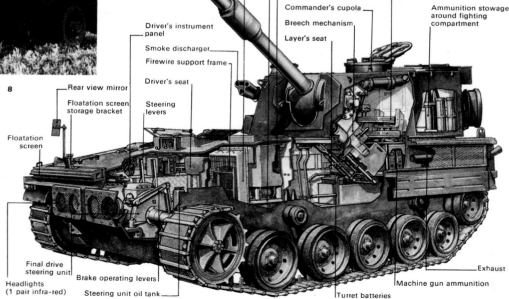

QF 105mm L13A1 gun
Engine support frame
Radiator
Fume extractor
Driver's instrument panel
Smoke discharger
Firewire support frame
Driver's seat
Rear view mirror
Floatation screen storage bracket
Steering levers
Floatation screen
Final drive steering unit
Headlights (1 pair infra-red)
Brake operating levers
Steering unit oil tank
Air cleaner
Commander's cupola
Breech mechanism
Layer's seat
Commander's seat
Ammunition stowage around fighting compartment
Exhaust
Machine gun ammunition
Turret batteries

8 The British Abbot is a self-propelled artillery gun. Artillery needs mobility to keep pace with the battle and to avoid counter battery fire. Armour protects it against such fire and against attacks from enemy who infiltrate the forward areas. Some self-propelled guns are outwardly similar to tanks, but there are many differences to suit their different role of sustaining fire against targets covering a large area deep in enemy territory, beyond the direct line of sight used by tanks. The Abbot uses the same automotive components as the tracked APC, FV432. It can "swim" using an extended screen while being propelled and steered by its tracks.

Fighting ships: the age of sail

Nations dependent on sea trade have always needed fighting ships to protect their sea lanes. Peaceful maritime commerce has usually been the result of an acknowledged naval supremacy on the part of one nation or another. By building up sufficient naval strength, interlopers have been able to achieve dramatic reversals of commercial power, as when the Arabs supplanted the Byzantines in the fifteenth century or when the Spanish were pushed aside by the Dutch and English in the sixteenth and seventeenth centuries. Some nations used ships for attacking other countries, as when the Vikings in their longships terrorized eastern Britain in the ninth century.

Galley warfare
An early reference to naval warfare in Egyptian temple reliefs of the second millennium BC shows a battle between the Egyptians and the "Sea People", with the former using oared sailing boats and the latter pure sailing boats. The Egyptians won, perhaps because of the superiority of their galleys.

The ancient Greeks, Phoenicians and Romans had galleys with up to three (and perhaps more) staggered layers or banks of oars [1]. These boats carried an auxiliary square sail on a single mast. Bows, arrows and catapults were not decisive weapons and an engagement usually had to be decided by boarding the enemy. Because of the sweeps (oars) this could be done only by head-on attack and the ram was a useful means of holding or capsizing an adversary.

Galleys survived until the early nineteenth century in some areas with remarkably little alteration, the major changes being the replacement of staggered sweeps by a single bank of longer sweeps powered by more men, and the replacement of the inefficient single square sail by triangular lateen sails on one or more masts [2].

The demise of the galley followed the development in northern waters (unsuitable for galleys) of manoeuvrable sailing men-of-war, which could fire devastating broadsides. The galleys could carry at most only five guns placed in the bows and were no match for the men-of-war after the sixteenth century. The last great galley battle was fought off Greece at Lepanto in 1571 when the Turkish fleet was defeated by a combined southern European force.

During the Middle Ages, there was no difference between the design of merchant and fighting ships; the former could be converted readily into the latter if soldiers were carried instead of cargo and wooden battlements were nailed on the bows and stern (the origin of the forecastle and quarter-deck). In fact, the distinction between the type of vessel used for trade, war and piracy was hazy until the seventeenth century and even up to the nineteenth century when privateers or corsairs were easily adaptable.

Importance of artillery
Light anti-personnel artillery was fitted on the rails of vessels during the fourteenth century and the great ships of the early Tudor period had gun ports pierced in the hull. Larger ships of this period were built primarily for naval purposes but would often engage in trade. The same was true of the Elizabethan galleon and in the Anglo-Dutch wars of the seventeenth century the Dutch

CONNECTIONS

See also
166 History of artillery
174 Modern fighting ships

In other volumes
96 History and Culture 2

1 A Roman war trireme of the first century BC can be drawn quite accurately from reliefs of the time. It was propelled either by three banks of oars or by a square sail, the mast being taken out (as shown) before battle. The lower-bank oars were manned by one slave each, the middle by two and the upper bank by three. This galley needed 144 rowers. Armament comprised a ram, a drawbridge (corvus) for boarding, a catapult (onager) and a carroballista which fired arrows and blazing darts. The buffer above the ram prevented the galley ramming too far into an enemy ship. The castle was a vantage point for the commander as well as a refuge in battle. On galleys which did not have a castle, there was an elaborate throne placed under an awning for shade.

Castle
Steering oar
Carroballista
Onager (catapult)
Tortoiseshell formation
Mast hole
Corvus
Buffer
Occulus
Ram

Captain's wine store
Rhythm keeper
Captain's tent
Coursie
Driver
One-man oar
Three-man oar
Two-man oar

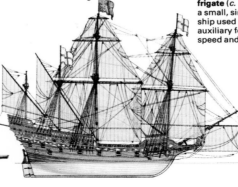

2 The 16th-century Venetian galley had a lateen rig, one bank of oars, a ram which doubled as a boarding bridge and five guns placed in the bows.

3 The Elizabethan galleon (c. 1585) was a forerunner of the ship of the line. The fourth mast disappeared in about 1625, but some 19th-century vessels had four.

4 The Napoleonic frigate (c. 1800) was a small, single-deck ship used as a fleet auxiliary for its speed and versatility.

fleets were essentially composed of the large two-decker trading vessels (known as "East Indiamen") of the Dutch East India Company. By that time, however, the fleet on the English side was purely naval and included some ships with three gun decks running the whole length of the ship. The upper decks such as forecastle and quarter-deck often carried guns as well.

Up to the time of the Anglo-Dutch wars, naval tactics remained fairly similar to those used by the galley; ships formed a line abreast and "charged" the enemy. But the increasing importance of artillery and its broadside placement led to the line-of-battle tactic [5] in which ships forming the line presented a formidable wall of guns. Only three- and two-decker ships were included in such lines and were known as "ships of the line". Two opposing lines would sail parallel to one another, firing broadsides in an attempt to create confusion in the enemy line so that it could either be pierced or divided.

Crossing an enemy's bow or stern not only ensured safety from his fire (which could be aimed only sideways) but also provided the opportunity to rake him. The bow and stern were the weak spots of a ship and shot ploughing through the whole length of a gun deck was particularly lethal. The side which achieved the windward line had the advantage of being able to disengage more easily in case of trouble. In addition, an enemy could be holed below the water-line (exposed as the ship heeled with the wind).

Close-range fighting

Attack at close quarters remained an important tactic and in the Napoleonic wars the British fleets made effective use of a large-bore, short-range gun called the carronade [10]. The British traditionally aimed at smashing the enemy's hull, while the French, firing on the uproll, concentrated on rigging and upperworks.

Beginning with small, paddle-wheel warships, navies experimented with steam engines early in the nineteenth century. Engines were fitted in ships of the line in the 1850s and in the next decade the appearance of the rifled shell gun rendered the "wooden walls" obsolete and ushered in the ironclad.

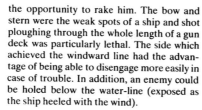
KEY

A

B

C

D

Men-of-war were classed into rates from the 1650s until the end of the era of sailing navies. Rates were based on numbers of guns carried (excluding the carronades) and were altered several times over the era of the line-of-battle ship. In Nelson's day first-raters [A] had 100 or more guns on three decks, second-raters [B] had 90 to 98 guns on three decks, third-raters [C] had 64, 74 or 84 guns on two or three decks and fourth-raters [D] had 50 on a single deck. Only the first three rates were classed as ships of the line. Fourth- and fifth-raters (32–44 guns) were frigates. Sixth-raters (20–28 guns) were sloops and brigs of war.

5 Naval fleet tactics of the 17th to 19th centuries were based on a line of battle with two- and three-decker ships forming a line ahead and frigates stationed outside to relay flag signals.

6 Deaths at sea during the age of fighting sail were common. Statistics for the British navy during the late 18th century show that accidents and diseases took many more lives than enemy action. Respiratory diseases like pneumonia and tuberculosis were rife, as were illnesses from bad food and drink. In the tropics, malaria and yellow fever (the "black vomit") were added hazards.

Enemy action 9%

Fire, sinking, wreck 10%

Accident 31%
Falls
Snapping cables
Wind
Storm hazard, lightning

Disease 50%
Scurvy
Tuberculosis
Pneumonia
Alcoholism
Typhus
Yellow fever
Malaria

7 Gunnery procedure changed little between 1560 and 1860. The powder cartridge was carried from the magazine in a box [A] to protect it from sparks. It was pushed down the gun's barrel with a rammer

[C] and pricked through the touch hole with a priming wire [E]. The barrel was swabbed, after firing, with a sponge [B] to extinguish embers. The worm [D] was used to extract unfired charges.

8 Naval cannon had a touch hole [1], cartridge [2], wads [3], shot [4], muzzle [5], breech [6].

9 Types of disabling shot included chain [A], bar [B], elongating [C], grape [D] and canister [E].

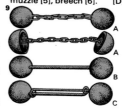

10 The carronade or "smasher", first fitted in 1779, was a short, stubby gun that fired a heavy shot and was deadly

at close range. Light in weight and more manoeuvrable than other guns, it was mounted on the upper decks.

11 HMS Victory (1765) was a typical "first-rate" ship of the period. She

carried 104 guns: 2 12-pounders on the forecastle, 12 12-pounders on the quarter-deck, 30 12-pounders on the upper gun deck, 28

24-pounders on the middle gun deck, 30 32-pounders on the lower gun deck and 2 68-pounder carronades. She carried 850 officers and men.

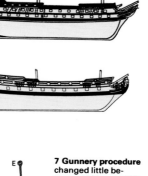

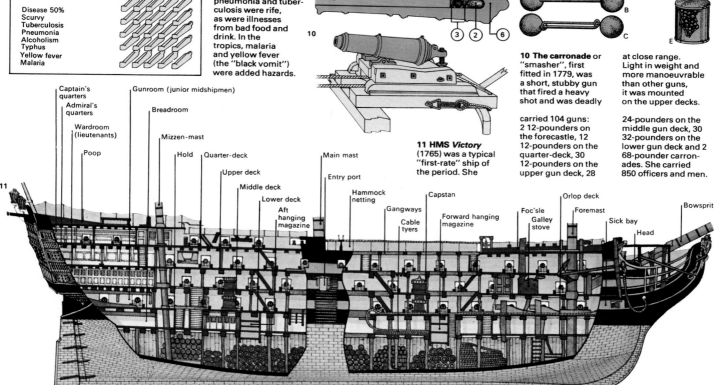

Captain's quarters
Admiral's quarters
Wardroom (lieutenants)
Poop
Gunroom (junior midshipmen)
Breadroom
Mizzen-mast
Hold
Quarter-deck
Upper deck
Middle deck
Lower deck
Aft hanging magazine
Main mast
Entry port
Hammock netting
Gangways
Cable tyers
Capstan
Forward hanging magazine
Foc'sle
Galley stove
Orlop deck
Foremast
Sick bay
Head
Bowsprit

Modern fighting ships

Navies are still used to protect shipping lanes, deter invasions or support military operations on land. But the nature of modern warships has changed radically in the past 30 years. Until World War II, battleships were the most important units of any fleet, armed with large guns and protected by thick armour plate. In essence, they had altered little from HMS *Dreadnought* [1] of 1906, just as most other types of warship, particularly cruisers, destroyers and submarines, were fundamentally similar to their predecessors of 30 years earlier.

Rise of the carrier

World War II changed the pattern of naval power. Aircraft carriers with dive-bombers and torpedo-bombers could strike enemy targets at 500km (300 miles) range, compared with the 32km (20-mile) range of a battleship's guns. Carriers had been under development since 1912, but were perfected as weapon systems only between 1939 and 1945, when it became clear that their mobility and striking power made them supreme. Today large attack carriers [5] are still

the most powerful surface warships, carrying about 80 bombers and interceptors, all similar in performance to their land-based counterparts.

Modern naval aircraft are too heavy to take off from a short deck, so they are launched by powerful steam-driven catapults. An incoming aircraft is arrested on landing by means of a wire across the flight deck which engages a hook under its tail. These special techniques enable carrier aircraft to use a shorter runway than that needed for normal take-off and landing. Naval aircraft carry the entire range of weapons and equipment used by land aircraft, including guided missiles, bombs and electronic counter-measures.

The development of Vertical Short Take-off and Landing (V/STOL) and Short Take-off and Landing (STOL) aircraft has led to a new type of hybrid carrier, whose main function is to provide air cover for a group of ships and helicopters for anti-submarine defence. Helicopters have now become effective weapons against submarines because they carry both detection gear and weapons.

Air attack is one of the greatest dangers

faced by surface warships and modern navies are equipped with a range of defensive guided weapons. Surface-to-air and surface-to-surface missiles are large and so destroyers armed with them are now as big as the light cruisers of World War II; many of the larger classes are rated as cruisers. American surface warships were the first to be armed with long-range surface-to-surface missiles, but Soviet ships now carry them as well.

Development of the submarine

Submarines have changed almost beyond recognition since the introduction of nuclear propulsion. By eliminating the need to use oxygen from the atmosphere for burning fuel in diesel engines, nuclear propulsion has given submarines not only great power for high underwater speeds but also unlimited endurance. When the tactical advantages of nuclear propulsion were combined with the awesome destructive power of nuclear ballistic missiles, submarines suddenly became the most deadly weapons in history [6, 7]. The Polaris missile, and its successors the Poseidon and Trident, are fired from under

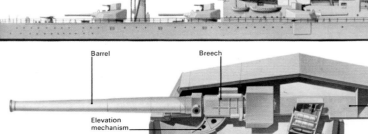

1 HMS *Dreadnought*, the world's most influential battleship, was completed in 1906. She had ten 12-inch guns and was driven at 21 knots by Parsons steam turbines. She made all previous warships obsolete and started a naval arms race which contributed to the outbreak of war in 1914. She was the first warship whose main guns were of equal calibre, essential for correcting fall of shot and firing salvoes at long range; she was also three knots faster than her contemporaries. She saw very little action during World War I, but she influenced design and gave her name to a new type of ship. The four-shaft Parsons steam turbines she carried were the first fitted in a large ship. They proved more reliable and economical than existing vertical triple-expansion steam engines. The ship's overall length was 160m (526ft) and her beam 25m (82ft).

2 The British motor torpedo boat (MTB) was first used in World War I but developed into a more potent weapon in World War II. The German equivalent was the Schnellboot, and the Americans called it the PT. Armed with torpedo tubes and light guns, MTBs could reach 40 knots. By the end of World War II, MTBs and MGBs (armed only with guns) had sunk 269 enemy ships for the British navy.

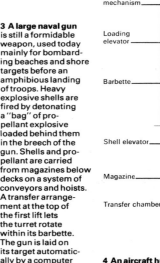

3 A large naval gun is still a formidable weapon, used today mainly for bombarding beaches and shore targets before an amphibious landing of troops. Heavy explosive shells are fired by detonating a "bag" of propellant explosive loaded behind them in the breech of the gun. Shells and propellant are carried from magazines below decks on a system of conveyors and hoists. A transfer arrangement at the top of the first lift lets the turret rotate within its barbette. The gun is laid on its target automatically by a computer that uses data from the ship's radar. Both radar and guns are stabilized gyroscopically to compensate for the ship's pitching and rolling.

Labels: Barrel, Breech, Armoured turret, Firing chamber, Elevation mechanism, Loading elevator, Barbette, Shell store, Shell elevator, Magazine, Transfer chamber

4 An aircraft hunting a submarine uses a variety of devices. Radar [1] detects anything above surface. Different types of sensors pick up fumes [2], temperature changes in the ocean [4] or magnetic fields [5] caused by the submarine. Underwater sonar buoys either record sound [6] or send out intense sound and broadcast any reflections [7]. An airborne computer or "tactical display" [3] gives an overall picture of the search being made.

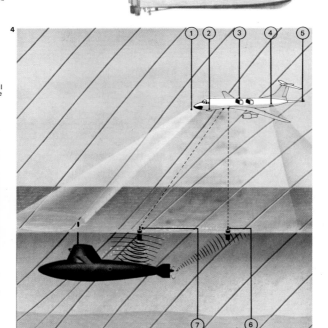

water and use the surface of the sea as a launch pad. Ballistic missile submarines do not reveal their position until the moment of firing and so there is no counter-measure short of finding and destroying all hostile submarines simultaneously. Even the possible interception of incoming missiles by anti-missile weapons is made more difficult in the Trident missile which has 14 independently targeted war-heads. For the foreseeable future, ballistic missile submarines remain almost invulnerable as a nuclear deterrent.

The submarine, together with the aircraft carrier, forms an important part of a fleet. However, to gain full advantage of the submarine, particularly the nuclear-powered submarine, it is never used against surface ships. The submarine acts independently, although remaining a vital part of a fleet.

In the last ten years, the fast patrol boat (FPB) has become more powerful with the introduction of light but destructive subsonic surface-to-surface missiles. These missile-armed FPBs are small and make fast, difficult targets. They are relatively cheap to build

and are highly effective in coastal waters. The success of a Soviet-built missile FPB used by Egypt against an Israeli destroyer in 1967 speeded up the introduction of similar weapons in larger ships, although light rapid-firing guns remain an effective defence against subsonic missiles.

Surface propulsion

Nuclear propulsion has been used in a few surface warships, but its extremely high cost and bulk offset many of its advantages. Only the American navy has shown any inclination to change from conventional forms of propulsion. Other navies, notably the British, have chosen gas turbine engines for their lightness and power, but these are frequently combined with diesel engines or steam turbines for greater economy.

Despite the threat from submarines and aircraft, surface warships have proved adaptable and remain the main vessels of navies. Electronic counter-measures and computer-aided weapon systems give better defence than ever before, but the ideal naval force combines aircraft and surface ships.

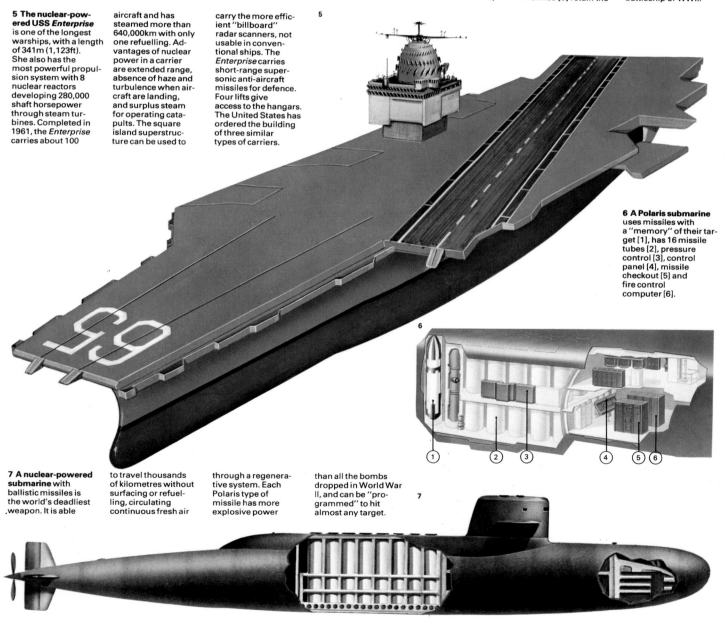

KEY

Warship profiles have remained remarkably constant since the days of HMS *Dreadnought* [3], which was completed in 1906. Today's guided missile destroyers [2], frigates [4] and minesweepers [5] are similar to vessels designed at the turn of the century (though more rakish) and even nuclear-powered ships like the USS *Enterprise* [1] and polaris submarines [6] retain the basic shapes of their predecessors. What has changed is fire-power – a single missile from a destroyer can hit as hard as a broadside from the biggest battleship of WWII.

5 The nuclear-powered USS *Enterprise* is one of the longest warships, with a length of 341m (1,123ft). She also has the most powerful propulsion system with 8 nuclear reactors developing 280,000 shaft horsepower through steam turbines. Completed in 1961, the *Enterprise* carries about 100 aircraft and has steamed more than 640,000km with only one refuelling. Advantages of nuclear power in a carrier are extended range, absence of haze and turbulence when aircraft are landing, and surplus steam for operating catapults. The square island superstructure can be used to carry the more efficient "billboard" radar scanners, not usable in conventional ships. The *Enterprise* carries short-range supersonic anti-aircraft missiles for defence. Four lifts give access to the hangars. The United States has ordered the building of three similar types of carriers.

6 A Polaris submarine uses missiles with a "memory" of their target [1], has 16 missile tubes [2], pressure control [3], control panel [4], missile checkout [5] and fire control computer [6].

7 A nuclear-powered submarine with ballistic missiles is the world's deadliest weapon. It is able to travel thousands of kilometres without surfacing or refuelling, circulating continuous fresh air through a regenerative system. Each Polaris type of missile has more explosive power than all the bombs dropped in World War II, and can be "programmed" to hit almost any target.

Early military aircraft

An Italian, Lieutenant Gavotti, is credited with dropping the first aerial bomb (on Turks, in Tripolitania, on 1 November 1911) and two Frenchmen, Sergeant Joseph Frantz and Corporal Quenault, with the first victory in aerial combat. On 5 October 1914, a bare 11 years after the first successful flight by the Wright brothers, they shot down a German two-seater with a machine gun mounted on the nose of their Voisin III biplane. The Voisin, with its "pusher" propeller at the rear, offered a clear field of fire forwards.

At the outbreak of World War I, the Germans with 285 aircraft, and the French and British, with a total of 219, were fairly evenly matched. The aircraft were intended mainly for reconnaissance and though some pilots and observers, like Frantz and Quenault, carried weapons, it was not until the first months of 1915 that the air war became serious. In April of that year Roland Garros of the French Air Force fitted steel plates to the propeller of his Morane-Saulnier monoplane. The plates deflected bullets fired through the propeller disc and for two weeks Garros was unbeatable. Then he was forced to land behind German lines. By this time, however, German engineers of the Fokker works had gone a step further and had developed an interrupter gear that synchronized the firing of the gun with the rotation of the propeller. Single-gunned Fokker Eindeckers fitted with the gear created a reign of terror on the Western Front for almost a year. By the summer of 1917 fighters like the Sopwith Camel [1] had twin synchronized guns.

Specialization

Gradually, as aircraft were made more powerful and reliable, it became possible to design them for specific tasks. Fighters became manoeuvrable flying guns for attacking other fighters, bombers and ground targets. Reconnaissance machines carried not only an observer but also cameras and radio sets. Bombers grew in size until by the end of the war the newly formed Royal Air Force had bombs that weighed 1,500kg (3,300lb) as well as aircraft – the Handley Page V/1500, successor to the 0/400 [3] – capable of bombing Berlin from airfields in England. Special torpedo carriers were in service, their crews rigorously trained to drop the 800kg (1,764lb) "tin fish" at exactly the correct height and speed.

Between the wars

There was little change in aircraft armament between 1918 and 1935. The fighter still carried two machine guns firing ahead through the propeller disc and the bomber was defended by the same two or three men who aimed their machine guns by hand. But there were dramatic changes in aircraft engineering. Engines had to be improved so that they could run not for the 20 or 30 hours that had sufficed in 1918 but for hundreds of hours, without failure. Airframes became both lighter and stronger, the fabric-covered skeleton of spruce or steel tubing gradually giving way to a monocoque (single shell) structure of light metal alloy.

After 1930 the increasing power of engines and ceaseless competition to outfly potential rivals led to the gradual abandonment of the trusty wire-braced biplane, which seldom reached 322km/h (200mph). Instead,

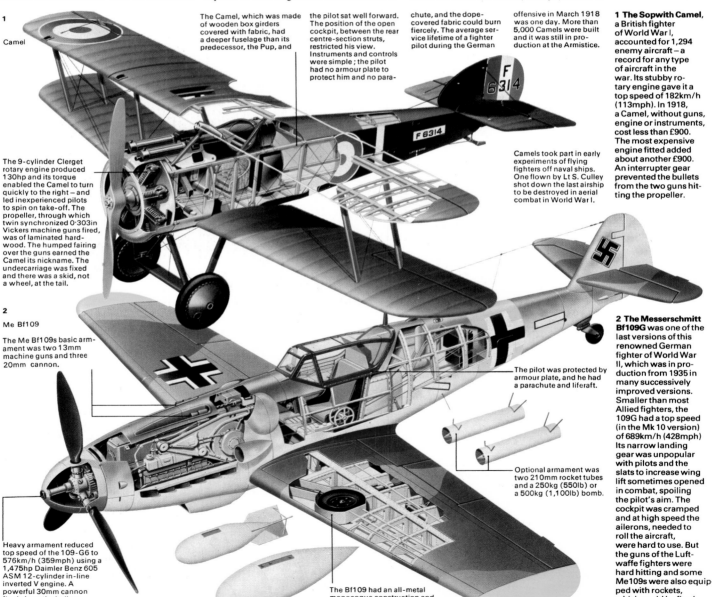

1
Camel

The Camel, which was made of wooden box girders covered with fabric, had a deeper fuselage than its predecessor, the Pup, and the pilot sat well forward. The position of the open cockpit, between the rear centre-section struts, restricted his view. Instruments and controls were simple; the pilot had no armour plate to protect him and no para-chute, and the dope-covered fabric could burn fiercely. The average service lifetime of a fighter pilot during the German offensive in March 1918 was one day. More than 5,000 Camels were built and it was still in production at the Armistice.

The 9-cylinder Clerget rotary engine produced 130hp and its torque enabled the Camel to turn quickly to the right – and led inexperienced pilots to spin on take-off. The propeller, through which twin synchronized 0·303in Vickers machine guns fired, was of laminated hardwood. The humped fairing over the guns earned the Camel its nickname. The undercarriage was fixed and there was a skid, not a wheel, at the tail.

Camels took part in early experiments of flying fighters off naval ships. One flown by Lt S. Culley shot down the last airship to be destroyed in aerial combat in World War I.

2
Me Bf109

The Me Bf109s basic armament was two 13mm machine guns and three 20mm cannon.

The pilot was protected by armour plate, and he had a parachute and liferaft.

Optional armament was two 210mm rocket tubes and a 250kg (550lb) or a 500kg (1,100lb) bomb.

Heavy armament reduced top speed of the 109-G6 to 576km/h (359mph) using a 1,475hp Daimler Benz 605 ASM 12-cylinder in-line inverted V engine. A powerful 30mm cannon fired along the hollow propeller shaft.

The Bf109 had an all-metal monocoque construction and a retractable undercarriage.

1 **The Sopwith Camel**, a British fighter of World War I, accounted for 1,294 enemy aircraft – a record for any type of aircraft in the war. Its stubby rotary engine gave it a top speed of 182km/h (113mph). In 1918, a Camel, without guns, engine or instruments, cost less than £900. The most expensive engine fitted added about another £900. An interrupter gear prevented the bullets from the two guns hitting the propeller.

2 **The Messerschmitt Bf109G** was one of the last versions of this renowned German fighter of World War II, which was in production from 1935 in many successively improved versions. Smaller than most Allied fighters, the 109G had a top speed (in the Mk 10 version) of 689km/h (428mph). Its narrow landing gear was unpopular with pilots and the slats to increase wing lift sometimes opened in combat, spoiling the pilot's aim. The cockpit was cramped and at high speed the ailerons, needed to roll the aircraft, were hard to use. But the guns of the Luftwaffe fighters were hard hitting and some Me109s were also equipped with rockets, which could be fired at distant bombers.

aircraft became monoplanes. The retractable undercarriage made its appearance – the first military aircraft to have one was the American Grumman FF-1 of 1933 – and guns (up to eight in aircraft like the Spitfire) were located in the wings outside the propeller arc.

The spur of World War II

Engines grew in power from 500hp in 1935 to 1,000hp by 1939 and by 1944 a 2,500hp engine was not uncommon. This meant that aircraft could be heavier and faster. Both Britain and the United States used bombers as strategic weapons. The aircraft were four-engined, carrying up to seven tonnes of bombs each and defended by power-driven gun turrets. The United States equipped them with turbo-charged engines that enabled aircraft like the B-29 Superfortress [4] to fly as high as 10,700m (35,000ft). German bombers, part of the *Blitzkrieg* (lightning war), were really close-support tactical machines. Shot out of the British daylight sky in 1940, they took to raiding by night. This spurred the development of large twin-engined night fighters equipped with radar.

Ocean patrol aircraft flew for 24 hours at a time, packed with new systems for detecting surface vessels and even submerged submarines. Reconnaissance aircraft brought back clear photographs taken from heights that ranged from 12,200m (40,000ft) to tree-top level. Specially built transports and gliders carried airborne forces and supplies, and naval combat aircraft were developed to operate from aircraft carriers.

By 1945, one or two outstanding machines designed before 1939 still survived in improved forms, the two greatest being the Spitfire and Bf109 [2]. But the way to the future was being shown by dramatically new developments. In 1944 Messerschmitt introduced the bat-like Me163 rocket interceptor and followed this with the formidable Me262 twin jet which could carry four devastating 30mm cannon as well as bombs. Entering service in the same week as the Me262, the British Meteor jet had only 20mm guns, but it was a much safer and more refined aircraft which began its combat career by destroying V1 flying bombs speeding towards London – the first operational missiles.

Boeing 747 "Jumbo"

B-29 Superfortress

Handley Page 0/400

Me Bf109

Camel

Even the large B-29 Superfortress, the most highly developed of World War II bombers, is dwarfed by a modern Boeing 747 "Jumbo" jet airliner. Yet compared with the Handley Page 0/400 of World War I, the B-29 was nearly four times as fast and could carry 12 times the bomb load. Fighter development was equally dramatic, 1945 aircraft flying up to six times as fast as their World War I predecessors.

3 The best British heavy bomber of World War I was the Handley Page 0/400, originally ordered by the Royal Naval Air Service. Its wings could fold to fit the small canvas hangars of 1918 and it operated from rough grass fields a few hundred metres across. With an endurance of 8 hours, the 0/400 was used against German cities in late 1918. After the war, some 0/400s were used as passenger aircraft.

4 The Boeing B-29 Superfortress was by far the most advanced bomber used in World War II. First flown in September 1942, it was in action over Japan little more than a year later. Before dropping the atomic bombs that ended the war, B-29s dropped 1,500,000 leaflets on Japanese cities, warning of heavy raids to come. Flying at 10,700m (35,000ft), the B-29 was almost impossible to intercept and it was as fast as most Japanese fighters.

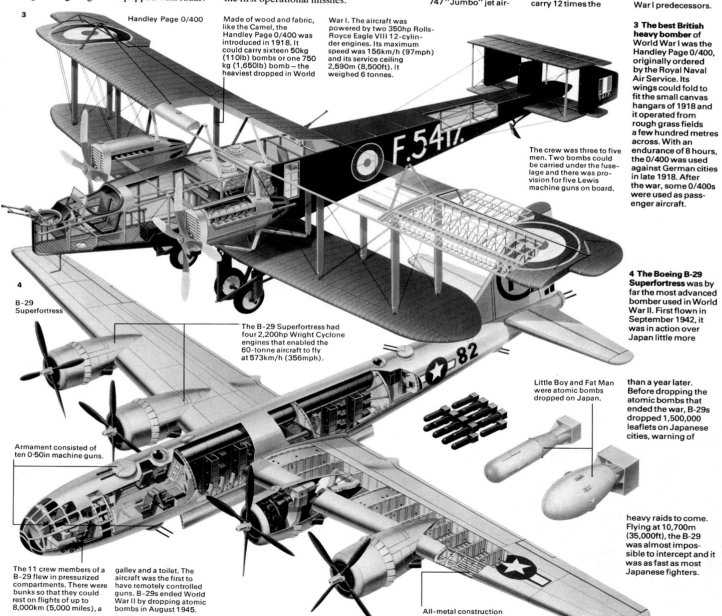

3

Handley Page 0/400

Made of wood and fabric, like the Camel, the Handley Page 0/400 was introduced in 1918. It could carry sixteen 50kg (110lb) bombs or one 750 kg (1,650lb) bomb – the heaviest dropped in World War I. The aircraft was powered by two 350hp Rolls-Royce Eagle VIII 12-cylinder engines. Its maximum speed was 156km/h (97mph) and its service ceiling 2,590m (8,500ft). It weighed 6 tonnes.

The crew was three to five men. Two bombs could be carried under the fuselage and there was provision for five Lewis machine guns on board.

4
B-29
Superfortress

The B-29 Superfortress had four 2,200hp Wright Cyclone engines that enabled the 60-tonne aircraft to fly at 573km/h (356mph).

Armament consisted of ten 0·50in machine guns.

Little Boy and Fat Man were atomic bombs dropped on Japan.

The 11 crew members of a B-29 flew in pressurized compartments. There were bunks so that they could rest on flights of up to 8,000km (5,000 miles), a galley and a toilet. The aircraft was the first to have remotely controlled guns. B-29s ended World War II by dropping atomic bombs in August 1945.

All-metal construction

Modern military aircraft

The largest single advance in aircraft propulsion was the invention of the jet engine. After World War II, two main groups of aircraft design emerged: those with jets in traditional airframes and those with new types of airframes that took better advantage of jet engines. By the outbreak of the Korean War in June 1950 Britain still had only obsolescent fighters, which were not as advanced as the Soviet MiG-15 [1].

The MiG incorporated the results of wartime German research and had wings and tail swept back to delay the onset of shock waves at regions of local supersonic airflow. The plane could fly 160km/h (100mph) faster than "straight-winged" jets. The American F-86 Sabre, although not as fast as the MiG mastered the North Korean-piloted MiGs because it was better equipped and flown.

Unpredictable development
Radical design advances were being made at an accelerating rate, calling for expert judgement in ordering new military aircraft. Unexpectedly, bombers of dramatic new design failed to enter service, whereas one that seemed almost conventional – the British Canberra [3] – was made in large numbers in Britain and the United States. Used Canberras were still in keen demand more than 25 years after this aircraft's first flight in May 1949. The Canberra had broad unswept wings for good performance and manoeuvrability at great heights, yet it was used mainly in low-level tactical roles for which its simplicity, flexibility, low costs and short runway requirements made it particularly suitable.

By the early 1950s the first supersonic military aircraft were being designed. This meant further major changes in aircraft design – smaller wing span, broader wings with thicker skins, greater body length, completely rearranged packing of equipment, ejector seats, greater fuel capacity and weight and considerably more powerful engines with afterburners to boost thrust for supersonic flight. Some designers chose swept wings, some a triangular delta shape (with or without a horizontal tail), some a stubby unswept wing and a few the canard (tail-first) layout. In each the wing had to be extremely thin and so most or all the fuel had to be in tanks inside the ever-bigger fuselage. Increasingly bombs were carried externally, even in the large B-58 Hustler of the US Air Force which in 1957 flew at twice the speed of sound (Mach 2). Overloading with equipment could cause problems: the Starfighter crashed frequently when it went into service.

Tactical requirements
Until the mid-1950s it had seemed sensible to strive for greater speed and altitude. But it was gradually realized that improvements in surface-to-air missiles would make high-altitude flights risky and attack aircraft had to be redesigned or modified to fly as low as possible to try to escape radar detection and give defenders less warning. A new generation of bombers was planned, some of them notable for variable-sweep wings that could be spread out for take-off, cruising flight and landing, yet folded sharply backwards for a low-level dash at high speed. The first such variable-geometry or "swing wing" aircraft was the American F-111 bomber (originally in production as a fighter). Later examples include the US Navy F-14 Tomcat, the Euro-

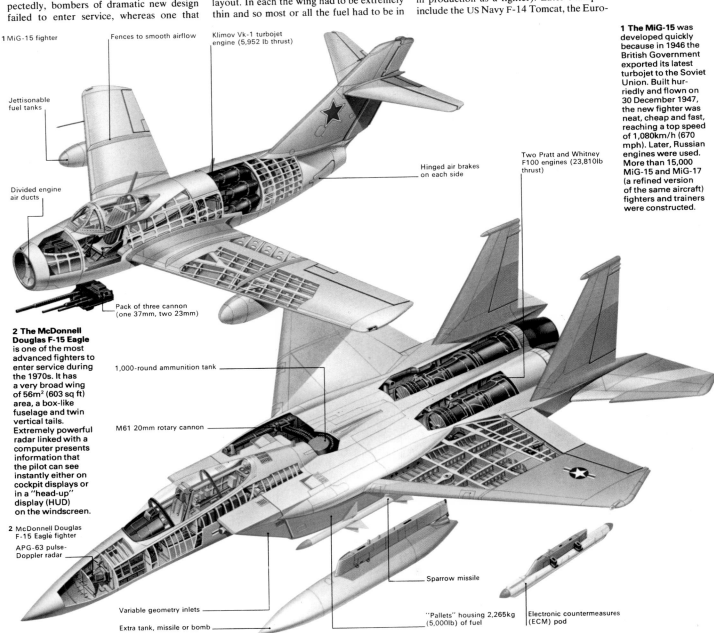

1 MiG-15 fighter

Fences to smooth airflow

Klimov Vk-1 turbojet engine (5,952 lb thrust)

Jettisonable fuel tanks

Divided engine air ducts

Hinged air brakes on each side

Two Pratt and Whitney F100 engines (23,810lb thrust)

Pack of three cannon (one 37mm, two 23mm)

2 The McDonnell Douglas F-15 Eagle is one of the most advanced fighters to enter service during the 1970s. It has a very broad wing of 56m² (603 sq ft) area, a box-like fuselage and twin vertical tails. Extremely powerful radar linked with a computer presents information that the pilot can see instantly either on cockpit displays or in a "head-up" display (HUD) on the windscreen.

1,000-round ammunition tank

M61 20mm rotary cannon

2 McDonnell Douglas F-15 Eagle fighter

APG-63 pulse-Doppler radar

Sparrow missile

Variable geometry inlets

Extra tank, missile or bomb

"Pallets" housing 2,265kg (5,000lb) of fuel

Electronic countermeasures (ECM) pod

1 The MiG-15 was developed quickly because in 1946 the British Government exported its latest turbojet to the Soviet Union. Built hurriedly and flown on 30 December 1947, the new fighter was neat, cheap and fast, reaching a top speed of 1,080km/h (670 mph). Later, Russian engines were used. More than 15,000 MiG-15 and MiG-17 (a refined version of the same aircraft) fighters and trainers were constructed.

pean MRCA and the large B-1 strategic bomber [4] built for the US Air Force.

In contrast some pure interceptors of the 1970s have seemingly old-fashioned fixed wings, as typified by the MiG-25 and its Western rival the F-15 [2]. Here the prime needs are tremendous engine power and a large wing area, for outstanding manoeuvrability. In 1955 fixed guns were considered obsolete and fighters became equippped with air-to-air guided missiles. But all new fighters now have guns for close-range dog-fighting (feasible only at much less than the speed of sound) as well as special close-range missiles.

Aircraft radar systems

Dramatic advances have been made in radars and infra-red and other systems for gaining a detailed picture of the whole battle scene. But several recent tactical aircraft, such as the US Air Force A-10, have no target-seeking radar and rely mainly on the pilot's eyesight to attack targets; again surprisingly, they are no faster than World War II aircraft, although they can carry an external weapon load of 7,270kg (16,000lb). In a class of its

own, the British-developed Harrier V/STOL (vertical or short take-off and landing) has a unique tactical role as it can carry weapons or reconnaissance gear out of a forest clearing or off the deck of a small ship.

In anti-submarine warfare the task is to pack inside one aircraft a versatile array of sensing systems for detecting a submerged submarine and weapons for destroying it. The aircraft may be a large land-based jet, a carrier-based aeroplane with folding wings or an amphibious helicopter. Modern anti-submarine aircraft, bombers and military transports are increasingly being designed for instant readiness and the ability to use short unpaved airstrips. Helicopters include heavy-load carriers lifting up to 40 tonnes and attack helicopters.

A major factor with modern military aircraft is their astronomic cost. The Soviet Union and United States build such large numbers that the price can be kept down (although a B-1 is likely to cost more than $75 million). Other nations have increasingly to collaborate on joint projects or buy from one of these two superpowers.

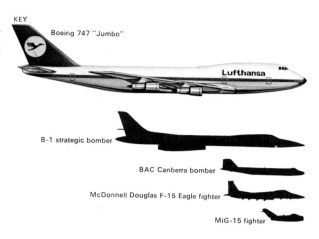

KEY
Boeing 747 "Jumbo"

Lufthansa

B-1 strategic bomber

BAC Canberra bomber

McDonnell Douglas F-15 Eagle fighter

MiG-15 fighter

Military aircraft tend to be smaller than civil transports, partly because the density of weapons and other military loads is very much higher than that of passengers and commercial cargo. Here the B-1 and Canberra bombers and the F-15 Eagle and MiG-15 fighters are compared with a Boeing 747 "Jumbo". Four-engined heavy bombers of World War II were about the same length as the Canberra and F-15 but it seems likely that future combat aircraft will be generally smaller.

3 The BAC Canberra (originally English Electric) was one of the first jet bombers and is still bought by many air forces. This version has a pilot offset to the left under a fighter-type canopy with a navigator in the nose and a battery of cannon under the belly. Other versions have a broad canopy, often with dual pilot controls, and the navigator in a third seat behind. Designed originally to bomb from 15,000m (50,000ft), Canberras now operate mainly at low levels.

4 The Rockwell International B-1, the only strategic bomber built outside the Soviet Union in the 1970s, has swing wings. The airframe is advanced in concept and takes the heaviest loads of fuel and weapons ever carried by a combat aircraft. Its speed at high altitudes is about Mach 1.6. Combat missions flown at 1,200km/h (750mph) would rely on decoy and counter-measure systems for protection.

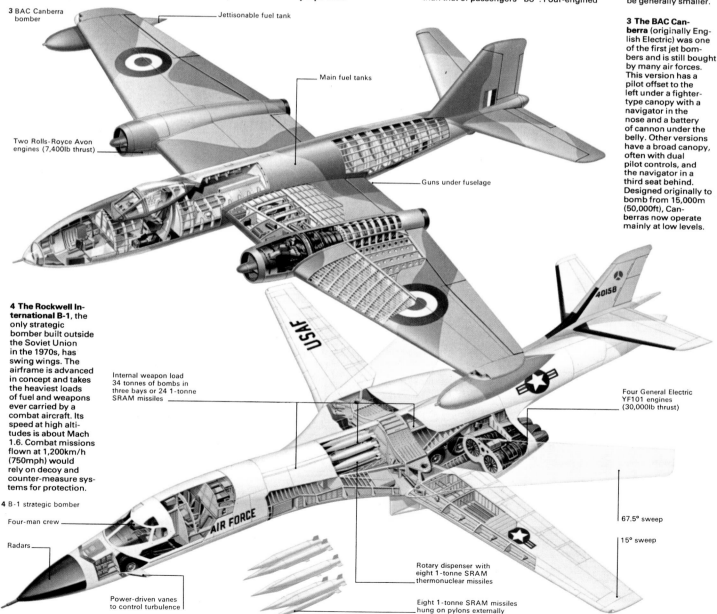

3 BAC Canberra bomber

Jettisonable fuel tank

Main fuel tanks

Two Rolls-Royce Avon engines (7,400lb thrust)

Guns under fuselage

4 B-1 strategic bomber

Internal weapon load 34 tonnes of bombs in three bays or 24 1-tonne SRAM missiles

Four General Electric YF101 engines (30,000lb thrust)

67.5° sweep

15° sweep

Four-man crew

Radars

Power-driven vanes to control turbulence

Rotary dispenser with eight 1-tonne SRAM thermonuclear missiles

Eight 1-tonne SRAM missiles hung on pylons externally

Nuclear, chemical and biological warfare

The nuclear bomb was born at the end of the most destructive war the world has known. The very first nuclear bomb, exploded in 1945, illustrated the awesome power of the new weapon; it was 2,000 times more powerful than any bomb used during the entire war in Europe. Since then, the power of the bomb [5, 6] has increased by another 3,000 times; the largest ever detonated, by the USSR in 1961, was equivalent to almost 60 million tonnes of TNT.

The first nuclear bomb dropped in a war, on 6 August 1945, killed 75,000 people in Hiroshima, Japan [1, 2], destroyed 62,000 out of 90,000 buildings, created a firestorm that lasted for six hours and burned out an area of 10.5 square kilometres (4 square miles). It was equivalent to 20,000 tonnes of TNT. Today the United States and the Soviet Union each has an armoury able to destroy more than 100,000 Hiroshimas [3, 4].

Atomic and hydrogen bombs
The first nuclear bombs derived their energy from fission – the splitting of the nuclei of certain uranium atoms (plutonium can also be used). When the atoms split the mass of the fragments produced is slightly less than the mass of the original heavy atoms and the difference in mass appears directly as energy according to Einstein's famous equation $E = mc^2$. The trigger that sparks off nuclear fission is a collision between the nucleus and a sub-atomic particle called a neutron. And for each uranium atom that splits, three more neutrons are produced. As long as the lump of uranium is big enough (larger than the "critical mass"), these neutrons go on to split more atoms, producing a chain reaction. In a piece of uranium of less than the critical mass the neutrons escape and fission stops.

The Italian physicist Enrico Fermi (1901–54) had recognized the possibility of a chain reaction in the late 1930s. In 1939 Albert Einstein (1879–1955) described the military application to the American president, Franklin D. Roosevelt (1882–1945) and within a year money was made available for atomic research. In December 1942 Fermi and his co-workers produced the first chain reaction in a primitive nuclear reactor at Chicago University. The first atomic explosion took place at Alamogordo, New Mexico, in July 1945 and three weeks later Hiroshima was bombed.

The discovery of atomic energy was yet another example, common in the history of science, of a discovery that could be turned into a weapon of war. But so terrible was this weapon that even the man who was largely responsible for its development – J. Robert Oppenheimer (1904–67) – spoke publicly against its use, and he was replaced as a government adviser.

Only two isotopes of uranium and plutonium are suitable for making bombs: uranium-235 and plutonium-239. Neither occurs naturally in a form from which it can readily be made into a bomb. Uranium-235 does occur in ores but it makes up only a small part of the total (0.7 per cent of the uranium). Expensive enrichment plant is needed to make a bomb from it.

The fission or atomic bomb was succeeded in the 1950s by the fusion or hydrogen bomb. This uses a fission explosion to trigger a fusion reaction in which isotopes of light hydrogen fuse to form heavier

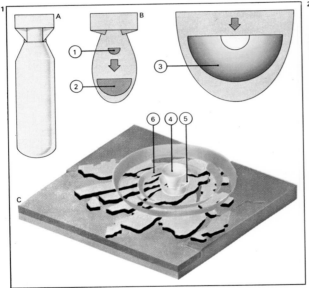

1 Atomic bombs, known as Little Boy [A] and Fat Man [B], were dropped by the United States on Hiroshima and Nagasaki, Japan, in 1945. For a nuclear chain reaction to lead to an explosion the "critical mass" of the fissile material must always be exceeded. An explosion forces a small piece of material [1] into a large, slightly sub-critical mass [2] to give a fissile lump [3]. The map of Hiroshima [C] shows the extent of the damage. At "ground zero" [4], directly beneath the explosion, everything was vaporized. All buildings were destroyed in [5] and the severe blast in [6] damaged everything. The bombings horrified mankind.

2 The ruins of Hiroshima, photographed after the atomic bomb blast, show the complete devastation. Hardly any buildings remain standing, for those that survived the initial blast were destroyed in the firestorm that followed. People who survived the blast and the fire were exposed to intense radiation, and many subsequently died of radiation sickness or other longer-term effects.

Unlike a fusion (hydrogen) bomb an atomic bomb produces no significant fallout of airborne, radioactive material downwind of the site of the explosion. An atomic bomb is limited in size.

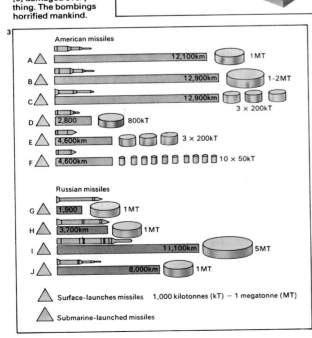

3 Ballistic missiles are the foremost weapons of the USA and USSR. America's Minuteman 1 [A] has been and 2 [B] is being replaced by Minuteman 3 [C]; Poseidon [F] is replacing submarine missiles Polaris A-2 [D] and A-3 [E]. The USSR has Sandal [G], Skean [H], Sasin [I] and Savage [J].

American missiles

A — 12,100km — 1MT
B — 12,900km — 1-2MT
C — 12,900km — 3 × 200kT
D — 2,800 — 800kT
E — 4,600km — 3 × 200kT
F — 4,600km — 10 × 50kT

Russian missiles

G — 1,900 — 1MT
H — 3,700km — 1MT
I — 11,100km — 5MT
J — 8,000km — 1MT

△ Surface-launches missiles   1,000 kilotonnes (kT) = 1 megatonne (MT)

△ Submarine-launched missiles

4 Global nuclear balance is controlled by five nations: USA, USSR, Britain, France and China. Eight others have the resources to develop them at short notice. Both the USA and USSR have enough nuclear force to survive an attack and still deliver an unacceptable counter-blow. Each symbol shows approximately 40 missiles or 30 bombers deployed in 1974.

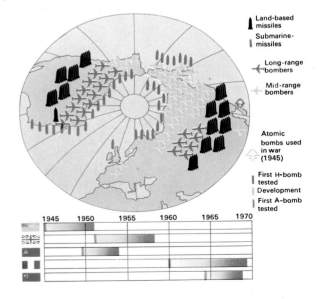

Land-based missiles
Submarine-missiles
Long-range bombers
Mid-range bombers
Atomic bombs used in war (1945)
First H-bomb tested
Development
First A-bomb tested

1945 1950 1955 1960 1965 1970

elements. In the process mass is "lost"and converted into energy. A fusion bomb is more destructive than a fission bomb.

Chemical weapons and their effects

Chemical warfare has been used in a major war only once – during World War I, when gas was used by both sides to poison enemy troops in the trenches [Key]. Since 1925 chemical warfare has been banned.

The most potent anti-personnel weapons in the chemical arsenal are nerve gases, first developed in Germany during World War II. Three such gases were developed, all derivatives of phosphine oxide; they were given the names Tabun, Sarin and Soman. If deposited as small drops on the skin they penetrate without blistering or irritation and act by inhibiting the action of an enzyme, cholinesterase, essential for muscle control. Death follows in as little as a minute or, in some cases, as much as an hour. The lethal dose for an adult is about 0.7 milligrams (about a forty-thousandth of an ounce).

In addition to nerve gases there are incendiaries such as napalm and white phosphorus

[7], "conventional" poison gases such as hydrogen cyanide, choking gases such as phosgene and defoliant chemicals that can be used against crops to create famine or against trees to destroy ground cover [8]. A further possibility is the use of incapacitating agents designed not to kill but merely to produce temporary mental disorder in the victim.

Biological warfare: the ultimate weapon?

Biological or germ warfare has never been used in human conflict although America was falsely accused of doing so in Korea. Like chemical warfare it has been the subject of intense development. During World War II, for instance, the United States perfected a means of isolating botulinus toxin, the natural product of a bacterium so toxic that 500gm (about 1lb) of the most toxic variety, properly dispersed, could extinguish all life.

There are growing doubts that the theoretical efficiencies of biological warfare could be achieved in practice and in 1972 it was agreed by convention to ban the production or use of biological weapons. Research, purportedly for defence, continues.

Gas was introduced as a weapon during World War I and used by both the Allies and Germany. Gases employed included the poisonous choking chlorine (which revealed its presence as a yellow-green cloud), the poisonous phosgene and the blinding mustard gas. An unreliable weapon, gas could rebound on the attacker if the wind changed. The only defence was to equip every soldier with a gas mask that absorbed the gas and allowed the wearer to breathe non-poisonous air. Lingering effects of gas poisoning are among the worst of all man-inflicted diseases.

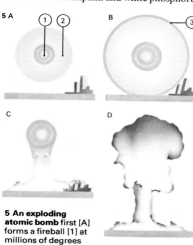

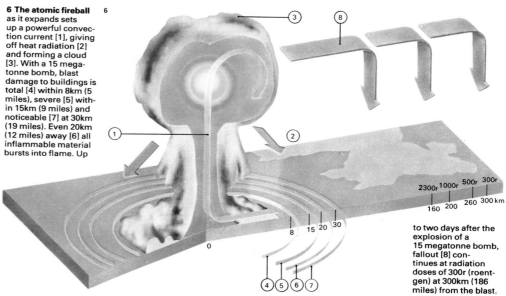

6 The atomic fireball as it expands sets up a powerful convection current [1], giving off heat radiation [2] and forming a cloud [3]. With a 15 megatonne bomb, blast damage to buildings is total [4] within 8km (5 miles), severe [5] within 15km (9 miles) and noticeable [7] at 30km (19 miles). Even 20km (12 miles) away [6] all inflammable material bursts into flame. Up

2300r 1000r 500r 300r
160 200 260 300 km

5 An exploding atomic bomb first [A] forms a fireball [1] at millions of degrees and gives out radiation [2]. Within a few seconds [B] it expands and creates a high-pressure shock wave [3]. The fireball rises [C], sucking up dust and rubble to form the familiar shape of a mushroom cloud [D].

to two days after the explosion of a 15 megatonne bomb, fallout [8] continues at radiation doses of 300r (roentgen) at 300km (186 miles) from the blast.

7 White phosphorus, packed round an explosive charge, is a new and horrific incendiary material. Dropped in cannister "bombs" from high-speed aircraft, it spreads and sticks to anything it touches, causing appalling burns.

8 In Vietnam the Vietcong sought shelter under the forest canopy for their men, machines and supply lines. Their American opponents removed this natural cover by stripping the leaves off the trees with defoliants sprayed from aircraft. Similar chemicals can be sprayed on an enemy's crops, destroying his food supply and reducing the civilian population, and the armed forces, to likely famine. Most defoliants affect trees for one season.

Road building

The first roads used by man were probably beaten out by the hoofs of animals as they made their way between feeding- and watering-places. Routes that began as mere cattle tracks or hunting and supply trails developed after the invention of the wheel into harder-wearing roads criss-crossing the ancient world and linking great trade markets. The Romans built a more permanent and extensive road network [Key] so that the empire could back up the authority of remote administrators by rapid troop movements.

Stone paving
Few authenticated records exist of pre-Roman road paving. But given the known skill of the ancient stonemason, we may assume that heavily used roads of the earliest civilizations were surfaced with slabs of cut stone. A drained earthen track becomes compacted by foot and hoof but cannot stand up to the wheel, which gradually cuts and breaks the surface.

The engineering of Roman roads [3] is well documented. Descriptions are detailed and some roads lasted so well that the original formation has been discovered, almost intact in places, by archaeologists. The Appian Way, started in 312 BC, which linked Rome with Brindisi, was typical. About 4.5m (14.75ft) wide across its two-way central lane, it was built in five layers and had three features to ensure drainage: it was constructed above ground level with a cambered surface and flanking ditches. The wearing surface was of crushed lava (plentiful in south Italy) on a gravel core. The larger stones of the base course of many Roman roads were mixed with lime mortar as a binder or with *pozzuloana* (natural volcanic cement), forming what was virtually a concrete footing. The surface of Roman roads varied with local materials and their availability.

The Romans employed cheap, expendable slave labour under military supervision for road building. After them, no administration could afford to repair their highways, let alone build new roads, until the introduction of tolls in the late Middle Ages.

Modern road technology evolved in eighteenth-century France. The Corps des Ingénieurs des Ponts et Chaussées was set up in 1720 within the French army. Twenty-seven years later the Ecole des Ponts et Chaussées was established as a state college at which civilians could study. It was Pierre Tresaguet (1716–96) who designed and built the first roads that combined good engineering practice with sound economics. He taught that there were two essentials for a lasting road – a firm foundation protected by a water-resistant surface [6].

Influence of McAdam
The French lead was soon followed elsewhere in Europe and the names of two British engineers became closely associated with improved road design. Thomas Telford (1757–1834), originally a stonemason, built roads similar in section to those of Tresaguet. But Telford's pavement was costly and it was a Scotsman, John McAdam (1756–1836), who found a way to cut costs without impairing efficiency. McAdam eliminated the deep foundation, recognizing that it is the soil that ultimately supports the weight of traffic, and that compacted soil, kept dry, will support any load. McAdam's road [4] was

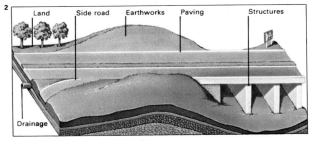

1 Road construction methods grew from simple compaction of early routes by foot and horse-drawn traffic. The Romans [A] made their routes more permanent by removing obstacles, laying foundations of gravel, surfacing with crushed stones or paving slabs and leaving drainage ditches at both sides. In the eighteenth century [B], major roads were built of tamped gravel on top of a foundation of large blocks. Roads in the nineteenth century [C] were similar but were sometimes raised and had an upper surface of rolled gravel. With the vast increase of traffic in the twentieth century and the introduction of heavy vehicles [D] roads had to be made more durable. Most modern roads have a lean concrete base with a surface composed either of reinforced concrete or of rolled asphalt.

2 Land | Side road | Earthworks | Paving | Structures

Drainage

2 A cost breakdown for a modern motorway shows that earthworks are the most expensive item.

Earthworks	25%
Paving	18%
Structures	18%
Drainage	7%
Land	4%
Side roads	4%
Engineers' fees and miscellaneous	24%

cambered for drainage and was surfaced with stone chips which were crushed and rolled by the steel-clad wheels of his time into a smooth water-resistant surface.

The McAdam pavement served its purpose well until the invention of the rubber tyre. The tyre no longer compacted the crushed stone surface, but seemed to suck the finer material from between the larger stones until the surface broke up. A binder was needed and the answer was found in natural tar. The new surface was called tarmac.

Concrete paving

The final stage in the development of the modern road followed the steady increase in the weight of heavy road vehicles. Lean concrete (which has a lower cement content) began to be used for the footing and fine concrete for the surface of the "rigid" pavement which spreads heavy axle loads over a greater area. The modern "flexible" or "black-top" pavement has a tar surface up to 10cm (4in) thick over a 25cm (10in) base layer, such as lean concrete, laid on a sub-base course of any locally available granular materials such

as stone or clinker. In the modern reinforced concrete road, a granular base course is topped by a concrete slab up to 30cm (12in) thick, with a mat of steel reinforcement 5cm (2in) below the surface.

The rapidly expanding need for more and more roads has led to the development of sophisticated road-making machinery. Giant scrapers, graders and other heavy earth-moving machines are used to prepare the roadbed and lay the footing. Machines for automatic laying of an asphalt surface to a predetermined thickness are commonplace; advanced models can lay and consolidate a 20cm (8in) asphalt layer in one pass, providing a black-top pavement that needs no further treatment.

The modern automatic concrete paver [7] spreads and tamps a continuous slab of concrete up to 30cm (12in) thick and 5m (16ft) wide, handling about 300 tonnes of material an hour, its hopper fed by tip-trucks. Reinforcement, in the form of mats of welded steel rods, is either sandwiched between two layers of wet concrete or pushed down into the body of a single slab.

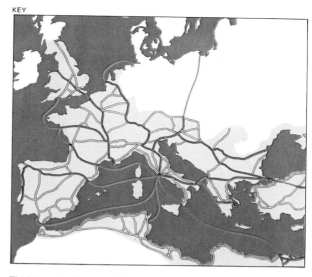

The Romans built 85,000km (53,000 miles) of roads to link Rome to its overseas centres of supply. Major land and water routes are marked on this map in orange; other routes shown are secondary.

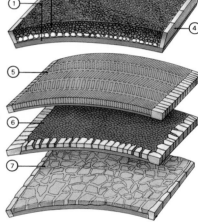

3 The Roman road pavement was based on a compacted earth footing [1] with a layer of small stones in mortar [2] and above this hard filling [3] and a slab surface [4]. At the sides were retaining stones [5] and ditches for drainage [6].

4 McAdam's pavement had a compacted, cambered earth footing [1], a 10cm (4in) base course of stones [2], a 10cm middle course of stones [3] and a wearing surface of small stones [4] which the steel wheels of the time crushed and rolled to a smooth surface.

5 A typical modern pavement has a granular sub-base [1], a 25cm base layer of lean concrete [2], a 6.5cm layer of tar or rolled asphalt [3], a 3.5cm wearing course of rolled asphalt [4], a concrete haunch [5] and a hard shoulder of asphalt on a concrete base [6].

6 Tresaguet's road in the latter half of the eighteenth century had a foundation of heavy stones rammed into an earth base slightly below ground level [1]. Above this was a 16cm (6in) layer of medium-sized stones [2] and an 8cm (3in) wearing surface of tamped small stones about the size of walnuts [3]. Retaining stones [4] were placed at the sides. The construction ensured good drainage. Different surfaces included herringbone [5], *opus testaceum* (brick and tile) [6] and cobbles [7].

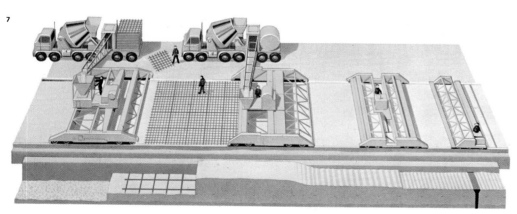

7 Road making is automated in the modern concrete "train". Ready-mixed concrete is fed into a paver which extrudes an even layer as it creeps along. This layer is compacted with poker vibrators. After a mat of steel reinforcing has been laid over the wet concrete a second paver places another layer of concrete over the steel. A vibrating beam is used to compact this thin layer without disturbing the steel. Surface irregularities are then eliminated with an automatic screeding machine carrying two transverse vibrating metal strips which spread a slight wave of excess mix over the surface. After final levelling by means of straight edges manipulated by hand from a travelling bridge, mechanical brushing is used to produce a non-skid surface. The concrete is protected with waterproof tenting.

Traffic engineering

Modern road systems evolved over many hundreds of years out of the network of routes that linked villages and townships across the length and breadth of every country. In early times, the farmer's cart threaded its way between fields, skirting woods, avoiding water and seeking the most convenient routes across hilly ground. In mountainous country, the principle of siting roads in the general direction of contours was developed to maintain reasonable gradients.

The Romans attempted to reduce the wastefulness of meandering roads by driving theirs, wherever possible, in simple, straight lines, using civil engineering techniques to carry the roadway over soft ground, natural depressions and rivers.

Development of road construction

By 1826, the British engineer Thomas Telford (1757–1834) had reconstructed the road from Shrewsbury to Holyhead to specifications which minimized curvature and gradient and provided perfect drainage. Now a part of Britain's A5 highway, his road was a model of its time, providing what is still a top-gear highway for the whole of its length.

Special highways for fast motor traffic appeared in America and Europe in the 1920s, notably around Milan where a system of single-carriageway *autostradas* was developed by private enterprise. In Germany, Adolf Hitler was impressed by the military potential of these roads and a vigorous programme of highway building began with the Frankfurt-Darmstadt *autobahn* in 1933–5. With subsequent design improvements, the modern high-speed freeway or motorway emerged. High speeds do not necessarily imply danger. Research has shown that a segregated freeway can handle much more traffic at faster speeds and much more safely than a typical arterial road with cross junctions, kerbside parking and other hazards.

The essentials of freeway design are relatively simple – no frontage development, separate up and down carriageways, no turns across the path of oncoming traffic, no sharp curves, and marked lanes of 3.7m (12ft) standard width. Simple though these principles are, old roads can rarely be upgraded to freeway specifications. Freeways are therefore built as new roads. A typical modern freeway with a design speed of 110km/h (70mph) meets the following specifications: three-lane carriageway width 11m (36ft); central reserve width 5m (16.4ft); crossfall grade for drainage 1 in 40; minimum curve radius 900m (2,950ft); super-elevation for 900m curve 1 in 22; uninterrupted visibility at 1.1m (3.6ft) above road surface 250m (820ft).

Urban freeways are designed for lower speeds with more frequent access for local traffic. Because it is impractical to drive urban freeways through existing cities at ground level, they are often built as elevated structures or in tunnels. Though tunnelled roads are more costly they may justify their expense by avoiding the damage elevated roads do to the environment.

Traffic controls

Some of the earliest attempts to control traffic were made in Rome with the banning of daytime cart traffic in the city during the first century AD.

With the expansion of trade and commerce during the Renaissance, several cities

CONNECTIONS

See also
182 Road building
192 Modern bridges
132 Cars and society
134 Trams and buses

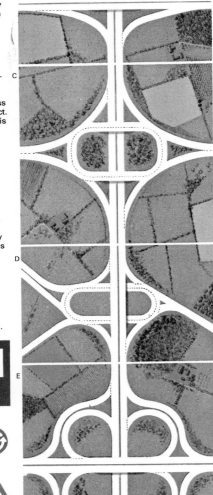

1 **Freeway interchanges** are designed to minimize traffic conflict. Different needs are met by a number of designs. [A] The Almondsbury interchange between Britain's M5 and M6 motorways allows traffic to flow safely towards any destination without much loss of speed. [B] The trumpet interchange provides a freeway T-junction with three-way access and minimum conflict. [C] Where a freeway is crossed by a major road, the major road is divided to form a roundabout either over or under the freeway itself. [D] Purpose-built junctions are designed to connect a country road network with a freeway. [E] The classic "cloverleaf" freeway interchange achieves safe access to and from all directions with little traffic conflict. Traffic that is changing direction must reduce speed in compact interchanges of this form.

2 **Electronic signs** on British motorways are operated by police at central control stations. They are normally blank but can be activated to show a variety of illuminated symbols. Surrounding these signs, coloured lamps flash to draw motorists' attention.

3 **Language problems** have been overcome in recent years by the introduction of a standardized system of road signs which can be recognized by motorists internationally.

Speed   Lane closed   Road clear

Change lane

Information

Instructions

Leave motorway at next exit

Warnings

Stop and wait until signal changes

Prohibitions

introduced rudimentary traffic regulations, including the use of one-way streets and parking restrictions. Leonardo da Vinci even envisaged separation of traffic on two levels.

The arrival of the motor age at the turn of the century saw the tentative beginning of scientific traffic engineering with the adoption of a traffic code for New York in 1903. Customary rules such as keeping to the left in Britain and giving way to the right on the Continent, began to be enforced by law. The introduction of safety road signs, automatic traffic lights at busy urban junctions and roundabout systems followed. Traffic engineering is a distinctively twentieth-century science, born out of the hazards brought to human societies by a torrent of motor vehicles. Its object is to provide safe, convenient and economic movement of vehicles and pedestrians.

More recently, traffic engineering has been defined as the science of fitting roads to traffic by planning and design, and traffic to roads by regulation and control, in order to achieve maximum capacity with safety. This depends on analysis of traffic flows, conges-

tion and accidents. On motorways, increased capacity at high speeds with safety is sought by minimizing the chance of traffic conflict. The road designer must achieve these aims with the least harm to the environment.

City traffic

The complexity of traffic engineering increases off the freeway as vehicles seek facilities to stop or park throughout the city. Expedients to reduce congestion include parking restrictions, designation of clearways where no stopping is permitted, bans on U-turns, extensive one-way systems and tidal flow systems of traffic lights to allow an unimpeded path along busy routes. In many cities, the policy is to reduce private traffic by both the improvement of public transport and the control and restriction of car parking. Already in some cities, traffic surveillance by remote-controlled TV cameras enables experienced traffic officers to operate electronic signs to reroute or restrict traffic.

Semi-automatic road systems now under study could conceivably lead to an electronically controlled flow of cars on major roads.

KEY

Today's road-user is served – or perhaps confused – by a proliferation of warnings, prohibitions and instructions applied in his own interests by the modern science of traffic engineering.

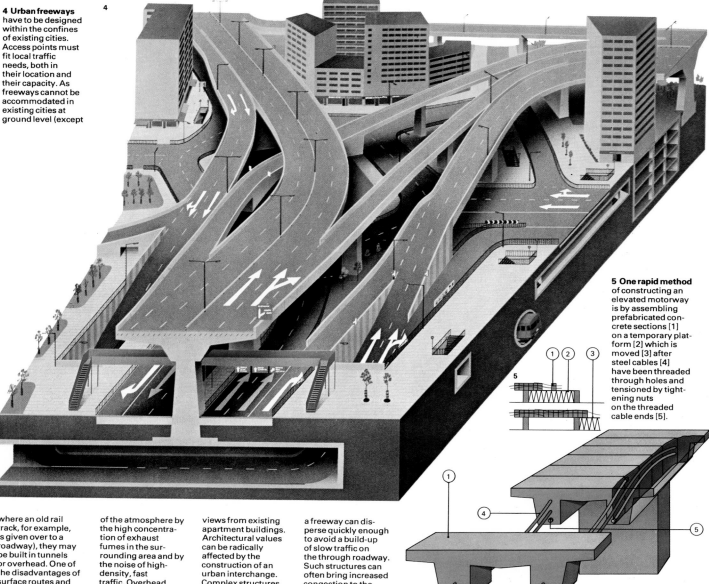

4 Urban freeways have to be designed within the confines of existing cities. Access points must fit local traffic needs, both in their location and their capacity. As freeways cannot be accommodated in existing cities at ground level (except

where an old rail track, for example, is given over to a roadway), they may be built in tunnels or overhead. One of the disadvantages of surface routes and elevated city freeways is pollution

of the atmosphere by the high concentration of exhaust fumes in the surrounding area and by the noise of high-density, fast traffic. Overhead roads may also look unattractive or spoil

views from existing apartment buildings. Architectural values can be radically affected by the construction of an urban interchange. Complex structures are needed to ensure that traffic leaving

a freeway can disperse quickly enough to avoid a build-up of slow traffic on the through roadway. Such structures can often bring increased congestion to the old roads near the freeway exit.

5 One rapid method of constructing an elevated motorway is by assembling prefabricated concrete sections [1] on a temporary platform [2] which is moved [3] after steel cables [4] have been threaded through holes and tensioned by tightening nuts on the threaded cable ends [5].

Airports and air traffic

When Orville Wright (1871–1948) and his brother Wilbur (1867–1912) made man's first powered flight on 17 December 1903 at Kitty Hawk, North Carolina, USA, they revolutionized transportation. In theory the new aircraft made possible the movement of men and goods from any point on earth to any other without the need for massive engineering projects on roads, bridges, tunnels and seaports. Landing facilities were necessary but before the 1920s any large, dry, level space sufficed for the purpose.

Runway size and strength
From the early 1920s the establishment of public air passenger services, and the increase in size, weight and speed of aircraft, resulted in the need for complex installations, including suitable runways and facilities for passengers and cargo. The rapid growth in the size of aircraft since their invention is illustrated by the fact that during the 1930s an aircraft carried on average 20 passengers, weighed about 12,000kg (26,450lb) and needed 600m (1,950ft) of runway while in the 1970s Jumbo jets weighing 372,250kg

(818,950lb) were carrying up to 500 passengers and needed more than 3,500m (11,480ft) of runway to take off. These lengths are for sea-level airports and at higher altitudes runways must be even longer. Typical commercial air speeds also increased from approximately 280km/h (173mph) to 1,000km/h (620mph).

Runway width varies today between 50m and 70m (160–230ft) with taxiways 25m (80ft) wide connecting the runways with the loading and unloading areas. The heavier airliners have a cluster of landing wheels to spread their weight; thus modern design practice specifies a runway strength sufficient to bear 100 tonnes per single wheel or 125 tonnes for a pair of dual wheels.

Most busy airports have at least two runways lined up in different directions [Key]. This enables the aircraft to take off and land into the prevailing wind . A good layout (neglecting noise considerations) is a six-pointed star providing a pair of runways in each of three directions separated by 120 degrees. But such a layout requires a clear area of 10 square kilometres (3.9 square miles), so it is

usual to economize on space by building all airport facilities and control, passenger, cargo and car parking in the centre and served by approach tunnels. An example of this design is London's Heathrow airport.

Airport location
Airports must be located as near as possible to major population centres if air transport is to retain its advantage of speed, especially on short-haul routes. London Heathrow to Amsterdam Schiphol [2], for example, takes about 45 minutes flying time. But the road journeys from the West London Air Terminal to Heathrow (24km [14 miles]) and from Schiphol to Amsterdam city centre (12km [7 miles]) can take up to 90 minutes and passengers can spend much time at the airports waiting for customs clearance and their flights to be called. Buenos Aires international airport is 50km (30 miles) from the city centre while Hong Kong airport is only 7km (4 miles) from the main centre.

Airports must be serviced by good road and rail facilities to enable the efficient inflow and outflow of passenger and freight traffic at

CONNECTIONS

See also
150 Modern aircraft
98 Earth-moving machines
144 Railways of the future

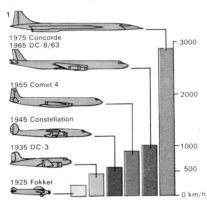

1 The technology of aviation has advanced much more rapidly than that of most other means of transport. Fifty years have seen aircraft speeds increase to as much as 2,335km/h (1,450 mph) in the case of Concorde. When the total area available limits runway length the only solution consistent with safety is to construct a new airport at a new location. It is doubtful, however, that aircraft speeds will increase in the near future due to the controversy over noise levels. Typical passenger transport aircraft and their maximum speeds are shown on the chart.

1975 Concorde
1965 DC-8/63
1955 Comet 4
1945 Constellation
1935 DC-3
1925 Fokker

3000
2000
1000
500
0 km/h

2 The development of Amsterdam's Schiphol airport illustrates the growth of a typical modern airport. In 1920 it consisted of a level landing field about 1sq km (0.4sq mile) in size [A]. By 1938 four relatively short runways had been built [B]. By 1967 [C] the airport boundaries had been greatly extended and the runways had been re-sited, lengthened and also interconnected with a network of taxiways and parking aprons. More extensions are planned.

3 The problem of aircraft noise near airports must be faced by all airport management authorities. In the enforcement of noise abatement programmes a balance must be struck between the demands of efficiency and the minimization of disturbance. Heathrow airport, London, is the busiest airport in Europe; because of its situation close to a number of suburban localities strict rules have been made to control noise. In 1958 noise limits were set at 110 PNdB (a measure of perceived noise) during the day and 102 PNdB at night, and there has been no relaxation since the advent of the larger more powerful and potentially noisier jet-engined aircraft of today. Noise abatement procedures include keeping within a defined flight path, making the approach for landing at a fixed angle of 3° from a height of 300m (1,000ft) and reducing power and the rate of climb as soon as the aircraft has reached 300m after take-off. The environmental battle over Concorde has been based as much on the question of exceeding accepted maximum take-off and landing noise as on the vexed question of sonic boom.

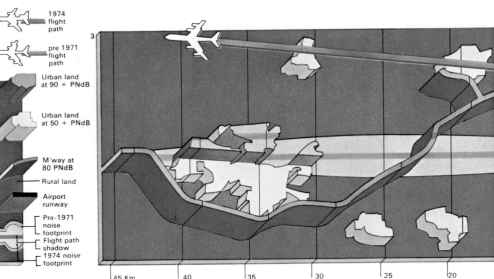

1974 flight path
pre 1971 flight path
Urban land at 90 + PNdB
Urban land at 50 + PNdB
M'way at 80 PNdB
Rural land
Airport runway
Pre-1971 noise footprint
Flight path shadow
1974 noise footprint

45 Km   40   35   30   25   20

densities equivalent to the airport's capacity. Some airports such as Kennedy airport, New York, can be reached by means of helicopter shuttle services.

Modern jet aircraft are noisy and airports must therefore be sited so that the take-off air lanes avoid populated areas as far as possible. Current legislation in most countries limits take-off noise [3] to between 110 and 120 PNdB (perceived noise decibels) at ground level in the flight path.

Passenger and cargo handling
Passenger facilities [6] at a modern airport have to be designed to deal with a wide variety of services: airline booking and checking in, baggage handling, open and duty-free shopping, departure and transit waiting and restaurant service, passport control, customs and short- and long-term car parking. These must be adequate to service the full loads that arrive with each giant airliner every few minutes. Cargo facilities include vehicle access, booking offices, warehousing, handling gear and customs. There must also be provision for airport police, and fire and ambulance services in case of an emergency.

The air traffic control facilities required at a modern airport are considerable. The control tower must be sited so that control staff can see all runways in fair weather and airport lighting must allow for comparable visibility during the night as well as providing pilots with adequate airport recognition and runway lighting for landing and take-off. There must be equipment to provide VHF radio-telephone contact between the control tower and pilots of all aircraft on the ground and in the air.

Landing aids [4], essential in low-visibility conditions, include radio beacons to provide control signals for blind landing. Full radar scanning is also necessary to enable staff to "see" all aircraft at night or in fog.

Airports must also be safe places. One of the new safety measures on the ground is the extensive search necessary to thwart potential hijackers. Airport personnel, aided by electronic devices, carry out these searches. The airport lighting system acts as a direction guide for a pilot, to make landing and take-off safe at night as well as in bad weather.

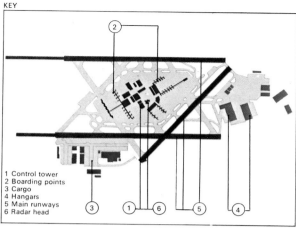

1 Control tower
2 Boarding points
3 Cargo
4 Hangars
5 Main runways
6 Radar head

The ideal airport provides landing and take-off directions into, or nearly into, the prevailing wind. This layout is the basis of London's Heathrow airport.

As the heavy aircraft of today are less affected by cross-winds, the full star is no longer necessary and runway extension has been confined to where it is most needed. The runways have been lengthened from time to time to accommodate these new aircraft. The longest runway today is 3,900m (12,800ft).

4 Aircraft navigational aids include VOR beacons (very high frequency navigational facility) [1] at flight path intersections to give bearings and distance.

At airports traffic plotted by a primary radar [2] is called up by surveillance radar [3], individually identified and guided to a holding stack [4] in which inbound aircraft fly in oval circuits at successively lower levels controlled by a radio beacon [5]. Stack control is essential at modern take-off and landing rates; those waiting their turn must remain in the stack. When the aircraft reaches the bottom the pilot locates the radio-defined glide path leading down to the runway, enabling a safe landing even in the worst visibility. Airport surveillance radar [6] "sees" all ground vehicles even in dense fog and the staff at the control centre [7] co-ordinate all movements by means of two-way radio. Near large airports there are additional stacks to hold extra aircraft if landings are temporarily stopped.

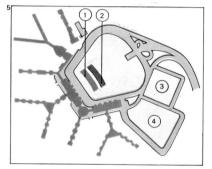

5 O'Hare International Airport, Chicago, is one of the busiest in the world. It has two domestic multi-berth terminals [1] in addition to an international one [2], and its other facilities include an hotel [3], car parks [4] and a restaurant as well as the usual passport controls, customs agencies and the means of handling air freight and other types of cargo.

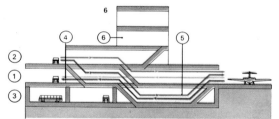

6 Passenger and baggage routes are kept separate in modern airport terminals to increase speed and efficiency. The diagram shows traffic flow to [2] and from [1] an aircraft; parking [3], customs and excise [4], baggage [5], and service [6] areas.

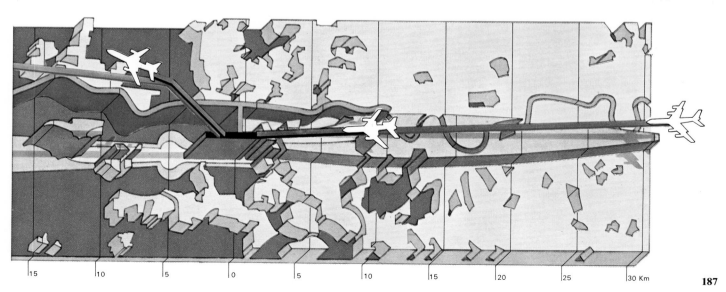

15   10   5   0   5   10   15   20   25   30 Km

Tunnel engineering

Tunnelling was probably one of man's earliest exercises in the field of civil engineering. The ancient Egyptians are known to have built tunnels for transporting water and for use as tombs. They also undertook mining operations, cutting deep tunnels to excavate copper ores. Today tunnels are used for road, rail and pedestrian transport, for carrying water – especially at dams – and in mining.

Some early tunnels

The first underwater tunnel was probably built in about 2160 BC by the engineers of Queen Semiramis of Babylon. The Euphrates had been diverted and the engineers dug a channel in the river bed. In this they built a brick-lined tunnel some 900m (3,000ft) long, waterproofed with bitumen plaster about 2m (6.5ft) thick. It connected the palace with a temple across the water.

Tunnels have often been used in warfare to penetrate enemy defences. Historians suggest that the walls of Jericho were almost certainly brought down by driving a tunnel beneath them and then lighting a fire to burn away the wooden props supporting the roof.

Tunnels, some cut through hard rock, were used extensively by the Romans in building their famous system of aqueducts. The Appian aqueduct, built in about 312 BC, ran as a tunnel for almost 25km (16 miles).

After the time of the Romans no large tunnels were built for more than a thousand years. It was the coming of the canal age in the seventeenth century that produced a new generation of tunnel builders.

Man's first great tunnel built for transportation was part of the Canal du Midi. It was completed in 1681 and ran across France from the Bay of Biscay to the Mediterranean. At Malpus, near Beziers, a 158m (515ft) tunnel was cut to carry the canal through a rocky ridge. It was the first tunnel built with the aid of explosives – gunpowder was placed in hand-drilled holes.

Rail and underwater tunnels

Throughout the eighteenth century canal tunnels were built both in Europe and America, but with the onset of the railway age in the early nineteenth century canals fell into disuse as a means of transport. The construction of railways, however, itself produced a huge increase in tunnelling. One of the most remarkable and difficult tunnels was the Simplon tunnel under the Alps, completed in 1906. It runs for 20km (12 miles) and connects Switzerland and Italy, with a 2.1km (1.3 mile) stretch below Monte Leone through rock.

Between 1825 and 1841 the British domiciled French engineer Marc Brunel (1769–1849) built the first major underwater tunnel – still in constant use by the London Underground – beneath the Thames at Rotherhithe. Brunel constructed his tunnel by means of a tunnelling shield [1] – a device consisting basically of a vertical face of stout horizontal timber baulks that could be removed one at a time to enable clay to be dug out, each baulk then being replaced farther forwards. Brunel's shield preceded the modern circular tunnelling shield.

Soon after Brunel's triumph plans were prepared to build a tunnel under the English Channel. In 1882 an army engineer, Colonel Frederick Beaumont, designed and built a tunnelling machine suitable for cutting a hole

1 Brunel's tunnelling shield consisted of 12 vertical cast-iron sections. The men were protected by horizontal timbers held against the tunnel face. These could be removed for digging, one at a time. When the face had been excavated about 30cm (12in) the jacks holding the timbers were released and the shield moved forwards by means of large jacks that acted against the tunnel shields.

2 A typical drilling pattern for blasting through rock involves a centre group of holes that are packed with a high explosive such as dynamite. [A] shows the holes in a vertical cross-section. [B] is a horizontal section showing the explosive placed in the rock face. After the explosive charges have been detonated a heap of rock lumps and rubble block the tunnel face. Engineers drill the hole-pattern in such a way that the rock is smashed into manageable lumps. The machine for the removal of blasted rock [C] includes a rock shovel near the face, which loads material into a long shuttle car with a steel-slat conveyor for moving the rock rubble back to trucks. The whole machine is gradually moved towards the face.

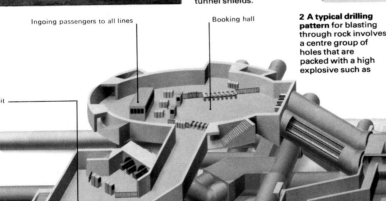

3

Ingoing passengers to all lines

Booking hall

Exit

3 London's reconstructed Oxford Circus Underground junction provides interchange facilities for three systems: the Bakerloo (brown), Central (orange) and Victoria (blue) lines. The new booking hall was built under a steel 'umbrella', temporarily above street level. The Victoria line was excavated mainly by automatic "drumdigger", which advanced up to 18m (59ft) a day. This machine consists of two steel drums, with the edge of the outer drum bevelled to cut through the London clay. At one Victoria line station water-bearing ground was frozen by pumping liquid nitrogen into tubes driven 1.6m (5.5ft) apart. Station tunnels on the Victoria line are 6.5m (21.5ft) in diameter.

2.3m (7.5ft) in diameter through chalk. Driven by compressed air, the machine advanced 12m (39ft) every 24 hours and had cut a hole 1.6km (1 mile) long eastwards from Dover when work was stopped for political reasons. A second tunnel, begun in 1973, was abandoned in 1975 because of the escalating construction costs.

Modern tunnelling methods

Tunnels through hard rock are usually built by drilling and blasting [2]. A pattern of holes is drilled into the rock face by using compressed air drills operated by men on a moving carriage or "jumbo" running on temporary rails. A drill tipped with a tungsten-carbide bit can penetrate 2 to 3m (6 to 10ft) in four to five minutes. When a round of holes is ready a high explosive such as dynamite is packed in and detonated. A mobile rock shovel lifts the shattered rock into dump cars which are hauled away by a locomotive.

Soft material such as sandstone, clay and chalk is cut through by automatic machines [Key]. One of the largest of these machines to be used cut five 9m (29ft) tunnels through

480m (1,600ft) of sandstone and limestone during the building of Pakistan's Mangla dam in 1963. Machines of this kind have a hydraulic cutting head that turns slowly and scrapes out the material at the face of the tunnel. This is lifted by a mechanical shovel on to a conveyor which carries it back to dump cars behind the machine. Mechanical arms lift and place in position huge prefabricated sections of concrete tunnel lining.

Shallow tunnels are sometimes built by the "cut and cover" method. A deep trench is excavated, the completely roofed tunnel lining is built at the bottom and this is then covered with excavated material.

In underwater tunnelling the working area may have to be pressurized so that the internal air pressure exceeds the pressure of water. After placing the tunnel lining engineers pump cement grout, a sealing mixture, round it to make it watertight. Another method sometimes used in water-bearing gravel is to sink tubes on each side of the path of the tunnel and pump in liquid nitrogen. The water in the gravel is frozen solid and the tunnel can then be cut.

The Mersey Mole, a new soft tunnelling machine, was used in 1967 to cut a 10.3m (34ft) hole through the soft rock under the River Mersey at Liverpool, England. The machine can cut away the tunnel face and convey the material back to skips running on rails. It also contains handling gear to position fabricated concrete lining sections.

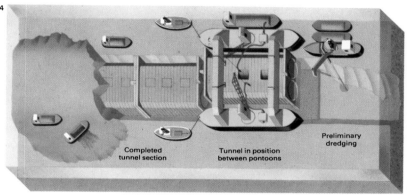

Completed tunnel section
Tunnel in position between pontoons
Preliminary dredging

4 Underwater tunnels are increasingly being built by the immersed tube method. Prefabricated steel or concrete sections of tunnel are sealed at the ends and floated into position. There they are sunk into a trench dredged into the bed of the lake or river. The sections are then joined together end to end with watertight joints and covered with previously dredged material. It is economical to make all the tunnel sections close to the tunnel site and factories for this purpose may be built on the shore; the tunnel sections are floated to the site on barges. Alternatively, prefabricated concrete or steel tunnel sections can be made in a floating dry dock that is moored alongside the site at which they are to be sunk into the dredged trench.

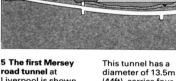

5 The first Mersey road tunnel at Liverpool is shown here in section. The tunnel falls steadily from its portals to a central point where pumps operate to remove any water that may seep in. This tunnel has a diameter of 13.5m (44ft), carries four traffic lanes (with space, never used, for two double-decker tramway lines below) and runs for 3.2km (2 miles), half of it under water. The tunnel was hand excavated in four stages. First two pilot tunnels were cut, one above the other, then the entire tunnel was opened out, top half first. The tunnel was completed in 1934.

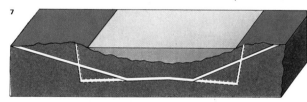

7 The Seikan tunnel, due to be completed in 1976–7, links Japan's main island of Honshu with the north island, Hokkaido. At 36km (22 miles) it will be the world's longest tunnel. Although 100m (328ft) of rock cover the tunnel under the deepest water, the danger of a leak at high pressure prompted the designers to extend the low section to a lower level at each end, providing a major draining facility to the high-power pumping plant. The tunnel runs for 22km (14 miles) through volcanic rock. The drilling and blasting method and rock tunnelling machines are both being used.

6 The Chesapeake Bay bridge-tunnel was one of the boldest prefabricated underwater tunnel projects. This 2km (1.25 mile) tunnel between two man-made islands passes under one of three shipping lanes along the remarkable 28km (17 mile) causeway that crosses Chesapeake Bay on the US east coast. There is a second, similar tunnel and at the east end of the structure, between two islands, there is a high-level bridge. The four man-made islands built of sand, stone and concrete are each 450m (1,480 ft) long and 70m (230ft) wide. The tunnels were built by sinking prefabricated double-skinned steel tubes into a dredged trench.

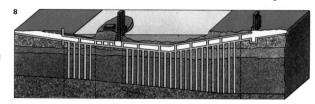

8 The four-lane road tunnel under the IJ river in Amsterdam, completed in 1967, was built mainly by the immersed tube method although one section, where the ventilation facilities are located, was constructed on site. The prefabricated reinforced concrete tunnel elements were from 70–90m (220–300ft) long and weighed up to 17,000 tonnes. They were designed as a flat box with two 7m (23ft) roadways side by side with service ducts between them and a fresh and foul air duct below each. As the river bed was very soft the tunnel sections were laid on sliding plastic bearings on a concrete raft supported by long piles. The tunnel sections were waterproofed with a bituminous membrane and a steel skin.

189

History of bridges

The earliest bridges were probably logs thrown across streams or the stone slab "clapper" bridges such as those still surviving in Devon, England. Another type of water crossing was provided by bridges formed of boats fixed together. The Greek historian Herodotus (c. 485–425 BC) provides the earliest record of a more permanent structure. This crossed the River Euphrates at Babylon about the eighth century BC.

Less durable perhaps, but certainly as remarkable technically, was the bridge of boats built for King Darius (548–486 BC) in 512 BC. It enabled the advancing Persian army to cross the Bosporus and invade south-eastern Europe. Darius's successor Xerxes (c. 519–465 BC) instructed his engineers to repeat the Bosporus exercise on the Hellespont (now the Dardanelles) [2]. On this occasion two bridges using 674 boats were built side by side spanning 1.4km (0.87 mile).

Rivers in mountainous country pose different but equally difficult problems and provoke some equally impressive solutions. Fâ-Hsien, a Buddhist monk, writing in AD 412, came across a 92m (300ft) bridge of ropes traversing a deep ravine while he was travelling in India. This form of primitive suspension bridge is known to have been widely used in South America, Central Africa, South-East Asia and China as well as in India. Jungle creepers served as ropes for many of these bridges, woven split bamboo being used in the construction of others. The Incas of Peru were still building such bridges to carry mountain roads over ravines as late as the sixteenth century [3].

Roman bridge building

The Romans developed the art of bridge building, like most things they tackled, systematically. The Pons Sublicius built over the River Tiber in Rome in 621 BC was 150m (492ft) long and famous for being defended by Horatius in 508 BC. Built entirely of timber, it was founded on timber piles driven deep into the river bed. The most remarkable Roman timber bridge was the 420m (1,378 ft) structure built across the River Rhine in 50 BC. The last plank was placed in position just ten days after Julius Caesar had ordered the bridge to be built.

The Romans' legacy to bridge building was the heavy masonry arch bridge, hundreds of which were built throughout Europe [5]. In this, large stone blocks were wedged against each other to form an arch. The central stone at the top of the arch was known as the keystone. The finest surviving example of such a bridge is the Pons Fabricius in Rome. Completed in 62 BC the bridge (now called the Ponte Quattro Capi), has two fine semicircular arches each spanning 24m (78ft). A small "relief" arch in the central springing of the two main arches releases excess water in times of flood.

Priests and professionals take over

So prolific and efficient was Roman building that it was hundreds of years before Europeans took to bridge building anew. Then, surprisingly, it was the Christian Church which, recognizing the advantages of good road communications in a developing society, took the lead. In France a group of interested priests formed a new order, the Frères du Pont, to design and build lasting bridges. Most famous of this order's works

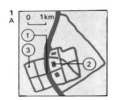

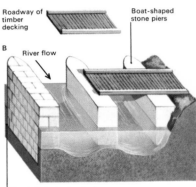

1 This drawing of the Euphrates bridge [B] at Babylon has been projected from ancient records. [A] The bridge [1] connected the old city [2] with a newer residential suburb on the west bank [3] of the river.

Roadway of timber decking

Boat-shaped stone piers

River flow

Piers built in dry river bed

2 A unit of King Xerxes' army of 480 BC crosses the Hellespont by a bridge of boats in this artist's impression. Alexander the Great crossed the Indus in India on a pontoon bridge.

3 Suspension bridges built from jungle creepers and stakes have been used for centuries by primitive societies in hilly regions. This Indian bridge in Peru is a typical modern example.

Heads of martyrs, traitors and criminals displayed over Great Stone Gate; collapsed and rebuilt 1437–40

Houses joined by gallery (hauptas) spanning street

Nonesuch House (1577) built on site of Newstone Gate

House and shop built over chapel remains

Remains of Peter of Colechurch's chapel

4 Old London Bridge was designed by a priest. Finished in 1209 after 30 years of toil, it had piers built on heaps of rubble dumped in the river and held in place by encircling rows of closely spaced wooden piles driven deep into the river bed. Over one of the piers stood a chapel dedicated to Thomas à Becket. The sixth span from the south was a timber draw span, which could be opened for tall ships to pass. Multi-storeyed houses and shops were built along the bridge with a wide enclosed corridor forming the roadway below. These buildings were destroyed by fire on several occasions and rebuilt each time in the current architectural style. The bridge's 20 pointed arches varied in span from 4.5m (15ft) to 10.5m (34ft), the piers occupying half the total river width. For many years London's only dry crossing, old London Bridge stood for more than six centuries before being replaced in 1831.

was the Pont d'Avignon built in 1177 over the River Rhône. It had 21 arches in all, the longest being 35m (115ft) in span. Likewise in England it was Peter de Colechurch, a chantry priest, who conceived, designed and eventually built the first stone bridge over the Thames – the famous multi-span London Bridge [4].

Until the late seventeenth century bridges continued to be designed and built largely by priests or architects with a flair for engineering. In Florence the local chamber of commerce commissioned the architect Taddeo Gaddi (c. 1300–66) to replace the Ponte Vecchio, which had been destroyed by flood. Gaddi's chief design innovation, incorporated in his bridge over the River Arno, was arches that were only part of a semicircle. Gaddi's idea was adopted by the architect-priest Giovanni Giocondo (c. 1433–1515). He used the segmental arch in Paris's first masonry bridge, built in 1507.

Such complex and essential work could not rest in the hands of gifted amateurs for ever. In 1716, following the work of French army engineers, France stole a march on the

rest of the world by forming the Corps des Ingénieurs des Ponts et Chaussées (Corps of Bridge and Road Engineers). Jean Rodolph Perronet, chief engineer of the corps, replaced the segmental arch with the even more daring and flatter eliptical shape [7].

Iron and steel bridges
In 1779 the first bridge to be made of iron was built across the River Severn, England, at Coalbrookdale. The iron ribs and plates were cast at the local works of the ironmaster Abraham Darby.

Wrought iron was the next major material for bridges and Thomas Telford (1757–1834) used it for the chains of his 178m (580ft) bridge of 1826, which spanned the Menai Strait to link Wales and Anglesey. Robert Stephenson's (1803–59) nearby tubular railway bridge (1850) also used wrought iron.

For long-span railway bridges carrying heavy loads, an arch or cantilever is the best design. An early bridge of this type was the steel arch over the River Mississippi at St Louis (1874) by James Eads (1820–87).

KEY

A

B

C

Bridges as a means of transporting people or goods have evolved to meet the needs of society.

In medieval times [A] road bridges were used in conjunction with ferries. In the 19th century [B] these

were supplemented by rail bridges and in modern times [C] by flyovers and multi-purpose bridges.

5 A most impressive Roman bridge was built over the River Tagus at Alcantara, Spain. Finished in AD 109, it was 204m (670ft) long with six stone arches and stood 52m (170ft) above the water.

The Romans built aqueducts in a similar way, to carry the water supply to Rome. Some of these also had a roadway. All Roman arches were semicircular, giving the name Romanesque to this type of arch.

6 This timber cantilever bridge is of a type that has stood for centuries in Srinagar, capital of Kashmir. Built on stone foundations these bridges have log piers spanned by Indian cedar.

7 The Pont de la Concorde, Paris, was completed in 1791 by Jean Perronet. He was the engineer who replaced the circular arch with the more graceful ellipse,

later to be widely copied by others. The arch under construction is shown supported by timber falsework, and a coffer dam enclosing the pier keeps out the water

8 The stayed suspension bridge at Niagara Falls, completed in 1855 by John Roebling (1806–69), had a rail track above and a road below. Spanning 250m (820ft), the main supports for the

rail and roadway were cables of wrought iron wire. The stays anchored the structure and reduced vibration in it in winds or when trains passed over it. The decks were hung on wires from the cables.

9 Eads' bridge over the Mississippi at St Louis was opened in 1874. It was the world's first major steel bridge and carried a roadway above a rail track. The centre arch is 156m (512ft) in span; the others 6m (20ft)

less. Eads sank the foundations down to rock using pressurized caissons to exclude water from the working area on the river bed. The main arch had to give sufficient room above the water-level for boats to pass.

Modern bridges

Two nineteenth-century inventions led to a revolution in bridge building. These were Portland cement and mass-produced steel. Cement is the vital ingredient of concrete, and mass concrete can be used to build piers, abutments (bank supports) and arches of "artificial" stone to any required shape. Well-made concrete is extremely strong in compression (when squeezed) but it has very little strength in tension (when stretched). On the other hand, steel can withstand great tension as well as compression, and can be used for building girders of far greater strength then the wooden trusses of early days. High-tensile steel wire cables will support immense suspension bridges.

Reinforced concrete bridges
These new materials, concrete and steel, can also be used in combination with each other. For example, a concrete structure does not have to be designed so that the material is entirely in compression, because steel rods can be used to carry the tension.

The French engineer Eugene Freyssinet (1879–1962) overcame the remaining weak-ness of reinforced concrete (the fact that steel in tension stretches, allowing the concrete immediately around the steel to stretch and frequently to crack) by using high-strength tensioned steel wires as the reinforcement. This technique permitted Freyssinet to "pre-stress" concrete (pre-load it in compression), so that it would never be subjected to tension at all. The result was a material so versatile that it could be used to make stronger, lighter and more architecturally satisfactory bridges.

Types of bridge
There are four basic types of bridge: beam, arch, suspension and cantilever [1]. The beam bridge is, in effect, a pair of girders supporting a deck spanning the gap between two piers. Such a beam has to withstand both compression in its upper parts and tension in its lower parts. Where it passes over supports, other forces come into play. A beam may be a hollow box girder or an open frame or truss.

An arch bridge can be designed so that no part of it has to withstand tension. Concrete is therefore well suited to arched bridge design. When reinforced concrete is used, a more elegant and sometimes less costly arch can be designed and most concrete arch bridges are, in fact, reinforced.

A suspension bridge consists, basically, of a deck suspended from two or more cables slung between high towers. The cables, of high-tensile steel wire, can support an immense weight. The towers are in compression and the deck, often consisting of a long slender truss (used as a hollow beam), is supported at frequent intervals along its length.

A cantilever bridge is generally carried by two beams, each supported at one end. Unlike a simple beam supported at both ends, the cantilever must resist tension in its upper half and compression in its lower.

There are also many composite forms of bridges. The bridle-chord bridge is a combination of a long beam (usually a trussed girder) partially supported by steel wires from a tower at one end, or from towers at each end. Most cantilever bridges are designed so that a gap remains between two cantilevered arms that reach out from their abutments; the gap is bridged by a simple beam. "Movable" bridges, like London's

CONNECTIONS

See also
190 History of bridges
184 Traffic engineering
96 Machines for lifting
170 Armoured fighting
vehicles

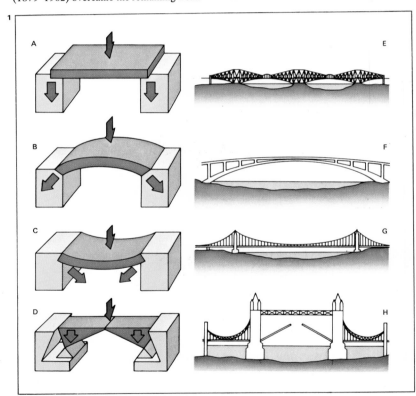

1 Bridges are of four main types: the beam bridge [A], the arch bridge [B], the suspension bridge [C] and the cantilever bridge. The three main structures of the Forth railway bridge [E] are cantilevers and the two steel trusses that connect them are end-supported beams. In [F] it is the lower member that forms the arch. The suspension bridge [G] has its roadway hanging from enor-mous steel cables passing over tall towers. The two rising arms of Tower Bridge in London [D, H] are cantilevers, whereas the slender truss which connects the top of the towers is a long beam.

2 The Howrah Bridge at Calcutta is the world's fifth longest cantilever, spanning 453m (1,500ft). The two main piers of this bridge were built hollow with twenty-one vertical shafts. By digging out the sand at the bottom of these shafts, the piers were made to sink steadily down until they reached impervious clay. The bridge was opened in 1943.

3 The Gladesville Bridge over the Parramatta River at Sydney, Australia, is the world's longest concrete arch. Completed in 1964, it spans 305m (1,000ft) and carries an 8-lane roadway.

4 The Medway Bridge in southeast England is the longest pre-stressed concrete cantilever bridge. Its 150m (490ft) main span is made up of two 60m (196ft) canti-lever arms linked by a 30m (98ft) sus-pended beam. It was opened in 1963.

famous Tower Bridge, have cantilevered arms or "bascules".

There is more to a bridge than the main span and without proper foundations for the piers or towers the whole structure would fail. Most modern bridges have reinforced concrete foundations, often keyed into bedrock. They may have to be designed to withstand the scouring action of tides, buffeting from pack ice and even mild earthquake tremors. If solid bedrock is too deep to be reached by excavating, foundations can be built on piles driven into the subsoil.

Theoretical limits of bridge spans

A bridge carries two loads. The useful load is the live load of crossing traffic. In addition it must carry its own weight, the dead load. The longer the span of a bridge, the greater its dead load: consequently there is a theoretical span limit for any given material and method of construction. Theoretical limits can be compared with current achievements using modern materials. The longest steel arches in existence are the 496m (1,652ft) Bayonne Bridge in New York City and the 503m

(1,650ft) Sydney Harbour in Australia [7]. The theoretical span limit for steel bridges of this type is about 1,000m (3,280ft). In theory engineers have considerable scope here. But the limiting factor is cost, and a long steel arch is not usually the most economical method of bridging a wide gap.

The longest steel cantilever is the 540m (1,800ft) Quebec Bridge in Canada. This was a great achievement considering it was completed in 1918. The theoretical cantilever span limit is about 750m (1,860ft).

The longest reinforced concrete arch yet built is the 305m (990ft) span Gladesville Bridge at Sydney, Australia [3].

The modern suspension bridge has the greatest span potential. The longest yet built is the 1,298m (4,240ft) Verazzano-Narrows Bridge at the mouth of New York harbour. The new Humber Bridge, under construction in Britain, will extend this record span to 1,410m (4,610ft). Suspension bridges already hold the span record, and their potential is even greater. Experienced designers consider a span of 3,000m (9,840ft) to be possible with present-day materials.

KEY

The Bosporus Bridge at Istanbul, opened in 1973, has a main span of 1,074m (3,520ft). An example of the modern trend in suspension bridge design, this bridge is far lighter and therefore more economical than many earlier suspension bridges. A similar bridge, now being built across the Humber at Hull, will break the world's long span record.

5

6

5 Gravelly Hill road junction near the city of Birmingham, England, seen here under construction is a complex of box girders with simple beams that carry the various links of the free-flow junction over each other. A large number of concrete columns support the many spans. Known popularly as "Spaghetti Junction", the complete work is a fine example of the combined skills of the civil and the traffic engineer.

6 Box girder bridges are increasingly used today because of their high strength-to-weight ratio. Building a bridge of this type often involves construction outwards from a support [1] in the form of long cantilevers. The method imposes stresses [2] on the structure which will not occur when it is complete, and it is vital for these to be adequately allowed for. Once complete, the bridge assumes its designed strength.

7

7 The Sydney Harbour Bridge is the longest steel arch bridge. Four urban rail lines and 6 motor lanes are carried by it over a clear 503m (1,650ft) span, 52m (170ft) above water. It was opened in 1932.

8 This complex of slip roads lead to an urban motorway junction at Copeley Hill, Birmingham, England. The elevated Aston expressway in the foreground is a good example of a reinforced concrete beam bridge.

8

Harbours and docks

A harbour is any place that provides seagoing ships or boats of any kind with a measure of protection from the wind and waves. To fulfil this role, a harbour must be deep enough to accommodate the largest ships that may use it, and have a bottom that will hold them at anchor. A port is a harbour with facilities for transferring passengers and cargo from shore to ship and vice versa.

Natural harbours

A harbour is often a natural product of geography [Key], whereas a port is a man-made communications facility and usually has piers, wharves, quays and docks. There may be cranes and other cargo-handling equipment, warehouses, customs and excise facilities and passenger facilities, including immigration control. Large ports have repair facilities for ships – perhaps including a dry dock [4] – and facilities for supplying water, food, oil and sometimes coal.

The Phoenicians used the natural eastern Mediterranean harbours of Tyre and Sidon from the thirteenth century BC until the Romans ended their power with the sack of Carthage (146 BC). The port of Alexandria (founded 332 BC) was an early example of a well-developed port. The natural harbour had land on all sides but the east where the deep-water pier was built within the protection of a line of reefs.

To guide ships safely into the harbour, Ptolemy II of Egypt built the 135m (443ft) Pharos lighthouse, one of the Seven Wonders of the ancient world, in about 280 BC.

The growing prosperity of Europe in the Middle Ages led to the establishment of extensive trade through the ports of Venice and Genoa, where docking and repair facilities were provided. Genoa was the terminal for sea communication with western Europe via the Straits of Gibraltar, whereas Venice was the link with Constantinople, which had direct overland trading contact with the countries of the Far East.

Among early British harbours known to have been developed by primitive civil engineering techniques were those built at Hartlepool (AD 1250) and Arbroath (1394), where the works consisted basically of protective breakwaters. Dover, formerly an exposed seaside town, was provided with an artificial harbour in the reign of Henry VIII by the construction of an enclosing breakwater of stone and timber.

Until the eighteenth century most harbours were natural. It was the Industrial Revolution in England that saw the founding of the science of harbour engineering by John Smeaton (1724–92), Thomas Telford (1757–1834) and John Rennie (1761–1821). One of the most famous harbour conceptions in recent times were the Mulberry Harbours, built by the Allies and used during World War II for the invasion of Normandy (1944). They were constructed of prefabricated concrete sections and floated to the French coast. There they provided instant harbour facilities for the landing of troops, vehicles, ammunition and supplies.

Harbour types

There are four main types of natural harbours. These are harbours in a coastal bay [1], river estuary harbours [3], inland harbours and the open roadstead harbour where a stretch of coast is relatively sheltered from

CONNECTIONS

See also
196 Canal construction
114 Modern ships
78 Oil and natural gas

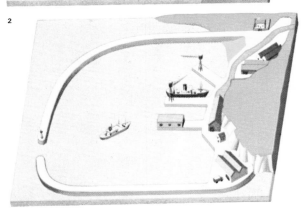

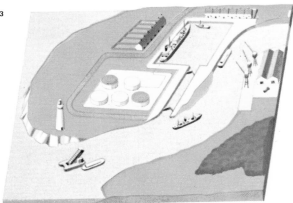

1 Natural harbours can be used without necessarily introducing improvements by engineering works. They are usually located in sheltered areas, such as enclosed bays or river estuaries. Those at Kingston, Jamaica, Southampton and San Francisco are examples. New York Harbour is one of the finest natural havens in the world with deep water, shelter, accessibility, small tidal variation and moderate current.

2 An artificial harbour, or shelter, can be formed in an exposed bay by the construction of a breakwater projecting from each shore. A single entrance is left open where the deepest water is found. The harbour at Colombo, situated in an exposed location in an embayment on the Sri Lanka coast, is typical. Others are the ferry port at Dover in the English Channel and the port of Monaco on the Mediterranean.

3 Estuaries often have wide entrances and then the river gradually narrows. Tides, therefore, are accentuated. The great tidal variation necessitates various devices to keep a relatively constant level between vessels and the shoreside facilities. At ports such as London, Liverpool, Le Havre and Antwerp large artificial basins are provided, separated from the tidal estuary by locks (wet docks), through which seagoing ships can enter.

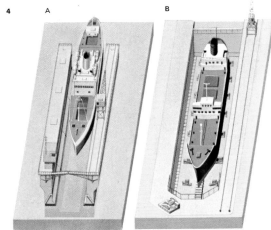

4 A floating dock, or a dry dock, is where underwater ship repairs are carried out. A floating dock [A] is partly submerged and the ship is brought in. A dry dock [B] is a solid, permanent structure.

5 The container dock is equipped with special devices for handling pre-packed cargo. A lorry [1] brings containers to the dock. The "Spider" or travelling crane [2] carries the containers from the lorry and places them in a suitable position for the crane [3], which loads the containers on to the ship. The crane's central cabin [4] follows the container as it is loaded.

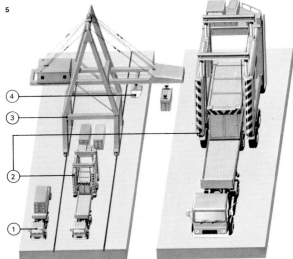
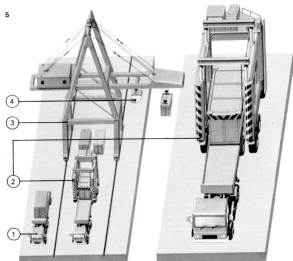

storms (moorings may be protected by breakwaters). Ports, too, are of four main types: commercial ports, naval ports, fishing ports and (a modern development) leisure ports, often called marinas, for yachts and power craft.

Commercial ports are vital to today's economy based on industry and trade, and they play a variety of specialized roles. There are highly mechanized cargo ports, including container ports [6]; passenger ports for ocean liners; short-haul passenger and drive-on ferry ports; and the new hoverports.

There are many natural harbours in the world, and among the finest in constant use are New York, San Francisco, Liverpool, Buenos Aires, Montevideo, Le Havre, Brisbane and Sydney. Of these New York has the advantage of enormous capacity providing about 720km (450 miles) of potential natural berthing along with 240km (150 miles) of man-made piers. The inner harbour is a perfect land-locked haven with adequate depth, a tidal range of less than 1.7m (5.6ft) and moderate tidal currents. The outer harbour, beyond the Verazzano Narrows, is protected

from rough weather by extensive sandbanks.

Inland ports are typified by Chicago, situated on Lake Michigan. It is 3,000km (1,900 miles) from the Gulf of St Lawrence, but connected to it. The channel runs via Lake Huron, Detroit, Lake Erie, and the 9m (30ft) deep Welland Ship Canal, which carries ocean-going ships 44km (27 miles) from Lake Erie to Lake Ontario (bypassing the Niagara Falls). From there it runs along the St Lawrence Seaway, via Montreal and Quebec, to the Atlantic Ocean.

Port maintenance
Most important of the maintenance operations of most river ports (and the majority of the world's ports are river ports) is dredging out the silt that accumulates in estuaries. Most modern ocean-going ships have a draught (depth below the waterline) of between 10 and 15m (33 and 49ft), and trailing suction dredges are kept constantly at work maintaining the depths of the channels. The new breed of supertankers need up to 25m (82ft) of water, so modern oil ports require unusually deep channels.

Falmouth's natural harbour on the south coast of Cornwall, England, has had a place in history for hundreds of years as a haven for ships seeking shelter from Atlantic storms, as a coastal port and as a mail-packet station. Today it is the headquarters of the Royal Cornwall Yacht Club.

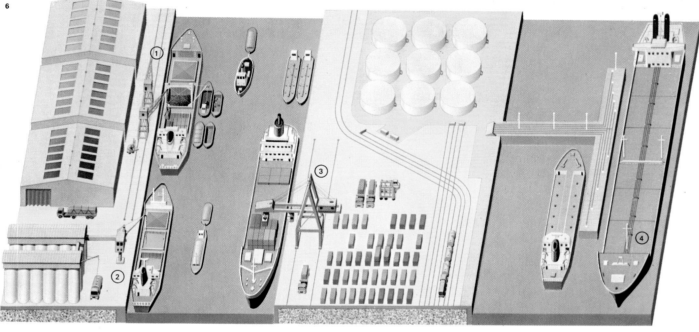

6 A multi-purpose port handles various types of cargo from different types of ships. When a normal cargo ship unloads at a typical dockside [1], the cargo is removed from the open hold by cranes and stored in warehouses. The cranes are generally mounted on rails so that they can be moved along the quayside. Some cargo may be unloaded by the ship's own derricks into barges or lighters moored alongside. A bulk grain ship [2] is loaded or unloaded by means of pumps connected with dockside silos. A container ship [3] provides maximum speed and efficiency for mixed cargo, as described in illustration 5. Today's large oil tankers [4] and supertankers have too great a draught to be able to dock normally in a port. They may moor offshore, unloading their crude oil by hose to land or into smaller tankers moored alongside.

7 The cargo-handling capacity of a modern port is in proportion to the capital invested in cranes and other heavy mechanical handling devices. Apart from the traditional luffing crane running on dockside rails, special grab cranes are used at industrial ports for the handling of coal and ore. Floating cranes operating from pontoons are used for unusually bulky and heavy lifts. Rotterdam currently handles more goods each year than any other port. Next in size are New York and Marseille. Each crate on this diagram represents 25 million tonnes of cargo in and out of the ports each year. The port of Rotterdam handles 268 million tonnes a year, New York 132 million, Marseille 83 million, Antwerp 67 million, London 66 million and the port of Tokyo 46 million. In the 20 years up to the mid-1970s the tonnages handled by some of these ports increased by seven times.

8 Harbour construction presents many problems. These necessitate building firm foundations and protecting against salt water corrosion. Quay construction varies. The solid type [A] has a paved apron [1] above the water level [2], supported by concrete blocks [3]. The bottom is dredged [4]. In one pile type [B] the timber apron [5] above the water level [6] rests on concrete piles [7]; it can be backed by sheet piling.

195

Canal construction

Modern canals provide important transport highways throughout Eurpe and, to a lesser extent, North America. The rivers Rhine and Moselle feed an extensive canal system in Germany that has connections to the Dutch, Belgian and French networks. A new major waterway is under construction to connect the Rhine with the Black Sea, via the rivers Main and Danube, and a great network has been developed around the rivers Volga, Ob, Yenisei and Lena across the USSR.

Canals ancient and modern
The four greatest achievements by canal engineers were undoubtedly the Suez Canal, opened in 1868 [8], the Panama Canal, completed in 1914 [6], the canals of the St Lawrence Seaway, opened in 1959 [2] and the 320km (200 miles) of canal linking the White Sea to the Baltic, opened in 1975.

One of the earliest reports of a man-made canal tells of one completed by Ptolemy II (died 247 BC) of Egypt to link the River Nile with the Red Sea. Natural inland waterways such as major rivers had been used as transport routes since earliest times and primitive single-gate locks are known to have been used in China from about 500 BC. These were openings in weirs through which water poured. Boats were carried downstream by the rushing waters or winched up against the current. A two-gate lock, built at Vreeswijk, Holland, in 1373, is believed to have been the first true pound-lock – a lock in which the flow of water is controlled by alternately lifting up or lowering the gates. The first to have swinging mitre gates of the kind used today was designed and built by Leonardo da Vinci (1452–1519) when he was engineer to the Duke of Milan. Mitre lock gates, when closed, form an angle pointing upstream so that water pressure holds them shut.

In 1681 French engineers made history when they completed the 250km (155 mile) long Canal du Midi linking the Atlantic Ocean with the Mediterranean Sea by a man-made waterway. It had many locks connecting the River Garonne, near Toulouse, with the River Etang de Thau, near Sète and included three aqueducts and a tunnel.

The greatest single obstacle to the industrial development of Europe and the United States in the eighteenth century was poor internal communication. Thus it was the needs of industry that brought about the dawn of the canal age on both sides of the Atlantic.

In the United States canals were built to link the Ohio and Mississippi basins [1] with the east coast ports, to provide routes from inland areas to the navigable rivers and to bypass natural obstacles on these rivers. The most spectacular early American achievement was the completion, in 1825, of the 580km (363 mile) Erie Canal linking New York City with Lake Erie. It took eight years to build. The lake was subsequently connected by canal to the Ohio River and so to the Mississippi and the port of New Orleans.

The British achievement
By 1850 England was traversed by more than 8,000km (5,000 miles) of navigable rivers and canals. Canals were also constructed throughout Europe, especially along the North Sea coasts of the Netherlands, Belgium and France.

During the nineteenth century British

CONNECTIONS

See also
194 Harbours and docks
200 Water supply

In other volumes
168 The Physical Earth
66 History and Culture 2

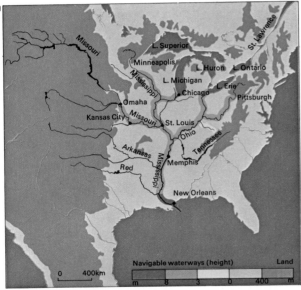

1 **The river systems** of the Mississippi and Ohio form one of the largest in the world. Their basins cover two-thirds of the USA. Many of the rivers are navigable and canals have been built to connect other waterways, extending the system to provide a continuous water route from New Orleans in the south to Chicago and the St Lawrence Seaway, and via the Ohio River direct to New York. Pittsburgh and Philadelphia were also in the 19th-century canal network, which connected with both the Columbia River and Chesapeake Bay. There is also a link from New York direct to the St Lawrence.

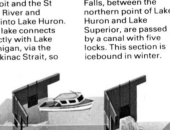

2 **The St Lawrence Seaway** provides a 3,830km (2,380 mile) inland route for ocean-going ships from the Atlantic to the heart of Canada and the USA. From Montreal it rises 51.5m (169ft) to Lake Ontario via two small lakes, three canals and seven locks. Between Lake Ontario and Lake Erie the waterway bypasses the Niagara Falls, rising 98m (322ft) in 45km (28 miles) through the eight locks of the Welland Ship Canal. From Lake Erie the route passes via Detroit and the St Clair River and lake into Lake Huron. This lake connects directly with Lake Michigan, via the Mackinac Strait, so linking the port of Chicago with the sea. The St Mary's Falls, between the northern point of Lake Huron and Lake Superior, are passed by a canal with five locks. This section is icebound in winter.

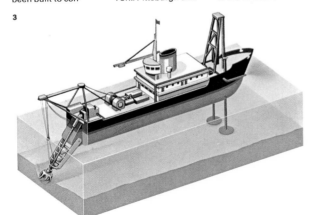

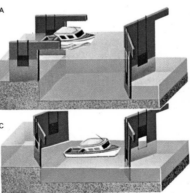
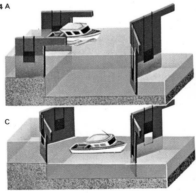

3 **The depth of a shipping canal** must be rigorously maintained to ensure that ships do not ground, causing delay not only to themselves but to all shipping along the length of the water channel. Dredgers of many types are used. This cutter dredge has a twin-leg bracing system to resist the action of the rotary cutter. A suction pipe close to the cutter draws off the debris and discharges it away from the vessel or into a barge moored alongside.

4 **Passing down a lock,** a vessel first enters it from the higher level [A] and the gates are shut behind it. Sluices in or around the lower gates are then opened [B], allowing water from the lock to pass to the lower level, lowering the water level and with it the floating vessel. When the lock level equals the lower canal level [C], the lower gates are opened [D]. Sluices in or around the upper gates are used to raise the lock level whenever a vessel passes up the canal.

civil engineers conceived and constructed a perennial irrigation system in northwestern India [5] in which huge masonry barrages were built across the rivers to divert part of the water into a great network of canals. By the time India and Pakistan gained independence in 1947 the subcontinent had more than 20 million hectares (50 million acres) of previously arid land under irrigation.

Technique of canal engineering

A canal, unlike a road, must be built in level sections and the canal engineer's first problem is the selection of a route along which a level can be maintained with the minimum of engineering work. This problem is approached by siting a canal to run as far as possible along the natural land contours. Where high ground must be crossed the cut can be correspondingly deeper or the canal can be built in a tunnel. Where there is an unavoidable depression in the ground the best solution is to construct the canal along a low embankment. There comes a point where the rise or fall of the land to be traversed requires so deep a cut, so long a

tunnel or so high an embankment or aqueduct that it is more economic to build a lock and continue the canal at a new level.

Wherever canals are built obstacles, such as roads, streams and rivers, have to be crossed. If they are roads or railways, bridges must be built; if they are waterways, then either the canal or the waterway must be carried on an aqueduct [9]. Where locks are introduced there must be a sufficient supply of water at the highest level to replace the water that flows when any lock is used [4].

A canal must not lose excessive water by seepage into the ground. Where the ground is porous, and where a canal runs along a manmade embankment, the engineer must waterproof its bottom and sides. In early canals this was achieved by lining the canal with puddled clay. Today there are alternative materials including bituminous materials, sheet polythene and concrete. Machines have been developed to lay a continuous concrete lining at minimum cost [7] and these are increasingly used for irrigation canals in the Middle East and other arid areas to carry water long distances.

A massive excavation 5.6km (3.5 miles) long was made in 1893 across the isthmus at Corinth to link the Ionian and Aegean seas. The Corinth ship canal has no locks, the whole waterway being at sea level to allow the direct passage of vessels.

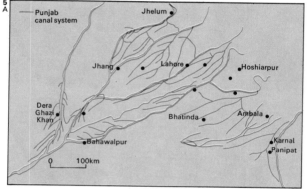

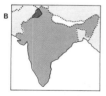

5 A system of irrigation canals in western Punjab [A], constructed by British engineers, was one of the major engineering achievements of the 19th century, These were running water canals fed from the River Indus and its tributaries by building great masonry barrages. They turned an arid area half the size of England into northern India's most fertile food-producing region [B].

6 The Panama Canal links the Atlantic and Pacific oceans, enabling ships that are less than 306m (1,000ft) long, 34m (112ft) in the beam and have a draught of less than 12m (40ft) to avoid the voyage around South America. From the Atlantic an 11km (7 mile) sea-level stretch ends at the Gatun Locks [1] leading to Lake Gatun [2], 26m (85ft) above sea-level. The route then runs 38km (24 miles) across Lake Gatun, along the Gaillard Cut [3] to the Pedro Miguel Lock [4] and Miraflores Lake [5], 16 m (53ft) above sea-level. Finally the Miraflores Locks [6] drop down to sea-level and lead to the Pacific Ocean.

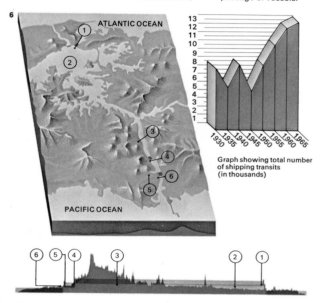

Graph showing total number of shipping transits (in thousands)

7 Construction techniques in canal engineering include the use of special lining machines, as in the Jordan Canal section of Israel's National Water Carrier. Wet concrete is fed on to a conveyor belt and, as the machine moves forwards on caterpillar tracks, an even layer 10cm (4in) thick is laid as a continuous strip. The Israeli system carries water 28km (17 miles) from Lake Tiberius southwards in an open canal to feed a pipeline extending much farther to the south.

8 The Suez Canal, opened after a decade of effort by French engineer Ferdinand de Lesseps (1805–94), was dug out of the desert by hand. It connects the Red Sea with the Mediterranean and covers a distance of 169km (105 miles). It has a minimum width of 150m (500ft) and a minimum depth of 10m (33ft). Britain bought a controlling share in the canal from the Khedive of Egypt in 1875 and retained control until Egypt's President Nasser nationalized it in 1956. The canal was blocked with sunken ships during the 1967 Arab-Israeli war and was cleared and reopened only in 1975.

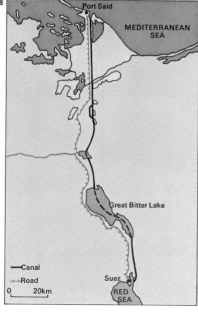

9 An aqueduct built over a road is often necessary in canal engineering for economic reasons, as in this case in the Netherlands. Because the canal is full of water it has to be built to one level throughout, or in level reaches connected by locks. This is achieved by judicious choice of route, by cutting deeper into high ground and by building on a raised embankment over low-lying areas. Most roads cross canals by bridges.

Building dams

Control of his life-sustaining water supply has always been one of man's primary concerns. For 5,000 years, dams have been instrumental in this control, being used to avert floods, divert rivers, store water and irrigate land. Many of today's constructions fulfil these same age-old functions; dams are still used for agricultural irrigation and domestic water storage and supply, as well as for the more sophisticated purposes of hydroelectric power generation, land reclamation, control of erosion by floods and the prevention of build-up of silt.

The use of barriers to divert part of a river's flow into irrigation canals was developed by British engineers during the nineteenth century, and used widely in the Punjab of northern India. French engineers built a similar barrage across the River Nile, which allowed the summer flood to pass over it but formed a reservoir from which the water was led, during the spring and early summer, into irrigation canals. In this way, huge areas of previously parched land were made productive.

The development of modern engineering materials and techniques have since made it possible not only to divert part of a river's water, but also to hold and store water by the creation of huge lakes behind a solid dam wall. A typical early example is the Hoover Dam, built across the Colorado River in the United States. Completed in 1936, this dam has a reservoir capacity of 38,000 million cubic metres (49,000 million cubic yards) and a power output, from the water released through its turbines, of 1,340 megawatts.

Dam design

There are two main types of modern dams: embankment dams, built from earth and rockfill, and concrete dams (which may or may not be reinforced). Embankment dams came first and today are generally the cheapest to build because they need fewer workers. They include the world's tallest dams, such as the 310m (1,017ft) Nurek Dam in the USSR, and the 234m (770ft) Oroville Dam in the United States. They are built from the natural earth and rock found near the site, consisting essentially of a central earth core to hold back the water, faced and supported on each side with more earth or rock. "Filter curtains" between the various layers prevent fine grains from one being washed into the voids of the next.

Embankment dams are suitable for nearly all sites because they can withstand some settlement of the foundation and do not need strong valley sides. A deep trench filled with compacted clay or concrete forms a curtain to prevent water from seeping under the dam. A line of wells under the toe of the dam collects any seepage water and leads it back to the course of the river.

The greater strength of rockfill allows it to be used with steeper slopes. One of the cheapest types of dam has rockfill with a waterproof skin of bitumen concrete on the reservoir side.

There are a number of designs for concrete dams. The simplest, the gravity dam, works rather like a bookend, its own weight preventing it from being turned over by the force of the reservoir water. Volume of material is reduced by a buttress construction.

An arch dam, a more sophisticated construction, is like an arch bridge lying on its

CONNECTIONS

See also
200 Water supply
46 Materials for building
84 Electricity generation and distribution

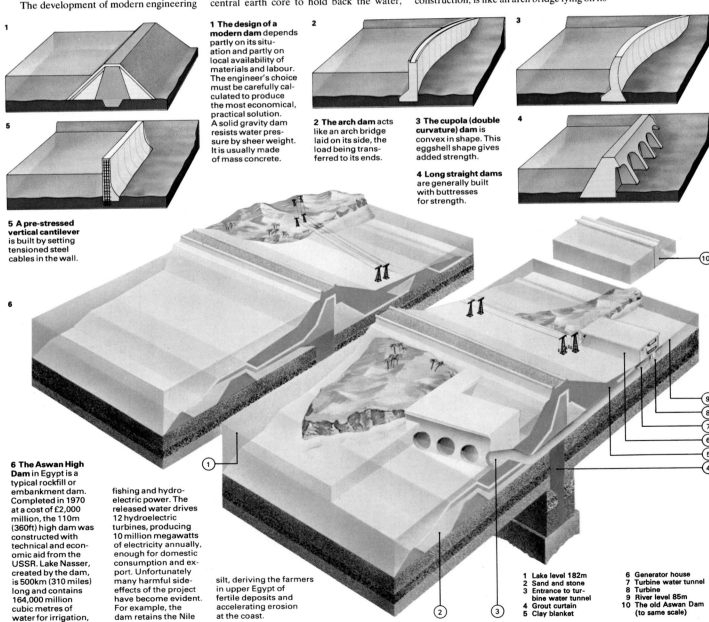

1 The design of a modern dam depends partly on its situation and partly on local availability of materials and labour. The engineer's choice must be carefully calculated to produce the most economical, practical solution. A solid gravity dam resists water pressure by sheer weight. It is usually made of mass concrete.

2 The arch dam acts like an arch bridge laid on its side, the load being transferred to its ends.

3 The cupola (double curvature) dam is convex in shape. This eggshell shape gives added strength.

4 Long straight dams are generally built with buttresses for strength.

5 A pre-stressed vertical cantilever is built by setting tensioned steel cables in the wall.

6 The Aswan High Dam in Egypt is a typical rockfill or embankment dam. Completed in 1970 at a cost of £2,000 million, the 110m (360ft) high dam was constructed with technical and economic aid from the USSR. Lake Nasser, created by the dam, is 500km (310 miles) long and contains 164,000 million cubic metres of water for irrigation, fishing and hydroelectric power. The released water drives 12 hydroelectric turbines, producing 10 million megawatts of electricity annually, enough for domestic consumption and export. Unfortunately many harmful side-effects of the project have become evident. For example, the dam retains the Nile silt, deriving the farmers in upper Egypt of fertile deposits and accelerating erosion at the coast.

1 Lake level 182m	6 Generator house
2 Sand and stone	7 Turbine water tunnel
3 Entrance to turbine water tunnel	8 Turbine
4 Grout curtain	9 River level 85m
5 Clay blanket	10 The old Aswan Dam (to same scale)

side. The curve of the arch is designed so that the concrete (weak in tension) is held permanently in compression. For success, this type of dam has to be firmly keyed into the rock of the valley sides at its ends.

The thinnest type of dam, also with its concrete kept in compression, is the double-curved cupola dam. It is shaped rather like half an egg, with the bulge facing into the pressure of the reservoir water.

Dam foundations

Engineers generally fill fissures in the rock foundation of a concrete dam with a curtain of grout. A drainage system leads away any seepage water, to prevent a build-up of pressure under the dam. There is often a row of drainage wells, drilled into the rock from a gallery running the length of the dam.

To construct the foundation of a dam while water is flowing, engineers generally drive a diversion tunnel through the rock of the valley wall around the dam site, before the construction work starts on the dam itself [7]. A temporary dam is then constructed to divert the water into the tunnel and the main

dam constructed on the dry downstream side of the temporary one.

Every dam construction must allow excess water to flow away in times of flood, without causing erosion in the process. The excess is usually dispersed in one of three ways: by a concrete-lined spillway at a level lower than the top of the dam; by an overflow shaft – usually funnel-mouthed – leading vertically down from a point well within the reservoir; or by an overflow channel or tunnel leading from a point at the side of the reservoir to below the dam.

Great modern dams

The world's greatest dams (in terms of reservoir capacity) are: Owen Falls, Uganda, built across the Victoria Nile in 1954 and holding a reservoir of 205,000 million cubic metres behind a gravity dam; Bratsk Dam, USSR, built across the River Angara in 1964 (169,000 million cubic metres), a concrete gravity dam; and Aswan High Dam [6], Egypt, built across the River Nile and completed in 1970 (164,000 million cubic metres), a rockfill embankment.

The Cubuk Dam, 10km (6 miles) north of modern Ankara, straddles the River Cubuk to form a reservoir with a capacity of 10 million cubic metres. This huge man-made lake provides Ankara with most of its water supply and there is a hydroelectric station to exploit the energy of the water flowing out. Set in an attractive wooded valley, this important engineering achievement has played a significant role in the industrial development of eastern Turkey. It has also been promoted as a recreation area and tourist centre, and has a fine casino and other forms of entertainment.

7 A river must be diverted before it can be dammed. A temporary coffer dam [1] is built and the water diverted via tunnels [2]. The permanent dam is then constructed behind the coffer dam. The rock core [3] of this gravity dam is first laid and compacted. Rubble [4] and an impervious facing [5] are added to the steep, concave upstream slope [6] and the spillway [7] to the downstream side. It is now vital that no subsequent settlement occurs. The impervious facing is grouted into the bedrock [8] to prevent seepage under the dam. The coffer dam is usually incorporated into the structure of the finished permanent dam.

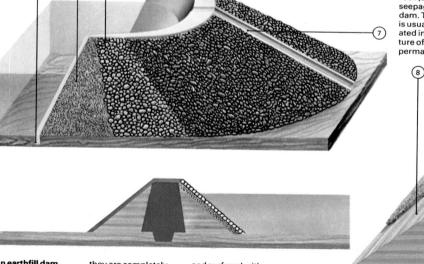

8 Laboratory testing of hydraulic models often precedes the start of work on the actual dam. These models give advance data on patterns of erosion which the water flow may cause. By this means the engineer who designs a river barrage or the outflow channels of a hydroelectric scheme can avoid the problems of water scour; he includes preventive measures in the design before construction begins.

9 An earthfill dam is begun by digging a trench until impervious rock is reached. This is excavated and the first layer of impervious core material (probably clay) is grouted into this trench [1]. Successive layers [2] of clay are added and rolled until they are completely compact. Once this foundation is ready only the core is formed from impervious material [3]. The rest of the dam is constructed with layers of any soil [4]. The upstream slope [5] is covered with gravel [6] and surfaced with rocks called rip-rap [7] to prevent water erosion. The downstream slope [8] is sown with vegetation [9] for stability. A spillway [10] is usually built into the slope to allow water to flow over the dam in times of flooding.

Water supply

The more sophisticated man becomes the thirstier he grows. In developing countries (the Third World), some communities manage with an average of 12 litres (2.5 gallons) of water per person per day. In most European cities the daily domestic consumption per head of population is nearer 150 litres (33 gallons). In the most prosperous urban areas of the United States the figure can be as high as 250 litres (55 gallons).

Demands and resources
Domestic consumption of water mounts steadily but even this is far outweighed by the demands of modern industry. The production of steel needs 300kg (660lb) of water for each kilogramme (2lb) of manufactured steel, although some of this is returned to its source. The total average daily water consumption in a Western city – domestic, commercial and industrial – may be as high as 2,000 litres (400 gallons) per head.

Agriculture's demands are even greater. To produce one kilogramme of wheat a farmer may "use" up to 1,500kg (3,300lb) of water from rain, irrigation or both.

Water is the world's most plentiful natural substance and is in constant circulation [Key]. But man's demand for water often exceeds the local natural supply. It is the engineer's job to transport water to where it is needed and to purify it as necessary.

Collection and storage
Water is normally collected from underground sources by drilling and pumping. The old farmhouse well and bucket provide an example of the extraction of flowing ground water by the simple expedient of digging until the natural water-table is reached. In the case of trapped "fossil" water, sometimes found under deserts, the water is not replaced by natural flow and the process of removing it is a form of mining – the extraction of a limited resource. Deep-lying water under pressure can be tapped by the bore-hole of an artesian well, in which the water is forced up the bore without the need for pumps.

There are three ways of exploiting surface water. The first is by pumping, from rivers [2] or lakes, the second is by building a barrage across a river and diverting its flow through canals or pipelines and the third is by constructing a dam across a valley at the lower end of a natural catchment area.

Apart from short-term local storage, water is normally stored in large, open reservoirs into which it is pumped, or in the huge, man-made lakes which form, by means of gravity flow, behind the walls of valley dams.

Water purification
Natural water, other than the rain, is rarely pure. Rivers in peaty moorland pick up traces of organic acids. Ground water takes up mineral salts including common salt (up to 0.1 per cent is tolerable in drinking water), calcium bicarbonate (above 0.02 per cent makes water unacceptably "hard") and fluorides (amounts above 0.0001 per cent, but not exceeding 0.00015 per cent are said to reduce tooth decay).

The Industrial Revolution produced a new problem. The discharge of effluents – factory waste products – into rivers resulted in chemical pollution and the development of water-borne sewage systems caused bacteriological pollution. Today many coun-

1 One of the first major water supply systems was built by the Romans. Their first nine aqueducts, constructed between 313 BC and AD 226, conveyed water by gravity feed from various sources to Rome. The longest, the Aqua Marcia, extended 90km (56 miles) and bridged low-lying land by a water channel of cut stone slabs raised on a succession of stone arches. The Roman aqueduct here is the 274m (900ft) long Pont du Gard in France. The top tier carries the water channel [1], the lower tier a footbridge [3]. So accurate were the Roman masons that the stones of the two lower tiers of arches, including the voussoirs [2], were laid dry without mortar.

2 As a city grows, so does its demand for water. The London Bridge Waterworks were opened in 1581 to serve the City of London. By the 19th century, the huge waterworks and pumps were capable of supplying almost four million gallons of river water daily. However the fall of water from the works was so great that it endangered navigation on the river and the works were completely removed in 1822. The river current turned the great wheel which rotated the crankshafts [3]. These operated the pumps [4] via levers [1], water being discharged through pipes [5]. The supply of water was controlled by turning the cranks [2] which raised and lowered the pumps.

3 The modern water treatment plant can be very complex as this diagram shows. Here raw river water is screened at the intake and drawn by pumps to an upward flow sedimentation tank. A flocculant (chlorinated ferrous sulphate) and softener (lime slurry) are added as the water enters, the lime also regulating the acidity. The sludge formed in the sedimentation tank is pumped to a sludge lagoon from which the clear water is recycled. Activated carbon is then added to absorb impurities of the kind that give an unacceptable taste, smell or colour to water. The water then flows into a rapid gravity filter in which any organic matter is decomposed by non-pathogenic-bacteria to form unobjectionable inorganic products. One of two chlorinators supplies chlorine gas as a sterilizer and additional lime is added as the water enters the contact tank. (The second chlorinator feeds the flocculant supply.) A sulphonator feeds the contact tank with sulphur dioxide, which dechlorinates the sterilized water. Finally the treated water is pumped to the mains supply system where it is stored in water towers or reservoirs before distribution to domestic consumers and industry.

Figure 3 labels: Intake; Pump; Sludge lagoon; Chlorinated ferrous sulphate; Lime slurry; Sedimentation tank; Activated carbon; Chlorine; Rapid gravity filter; Contact tank; Sulphur dioxide; Pump

tries have laws to control chemical pollution, and bacteria are eliminated by treatment.

The principal processes of water treatment [3] are sedimentation, filtration, aeration and sterilization. Sedimentation is merely settling, carried out by allowing water to stand in large, shallow basins; solid particles sink slowly to the bottom. Sedimentation is aided by the addition of a flocculant, such as alum, which causes the smallest particles to clump together. Filtration is carried out by passing water through a sand bed – 100m × 40m (328ft × 131ft) is a typical size - in which harmless bacteria (in a layer up to 30cm (12in) deep) decomposes organic matter in the water passing through, forming unobjectionable inorganic substances. A rapid sand filter may be only 8m × 5m (26ft × 16.4ft) in size and 4m (13ft) deep. It is cheaper and filters 20 times faster, but the filtered water is never entirely bacteria free.

Aeration (usually effected by passing water over a cascade) increases the amount of dissolved oxygen in the water, reduces the carbon dioxide content by as much as 60 per cent and aids natural inorganic purification

by aerobic bacteria. Sterilization (killing harmful micro-organisms) is achieved, where bacteriological pollution has been high, by the addition of small quantities of chlorine or ozone. A dose of 0.0001 per cent of chlorine destroys all germs within four minutes.

Where exceptionally pure water is required, demineralization (softening) may be carried out. Here the exchange of the "salt" part of soluble mineral salts forms easily eliminated non-soluble compounds.

Where natural pure water is scarce and the sea near, the process of desalination (removing salt) is used today to produce water for both human and agricultural use. The main processes for eliminating the dissolved salts are called distillation, electrodialysis, reverse osmosis and freezing. Multi-stage flash distillation [5] is used in most modern desalination plants, using steam as the heat source.

Today nost water is distributed by pumping it to a local storage facility (usually a water tower), which provides sufficient static "head", or pressure, to force the water through a network of pipes.

The hydrologic cycle, powered by the sun's heat, keeps the world's stock of water in constant circulation. Enormous quantities are moved by evaporation from the sea, lakes and rivers and by transpiration from trees, crops and other vegetation. It moves through the atmosphere, condensing as clouds, then returns to the land as rain or snow, forming streams and rivers leading back to lakes and seas. It sinks into the soil forming aquifers, reappearing as springs.

4

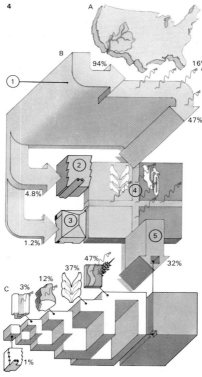

4 Growing salinity can pollute natural water. Man's "control" of the Colorado River in the United States has caused an excessively high salinity. The map [A] shows the location of the Colorado River. [B] shows how the water obtained from the river is used. The total capacity of the Colorado River [1] is approximately 60 million million litres per day; 94% is supplied to agriculture, although 16% of this is lost in evaporation before it reaches the farms.

Of total capacity 4.8% [2] is supplied to industry and 1.2% [3] for domestic use. Run-off from all uses is 32% [5]. 47% of total capacity [4] is actually used by agriculture. The run-off is highly saline when it returns to the river. Salinity [C] in the water is caused by evaporation from the river (47%), evaporation from land and transpiration of plants (37%), evaporation from reservoirs (12%), evaporation from canals (3%) and from industry (1%). In highly developed

industrial countries industry has been responsible for some other forms of river pollution. Industrial effluents can cause serious chemical pollution and the disposal of water-borne sewage can be responsible for serious biological pollution. Legislation has been passed in many countries to help keep river water clean.

5

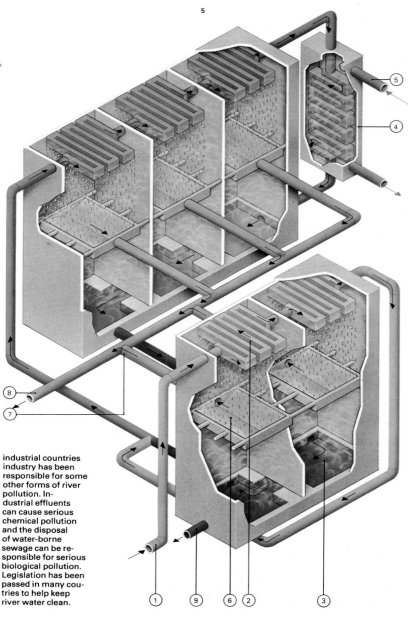

5 Desalination by flash-distillation is a widely used process. Raw seawater [1] is fed through the condensing coils [2] of the first two flash chambers. It is then mixed with strong brine from the brine pans [3] of these two chambers before passing on through the condensing coils of the third, fourth and fifth flash chambers. From there it passes through the heat exchanger [4] where it is heated by steam [5]. The seawater, then at 80°C (176°F), next flows through the brine pans of the five chambers in reverse order. Water vapour rises from the hot brine, condensing on the much cooler coils above. The condensate drips on to the fresh water catchment troughs [6] from which it is piped [7] into the main fresh water outlet [8]. Meanwhile the hot brine, on its way through the five chambers, grows progressively more concentrated and cooler. When it reaches the first chamber part of it is recycled by mixing with the sea water passing between the coils of the second and third chambers, the rest being discarded as waste [9]. Modern practice achieves economy in the desalination process by locating the plant, wherever possible, near a nuclear power station. The exhaust steam from the power plant can then be used as the main energy source for the distillation process.

Sewage treatment

As soon as early man established settlement he had the problem of getting rid of sewage – the organic wastes of man and, sometimes, domestic animals. Nature disposes of surplus organic material in four principal ways – by dilution, oxidation, putrefaction and filtration. Early man relied on these natural processes and generally dumped sewage in the fields. There its moisture seeped into the land, becoming filtered and purified. When rain fell on the solids they dissolved and were oxidized into natural nutrients for the soil.

The Roman system of aqueducts, built from 312 BC to AD 226, provided a water-borne sewage system that drained into the River Tiber. Today the large volumes of sewage from towns and cities need the resources of modern technology for efficient treatment and disposal. Also industrial wastes present additional problems for treatment plants if pollution is to be avoided.

Treatment by dilution

Sewage treatment by dilution works because water contains dissolved oxygen. When sewage contaminates a small volume of water, as in a lake or stream, aerobic bacteria (which oxidize the organic material) absorb the dissolved oxygen at a rate greater than it is naturally replaced from the air. As a result fish cannot live and the water is no longer self-purifying. But if the surface area of the water is large enough to dissolve oxygen faster than it is absorbed by the bacteria the water remains pure and unpolluted.

The River Thames in London provides an illuminating case history [Key]. In 1750 the city's population was 750,000 and the river was teeming with fish. It had long been used as a sewer but this had not seriously polluted the water. By 1840 the population was over two million and sewage, swollen in volume by industrial effluents, then exceeded the river's capacity for self-purification. Only eels had survived the advent of industrialization.

London's first sewage treatment works were commissioned in 1889. By 1900, with the population at over six million, six of the less sensitive species of fish had returned to its waters. Between the wars London's population grew to eight million people and industry increased. The sewage works could

not handle the greater volumes of domestic and industrial effluent and by 1945 there were no fish in the river, which had become more polluted than ever. In the 1950s there was a drive to clean the river. New sewage works were built and by 1970 fish of many kinds had returned to the Thames.

The dangers of pollution

Health authorities today agree that the disposal of diluted sewage into lakes or rivers is satisfactory – if it contains not more than 30 parts per million of suspended solids and does not absorb more than 20 parts per million of dissolved oxygen in five days. The latter figure, called the "biochemical oxygen demand" (BOD), provides a means of measuring the degree of pollution.

Where there is sufficient water for effective sewage treatment by dilution, as on coasts, on the shores of major lakes, or near large rivers, sewage is sometimes discharged without prior treatment [4]. However, the growth of industry and high population densities usually require sewage treatment by processes designed to reduce its BOD to a

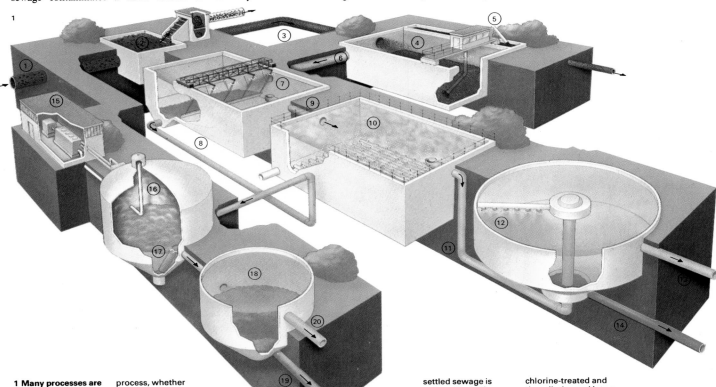

1 Many processes are used in a typical modern sewage works. The first stage consists of screening to remove solid material including heavy grit. Screenings are burnt to destroy micro-organisms and render them inoffensive. Fine grit is next removed by gravity settlement. Screened, de-gritted sewage is then led to sedimentation tanks. The remaining suspended solids are eliminated. This reduces the liquid sewage strength by up to one half. As gravity precipitation is a slow

process, whether still or continuous flow is used, the tanks provided for this purpose are usually duplicated and made large. The sludge and clear sewage are then treated separately. Clear sewage passes to aeration tanks where the organic content is destroyed by aerobic bacteria. The effluent from the aeration tanks is once again settled, the clear liquid chlorinated and discharged into a river or the sea. Sedimented sludge is fed to digestion tanks where it is warmed and allowed

to putrify in the absence of air. This produces gas, to power the works itself and to heat the sludge tanks, and a nitrogen-rich "clean" sludge which is dried to form an excellent fertilizer. A typical process operates in the following way. Raw sewage [1] enters the works where coarse screens [2] remove solid trash (wood, rags and so on) from the sewage so that the machinery is not damaged or pipelines

blocked. The screened sewage [3] is then pumped to grit-settling tanks [4] where grit and sand settle out and are dredged up, washed and then used as filling material such as aggregates for road-bed construction. The settled grit and sand [5] are removed to allow the grit-free sewage [6] to be pumped to the primary sedimentation tanks [7]. During this process 50 per cent of the suspended solids settle out to form sludge and the BOD of the

settled sewage is reduced by half. The sludge [8] is then pumped to digestion tanks and settled sewage [9] pumped to the aerator [10], in which the sewage is mixed with bacteria-rich activated sludge. As the sewage is aerated the bacteria transform organic matter into harmless by-products. The aerated sewage [11] is pumped to a final settling tank or secondary sedimentation tank [12]. At this stage of the process activated sludge settles out, leaving a clear effluent. The upper liquid part [13] is filtered,

chlorine-treated and then discharged into a lake or river. The activated sludge [14] is removed and reused in the aerator with incoming settled sewage. In the power house [15], gas from the collector [16] – containing about 70 per cent methane – is burned to generate power for pumps and air compressors; some gas is used to heat the digestion tanks. The sludge that settles out is then pumped to the primary digestion tanks [17] which are kept at a temperature of 30°C (86°F). The temperature speeds the action of micro-

organisms which, in the absence of oxygen, digest the sludge rapidly, producing gas and a relatively inoffensive sludge. The secondary digestion tanks [18] are where the digestion is completed (unheated) producing concentrated, nitrogen-rich sludge and relatively pure but bacteria-rich water. The sludge [19] is then removed and after going through a drying process is used as a fertilizer. Finally, water [20] is drawn off and discharged into a lake, river or the sea.

very low figure before discharge and sometimes to eliminate all bacteria.

Sewage treatment by putrefaction is a natural process in which anaerobic bacteria destroy organic matter by breaking it down into simpler substances. The products are nitrogen-rich humus and a mixture of gases in which methane predominates.

A modern sewage works [1] uses both the natural oxidation and putrefaction processes in treating sewage. It also uses several other processes that may include screening, sedimentation, flocculation, digestion, aeration, filtration and chlorination. Industrial effluents often require other special treatments to eliminate toxic substances before they are discharged into the disposal system.

Modern treatment processes

Screening is simply the removal of large, solid particles by passing the sewage through a wire screen or other form of mesh. Quiescent sedimentation is natural settlement of sediment by gravity in undisturbed tanks. Continuous-flow sedimentation achieves the same result by passing the fluid slowly, without turbulence, through long, relatively shallow tanks and out over a weir. Flocculation, mechanical or chemical, makes non-settling material coagulate into particles of sufficient size to settle by gravity. Alum is an efficient flocculent but is too expensive for general use in sewage works. Mechanical flocculation is achieved by slow stirring.

Digestion uses natural aerobic putrefaction, producing gas and a sludge which, when dried, forms a useful fertilizer. Some sewage works use the gas to produce power and light, and often warm digestion tanks by passing gas-heated water through coils fixed inside them. Heating speeds the digestion process but it is generally too expensive because it consumes too much energy.

Aeration employs natural oxidation to reduce the BOD of clear sewage, following sedimentation. It is commonly achieved by pumping air bubbles through the lower part of the tank and redistributing the clear sewage over the surface in the form of a rotating spray. Filtration and chlorination remove the final effluent from the water, leaving it totally free from bacteria.

In 1858 the Thames was so polluted that *Punch*, England's weekly humour magazine, printed this savage caricature of Death rowing through its flotsam and jet-sam. The flow of sewage, a product of the Industrial Revolution, into the Thames created an enormous health hazard. *Punch* called its cartoon "The 'Silent Highway' – Man" and gave it the sub-title "Your MONEY or your LIFE!" In 1889, 31 years later, London commissioned its first sewage works.

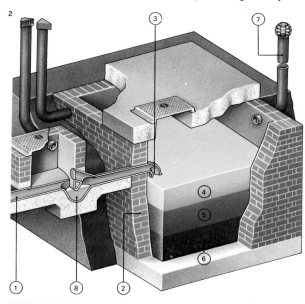

3 The sludge from a septic tank must be pumped out once every three to six months. A custom-built tanker, the sludge gulper, is used by local authorities for this purpose. A sludge removal vehicle must have a tank large enough to empty the largest septic tanks in its area in one lift, for partial emptying results in the scum being left behind. The vehicles pump out liquid sludge through a long flexible hose.

2 A septic tank can process sewage from buildings not connected to the local authority's drainage system. It is a watertight, airtight underground tank that serves the triple purpose of sedimentation, digestion and sludge storage. Digestion is automatic, being effected by anaerobic bacteria. Effluent should be led to a deep soakaway and sludge pumped out every 3 – 6 months. The tank's walls are waterproof [2], usually of cement-rendered brick. Sewage enters through the inlet pipe [1] and is delivered low in the tank [3] without stirring the contents. Relatively pure water [4] separates above the sludge, anaerobic bacteria digest the sewage [5] and the sludge sinks [6]. Gas escapes up the vent pipe [7] and a gas trap [8] prevents it passing back into the inspection chamber.

4 Untreated sewage used to be discharged into rivers, estuaries or the sea [A], leading to pollution and the possible spread of disease. Today most industrialized countries process sewage in treatment plants [B]. One of the problems of modern sewage treatment is the presence in urban waste of ever-increasing quantities of detergents. Domestic detergents must by law be bio-degradable, so that they are digested by bacteria, but most sewage treatment plants are unable to remove the phosphates usually found in detergents. These act as nutrients to green algae, which grow too quickly and upset the natural biochemical processes that keep water pure in lakes and rivers.

5 The digestion of sedimented sewage sludge in a modern sewage works results in the production of a relatively inoffensive nitrogen-rich sludge which, after drying, forms an excellent general fertilizer ready for immediate use on the land.

History of printing

Movable metal type was probably first produced in the Royal Type Foundry of Korea in 1403 and a book was printed from this type six years later. But not until 1439 is there evidence of printing, as we know it today, in Europe. It was a German, Johannes Gutenberg (1400–68), working in Strasbourg, who developed printing using movable type.

Early printing methods
In the year 1456 the first substantial printed book appeared. This was a Latin Bible printed in Mainz, almost certainly by Johannes Gutenberg and his associates [2]. How Gutenberg manufactured his type is not known and it was not until 1540, in Vanoccio Biringuccio's book *De La Pyrotechnica*, printed in Venice, that there was a description of typefounding. Type was made by pouring molten metal into a copper matrix or mould formed by punching an engraved steel character into a piece of copper.

From the days of Gutenberg there were, for many centuries, no significant changes in the basic methods used for printing. Metal type was set by hand into pages known as formes and these were inked and printed on to single sheets of paper in a hand press [Key]. The first change came in 1795 when Firmin Didot (1764–1836) tested ways of making duplicate printing plates (stereotypes) from set type.

Three years later lithography was invented by Aloys Senefelder (1771–1834) of Munich. While seeking a practical method of printing musical scores he tried drawing the music in reverse on a flat slab of stone, using an ink made of wax, soap and lampblack. His original idea was to etch the stone with acid but his experiments led to an entirely new printing process based on the mutual repulsion of oil-based ink and water.

From plates to printing machines
Didot's stereotype process was perfected in 1800 by Charles Stanhope, third Earl Stanhope (1753–1816), who used plaster of paris to make the moulds of the set type. Molten metal was then poured into the plaster moulds to produce solid printing plates, a whole page at a time. In 1806 Anthony Berte of London invented a mechanical device for typesetting, using a pump to force molten metal into the matrix.

In 1811 Friedrich König (1774–1833), a German printer who moved to England in 1806, built the first successful printing machine in which a series of leather-covered rollers, fed with ink from a container, automatically inked the type as it travelled to and from the printing platen. Except for the laying on and removing of the paper from the platen, this steam-powered press was automatic. In the following year König designed a cylinder printing machine in which the type forme was fixed to a bed which moved first under leather-covered inking rollers and then under a cylinder around which the paper was held. Soon after the first single-cylinder machine was successfully demonstrated in London two machines, each with twin-cylinders, were manufactured.

Nineteenth-century progress
In 1816 an Englishman, Edward Cowper (1790–1852), secured a patent for a method of bending stereotype plates for rotary printing, the plate first being cast flat in a

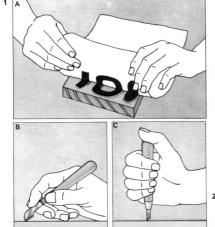

2 The first substantial book printed from movable type was the Latin Bible published in Mainz, Germany, in 1456. This remarkable book had 643 leaves, each page printed in two columns of 42 lines. Known after its celebrated printer, Johannes Gutenberg, as the Gutenberg Bible, it is not only the first major product of modern typography, but is still among the finest examples of the printer's art.

This illustration, taken from a page of the Bible, shows the typeface used by Gutenberg. This face was known as Textura and it was cut to resemble the manuscript hand used in the 15th century in Germany to prepare handwritten Bibles and church service books. The Gutenberg Bible was printed in ten sections on six presses at once, the edition running to 150 copies on heavy paper and 30 on fine vellum.

1 The woodcut was developed from the early 15th century onwards to produce devotional prints and playing cards. The artist cuts away the part of the design that is to appear white, so that only the parts to be inked appear in relief [A]. The block is cut along the grain using well-seasoned wood from apple, pear, cherry, sycamore or oak trees. Some early woodcuts were hand-coloured but prints can be produced by using a series of blocks. A variety of tools were used, such as the English knife [B] and the Japanese knife [C].

3 The first self-inking treadle platen press was designed and built by an American, Stephen Ruggles, in 1839. This improved design built by him 12 years later became the model for the popular jobbing press. In this machine the forme of type is clamped in a near-vertical position. Above the forme is a rotating disc that serves to distribute the ink. Three composition rollers roll down over the disc, picking up ink and then down over the type forme. When the rollers have returned, the platen (on which the paper has been laid by hand) moves up against the type forme.

4 Typefounding is the process of making individual pieces of type by casting molten type metal into moulds. Type metal is an alloy of tin and lead, with added antimony for hardness. In this Victorian typefoundry dozens of workers operate individual machines belt-driven from a shaft running the length of the room. Single casting of individual type was superseded by the linotype machine, which casts a whole line at once.

5 Letterpress printing includes the platen press [A] in which paper is pressed up against the inked image. The paper in the flat bed press [B] is laid on the image and pressed down by roller. The rotary press [C] has a curved image on one roller, the paper passing between this and a pressure roller. Metal type [D] has a face [1], a shoulder [2], a body [3], a foot [4] and a nick [5]. Zinc blocks [E] have unwanted metal [6] etched away from round the design [7].

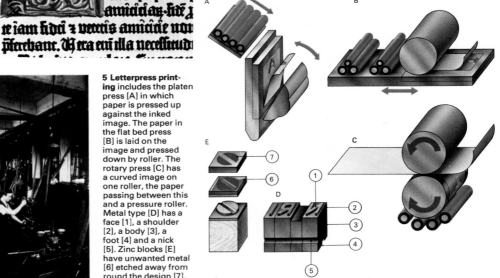

plaster mould of a type forme. In the same year Friedrich König and Andreas Bauer (1783–1860) built the first perfecting machine – which could print on both sides of the paper.

A traditional printing press uses mirror image type. In 1817 an Englishman, called Augustus Applegath (1788–1871), designed a machine to print bank-notes with the same design on each side of the paper and with each colour in perfect register with the others. In this machine a curved stereotype first printed on to a leather pad fitted around the printing cylinder. When one revolution was complete the paper was fed between the stereo and the leather pad so that the metal printed on one side of the paper and the inked leather pad on the other. The idea was to produce bank-notes that could be forged only with difficulty, and Applegath's machine in fact printed the lower side of the notes by what is now called the offset method. The leather pad was printed with a mirror image of the original type matter which it then transferred to the paper. As a result the printed image on the lower side of

the paper was identical to the original stereo. This principle was later made use of in offset printing, in which the type matter is identical to the finished print.

The next 70 years saw numerous advances in the art of printing. Stereotype preparation from papier mâché moulds [9] was originated by Claud Genoux of Lyons. These moulds or "flongs" were strengthened with clay and glue. In 1845 the French printing firm of Worms and Phillipe patented the idea of casting curved stereos direct from a curved flong. This is the method still used in most newspaper printing today.

In 1838 the American David Bruce, Jr (1802–92) built the first commercially successful mechanical typecasting machine, which made 100 characters an hour.

In 1852 the first printing by photolithography was carried out experimentally by Alfred Lemercier, a Frenchman. Lemercier coated his printing stone with a light-sensitive substance which was exposed through a paper negative. After washing with turpentine the design on the stone could then be inked for normal lithographic printing.

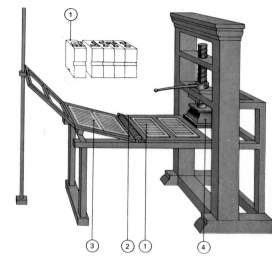

KEY

Early printing presses resembled 15th-century linen presses. The type matter [1] was wedged in a sliding tray [2] and then inked by hand. The paper was placed on a parchment covered "tympan" [3] which hinged into position over the type tray. This slid under the screw-down press [4].

6 A

6 Lithography is based on the mutual repulsion of greasy ink and water. The image to be printed is drawn, in reverse, on stone using a greasy pencil or ink. The stone is then soaked in water and inked. The ink adheres to the image but not to the wet stone. The machine [A] presses paper from the holder on to the stone, thus making the print. The process was used originally for printing hand-drawn music [B].

B

7 A

8 When an etched plate has been inked and wiped, ink remains only in the grooves made by the acid and can thus be printed. This is the principle of intaglio, in which the image is below the surface of the printing plate and so retains the ink when the polished surface is wiped clean. James Whistler (1834–1903), who used etching as an art form, made this etching of Black Lion Wharf, Wapping, London.

8

7 In an etched printing plate the ink fills the grooves made by acid in a polished metal plate (copper, zinc, aluminium or steel). The plate is coated with acid-resisting wax. The image is drawn in the wax with an etching needle [A] exposing the metal. The scribed plate is then put in acid [B] which eats into the metal where the wax has been removed. After cleaning off the wax the plate is inked, then wiped.

B

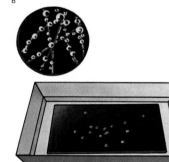

9

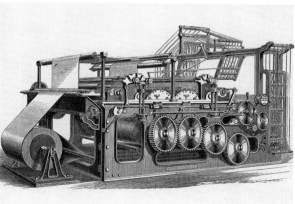

9 The principle of printing on paper fed from a roll with the sheets being cut after printing was invented by Rowland Hill (1795–1879), who later introduced the penny post in England. Rotary machines were developed for newspaper printing from 1846, using curved stereotypes cast from curved papier mâché moulds (flongs) of the original flat type formes. This early Victory press printed at speed from a paper roll (web) and folded the cut sheets.

10

10 The linotype composing machine, invented by Ottmar Mergenthaler (1854–99), an American of German parentage, was probably the greatest single printing innovation of all time. The machine was first used to typeset the *New York Tribune* in 1886. The operator "types" the copy on a keyboard and the machine sets letter moulds (matrices) in the correct order. At the end of each line a complete line of type (slug) is cast in one piece.

Modern printing

There are three main kinds of modern printing processes: relief, intaglio and planographic. In relief printing the ink-bearing surface of the type or engravings (for illustrations) stands out above the surrounding non-printing area. Letterpress printing is an example of relief printing and includes printing from one-piece stereotype plates and from formes made up of both type matter and illustrations, which may be metal line blocks, halftone engravings (which have tonal shades formed from minute dots), woodcuts or linocuts.

In letterpress printing the platen (flat plate) press and various kinds of cylinder presses are widely used. Paper is fed into the machine (by hand or automatically). The mechanism picks up single sheets from a stack (usually by means of suction pads) and feeds them into the machine.

Intaglio printing is relief printing "in reverse". In this case the ink is trapped in cavities in the printing plate surface, the polished non-printing area being "wiped" clean of ink before printing. Photogravure is the principal intaglio process of today;

formerly it was used for printing artists' engravings made on steel and copper and for aquatints and etchings (in which lines were etched in the metal by acid).

In planographic printing the printing and non-printing surfaces are completely flat. The printing plates are treated in such a way that the mutual repulsion of oil and water keeps the ink from the non-printing area. Lithography is the principal planographic process of printing.

Rotary printing

Modern high-speed printing is carried out on a rotary press [1] using continuous paper from a roll. The printing formes are prepared flat, using type (text) and blocks (illustrations). A papier-mâché "flong" mould is made, again flat. The mould is placed in a casting box which has been bent to a half circle, so that when molten metal is pumped between the flong and the circular backing of the mould a semicircular stereotype plate (stereo for short) is produced. The edges are cleaned up, the white (non-print) areas routed out, the inner side shaved to form a

perfect semicircle and two stereos are fitted around each printing cylinder in the press. This forms a circular printing surface.

Modern photogravure, developed from hand etching, uses photographic techniques to produce copper printing cylinders or plates. Positive prints of the type matter and illustrations are printed photographically on to a gelatin-coated carbon tissue. When developed and laid on a copper sheet this forms an etch-resistant pattern (called a resist), which allows acid or other etchant through the image area, leaving the copper surface untouched in the non-image areas. The etched copper cylinders are then fitted to a high-speed rotary gravure press, which operates in much the same way as a letterpress rotary. Rotary photogravure is widely used for the production of high-quality large-circulation magazines.

Web-offset lithography

Lithography, originally carried out by forming the image to be printed on a porous stone slab, was soon developed so that the printing surface could be prepared

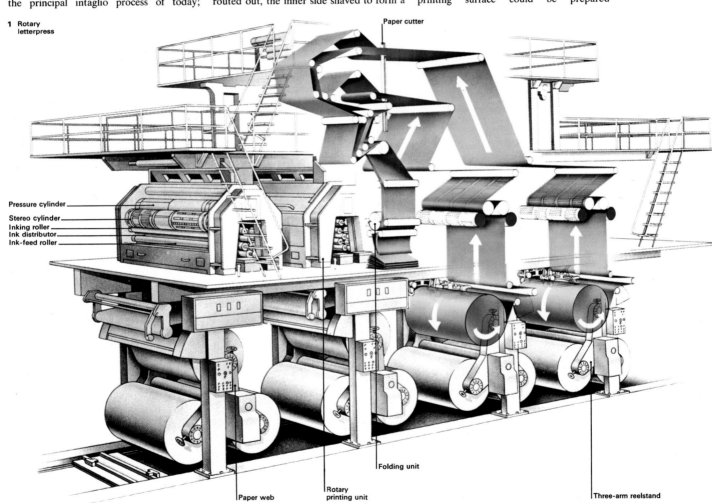

1 Rotary letterpress

Paper cutter

Pressure cylinder
Stereo cylinder
Inking roller
Ink distributor
Ink-feed roller

Paper web

Rotary printing unit

Folding unit

Three-arm reelstand

1 The newspaper industry has improved and developed its methods of printing over the last 100 years. A modern machine prints both sides of a paper web four pages wide, cutting and folding to make up a finished eight-page newspaper section. Shown here is a typical installation with four presses that can produce a complete 32-page newspaper at once. The three-arm reelstands allow the machine to run without interruption by pasting the paper from the following reel on to the tail of an exhausted web. Colour printing can be achieved by re-routing the paper web so that it runs through two machines in sequence, one using black, the other a coloured ink, or by adding multi-colour press units in the paper path following one of the single-colour presses. They then colour-print in the white space left by the normal press. Presses of this type are designed to deliver at least 50,000 copies per hour. Presses working at such high speeds are fitted with automatic devices to maintain correct paper tension and flow through the mechanism and to apply brakes if there is some failure such as a break in the paper web. Newspapers and magazines are printed increasingly frequently by photolithography, but many large-circulation magazines, printed in black and one other colour, are produced on automatic high-speed letterpress machines similar to newspaper presses. In Britain telephone directories and similar publications are printed on rotary letterpress machines at a rate of 15,000 copies per hour. Rotary presses sacrifice quality for speed; in eight hours several presses can print 2.5 million copies of a paper.

photographically on a sheet of metal such as zinc or aluminium. The methods used for offset platemaking vary widely, but the result is the same – a flexible metal plate that can be fitted around the printing cylinder of a high-speed rotary press [2]. Web-offset presses, printing on a continuous paper roll (the web), have been developed in recent years. As a result many small-circulation newspapers and magazines use this system of printing.

Modern printing techniques

In letterpress printing and lithography, the reproduction of continuous-tone illustrations such as photographs is achieved by photographing the picture through a "half-tone screen", which is a fine grid of crossed parallel lines. This breaks the picture into tiny elements which are printed separately as dots of varying size [3].

Linotype machines (which cast a whole line of type at a time) continue to be used widely for newspaper production and the Monotype machine [4] (which casts one letter at a time), with its ability to set, say, symbols and complex mathematical equa-

tions, is still used extensively for book production. But photosetting, introduced in a practical form in 1955, is superseding the metal type processes in many fields.

Photosetting is an entirely new process in which the alphabet and other characters of each type style are stored on film or as tape-recorded instructions. The photosetting machine [5] produces positive or negative copy suitable for photographic platemaking processes and has the advantage that the size of any typeface can easily be reduced or enlarged photographically.

A modern photosetting machine, with appropriate lens systems, also produces condensed and expanded faces, and italics, from the same master negatives. It operates at high speed – computers control spacing and produce proofs as fast as the operator can use his keyboard. In gravure printing and lithography printing plates are made directly from the positive image produced by the photosetter. In letterpress work a negative image is made photographically into a metal printing plate in a similar manner to that used in the production of halftone blocks.

The essence of modern printing is speed. Millions of

copies of daily newspapers appear every morning filled

with news stories that were written the previous night.

2 Offset lithography

3 A

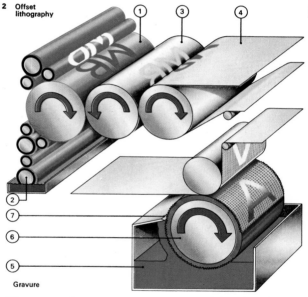

Gravure

3 Yellow, red and blue are used in various combinations to print a wide variety of shades. Where all three overlap the result is almost black. In colour reproduction a greatly improved result is achieved by a fourth, black printing. More accurate colours are obtained by using magenta and cyan in place of red and blue. An image is printed in each colour [A-D] to give the combined effect, seen here magnified [E].

2 Image areas of offset lithographic printing plates [1] accept a grease-based ink and non-image areas, damped by rollers [2], reject the ink. The inked image is offset on to a rubber roller [3] and then on to the paper [4]. The

copper gravure printing cylinder [6] has the image etched as "cells" (small depressions beneath the surface), the volume of cell determining the quantity of ink and hence the tonal value. The cylinder is covered with ink [5] and the surface

is then cleaned by a "doctor" blade [7], leaving ink only in the etched depressions. When the cylinder contacts the paper the ink in the cells is transferred to it. The gravure process is used principally for high-quality pictorial work.

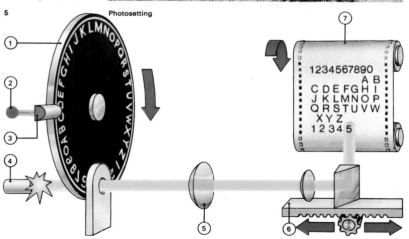

5 Photosetting

4 The Monotype machine, invented in 1887 by Tolbert Lanston (1844–1913), produces three characters of set type each second, using matrices (moulds) and molten metal. It is widely used for books. The operator types the copy on a keyboard, producing a punched paper tape. The tape is later fed into a second machine which casts individual letters, setting the type in lines of equal length ready for printing.

5 A typical photo-composing machine has a continually rotating disc [1] which bears negative images of all the letters, numerals and punctuation

marks in a given type face. An exciter lamp [2] and photo-electric cell [3] sense the exact orientation of the rotating disc. When the chosen character

is in line (instructions come from a computer tape by "typing" copy on a keyboard) a micro-flash unit [4] fires. The lens system [5] focuses an image on

film or paper [7]. The transport system [6], also computer controlled, sets the images in lines with justification and hyphenation where necessary.

Copying and duplicating

Reproduction of the written word and of drawings and photographs is not exclusively the art of the printer. An increasingly large volume and variety of reproduced material is prepared quickly and relatively cheaply in offices found in most parts of the world by duplicating and copying processes developed for this purpose [1].

The hectograph and rotary duplicator

The oldest of office copying processes is the once familiar spirit duplicator [2]. The original material – the master – was drawn by hand using synthetic, aniline inks (purple, red and green were the most usual colours, although black, blue, yellow and brown inks are also available today) or by typing with special aniline dye ribbons. The master was then rolled face down on to a gelatin bed, which absorbed a proportion of the ink, to form a mirror image on its surface. Plain paper moistened with spirit was then pressed on to the gelatin and peeled off to leave a positive image on each sheet. This convenient process, which can print up to 50 or even 100 copies from a good master, has

been modernized and is the basis of the rotary duplicator. A "transfer" sheet surfaced with a wax film containing aniline dye is placed behind the master sheet and material is drawn or typed on to both sheets. In this way a mirror image in aniline ink is formed on the back of the paper, just as though a sheet of carbon paper had been placed below it, black side up. The master so made is fitted on to a cylindrical drum which presses against sheets of paper, moistened with spirit, to provide copies quickly and cleanly. A master can be made using a number of different coloured transfer sheets, the process printing the different colours simultaneously.

Modern copying processes

The ink duplicating machine, or mimeograph, has the advantage of being able to produce a thousand or more good copies from the master. This master, once made of wax on porous paper, is now generally made of a pressure-sensitive plastic composition. The image is either hand-drawn with a sharp stylus or typed on any machine with a fine typeface. The lines or letters cut into the sur-

face and break the skin so that ink can penetrate. The master so formed is a negative stencil. The modern mimeograph is a high-speed electrically powered machine and produces good quality copies at speed.

The blueprint [Key] is no more than a negative prepared photographically by contact with a translucent original. The modern process, which has superseded the blueprint, uses special paper sensitized with diazo compounds. This is placed in contact with the original and exposed to what is known as actinic radiation (mainly blue and violet light, together with ultra-violet rays). The radiation passes through the clear paper of the original and chemically alters the diazo compound. But where it is masked by the lines of the image, the diazo compound remains unchanged. "Dry" development by ammonia gas can produce a black or blue image; wet development can produce prints in red, sepia and some other colours. Variations of this process are used in the many dry, "semi-wet" and heat development azo copiers available today. The process has the advantage of producing copies of large drawings more cheaply

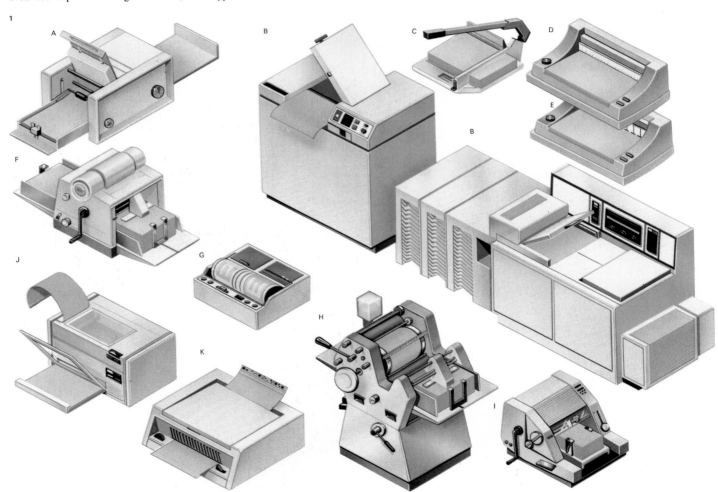

1 The in-house print room of a modern commercial office, research establishment or other organization where reproduction of printed matter and illustrations is required quickly and often, contains a variety of machines, each serving a specific purpose. Among typical machines shown here there is a spirit duplicator [A] for short runs, a mimeograph [I] for medium runs of non-illustrated material and an offset-litho press [H] for high-quality printing that may include colour work as well as drawings and photographs. For rapid copying and high quality, a type of photostatic copier [B] is generally used. Large drawings can be quickly reproduced on a diazo copier; "semi-wet" and dry working types are available. Stencils for a mimeograph can be typed by hand on an ordinary typewriter or produced on a thermal stencil maker [K]. A high-speed stencil duplicator [F] can reproduce these much faster than can a mimeograph. Litho plates can also be made on a typewriter, and an electrostatic litho-plate maker [J] can produce them from drawings and photographs. Also shown here are an electronic machine for making mimeograph masters [G], a guillotine [C], paper punch [D] and binder [E]. A well equipped print room can copy any documents, drawings, pages from books or any other originals, and it can print circulars, memoranda, publicity material, invitations, house journals, technical reports and even good quality letter headings. Among other methods of graphic reproduction (not shown) are microfilm systems. There are also titling systems such as adhesive transfer lettering and machines for printing serial numbers.

than any of the other copying processes.

The introduction of photostatic copying machines in the mid 1960s [4] has resulted in a minor revolution in the copying field. The term xerography, which was based originally on a proprietary name, has come into general use for processes marketed by many companies. Photostatic copying depends on the phenomenon of photo-conductivity [5].

Offset lithography is another process widely used in modern offices. The office offset machine uses the principle employed in commercial lithography – the fact that water and an oil-based ink will not mix. The offset-litho masters can be prepared in several ways: photographically, as in commercial printing; by the use of a suitable toning powder in a photostatic copier; or simply by typing or drawing with a special ribbon or ink direct on to plastic or thin metal plates.

Preparation of copy
The first step in the traditional printing process is the setting of movable type – a complicated procedure ill-suited to the office print room. It was the possibility of preparing printing masters on standard typewriters that made the hectograph and the mimeograph, and later the offset-litho press, practical alternatives to commercial printing.

The standard typewriter, however, has three drawbacks. The quality of the typed copy is variable in intensity and clarity, the typeface cannot be altered and the letter spacing is constant, unlike that of movable type where "m"s and "w"s, for example, occupy much more space than "i"s or "l"s.

All three weaknesses have been overcome with the invention of more sophisticated typewriters. The modern electric machine, with a carbon or plasticized ribbon used only once, produces uniformly clean and clear jet-black letters. Machines with interchangeable typefaces have also been introduced. Finally machines with "proportional" spacing were invented, so that narrow, medium and wide letters do not have to occupy the same space. The most advanced machines incorporate the alternative type facility as well as proportional spacing and have special spacing facilities for the typing of fixed-width columns.

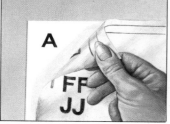

The original blue-print used by architects and engineers for the reproduction of plans and drawings is a photographic negative, white on blue, made on paper sensitized with a ferric (iron) salt and potassium ferrocyanide.

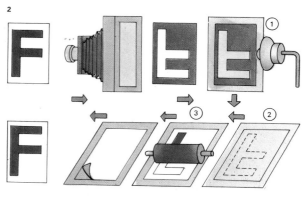

2 The original spirit duplicator used a gelatin or clay bed which absorbed some of the aniline ink with which the master copy [1] had been prepared. Some of the ink mirror image in the surface of the "jelly" bed [2] was then transferred back to sheets of paper that had been moistened with spirit and pressed on to it by means of a rubber roller [3]. As aniline inks of various colours could be used on the master the flat bed hectograph could reproduce several colours simultaneously. Its disadvantage lay in the fact that only short runs – fewer than 100 copies – could be printed.

3 Adhesive transfer lettering, which is available in a wide range of typefaces and sizes, has made possible the preparation of titles quickly and cheaply, complementing the modern electric typewriter's function of producing near-perfect printed material. Line borders, human figures in various attitudes and sizes for use in simple artwork, electronic symbols, mathematical signs and a wide variety of other symbols and colours are produced in the form of adhesive transfers. Transfer lettering is produced on a transparent sheet; letters are placed in the position required and rubbed on to the paper below where they remain fixed after the sheet above is removed. "Instant" lettering is ideal for small-scale print operations at a relatively low cost.

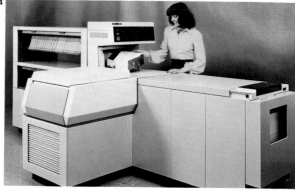

4 Modern photostatic copying machines are produced in a wide range of sizes and types. Table-top machines, designed specifically for use in offices where noise must be kept to a minimum, require no expertise on the part of the operator. A dial is set to specify the number of copies and the original is then fed in through the slot. The sheet trips a micro-switch which sets the process in motion. The copies emerge singly, followed by the original. Floor-standing copiers usually have a flat bed for the original, making reproduction from open books possible without the removal of separate sheets. Enlargement or reduction of the image is possible on the most comprehensive copiers.

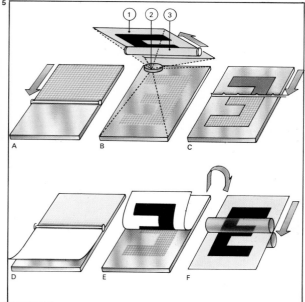

5 The photostatic copying process uses a metal plate (which is often wrapped around a cylinder) coated with a material that conducts electricity only when exposed to light. In this process: the plate is first given an overall charge [A]. Then an image [B] of the original [1] is focused on to the plate using a lens system [2] and a light source [3]. Where the light falls the charge is conducted away. [C] Toning powder is now dusted over the plate. This adheres to the charged image area but not elsewhere. [D] Resin-coated paper is pressed in contact with the plate. [E] The powder adheres to the paper. [F] The image is fixed by heat.

Newspapers and magazines

Newspapers that can be properly so called – because they contained topical information with an attempt at regular appearance – date from the early 1600's. Perhaps the first was the *Nieuwe Tijdingen*, published in Antwerp from 1605.

Early English newspapers suffered heavily from censorship. A government-issued licence to print, which could be revoked if a newspaper offended the government of the day, acted as an effective curb until the Licensing Act was allowed to lapse in 1694. The relaxation led to a host of papers and journals and Britain's first daily newspaper appeared on 11 March 1702, priced at one penny. Papers were printed by hand and circulations were small: in 1711 total daily sales were only 7,500 papers. Not until *The Times* of London introduced a steam printing press in 1814 could the modern newspaper be said to have arrived.

Birth of popular papers

Stamp tax, introduced in 1712, also severely restricted newspaper circulations. In 1836 it was reduced from 4 to 1 old pence (d) a copy and new Sunday papers, carrying accounts of crime, sex and sport, and appealing to a new type of reader, began to make their appearance. The *News of the World*, the largest selling Sunday newspaper in the Western World today, dates from this period. In 1855 the tax was abolished and several more "middle-class" dailies were established.

The 1870 Elementary Education Act, which increased literacy among the British working and lower-middle classes, stimulated further growth. Alfred Harmsworth, later Lord Northcliffe (1865–1922), launched the *Daily Mail* in 1896, priced at ¹/₂d. The first issue sold 390,000 copies, more than any other paper had ever sold in one day. The "popular" paper had arrived.

Editorial content

Modern newspapers should not be regarded merely as a public utility. Many are manufactured "goods", produced to make a profit. They get their income from selling the paper itself [Key] and by selling space in it to people who wish to advertise goods or services. Some, however, are produced at a financial loss as a public service, supported by other activities or by the proprietor.

The editorial content of all newspapers can be divided into two major sections: news and features. Most news is predictable: it comes from events that are known about in advance – the so-called "diary" events such as sessions of Parliament, court cases, athletic meetings and so on. But the news that the reader often finds most fascinating, and that may lead him to buy a newspaper because he catches sight of a headline, is "spot" news – the event, such as a presidential assassination, that cannot be predicted. Leading articles ("leaders" or "editorials") are generally based on a principal news item.

Features include articles giving the background of news; gossip columns; articles on specialist subjects such as gardening, fashion and cooking; strip cartoons and horoscopes.

All newspapers report "spot" news, but the quality newspapers carry much greater coverage of serious diary news – especially politics – than popular papers and supplement it with articles by specialist writers on science, medicine, education and current

1 Birth of a story in a newspaper.
7 pm: News of a fire at a politician's home comes over the teleprinter from the Press Association, an agency with a nationwide network of reporters. The story, or "copy", is assessed or "tasted" by the "copytaster" [1] who informs the news editor [2] and night editor [3]. They send a reporter [4] to the scene and the picture editor [5] sends a photographer [6]. 7.15: A reporter telephones that the man has died. Two reporters and a photographer are sent to assist in covering events. The "back bench" – the night editor and assistants [7] – decide that the story is the main front page news, the "splash". A staff writer [8] is briefed to compile the man's obituary, using reference books, contacting his friends and studying newspaper cuttings from the paper's own library. Pictures of the man are obtained from the picture library and an artist [9] draws a map of the house and grounds. 8.00: A reporter returns to write his story. It is a "running story" – as he writes, his colleagues telephone further information. A photographer rushes to the darkroom to develop his film. 8.15: With the designer [10] the back bench choose a picture and draw, or "lay out", a page, showing sizes of illustrations and headlines, and lengths of stories. 8.40: The reporter's copy is scrutinized by news and night editors. 8.50: Copy and scheme go to the chief sub-editor [11] who briefs a sub-editor [12]. He shortens the reporter's copy to the number of lines allocated, adds in late news, rechecks facts and writes a headline of the prescribed number of letters. His work is vetted by the revise sub [13] before it goes to the printer.

affairs. They also carry several pages of financial and industrial news. Popular newspapers report only the most arresting diary news, such as Parliamentary disagreements.

The task of a newspaper editor is to direct his journalists to produce stories and pictures that together strike a suitable balance between information and entertainment for his particular readers. On a mass-circulation newspaper he has assistant editors in charge of the various sub-sections – news, foreign news, sport, women's pages, leisure, business and general features. Each assistant has his own team of journalists.

The total number of pages in each issue of a paper is determined by the amount of advertising matter. Each morning the editor studies a "mock-up" of the paper that indicates the pages and positions of the advertisements. Knowing the space available, he then holds a news conference to deal with stories that are expected to develop during the day. Information comes from staff, independent or "freelance" journalists, or, by way of national and international teleprinter networks, from news agencies. At a later confer-

ence the editor receives progress reports and finally decides which items deserve prominence and which can be put on one side.

The editor holds features conferences to discuss immediate ideas arising out of the news and those for future issues. At a leader conference (particularly on quality papers) he discusses with senior staff the attitudes to be taken to current affairs.

The magazine world

Periodicals focusing on narrow fields – such as gardening, homes, stamp collecting, motoring and chemical engineering – are prospering, as are low-priced women's magazines. But general magazines have suffered from competition from television and from magazines given away with newspapers. In particular, magazines of photo-journalism – features or news stories told in pictures – have all but vanished in all Western countries. Successful and comparatively new fields are glossy magazines for the newly rich teenagers and prestige journals distributed free to professional people such as doctors in order to attract lucrative advertisements.

Most sales of British newspapers and magazines are through newsagents. Between 50 and 80 per cent of newspaper sales are home deliveries – the rest are counter sales. A newsagent's profit is up to 28 per cent of the selling or "cover" price. Some local newspapers deliver direct to newsagents but national papers go first to area wholesalers who take about 10% of the cover price. Newsagents operating on "sale-or-return" send back unsold copies to the wholesaler. To lessen this possible loss, some newspapers ask agents to place firm orders and bear any loss themselves – but this can lead to under-ordering, a shortage of papers and sales for rivals.

2 Linotype machines convert stories into lines of metal type. The operator sits at a keyboard like a typewriter and re-types the edited copy. The keys release tiny moulds – matrices – of the letters, which form into a line. Finished lines move to a slot into which molten metal is pumped. A strip of metal – a "slug" – is formed with raised letters on its surface. Lines gather in a tray until the copy ends.

3 Pages are assembled on a metal workbench called a "stone". Following the editorial scheme, a compositor arranges the columns of type with the pictures (etched into metal blocks). The whole page is then locked into a "chase" and a proof copy taken. A journalist – the "stone sub" – reads proofs of each page and indicates which lines can be omitted if the metal does not fit. He also "passes" the final page.

- - - - Fold of spread
☐ Pictures
⊠ Advertisements
☐ Headlines
☐ Text

4 A typical tabloid newspaper contains 30 to 45% advertising and 55 to 70% editorial matter. Half the editorial is text (nearly a quarter headlines) and the rest pictures. Most pages carry an advertisement, except for information pages such as television or racing; without them pages look dull. Pictures and headlines are placed to counterbalance advertisements and some pages contain nothing but advertising.

5 A newspaper's income [A, B] is derived from advertisements and selling the paper. This chart assumes no profit. It shows [C] paper (called newsprint) and ink as the main costs and demonstrates how popular papers selling millions of copies depend more on sales, whereas quality papers selling perhaps 400,000 copies a day rely on advertisements (particularly classified). In local "giveaway" (free) newssheets, profit is derived solely from advertising content.

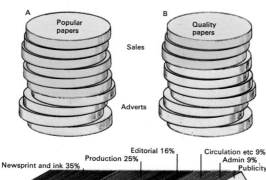

Popular papers
Quality papers
Sales
Adverts

Newsprint and ink 35%
Production 25%
Editorial 16%
Circulation etc 9%
Admin 9%
Publicity 6%

6 Future newspaper production will be by computer, with typesetting machines phased out. Writers sit at visual display units (VDUs) which are keyboards with cathode-ray tubes linked to a computer. By tapping a key a letter shows on the screen. The finished stories are kept in the computer under code numbers until an editor calls one on to his own screen for amendments. The computer can then instantly produce a tape that works a typesetting machine.

Books and publishing

"Another dammed thick, square, book! Always scribble, scribble, scribble, eh, Mr Gibbon!" So observed the Duke of Gloucester to the greatest of the 18th century historians, Edward Gibbon (1737–94). A hundred years later the greatest 19th century historian, Thomas Carlyle (1795–1881) refuted the implication: "A good book is the purest essence of a human soul."

The United Nations Educational, Scientific and Cultural Organization (UNESCO) defines a book as "a non-periodical printed publication of at least 49 pages, exclusive of the cover pages". In 1976 Britain alone published more than 30,000 books. They included everything from cheap, read-and-throw-away paperback novels to finely produced, limited editions of the classics. In the total were more than 1,000 works on religion, 2,000 on politics, 2,500 on natural sciences and 7,000 on literature. School textbooks and children's books accounted for more than 4,000 titles.

In spite of television, books are still the greatest medium for education, influence and entertainment that the world has ever known: probably as many square miles have been conquered by books such as Karl Marx's *Das Kapital* as by guns and tanks. And when the vision of an ideal state of society called communism contained in that book became perverted and distorted, one of the greatest instruments in the fight against it was another book, *Animal Farm* by George Orwell [2].

The business of publishing

Books like Gibbon's histories and Orwell's satires are grouped by publishers under the general term "trade books", that is books sold to the general public. They are a product of which the publisher is the manufacturer and the bookseller the retailer [Key]. To a businessman, they are exactly the same as mousetraps or motor cars and their sales are promoted in similar ways: just as the car industry has its annual trade shows, so the book business has its books fairs [6] at which both publishers and public can inspect the latest product.

Samuel Johnson (1709–84) said that nobody except a blockhead ever wrote except for money and with some modification the same could be said of publishers. But most established publishers are prepared to publish unknown authors – and make a financial loss – if they feel that the authors will one day become well known and therefore profitable. The initial losses are offset by profits from other books they publish. Publishers are in a way schizophrenic: Raymond Chandler (1888–1959) said of the publisher that "the minute you try to talk business with him he takes the attitude that he is a gentleman and a scholar, and the moment you try to approach him on the level of his moral integrity he starts to talk business".

The cost of producing a trade book is generally multiplied by about four to get the retail price. This "mark-up" allows for the publisher's profit, the author's royalty, and the bookseller's margin (which is normally 35 per cent of the retail price). Authors' royalties normally begin at 7.5 or 10 per cent and may increase on a sliding scale against the number of sales to a maximum of 15 per cent. Most of a publisher's outlay – about 80 per cent – is spent on typesetting, paper, printing

1 *The Godfather* by Mario Puzo was written originally in English and has been translated into most major languages. It has become one of the most successful fiction books of recent years.

2 George Orwell (1903–50), whose real name was Eric Blair, was a prolific and successful British author of more than ten best-sellers, including *Animal Farm* and *Nineteen Eighty-four*, as well as many essays.

3 Among the most successful books in terms of numbers printed and sold are textbooks and other books used in schools and colleges. An elementary book on mathematics, for example, may have several successive editions and reprints and can sell in large numbers for many years. In the past, books for classrooms were "hardback" with a stiff cloth-covered binding. But today the rising cost of books has forced publishers increasingly to produce paperback books for schools. Many advanced textbooks are available in both hardback editions (favoured by libraries) and as paperbacks (generally bought by students). Some never appear in hard covers.

4 High-quality books are produced by hand one at a time. Here the British craftsman Grahame Clarke is seen inking an etched metal plate [A] for an illustration, then [B] inking the type matter using a roller before [C] arranging the plate and type on a forme prior to printing. Sheets of paper for the book's pages are printed [D] in a flat-bed press and then any hand lettering carefully drawn in position [E]. Finally the book is bound; the pages are stitched together [F], the boards forming the covers are added and the cover bound in cloth or leather. Few people have all the skills needed to make books in this way and their products command very high prices.

and binding. The publisher takes a risk: if his estimate of sales proves wrong, he makes a loss. The author of this type of book is also at risk; if the book does not sell, many months or even years of work yield no income.

Compared with the production of illustrated books, publishing "text only" books is relatively straightforward. The author's manuscript is first edited for house style (the way in which the publisher chooses to present, say, numbers – that is, to print them as figures or to spell them out). It is then cast off – the process of estimating how many pages it will occupy when printed in the chosen typeface. The typeset "galley proofs" are checked for errors and marked up to indicate for the printer how the pagination of the text should be arranged. The printer "imposes" the pages in such a way that when a large printed sheet is folded [5] the pages fall into the correct sequence. These folded sheets (called "signatures" and usually 16, 32 or 48 pages long) are gathered together, sewn and bound. Most paperback books are not sewn but instead "perfect bound" – the folds are trimmed off and the back of the book held together, in its cover, with a special glue.

A successful textbook [3] can be highly profitable to both author and publisher. Britain's bestselling textbook author, Ronald Ridout (1916–), sold 50 million copies of his works in 25 years. Some novelists do even better: the Belgian writer Georges Simenon (1903–) has sold 300 million copies of his books. Story books for children also command large markets, generally about 8 per cent of all titles. Other main categories are fiction (25%), natural sciences (8%), history (7.9%), the arts (7.6%) and politics (7.4%). But the Bible is the best seller of all time: since 1800 about 1,500 million copies have been printed in various languages.

Home-made books
At the other extreme of the publishing business are those "cottage industry" publishers [4] who put as much art into production as the author puts into his text, making their own paper, cutting their own type and printing on small hand presses. Their books may make a reasonable profit, but they are not "economic" in the trade publisher's sense.

A large general bookshop, such as Foyles in London, may occupy several floors and have up to 750,000 volumes on display. Smaller shops may specialize in books on only one subject.

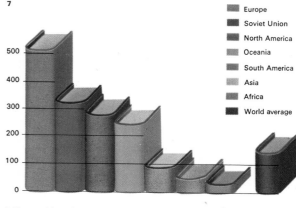

5 Modern bookbinding is carried out almost entirely by machines. The pages of a book are printed on flat sheets of paper, often eight pages at a time on each side of the sheet. Each sheet is folded [1] to give a "signature" (in this example, of 16 pages). All the signatures for a complete book are collated (put in page order) and sewn [2]. A heavy clamp [3] expels air from the sewn book, which is then trimmed to the finished size [4]. Net fabric called mull is glued to the spine [5] and the back of the book is rounded [6]. Colour is added to the edges of the pages at this stage. Endpapers are attached and the book fixed into its covers or "case" [7]. The title may be blocked – impressed into the cover [8] – and often a printed dust jacket is added to the finished book. Books are packed into boxes for distribution.

6 The Frankfurt Book Fair is held annually in the autumn. It is the principal market place at which publishers display existing and planned products and try to sell them to the book trade. International co-operation between publishers at the planning stage enables one book to be produced at the same time in several co-editions that differ only in the language in which the text is printed. And the same book that is printed as a single volume for one country may be produced in several volumes for another and as a weekly partwork for yet another; or a hardback book in one country may be a paperback in another, depending on the particular requirements of each market.

7 The world market for books can be stated in terms of the number of new titles published each year. This diagram shows the number of titles per million inhabitants in each of the major continents in 1972. More books were produced in Europe than in any other continent, with the Soviet Union and North America the second and third largest producers. The Russian total (325 per million), however, changed little over a period of ten years while those of Europe and North America increased steadily. The figure for Asia, about 48 new titles per million inhabitants, has remained virtually unchanged for a period of 20 years.

Europe
Soviet Union
North America
Oceania
South America
Asia
Africa
World average

500
400
300
200
100
0

Reference books and encyclopaedias

One effect of the explosion of human knowledge during the twentieth century has been to make a good reference library a necessary part of civilized life. The essential job of encyclopaedias, dictionaries and other reference books is to summarize knowledge in an easily accessible and comprehensive form. Encyclopaedias are about facts; dictionaries about definitions. Some encyclopaedias – and *The Joy of Knowledge* is one – also try to explain the more complex words they use, thus entering the intermediate category of the encyclopaedic dictionary.

Early encyclopaedias

The desire to summarize all human knowledge has a long history; the earliest surviving encyclopaedia in the West was compiled in Rome by Varro in the first century BC and in China encyclopaedias were in existence a thousand years before that. Varro's work was followed a century later by the *Historia Naturalis* of Pliny the Elder (AD 23–79). Like all early encyclopaedias, Pliny's work is arranged by subjects, as is the *Etymologiae* [1] of Isidore of Seville (560–636).

Pierre Bayle's *Dictionnaire Historique et Critique* (1695–97) attempted to give a comprehensive summary of human knowledge and appeared in many editions and translations. In 1704–10 John Harris (*c.* 1667–1719) brought out his *Lexicon Technicum*. The first true encyclopaedia in English, it was compiled at the request of the Royal Society and has a noticeable scientific bias. It was followed in 1728 by the *Cyclopaedia* of Ephraim Chambers (died 1740). This fully cross-referenced work was the model for all subsequent alphabetical encyclopaedias, including the largest European encyclopaedia ever published – the *Universal Lexicon* [2] of Johann Heinrich Zelder (1706–70).

The most famous foreign successor to Chambers' work was the *Encyclopédie* [3] (1751–65) of Denis Diderot (1713–84) and Jean le Rond d'Alembert (1717–83). Their approach was unashamedly radical and their enthusiastic humanism contributed to the undermining of the *ancien régime* in France. The original three-volume *Encyclopaedia Britannica* (1768–71), was conceived in part

as a more objective answer to the *Encyclopédie*. Always a monumental work, *Britannica* has grown through 15 editions to the completely reorganized 1974 edition.

Alphabetical order of subjects did not become standard practice until the late Middle Ages; even today some valuable encyclopaedias, such as the *Encyclopédie Française* (1937–) and the *Oxford Junior Encyclopaedia*, arrange their material in a logical sequence from subject to subject.

Advent of dictionaries

Without alphabetical order, dictionaries would be impossible to use. They are comparatively recent innovations. John Florio's *World of Words* (1598) was the first Italian-English dictionary, and Robert Cawdrey's *Table Alphabeticall* (1604) explained some of the learned "inkhorn" terms that had come into English from Latin.

In the eighteenth century the first modern dictionaries saw the light of day. In 1721 Nathan Bailey (died 1742) issued his *Universal Etymological English Dictionary*, which remained a bestseller for more than a

1 **Isidore's *Etymologiae*** was a compilation of seventh-century knowledge. This "Tree of Knowledge" is taken from an illustrated manuscript copy. The book was arranged thematically and included liberal arts, medicine, law, a timechart, the Bible, the Church, people, language, statecraft, a Latin dictionary, man, zoology, heaven, air, seas and oceans, geography, cities and towns, building, geology, weights and measures, agriculture, ships, houses, dress and costume, food and drink, tools, and furniture. Isidore remained a standard work of reference for 1,000 years. Today it provides us with the best view available of the life and thinking that prevailed during the Middle Ages.

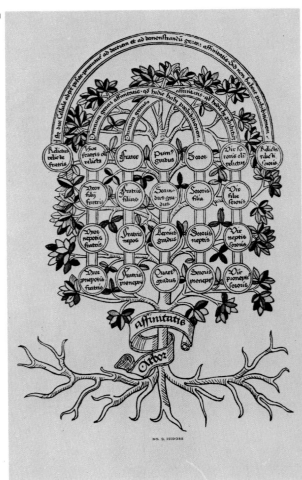

NO. 9, ISIDORE

2 **Zedler's *Lexicon*** or *Great Complete Universal Dictionary of all the Sciences and Arts* was particularly strong on biography, genealogy and topography. Zedler a Leipzig bookseller, claimed that his work was more comprehensive and complete than any previous encyclopaedia. His work prompted rival ones that later put him out of business.

3 **The *Encyclopédie*** was first planned as a French edition of Chambers' *Cyclopaedia*. Following disagreements over rights, the publisher hired Denis Diderot, who had just edited the *Dictionnaire de médecine* to compile a new work. Diderot enlisted the aid of his friend Jean le Rond d'Alembert, a brilliant and famous mathematician who wrote the "preliminary discourse" in 1751. They recruited the best scholars and the most distinguished *philosophes* and gained support from the leading *salons*. Each volume as it appeared caused a sensation throughout Europe. The Establishment was outraged and the number of subscribers rose from 1,000 to 4,000. In 1759, a year after d'Alembert withdrew, the work was banned by the French attorney-general, but the publisher, Le Breton, continued it as an "underground" publication. There were numerous pirate editions.

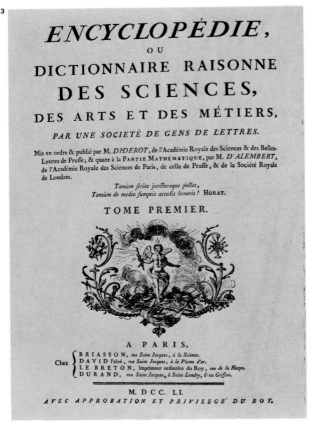

century. It was the first attempt to collect and define *all* English words, as opposed to just "difficult" words, and it was the foundation on which Samuel Johnson (1709–84) based his more famous English dictionary. Johnson's chief innovations were the introduction of illustrative quotations to demonstrate shades of meaning and styles of usage, a willingness to make judgments about levels of usage ("clever, a low word..."), and some famous jokes among the definitions ("lexicographer, a harmless drudge").

Among Johnson's successors was the cantankerous American patriot Noah Webster (1748–1843). His *American Dictionary of the English Language* (1828) first listed most of the spelling differences between British and American English.

The best modern encyclopaedias expend great efforts to keep abreast of developments in all fields of knowledge and to present information in a format that will be most useful to the reader. Major dictionaries now adopt one of two approaches to language. The first, exemplified by the great *Oxford English Dictionary* (1884–1933), is the

historical approach: through a systematic reading of the literature of a language, the history of each word is given – the *OED* also gives dates and examples of uses. In the other approach – the "synchronic" method – the lexicographer sets out to observe and record the language in its contemporary form. The most extreme example of this method – *Webster's Third New International Dictionary* (1961) – upset many people by its including such words as "ain't".

New works of reference
In the future, printed reference books will undoubtedly be supplemented by more technologically advanced ways of storing and retrieving information. Computer data banks can hold an enormous amount of information in a way that enables it to be instantly available and readily updated. Audio-visual techniques are now so far advanced that *The Joy of Knowledge Library* was itself carefully planned so that its printed pages could be related to general and educational films and audio-visual discs that could be shown on domestic and school television screens.

A general encyclopaedia is the ideal reference work for helping young people with school work.

4 *The Joy of Knowledge Library* was produced in an unusual way, and text and pictures were assembled in various stages. After an author had been asked to write about a topic, he attended a briefing [A] at the publisher's office, together with the section editor and art editor. The artist drew a rough layout indicating the size, position and content of all the pictures. The author then researched and wrote the text [B] while an artist prepared a page layout [C]. A picture researcher obtained photographs [D] from agencies and photographers, or photographs were taken specially for the book. Artwork – paintings and drawings – was produced by a team of artists [E]. An editor prepared the article for publication and ensured that it was consistent with other articles. Each author's edited manuscript was then submitted to a leading authority in the field, to make absolutely sure of the accuracy of all the facts. Printed proofs were checked [G] and assembled into a "paste-up" of the page [H]. The text was "married" with the colour pictures and a final full-colour proof was checked for accuracy and quality [I]. This process was repeated for every one of the major topics in the *Library*. Finally, cross-references were added to link various topics in related articles.

Information retrieval

In the Middle Ages, man's knowledge was restricted mainly to philosophy, history, medicine, astronomy and, to a very limited extent, geography. The boundaries of scientific and technical information were limited. Disregarding guesswork and unproven theories, it would have been possible at the time of Magna Carta to record all knowledge in a few hundred books.

The early recording systems

The volume of recorded knowledge grew slowly at first but, after the second half of the eighteenth century, men began to conduct systematic scientific research in many fields. Soon no one person could hope to read, understand and remember all recorded knowledge and instead people began to specialize in particular fields of study. The expansion of scientific and technical knowledge into new areas, each producing its own "literature", led to a rapid explosion of scientific information. Even modest libraries found it impossible to cope with the flood of new information.

The new problem led to the evolution of a new expertise called information science – the science of storing, organizing, retrieving and disseminating information in such a way that anyone who knows the system is able to identify and retrieve all available information on any specific subject as and when it is needed. There are now many aspects of information retrieval including market research, mail-order selling and population censuses. Its principles can best be illustrated by considering the problems involved in a large library of books.

The first successful system of information classification was the decimal system invented by an American. Melvil Dewery (1851–1913), and first published in 1876.

The Dewey classification system

Dewey divided "knowledge" into ten main numbered classes: 100 Philosophy; 200 Religion; 300 Social Sciences (including Economics); 400 Language; 500 Natural Science; 600 Useful Arts (now Technology); 700 Fine Arts; 800 Literature; 900 History (including Travel and Biography); and 000 General Works (covering subjects not fitting into the other nine categories). Each main class was subdivided into ten sub-classes (indicated by the second digit of classification number), and each of these into ten further sub-classes (third digit). Dewey made provision for even greater flexibility by adding a decimal point after the first three figures. The virtue of the system lies in the ease with which books can be given a numerical order on library shelves, so that related information is grouped together.

The Dewey system was widely accepted, but had one major drawback. Many books can be classified in more than one way and the Dewey system made little provision for this. To make it possible for a book or document to be given a classification number that indicates its subject-matter more precisely, the Universal Decimal System (UDC) was developed. This used the Dewey system as its basis and was first published by the International Institute of Bibliography, in French, in 1905. The Institute (which later became the International Federation of Documentation) has developed and widened the scope of the UDC progressively ever since.

CONNECTIONS

See also
108 What computers can do
214 Reference books and encyclopaedias
232 Sound recording and reproducing
234 Video recording and reproduction

In other volumes
44 Man and Society

1 This carpet loom is controlled by punched cards. It is based on a silk loom invented in 1801 by the French engineer Joseph Jacquard (1752–1834). Compressed air passes through the holes in the cards, joined to form an endless loop, as they pass over a perforated roller. Depending on the positions of the holes, the air operates mechanisms that lift the lengthwise warp threads of different colours. In the modern machine, push rods are used instead of compressed air to "read" the positions of the holes. Single punched cards, "read" by a lamp and photocell arrangement, are part of many modern data processing and information retrieval systems.

2 The punched card is the basis of most electro-mechanical data processing systems. Information is stored on the card by punching holes in specific locations according to a predetermined code [A]. In a population census, for example, there would be one card punched for each completed census form. The card sorter [B] uses electrical contacts to sense the holes and thereby sort cards into predetermined groups at high speed. For example, the sorter could sort cards into ten different age groups by sensing the "age" hole on each card, this being carried out in a fraction of the time possible by hand.

3 A computer display screen can be used to provide information about the stores of an engineering works. Suppose the store-keeper takes delivery of twenty 2m lengths of steel rod and wants to correct the records. The computer gives the "Stores" display [A], from which he selects "Metal" by pushing button number 1. On the "Metal" display [B] he selects "Rod"; from "Metal rod" [C] he chooses "Circular" and from the available types of circular metal rod [D] selects "Steel" by pushing button number 3. From the next display [E] he selects the 10mm size, and the following one [F] informs him that the present stock is 9 2m lengths. He feeds in the new stock [G] and checks the total stock situation [H].

3

A STORES	B METAL	C METAL ROD	D METAL ROD, CIRCULAR
1 Metal	1 Sheet	1 Hexagonal	
2 Plastic	2 Bar	2 Circular	1 Brass
3 Fixings	3 Rod		2 Aluminium
4 Electrical	4 Angle		3 Steel
5 Tools	5 Tube		4 Copper

(1) (2) (3) (4) (5)   (1) (2) **(3)** (4) (5)   (1) **(2)** (3) (4) (5)   (1) (2) **(3)** (4) (5)

E METAL ROD, CIRCULAR STEEL	F METAL ROD, CIRCULAR STEEL, 10mm	G METAL ROD, CIRCULAR STEEL, 10mm	H METAL ROD, CIRCULAR STEEL, 10mm
1 5mm	1 Update	1 Update	1 Update
2 10mm	2 File	2 File	2 File
3 25mm	3 Read	3 Read	3 Read
4 50mm	9 × 2m	20 × 2m	29 × 2m

(1) **(2)** (3) (4) (5)   **(1)** (2) (3) (4) (5)   (1) (2) **(3)** (4) (5)   (1) **(2)** (3) (4) (5)

When the US Library of Congress moved to new premises in 1897 it was already 97 years old and held about 1.5 million volumes and documents. The directors considered using the Dewey system but finally developed its own. In this, the LC system, subject-matter is first divided into 21 classes bearing the letters of the alphabet from A–Z, excluding I, O, W, X and Y. Each main class is then subdivided, each sub-class again bearing a letter. In this way the classification of every book or document begins with two letters and these give a fairly clear indication of the nature of the contents. The two letters are followed by figures, usually three or four, which are so devised to give the maximum detailed information on the contents of the book. Numerical sub-classes, where appropriate, are further subdivided by topics. Many advantages are claimed for the LC system; for example, it is easy to use.

Other classification systems

Other forms of information classification have particular advantages and are used in modern retrieval systems. Faceted classifica-tion recognizes that most subjects are com-pounds of several elements and seeks to enumerate the principal elements in every classification label. This system has not only been shown to be flexible but also to provide, if well conceived, a precise indication of, say, the contents of every book or document. Faceted classification is also easily adapted to use in punched-card information retrieval systems [2]. By defining all the "facets" of a subject on which information is required, a punched-card sorting machine can quickly identify all information cards bearing those facets in its classification label.

Information retrieval is an area in which computers are increasingly being used. As long as the "information" can be coded in a form that can be stored in a computer's memory banks, it can quickly be retrieved merely by calling it up using an appropriate code addressed to the computer. In addition, computer information can be continually updated, cancelling outdated material and replacing it with fresh data. The updating process in books and most other publications has to await complete new editions.

A record storage corridor in a US Patents Office, photo-graphed here, gives a vivid impression of the steadily growing problem of information storage and retrieval. When an application for a patent is made, the authority must be sure that the idea is genuinely new. A century ago the descriptions of all existing patents could easily be filed in a single room and classified in such a way that the validity of new applications for patents could be quickly established. Today the number of current patents is so large that the problem of validating these applications in the old way has now become excessively time consuming.

4 Libraries of rare books must guard against losses and theft. The Victoria and Albert Museum Library in London [A] uses a system under which all books are guarded [B]. A user enters his re-quest (author and title) on a form in duplicate, and hands this to an attendant. The latter files one copy, takes the book from the shelf (where the other copy is lodged in its place) and gives the book to the reader.

5 Large airlines operate 50 or more airliners flying daily on routes all over the world, and passengers may join or leave flights at 20 or 30 different major cities. The airline's inform-ation problem is to know instantly, at all booking offices, how many seats are available on each plane along each section of the route. This is overcome by having a computer at the head office with terminals at every booking office. The computer memory has a separate sub-store for every sec-tion of every flight and keeps a running total of the number of passengers booked for each section. If a proposed booking will exceed seat capacity on any section the computer informs the booking clerk instantly.

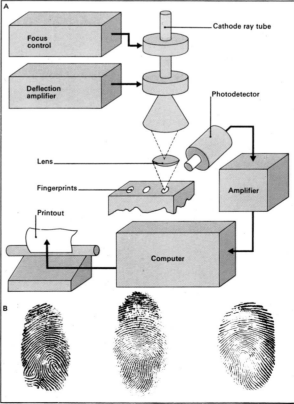

6 In the early days of police science the criminal invest-igation departments of even the largest countries held only a few thousand sets of fingerprints. Today the number filed at London's Scotland Yard is counted in millions. When a crime has been committed and fingerprints found at the scene, it may take days of search-ing to establish whether the prints belong to a known criminal. To solve the problem, scanning devices [A] are being developed that will provide information that can be read by a computer. When this new technique has been perfected, all fingerprint sets held by a national CID will be filed in a single computer store and it will be possible for the computer, if "shown" a print found at the scene of a crime [B], to compare it with every print in its store within a minute or two and come up with references to any matching fingerprints.

Photography

In just 150 years photographs, and allied processes such as photocopying, have become almost indispensable. Where words struggled to convey reality, and paintings could not capture the fleeting moment, the photograph came to achieve both, presenting in one dramatic image instant history, technical detail, or profoundly moving emotion. "Photography" comes from Greek words meaning "writing with light".

Development of photographic processes

The first photograph was taken by Nicéphore Niepce (1765–1833) in 1826 in a "camera obscura". This was a small dark room with a lens let into one wall: the scene outside was projected on to the opposite wall. It took about eight hours to expose a pewter plate covered with chemicals that reacted to light. In 1837, a Frenchman, Louis Jacques Mandé Daguerre (1789–1851), invented the daguerreotype process in which silver-plated copper sheets treated with iodine vapour replaced Niepce's pewter plate.

Development of the latent image (the picture held invisibly in the chemicals) required treatment with mercury vapour. The visible image produced was reversed, with the black parts white and the white parts black. The process became popular, but produced only one picture at a time. The multiple prints of today were impossible until William Fox Talbot (1800–77) developed the negative in 1839. But his calotype (later talbotype) was not as clear as the daguerreotype.

Talbot's paper negatives could be made transparent by wax or oil. They were soon replaced by glass negatives and in 1861 the first colour process was demonstrated.

The modern camera

In 1888 George Eastman (1854–1932) introduced the Kodak camera and brought photography to the man in the street. In 1924 the Leica was introduced as a miniature camera originally designed for testing 35mm motion picture film, and in 1925 the invention of the flash bulb released photography from dependence on sunlight or special artificial lighting.

The modern camera [Key] is a light-tight box with a mechanism that holds a piece of film flat and opposite a lens. The lens focuses on to the film a sharp upside-down image of the scene before the camera. A shutter, situated between the film and the lens, stops light reaching the film until the user decides to take a photograph. It then opens, usually for only a fraction of a second. The correct exposure is obtained by regulating the relationship between the shutter speed and the diameter of the lens – a factor that can be varied by adjustment of a diaphragm.

The diaphragm controls the amount of light passing through the lens and the shutter determines how long the film is exposed to the light. A fast shutter speed and a wide diaphragm opening will give the equivalent exposure of a slow shutter speed and a small opening. All cameras have a viewfinder – from a simple wire frame to a complex optical system – that enables the user to see what the camera "sees".

These features are common to all cameras, but there is great variation in their complexity and operation. The simplest cameras [3] have a single shutter speed and fixed diaphragms, both so chosen that on a

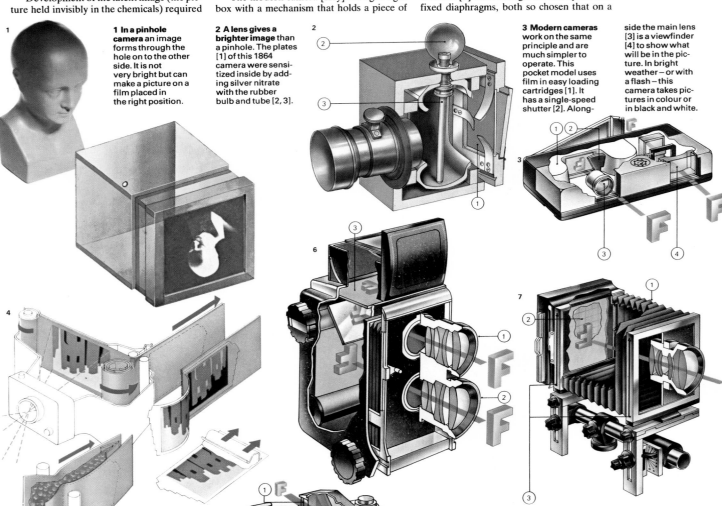

1 In a pinhole camera an image forms through the hole on to the other side. It is not very bright but can make a picture on a film placed in the right position.

2 A lens gives a brighter image than a pinhole. The plates [1] of this 1864 camera were sensitized inside by adding silver nitrate with the rubber bulb and tube [2, 3].

3 Modern cameras work on the same principle and are much simpler to operate. This pocket model uses film in easy loading cartridges [1]. It has a single-speed shutter [2]. Along- side the main lens [3] is a viewfinder [4] to show what will be in the picture. In bright weather – or with a flash – this camera takes pictures in colour or in black and white.

4 Polaroid cameras take composite "film and paper" packs which incorporate the necessary processing chemicals.

5 The single-lens reflex is a versatile camera design. The viewfinder [1] reveals an image formed via a mirror [2] by light from the main lens [3], showing exactly what is being focused upon.

6 The twin-lens reflex camera is really two cameras, one for focusing and viewfinding [1] and one for photography [2]. The lenses focus with a focusing screen [3].

7 The technical camera does not have a viewfinder. The image is focused by means of bellows [1] on a ground-glass sheet [2] at the back. Just before exposure this is replaced by a piece of film held in a dark slide. The front and back panels can both be tilted, shifted or swivelled independently [3]. This allows the professional photographer to manipulate subject angles and planes of sharp focus. It is ideal for static subjects.

sunny day the correct amount of light is admitted. Complex cameras, designed to take perfect photographs in all kinds of lighting conditions, have shutters with a wide variety of speeds, from hours to perhaps 1/2000th of a second; lenses that can admit a much greater amount of light (and still focus precisely and clearly); and built-in accessories of many kinds.

Generally the principal built-in accessory is an electronic exposure meter, which can automatically adjust shutter and diaphragm to the correct relationship. In many miniature (and some other) cameras the shutter is not built into the lens, or set just behind it, but lies almost against the film. These shutters are known as focal-plane shutters.

Film, developing and printing
Film may be cut into sheets, loaded into a light-tight cartridge or cassette, or wound (backed with paper) on to a metal spool. It is thin transparent plastic coated with a photographic emulsion made of grains of silver salts suspended in gelatine. The grains are relatively large in highly sensitive (fast) film

and small in slow film. Correct exposure is affected by film speed as well as by shutter speed and diaphragm opening.

The latent image formed on the film [13] is made visible and permanent by chemical processing in four main stages – developing, stopping development, fixing and washing. With negative film, developers darken the film in proportion to the light that reached it, so that the brighter parts of the scene that was photographed appear dark. In colour negative film [14] the colours of the original subject are also reversed, each colour being represented by its complement. Thus yellow appears as blue and red as cyan.

Fixing simply removes from the emulsion all the chemicals not affected by light, leaving areas of clear film. These portions print black in the final photograph. A "stop bath" between developer and fixer stops development at the correct point. Washing removes unwanted fixer, which would otherwise eventually spoil the negative. The printing process [10] is identical, except that light-sensitive paper is used (although film may be used if the photograph is to be projected).

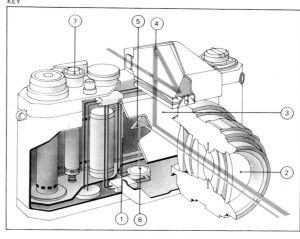

When the shutter [1] is released light reflected from a scene is focused on to film. Before the photo is taken, light entering the lens [2] is split along two paths. Most is reflected upwards from a mirror [3] into a viewfinder [4]. A little is either reflected downwards from a smaller mirror [5] into the photoelectric cell of an exposure meter [6], or directed to a cell in the base of the viewfinder. The release [7] opens the shutter.

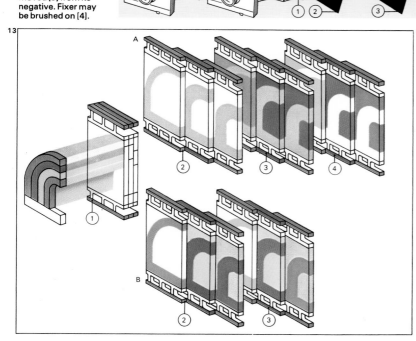

8 A black-and-white film is processed by immersion in a suitable developer [1]. The image is then washed [2] and made permanent by "fixing" [3], and the negative is washed again [4] and dried [5].

9 After drying, the negative is seen as a picture with reversed tones; the blacks are recorded as clear, the whites as black, with the intermediate tones as their complementary tones.

10 To make a print, light-sensitive paper is exposed to a same-size or enlarged image of the negative [1], then processed as is a film [2–6]. The paper is not as sensitive as film.

11 When it is washed and dried the paper has a permanent black-and-white image of the subject. The exact range from black to white depends on the paper's contrast and the surface.

12 The processing chemicals are spread on to Polaroid film as it is pulled from the camera [1]. After 15 seconds [2] the paper print is removed [3] from its negative. Fixer may be brushed on [4].

14 When light falls [1] on the light-sensitive emulsion of a film [2] it forms a latent image [3]. Grains of silver halide are slightly altered; the more light reaching the film, the more grains are affected [4]. In developer the grains are converted back to black metallic silver. At first a few are changed [5]; later a denser image forms [6, 7] consisting of clumps of silver grains [8].

13 Colour films have three layers sensitive to blue, green and red light respectively; each forms a latent image on exposure [1]. [A] Most transparency film is first developed to a black-and-white negative [2]. All remaining silver halide is then exposed and colour developed. Now the film is opaque, as all silver has been blackened [3]. The silver is bleached out leaving a naturally coloured dye image [4]. [B] Dyes form in the first development of colour negatives [2]. After bleaching the negative image is in complementary colours [3], and it also has residual yellow and orange dyes. This ensures the correct colours in the subsequent printing process.

Taking pictures

In the modern world, photography has replaced painting and drawing as the chief means of making pictures. It lets men capture images without the need for graphic skills.

Using a camera

Even with the simplest camera there are rules to be followed. For example, a suitable film must be used, and the camera must be properly positioned so that the subject is not too close and out of focus, nor too far away and too small in the finished picture. A live subject must keep reasonably still, the camera must be held level and steady, and the shutter release pressed gently.

With a more complicated camera the right exposure must be selected. On some cameras the photographer simply sets a pointer to the appropriate weather symbol. With others, the camera automatically sets its own aperture or shutter speed, or both. In the most sophisticated equipment the exposure is set manually, or by overriding the automatic system, to give creative control.

Lens apertures, or "stops", are normally expressed as *f*-numbers [2]. These are so calculated that the light-passing ability of any lens is the same at the same *f*-number, within the limits set by manufacturing tolerances and lens efficiency. The smaller the number, the more light reaches the film. Most cameras have apertures indicated by numbers from the scale: 1, 1.4, 2.8, 4, 5.6, 8, 11, 16, 22, 32. Each number symbolizes half the light-passing ability of the one before it on the scale. Some cameras have maximum apertures that fall between the scale figures.

Shutter speeds [1] may also be set on a scale that halves the light reaching the film with each step. The commonly used speeds are all fractions of a second, usually on the scale: 1, $^1/_2$, $^1/_4$, $^1/_8$, $^1/_{15}$, $^1/_{30}$, $^1/_{60}$, $^1/_{125}$, $^1/_{250}$, $^1/_{500}$, $^1/_{1000}$. Any one of a number of combinations of aperture and shutter speed can give the same exposure. For example, many colour films need an exposure of $^1/_{125}$ at *f*11 in bright sun. They receive the same amount of light with $^1/_{30}$ at *f*22, and $^1/_{500}$ at *f*5.6.

Filters and focusing

The way in which a film records colours as monochrome tone values can be modified by using filters. They are discs of coloured glass and when placed in front of the camera lens they transmit light of their own colour and absorb light of other colours. A yellow filter can be used in landscape photography to prevent over-exposure of the blue sky and to highlight detail in cloud formation.

The chosen combination of aperture and shutter speed is decided by the amount of light available. The aperture determines how much of the picture is in sharp focus. The distance between the nearest and farthest parts of the subject that come out sharp is called the depth of field. It is not only affected by the *f*-number. It is smaller with longer-focus lenses, often called telephoto lenses, and greater with shorter-focus (generally wide-angle) lenses. The depth of field also decreases or increases as the distance between camera and subject is decreased or increased [7].

Shutter speed determines how sharply moving subjects come out. Slow speeds carry the risk that the picture will be spoiled by camera shake. Few people can hand-hold a camera steadily enough to use times longer

1 **Shutter speed** determines how sharply moving subjects come out. At a slow speed (such as $^1/_{15}$ second), moving things are blurred. This gives emphasis to movement, as in this waterfall [A]. At intermediate speeds, $^1/_{125}$ or $^1/_{250}$, ordinary movements do not affect the picture, but really fast objects still blur. Fast speeds, such as $^1/_{500}$ or $^1/_{1000}$ second can freeze most movement and reveal details that are normally unseen, as shown in [B].

2 **When the lens** is set to a large aperture (such as *f*4) it has little depth of field. Only part of the subject is in sharp focus. The actual plane of the sharp image depends on how close [A] or distant [B] the lens is focused. At small apertures (large *f*-numbers such as *f*16) the depth of field is much larger and most of the subject comes out sharp [C]. Choosing the aperture can concentrate attention on selected parts of the subject.

than 1/30 second, and for hand cameras 1/60 or 1/125 is preferable.

Simple cameras are pre-set to give about 1/60 second at f11. They have enough depth of field to take sharp pictures of anything more than about 1.5m (5ft) from the camera.

The professionals
Professional photographers work in a wide range of fields. Some record life as it is for newspapers, magazines and similar publications. Others set up the situation they want in a studio or on a carefully chosen location. Their pictures are usually used as decorative illustrations or in advertising to show a product to its best advantage.

Photo-journalists and other reporting photographers have to take scenes as they find them. They create their pictures by choosing a suitable viewpoint and manipulating the shutter speed and aperture. They may have some simple portable lighting, such as an electronic flash gun. But this is normally used merely to illuminate a subject that is not well enough lit.

On the other hand, advertising, portrait and industrial photographers manipulate their lighting to make the most of the subject. A studio set may be lit by several electronic flashes [6], or by a series of theatrical spotlights and floodlights [3]. The first consideration is the angle of the main or key light, followed by the amount of fill-in light used to keep the shadows from being too dark. There may also be background lighting, or special effects, such as a backlight to make a model's hair glow.

Professional photographers understand how to get sharp, grain-free, perfectly lit and accurately coloured pictures. But much of the impact of good, exciting photography depends occasionally on breaking the rules to create something more than a mere record of what is in front of the camera.

Whether he breaks rules, or follows them, has sophisticated or simple equipment, a photographer creates his picture in the viewfinder the instant he presses the shutter release. Before this vital moment arrives the most important lesson is to make a critical examination of the whole composition of the prospective picture.

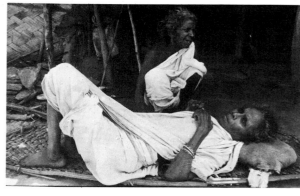

One good picture is worth 1,000 words – this maxim is known and understood by every newspaper editor. For example, this photograph, taken by Chris Steele-Perkins, shows all the agony and despair of the terrible conflict in Bangladesh. It says far more about human suffering than the most eloquent of reports could achieve with descriptive prose. The recent history of war and revolution in Asia, Africa, the Middle East, South-East Asia and Northern Ireland has brought to the fore photo-journalists able to capture the violence and compassion of the modern age, often risking their own safety in order to take good pictures.

3 The angle at which light falls on the subject changes its appearance in the photograph. Direct lighting from the front gives a flat look. If the key light is moved to the side [A], above [B], or below [C] it creates a more interesting picture. The effect can be harsh with a "hard" light source. To soften it, a fill light is used. It is placed in a more-or-less frontal position.

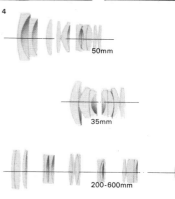

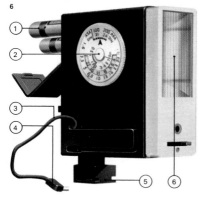

4 Camera lenses consist of glass or plastic elements in a tube. Number, shape and position of the elements determine the focal length and maximum aperture. On a 35mm camera, a 35mm lens covers a wide field of view; a 50mm lens covers a standard field. Some lenses can zoom through a range of focal lengths between two fixed points, as in an extremely long focus or telephoto lens of 200–600mm.

5 Simple cameras often take battery-fired flash cubes, with disposable flash bulbs mounted in reflectors, which rotate between flashes.

6 Electronic flash guns give amateur photographers great flexibility. Many of them have built-in automatic exposure circuitry. They give uninteresting, but accurate, lighting when mounted on a camera. They can be used off the camera to give greater variety. The chief parts are the batteries [1], exposure calculator [2], connecting lead [3], camera socket plug [4], mounting shoe [5] and flash tube [6].

7 Altering the focal length alters the subject's size, pictured from a fixed camera position. (If the camera is moved so that the subject stays the same size, it comes out a different shape with different angles of view.) A wide-angle lens [A] enlarges the nose and chin, while a narrow angle, long focus, lens [C] tends to flatten them, but is preferred for portraits. A standard lens is in between [B], and intended to give a normal view.

Cine photography

A movie camera [Key] works just like a still camera, with one major exception: instead of taking just one picture of a scene, it rapidly takes a series of pictures or frames [7]. The usual rate for taking and projecting amateur movies is 18 pictures a second and for professional movies 24. Older silent movies were shot at 16 pictures a second. Each exposure produces a static view of the scene; but anything that moves is in a slightly different place in each succeeding frame. When the pictures are projected in sequence the movement is re-created. If pictures are taken at a faster rate, such as 48 or 72 a second, they produce slow motion when projected at normal speed; pictures taken at longer intervals, such as 16 or 8 a second, produce a speeded-up motion at normal projection speed.

The camera
All movie cameras have a motor – usually electric, occasionally clockwork – to drive the film from one spool to another; and they all have a lens, shutter and gate mechanism. Apart from these features, they vary enormously. The simplest cameras have a fixed-focus lens and simple automatic exposure systems – ensuring acceptable results in most outdoor conditions. More versatile cameras are fitted with zoom lenses that allow a range of image sizes. These lenses generally have a focus control and a built-in rangefinder system. Only in the most sophisticated cameras can the lens be removed and replaced with another. Super 8 cameras range in sophistication from the movie equivalents of box-cameras to complex machines hardly distinguishable from 16mm professional equipment.

Cine camera film
Cine camera film is used in five widths – 70mm, 35mm, 16mm, 9.5mm and 8mm. The 70mm and 35mm sizes are used by major film companies in cameras that are highly complex and expensive. Such cameras are often used with special image-squeezing "anamorphic" lenses to produce wide-screen pictures. 16mm film is used for smaller-budget commercial pictures, documentary films, much television filming and by serious amateurs. A special format with a larger picture area, Super 16, is made for wide-screen filming to be printed on to 35mm film stock. Although apparently superseded for a quarter of a century, 9.5mm still has its adherents. 8mm is the popular amateur format and is beginning to be used for low-budget television work. There are two picture sizes – Standard 8 and Super 8. With Standard 8 the camera is loaded with special 16mm film either on simple spools or in reloadable magazines. After half the width (that is, an 8mm-wide strip down one side) has been exposed, the film is turned over to expose the other half. It is cut along the middle after processing. Super 8 film in plastic cartridges has virtually superseded Standard 8. It has a larger picture area than Standard 8 and is not as subject to light leakage.

Because the same piece of film is normally used both for filming and projection, almost all amateur movies are shot in colour reversal film – usually rated at 40 ASA (17 DIN) in artificial light or 25 ASA (15 DIN) in daylight. ASA and DIN ratings are a measure of the "speed" of the film – that is, its sensitivity to light. Super 8 cameras have a built-

1 **Moving subjects** were first pictured as a series of still photographs by Eadweard Muybridge (1830–1904) in 1877. He set up 12 and later 25 cameras fitted with high-speed shutters. As a horse galloped or trotted by, it set off each camera in turn by breaking a string or through electrical contacts. Later Muybridge extended his technique to a wide range of subjects, using the principles of moving picture toys (such as the zoetrope) to produce moving images.

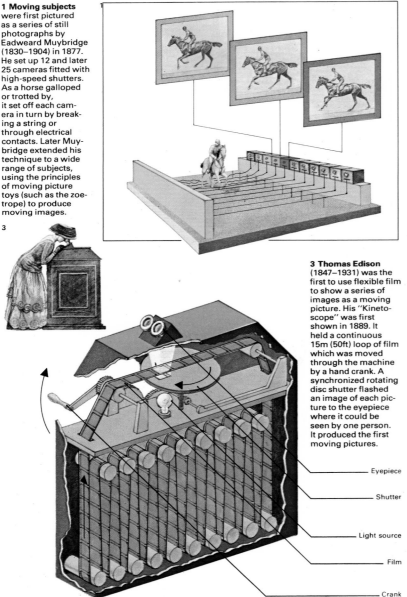

3 **Thomas Edison** (1847–1931) was the first to use flexible film to show a series of images as a moving picture. His "Kinetoscope" was first shown in 1889. It held a continuous 15m (50ft) loop of film which was moved through the machine by a hand crank. A synchronized rotating disc shutter flashed an image of each picture to the eyepiece where it could be seen by one person. It produced the first moving pictures.

Eyepiece

Shutter

Light source

Film

Crank

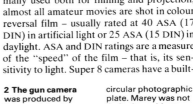

2 **The gun camera** was produced by Etienne-Jules Marey (1830–1903) in 1882. It was the first single camera to take a series of photographs. It was sighted like a gun and, when the trigger was pulled, took 12 small pictures in one second on a revolving circular photographic plate. Marey was not limited to subjects which could trip the camera shutter, and made numerous series of photographs of birds.

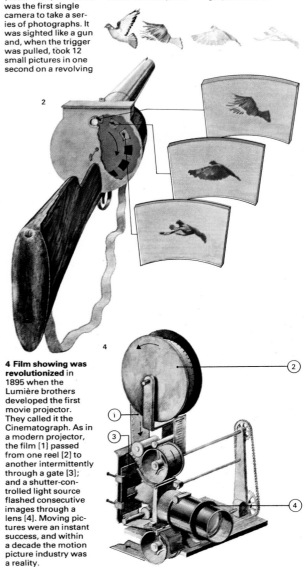

4 **Film showing was revolutionized** in 1895 when the Lumière brothers developed the first movie projector. They called it the Cinematograph. As in a modern projector, the film [1] passed from one reel [2] to another intermittently through a gate [3]; and a shutter-controlled light source flashed consecutive images through a lens [4]. Moving pictures were an instant success, and within a decade the motion picture industry was a reality.

in filter and normally use the 40 ASA (17 DIN) artificial light film with this filter in daylight. 160 ASA (23 DIN) films are available for poor lighting conditions and, with specially designed lowlight XL type cameras, can make films in ordinary domestic lighting.

Most professional moviemakers use negative film – either colour or black-and-white. The colours or tones of the final copies are determined in the processing laboratory. The films for projection are always positive – like colour transparencies, but negatives are used for some television transmissions and they are reversed electronically.

Projection and editing

For normal viewing, movies are projected at a suitable speed. The film is projected as a series of static images. If the images follow one another fast enough, the effect is a moving picture. To reduce flicker, the projector shutter closes and opens again during each frame. So at the cinema we see 48 pictures a second, each of which is repeated with a moment of blackness intervening.

For home movies, the original camera

film is projected. In professional work this hardly ever happens. Prints are made from the camera negatives and are edited and "cut together" to make a complete film. The camera negatives are then cut and joined together to match. Any optical effects are added at this stage. Such effects include "fades" to black or white, "dissolves" from one scene to the next, and "wipes" in which a new scene progressively displaces the old one. A master negative is then made from the complete film and used to produce all the prints for distribution. The prints need not be on the same size of film as the camera original. For the largest cinemas, 70mm film is used, although 35mm is more usual. 16mm copies are made for small cinemas, clubs and educational establishments.

Soundtracks on movies may be either magnetic or optical. Magnetic tracks are recorded on magnetic stripes coated on the film, just like a normal tape recording. Optical tracks record the sound as brightness variations in a dark stripe down the edge of the film. It is replayed by shining a light through it on to a photoelectric cell [7].

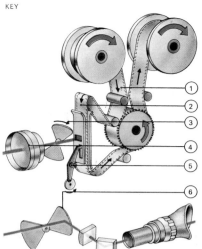

In a movie camera, fresh film [1] is fed from a spool to the gate [2]. A revolving shutter [3]

controls exposure to light focused through a lens [4]. This produces an upside down image like that

in a still camera. While the shutter is closed, a claw mechanism [5] moves the film through the gate in measured steps. Between steps the film remains stationary and the shutter opens and closes again. So the scene is recorded as a series of pictures taken one after the other. The reflex viewfinder uses a mirror shutter [6]. When the shutter is closed, the image through the camera lens is reflected from its mirror surface to the viewfinder. As a result, the image directed through the prism and lens system is exactly the same as that passing through the lens. In the viewfinder, the cameraman sees exactly what the lens "sees".

5 Floor-mounted movie cameras are used in studios; lighter cameras such as this 35mm model are used outdoors. The camera can take most lenses and accessories. It can be carried on a cameraman's shoulder, but gives steadier pictures when fixed to a tripod. Film spools are mounted coaxially, so the balance remains constant; and to keep the noise to a minimum, the camera is sealed in a rigid

housing with a window in front of the lens. The gate is fitted with register pins to ensure that each frame is accurately located on the film.

Key
[1] Focus control
[2] Aperture control
[3] Mirror shutter
[4] Viewfinder eyepiece
[5] Take-up spool
[6] Feed spool
[7] Gate
[8] Sprocket drive
[9] Pull-down claw
[10] Register pins
[11] Indicator
[12] Viewfinder
[13] Footage indicator
[14] Lens

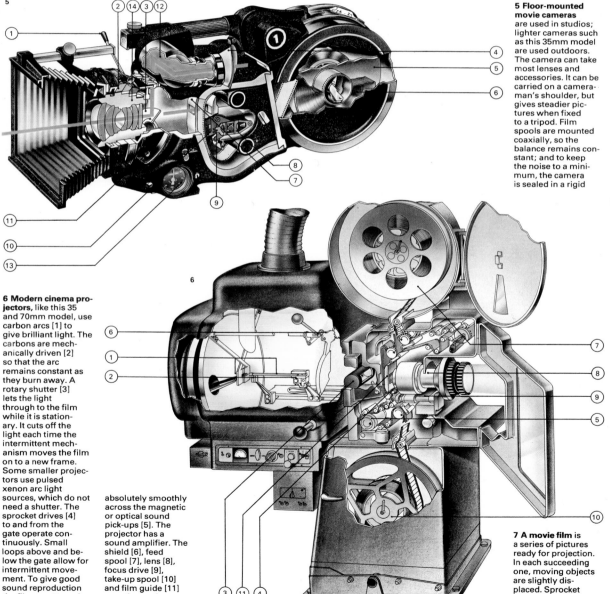

6 Modern cinema projectors, like this 35 and 70mm model, use carbon arcs [1] to give brilliant light. The carbons are mechanically driven [2] so that the arc remains constant as they burn away. A rotary shutter [3] lets the light through to the film while it is stationary. It cuts off the light each time the intermittent mechanism moves the film on to a new frame. Some smaller projectors use pulsed xenon arc light sources, which do not need a shutter. The sprocket drives [4] to and from the gate operate continuously. Small loops above and below the gate allow for intermittent movement. To give good sound reproduction the film must travel

absolutely smoothly across the magnetic or optical sound pick-ups [5]. The projector has a sound amplifier. The shield [6], feed spool [7], lens [8], focus drive [9], take-up spool [10] and film guide [11] are other main parts.

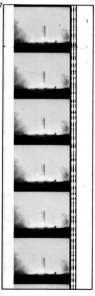

7 A movie film is a series of pictures ready for projection. In each succeeding one, moving objects are slightly displaced. Sprocket holes along the edge

ensure that each frame is held in just the right place in the projector. A variable area stripe along one edge carries the soundtrack. As this passes a light-sensitive cell, it modulates the light from a lamp and produces sound signals.

Communications: telegraph

Long-distance communications, which for thousands of years had depended on the slow and unreliable travel of messengers by foot, horseback or ship were transformed by telegraphy. Simple forms of semaphore had long been used, but the transmission of messages beyond the range of sight had to wait until the discovery of electricity. The idea was foreshadowed as early as 1753 by a Scottish doctor, Charles Morrison, in a letter to the *Scots Magazine*. In 1764 Georges Louis Lesages built and operated an experimental electric telegraph in Geneva using static electricity and an electroscope. The mutual repulsion of a pair of pith balls indicated the presence of an electric charge in a wire connected to them and a separate wire was used for each letter of the alphabet.

Single-wire telegraphy

At the end of the eighteenth century, Napoleon Bonaparte (1769-1821), who was the first to make use of a systematic telegraph, still had to rely on a visual system invented by a French merchant, Claude Chappe (1763-1805), to receive intelligence reports and send orders to his army [1]. It was not until 1816 that a single-wire telegraph was invented by an Englishman, Francis Ronalds (1788–1873). He set up discs turned by clockwork at each end of the wire. Each disc was initialled with the alphabet round its rim and, as the desired letter aligned with a pointer, the sender's wire was connected to an electroscope. With the discs synchronized, a man at the other end of the wire could note the letter indicated each time the wire received a charge.

Ten years later an American, Harrison Gray Dyer, built the first practical electrical telegraph by using the recently invented voltaic cell (battery) and a chemical solution that indicated the presence of an electric current by the formation of bubbles at two electrodes. Dyer sent messages along 12.5km (8 miles) of wire laid in Long Island, New York, using the earth to complete the circuit.

The final step in the evolution of the electric telegraph came in 1831 when another American, Joseph Henry (1797–1878), replaced Dyer's electrolytic indicator with an electric bell, using the principle of electromagnetism discovered in 1819 by the Danish physicist Hans Christian Oersted (1787-1851). Henry used a code to indicate the different letters of the alphabet.

Two Englishmen, William Cooke (1806-79) and Charles Wheatstone (1802-75), designed and installed the world's first commercial telegraph [2]. This was a five-wire system, first tested in 1837 between Euston and Camden Town on the London-Birmingham railway and installed two years later, between Paddington and West Drayton, for the Great Western Railway.

Morse code and later developments

Up to this time each inventor had devised his own means of coding messages. It was an American inventor and painter, Samuel Morse (1791–1872), who first recognized the practical and commercial inportance of devising a standard code. He demonstrated his own code in 1837 and it was the revised Morse code that eventually made possible the development of electric telegraphs throughout the world. Consisting essentially of a single wire, the telegraph circuit was

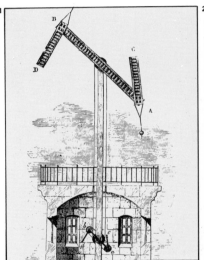

1 The earliest form of long-distance communication, apart from jungle drums, smoke signals and similar crude devices, was the visual semaphore invented by Claude Chappe and used by the French army from 1794. The Chappe telegraph consisted of pairs of hand-operated semaphore arms located on a chain of towers built on the tops of hills within sight of each other.

2 The world's first commercial telegraph was the five-wire system set up by Charles Wheatstone and William Cooke in England in 1839. Switch pairs made each needle deflect in either direction.

3 A Morse key is basically a switch. This early two-pole key has a closed circuit between points 1 and 2 when at rest. When the key is depressed contact [1] is broken and the circuit made instead through 3. Current entering via the black wire is switched from blue to red.

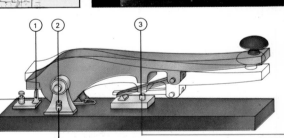

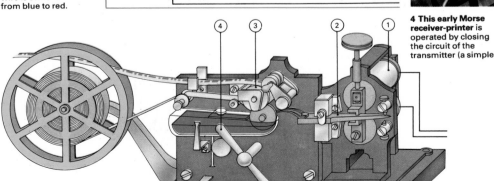

4 This early Morse receiver-printer is operated by closing the circuit of the transmitter (a simple key). The current energizes the coils [1]. The lever [2] is moved by magnetic attraction, bringing the printing disc [3] into contact with the paper tape for the duration of the current. The disc is linked by contact with a roller immersed in a bath of printer's ink. The movement of the paper tape is effected by clockwork wound by the handle [4]. It was found that a skilled operator could "read" a buzzer quicker than an inked tape.

5 Emile Baudot's multiplex system enabled several telegraph operators to send messages over the same line simultaneously. Each operator's set was connected in turn to the line by a distributor for just enough time to allow transmission of a single letter in the form of a five-unit code. By exact phase synchronization outgoing signals could be correctly separated and "read" on receivers at the other end.

completed by a battery and key between the wire and the earth at the sending end [3], and an electromagnetic sounder between the wire and the earth at the receiving end.

The laying of submarine cables [6, 8] was a natural development of the electric telegraph. A connection across the English Channel in 1850 encouraged engineers to attempt the more difficult task of laying a cable across the Atlantic. Success came in 1858 when a cable was laid from Ireland to Newfoundland, although a reliable link was not established until 1866 [Key].

Because wire is expensive inventors soon turned their minds to ways of sending a number of messages simultaneously along a single wire. The breakthrough came in 1874 when a Frenchman, Emile Baudot (1845–1903), designed an instrument which, when fully developed, could interlace six messages and unscramble them [5].

Automatic telegraphy
Baudot's system, in which every letter consisted of five pulses (or absence of pulses), was used successfully for about 50 years until replaced by a more sophisticated frequency division multiplex, invented in the 1890s by an American, Elisha Gray (1835-1901).

To reduce the time spent coding and decoding messages much effort was given, over many years, to the invention of an automatic printing telegraph. David Hughes (1831-1900), an American professor of music, built the first practical printing telegraph in 1854 [4], but it, too, was slow. In 1921 a Russian, N. P. Trusevich, invented what is called the "start-stop" system and the modern teleprinter became possible. The new invention solved the problem of keeping the receiving machine perfectly synchronized with the sending machine when the operator's typing speed varied slightly from letter to letter [9]. The modern teleprinter, using the five-unit code, can transmit up to 13 characters a second when working from punched paper tape which the operator first prepares, typing at his own speed. Telex [7] enables up to 26 teleprinter messages to be carried on a single telephone cable. Photo-telegraphy [10] completes the range of modern telegraphic communications.

KEY

A cable-laying ship is specially designed so that miles of submarine telegraph cable can be fed out over sheaves in the bows. Today cable-laying is a routine task, but it took the American businessman Cyrus W. Field (1819–92) nine years and five attempts before the first successful transatlantic telegraph cable was laid in 1866. For the laying the engineers used the largest ship then afloat, the steam screw and paddle ship *Great Eastern* built by the British engineer Isambard Kingdom Brunel (1806–59).

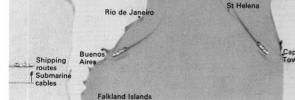

6 A typical early submarine cable was made up of a stranded copper conductor [3], usually of wires or copper tapes woven round a stout central strand; an insulating layer [2], originally of gutta percha; a layer of jute fibre [4] with galvanized steel wires embedded in it (for strength); another layer [1] of compounded jute; and an outer waterproof layer [5] tough enough to withstand chafing.

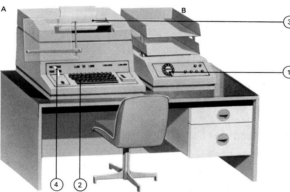

7 A modern telex installation comprises a teletypewriter [A] and a dialling unit [B], both often built into a single console. The operator uses the dial unit [1] to call the receiver's telex number and then types the message using conventional typewriter keys [2]. Alternatively, the message can be pre-typed on a punched tape [4] and fed in for extra speed in transmission or when a line is free. For checking, the transmitted message is automatically printed out above [3]. When not sending, the equipment is left ready for receiving. A buzzer can alert the receiving operator to an incoming message. If the operator is absent the teletypewriter is automatically activated to reproduce the message on the paper [3]. Incoming messages can thus be "stored" in typed form on the machine until the operator returns.

8 Telegraph cabling across the Atlantic first became fully successful in 1866. After this the world's oceans were soon crossed by a network of such cables. The map shows today's transatlantic routes. Cable communications were supplemented by radio in the 1920s and by satellite systems in the 1960s. Cables retain the advantage of privacy because, unlike radio signals, the messages they carry are difficult to intercept.

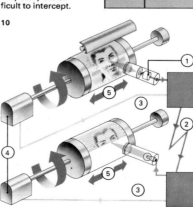

Letter	Morse code	Morse code electrical signals	Five-unit code	Five-unit electrical signals
A				Start Stop
E				
O				
Y				

9 The teleprinter has a seven-unit code consisting of a start signal, a five-unit character signal and a stop signal. The start and stop signals enable the equipment to keep the five-unit signals in step for each transmitted letter, despite the fact that no two typists operate at the same precise speed or perfectly evenly. Morse signals, being of unequal length, cannot be so used.

10 In photo-telegraphy, used widely by newspapers, the picture is "scanned" by a spot of light that covers its area in a series of parallel lines. The brightness of each element is converted by the cell [1] into an electrical signal that is transmitted [2] by telegraph. A machine at the receiving end converts the signal back into a picture by printing it dot by dot according to the incoming signal. Pulses [3] synchronize the motors [4] and traverse [5].

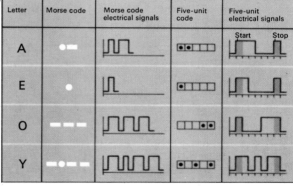

Communications: telephone

Inventors working independently often arrive at very similar solutions to a problem. So it was in 1876 when Alexander Graham Bell (1847–1922), a Scottish professor of vocal physiology living in the United States, applied for a patent for his electric telephone only a few hours before an American from Chicago, Elisha Gray (1835–1901), filed a similar application. Bell was granted the patent and has since generally been credited as the sole inventor of the telephone.

Bell's instrument [2] was used both as transmitter and receiver and needed no battery. But the current generated when sound vibrated the diaphragm in its "microphone" was small and the instrument was therefore unsuitable for telephone communications over long distances.

The microphone and telephone exchange
In 1877 the American Thomas Edison (1847–1931) invented the carbon microphone. By another coincidence a similar microphone was developed independently a year later by the English scientist David Hughes (1831–1900), who is now generally

credited as the inventor. The carbon microphone [4] modulates an electric current from a DC source creating a voltage that varies in step with the sound waves. It is used to this day as the transmitter in modern telephones, the receiver being an electromagnetic earphone similar to that patented by Bell. A varying voltage in the earphone's coil causes a metal diaphragm to vibrate and produce sounds at low volume.

For the telephone to become a practical proposition it was necessary to find a means of interconnecting any pair of a number of instruments. The first telephone exchange was opened at New Haven, Connecticut, in 1878 and a similar eight-line exchange was set up in London a year later. In the early exchanges, an operator used plugs and sockets to connect callers.

In 1889 an American undertaker, Almon Strowger, annoyed by the inefficient service from his local exchange, designed an automatic selector. The first automatic exchange was opened in La Porte, Indiana, in 1892. The Strowger electromechanical selector [5] became standard equipment for

telephone exchanges throughout the world during the next half century.

Since 1926 an American invention, the crossbar switch, has replaced the Strowger selector in Sweden. It is in use today in both the United States and Great Britain, although an even more efficient all-electronic exchange was developed in 1960.

Telephone cables
The largest single expense in any long-distance telephone system is the cost of the wires that connect subscribers to each other. Research into the problem of using a single cable for more than one simultaneous telephone call bore fruit in 1936 when the first 12-channel coaxial cable was laid between Bristol and Plymouth in southwest England. The system uses electric carrier waves of different frequencies, which can be transmitted simultaneously along the metal core of the cable and separated at the other end by a series of electronic "filters", each of which accepts signals of one frequency only.

The longer a telephone line, the weaker the electrical signal reaching the end; this is

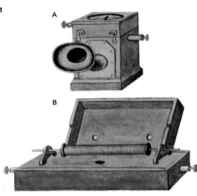

1 An early telephone designed by a German, Philipp Reis, in about 1861 has a transmitter [A] with a metal point in light contact with a metal strip fixed to a membrane. Reis believed that the intermittent circuit caused when the membrane vibrated would produce a varying electric current which could be reconverted into sound. The receiver [B] was based on – the change in length of an iron needle in a magnetic field.

2 Bell's first telephone used a parchment drum, which vibrated when sound waves reached it. A piece of iron was supported by a short length of clock spring so that it rested lightly on the parchment [A]. An electromagnet [B] was placed so that one pole was close to the iron piece. When the parchment and the iron vibrated a small varying electric current was induced in the coil. When two such instruments were connected the current produced by one energized the magnet of the other, causing the iron piece and the parchment to vibrate in step with the first. By this means a voice or any other sound that vibrated one parchment diaphragm was reproduced by similar vibrations of the other. Alexander Bell obtained the publicity he needed for his invention when it was seen by the Emperor of Brazil.

3 An exchange network allows telephone subscribers to dial any number in a national system. The lines and numbers indicate the direction in which calls can be routed and the dialling code for each route. The squares, always in pairs, are main exchanges, one part for local, the other for long-distance calls. The circles are sub-exchanges – those top right being directly interconnected exchanges in a city. All city, local and main exchanges have lines to subscribers. A caller on exchange P can call numbers on exchange N by first dialling the code "9", or numbers on exchange Q by dialling code "987". To call exchange S he must first dial code "991", and to call a number on exchange T the code is "99186".

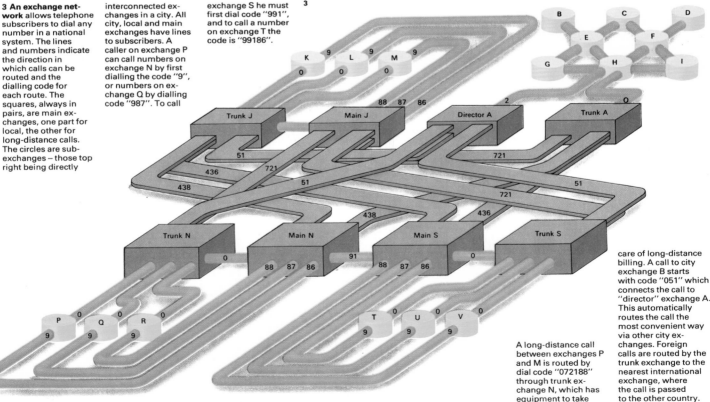

A long-distance call between exchanges P and M is routed by dial code "072188" through trunk exchange N, which has equipment to take care of long-distance billing. A call to city exchange B starts with code "051" which connects the call to "director" exchange A. This automatically routes the call the most convenient way via other city exchanges. Foreign calls are routed by the trunk exchange to the nearest international exchange, where the call is passed to the other country.

due to the resistance of the wire. Automatic amplifiers (called repeaters) overcame this problem and today these are incorporated in multi-channel cables every 16km (10 miles). The first transatlantic multi-channel telephone cable, called TAT 1, was laid in 1956. It was a twin coaxial cable – one for speech in each direction – running from Scotland to Newfoundland and included 51 repeaters in each cable. Today, submarine telephone cables with two-way repeaters encircle the world. The repeaters are powered by alternating current from the same cable core that carries the modulated carriers. The power frequency is much lower than the carrier frequencies, so there is no interference.

For many years automatic dialling could be used only for calling a subscriber connected to one's own exchange. The problem of extending the system was largely one of designing a means of automatic billing that would charge according to distance and time.

Today, when a call is dialled the equipment connects the line to an electronic pulse generator. Each pulse records a fixed unit charge on the subscriber's account. The pulse generator selected depends on the distance of the called exchange.

Modern coaxial cables can carry an increasingly large number of simultaneous conversations – current research is producing multiple cables capable of handling up to 3,000 channels – but microwave radio has been playing a rapidly growing part in telecommunications since the early 1960s.

Microwave telephony

Modern microwave systems use relay towers at 40–50km (25–30 mile) intervals. Telephone signals are used to modulate microwave radio carriers instead of the electrical carriers used in coaxial cables [6]. The British microwave system [Key], which has about 120 relay stations throughout the United Kingdom, is typical. It operates with 132 separate microwave carriers, each able to accommodate 2,700 simultaneous telephone conversations (many of the channels are used to relay television signals). For intercontinental telecommunications microwave carriers are beamed up to satellites, amplified and beamed down to ground stations.

KEY

London's Post Office Tower is the hub of the United Kingdom microwave network. This provides thousands of telephone circuits and up to forty television channels connecting all parts of the British Isles, and links them with the international satellite station at Goonhilly Downs in Cornwall and a cross-Channel link station near Dover. The tower is 189m (620ft) tall, to the top of the aerials, and has a mass of radio equipment on most of its lower 16 floors. Above is a series of open galleries around which stand an array of parabolic dish and horn aerials for microwave signals. Above are public observation galleries and a restaurant.

4 The heart of a carbon microphone is a small insulating cylinder packed with carbon granules. The centre of a metal diaphragm presses against the open end of the cylinder. When the diaphragm vibrates, the pressure on the granules, and thus the electrical resistance through them, varies. With a DC source connected, a variable current passes. This current will operate a magnetic earpiece.

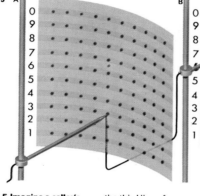
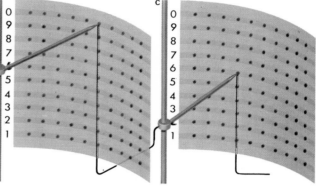

5 Imagine a caller's telephone is connected to a wiper [A] of a Strowger selector. He dials 3064. In response to the three electrical impulses produced by dialling the digit 3, the wiper moves up to the third line of contacts and then moves along until a free line is found. (In the illustration the wiper has stopped at the fifth contact because the first four lines are already in use by other callers.) The fifth contact is connected to wiper [B] of another selector. This moves up to the tenth row of lines and then moves round until it reaches a free one – once again the fifth, the first four being busy. This line is connected to wiper [C] of a third Strowger selector. This, unlike the first two, which are "number-line" selectors, is a "double-number" selector. The wiper moves up to the sixth row of contacts in response to the dialled third digit 6 and then waits until the final digit 4 is dialled. It then moves round to the fourth contact leading to telephone number 3064.

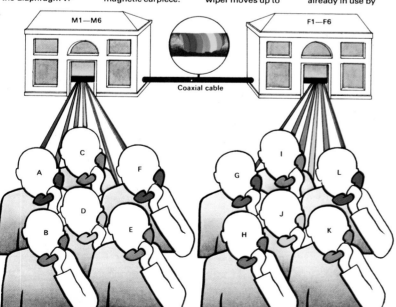

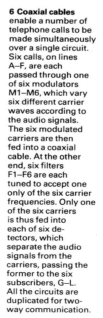

6 Coaxial cables enable a number of telephone calls to be made simultaneously over a single circuit. Six calls, on lines A–F, are each passed through one of six modulators M1–M6, which vary six different carrier waves according to the audio signals. The six modulated carriers are then fed into a coaxial cable. At the other end, six filters F1–F6 are each tuned to accept one only of the six carrier frequencies. Only one of the six carriers is thus fed into each of six detectors, which separate the audio signals from the carriers, passing the former to the six subscribers, G–L. All the circuits are duplicated for two-way communication.

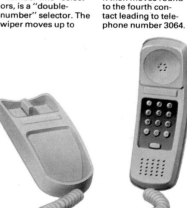

7 An up-to-date telephone has a set of numbered push buttons in place of the traditional dial. A number can be selected more quickly by push button and cross-bar and electronic exchanges can work as fast as a user can "dial". If a push button telephone is used with a Strowger exchange, the number has to be stored in a memory and converted into pulses at a speed the selectors can handle. Despite this limitation the push button system is more convenient.

227

Communications: radio

Radio waves were predicted before they were discovered. In 1865, James Clerk Maxwell (1831–79), a Scottish theoretical physicist, argued the existence of an unseen form of radiation. But his mathematics was so complex that his theory was at first rejected by some scientists. About 25 years later experiments showed that electromagnetic waves, which include gamma-rays, X-rays, and visible light and radio waves, all conformed to his formulae.

The first practical demonstration of what we now call radio waves took place in 1879, when the Anglo-American inventor David Edward Hughes (1831–1900) built a crude radio transmitter and receiver and passed signals without wires along Great Portland Street in London. Hughes failed to realize the full significance of his experiment and did not publish his findings for 20 years.

Successful transmission

In about 1887 the German scientist Heinrich Hertz (1857–94) built a spark generator [Key] that produced radio waves and a receiver that detected their presence at a dis-

tance. In a series of experiments he proved conclusively that energy could be transferred over a distance in a way that could not be accounted for by induction and he is generally credited with the discovery of radio.

Oliver Lodge, an Englishman, was the first to build a radio receiver – more sensitive than Hertz's coil and spark gap – that could be used for practical radio communication. It used a coherer [2], a device invented by a Parisian Edouard Branly (1844–1940). In an experiment in 1894, Lodge used his radio to operate at a distance of 137m (450ft).

The man who was to take wireless out of the purely experimental field was the Italian Guglielmo Marconi (1874–1937). After failing to interest the Italian government in his work he moved to England. In 1898 he set up a radio link between the mainland near Dover and the East Goodwin light vessel moored 19km (12 miles) offshore. A year later Marconi fitted the American liner *St Paul* with radio. The first message received was transmitted from 97km (60 miles) away.

In 1901 Marconi astonished the world by transmitting a radio signal across the Atlantic

Ocean [1] in Morse code. In 1906 a Canadian, R. A. Fessenden (1886–1932), transmitted from Brant Rock, Massachusetts, a signal in which operators at sea heard a voice and music in their headphones. This was the first audio-modulated transmission.

The nature of radio waves

When electrons oscillate in an electric circuit, some of their energy is converted into electromagnetic radiation. The frequency (the rate of oscillation) has to be very high to produce waves of useful intensity, but once formed they travel through space at the speed of light – 300 million metres a second (186,000 miles a second). When such a wave meets a metal aerial, some of its energy is transferred to free electrons in the metal, causing them to flow as an alternating electric current having the frequency of the wave. This, in the simplest terms, is the principle of radio communication. A radio transmitter produces concentrated electromagnetic radiation of a chosen frequency. The waves so generated are picked up by an aerial. From all the waves that come into contact with its

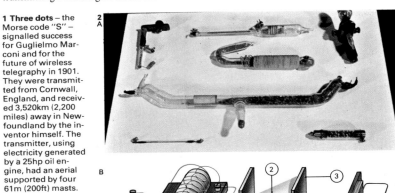

1 Three dots – the Morse code "S" – signalled success for Guglielmo Marconi and for the future of wireless telegraphy in 1901. They were transmitted from Cornwall, England, and received 3,520km (2,200 miles) away in Newfoundland by the inventor himself. The transmitter, using electricity generated by a 25hp oil engine, had an aerial supported by four 61m (200ft) masts. Marconi's receiver was connected to a 122m (400ft) aerial supported by a kite. Marconi had proved to doubting science his firm belief that radio waves were capable of travelling around the curvature of the earth.

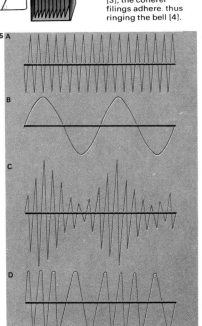

2 The first practical radio detectors were coherers [A], developed by Oliver Lodge (1851–1940). Each of those shown consists of a long glass tube with iron filings packed between two metal plates. In their loose condition the filings did not conduct electricity but when subjected to a vibrating electric wave they cohered (adhered to each other), making a conducting path between the plates. In the circuit shown [B] the coherer [1] acts as a switch in an electric bell circuit. As soon as streams of radio waves [2] reach the aerial plates [3], the coherer filings adhere. thus ringing the bell [4].

3 A 1920s crystal set, forerunner of the modern transistor, used a crystal of carborundum or lead sulphide, a semiconductor that rectified a radio carrier wave. This crystal produced an alternating current of the same frequency as the sound "carried" on the wave. The low frequency electric current had sufficient power to produce sound from sensitive earphones.

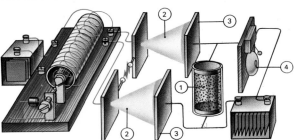

4 The modern transistor radio, like the old crystal set, "detects" (or demodulates) the radio carrier wave, although the process it uses is far more sophisticated. Having created an electrical analogue of the original sound waves, a series of amplifying circuits then produce a signal that has sufficient power to drive a small loudspeaker. Whereas radio wave energy was enough to operate earphones, the transistor set needs a battery.

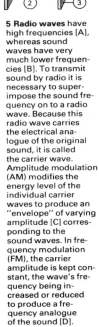

5 Radio waves have high frequencies [A], whereas sound waves have very much lower frequencies [B]. To transmit sound by radio it is necessary to superimpose the sound frequency on to a radio wave. Because this radio wave carries the electrical analogue of the original sound, it is called the carrier wave. Amplitude modulation (AM) modifies the energy level of the individual carrier waves to produce an "envelope" of varying amplitude [C] corresponding to the sound waves. In frequency modulation (FM), the carrier amplitude is kept constant, the wave's frequency being increased or reduced to produce a frequency analogue of the sound [D].

aerial, the radio receiver amplifies those of a selected frequency to which it is "tuned" and eliminates all others.

To transmit voice and music by radio the waves of a regular "carrier" signal must be modulated (varied) by the audio signal [5]. The waves may vary either in strength (amplitude modulation, AM) or in frequency (frequency modulation, FM). The receiver is then able to eliminate the high-frequency carrier waves, leaving only electrical waves of the same frequencies as those of the original sound. Finally, after amplification, the electrical waves are fed into earphones or a loudspeaker, which are able to convert electrical vibrations back into sound waves.

Frequencies, wavelengths and channels

Electromagnetic radiation can vary enormously in frequency. It includes gamma-rays, X-rays and ultra-violet, visible and infra-red light rays, all of which have very high frequencies. Electromagnetic radiation of lower frequencies is radio waves. Those waves next in frequency to infra-red rays are known as microwaves and are used mainly

for telecommunications between towers within visible range, but also for communication with satellites. These waves are followed in order of lower frequency (and so of longer wavelength) by ultra high frequency (UHF) – used for television broadcasts; very high frequency (VHF), for radio broadcasting and for local communication, such as between aircraft and ground control; short waves that, at high power, are used for worldwide broadcasting; medium waves for regional broadcasting; and the relatively little used long waves. The whole radio spectrum is divided by international agreement into bands reserved for specific uses and each band is generally further subdivided into channels spaced so that they do not overlap.

Stereophonic sound requires the reproduction of two separate sound signals that correspond to those received by the two ears of the listener. In radio this could mean doubling the channel width for every stereo transmission. Because radio space is already congested, engineers have devised a method of transmitting two separate audio signals over one radio channel.

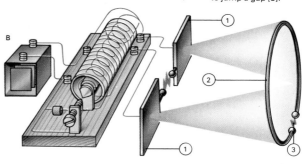

The first radio transmitters [A], as used by Heinrich Hertz and Oliver Lodge, made use of the radio waves generated when a high-voltage spark jumped between contacts [B]. Hertz beamed the waves from aerial plates [1] and detected them with a loop of wire [2] in which they caused a small spark to jump a gap [3].

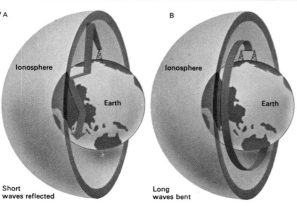

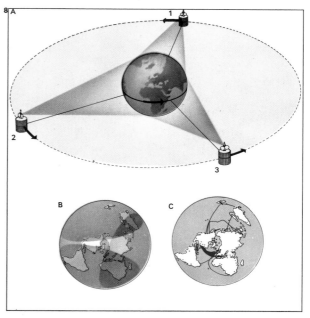

6 The non-stop trend towards miniaturization in electronics is shown in this comparison of a typical wireless set of the 1930s [A] and a transistor radio of today [B]. A loudspeaker must still be large if it is to reproduce the low-frequency components of sound, but now modern technology (much of it space technology "spin-off") has led to the design of smaller and smaller components of nearly all other kinds. The transistor [C, left] has almost entirely superseded the old thermionic valve [right]. Individual electronic components have been growing steadily smaller; more-over new developments have taken the process further, with the introduction of the integrated circuits (IC) in which a complete wired set of components is replaced by one minute IC.

7 The ionosphere (a layer of ionized gas in the earth's upper atmosphere) and the curved surface of the earth below act together as a kind of "waveguide", which bends the path of long radio waves [B] around the earth. The path of waves of the medium band is not as bent, which is why they cannot normally be received more than a few hundred kilometres from the transmitter. All radio waves travel strictly in straight lines, but short waves are used for round-the-world communication because they are reflected by the ionosphere and also by the earth's surface, as though these were mirrors [A]. Even shorter waves pass through the ionosphere and so are used for space communications.

Short waves reflected

Long waves bent

Ionosphere
Earth

8 Communications satellites, owned by an international consortium of over 80 nations, provide a major part of the world's global communications. Placed in synchronous orbit [A] 35,800km (22,375 miles) above the Equator over the Pacific [1], Atlantic [2] and Indian oceans [3], the Intelsat IV satellites remain in fixed positions in relation to the earth, each capable of relaying thousands of VHF radio signals to and from approximately one-third of the earth's surface. Together they cover most of the earth with some overlapping [B]. There are more than 70 earth stations (dots in [C]) capable of communication via the satellites, although submarine cables (red lines) are still used.

Communications: television

Unlike the telegraph, telephone and radio, television is unique among forms of tele-communications because it was originally developed purely as an entertainment medium. Today it has many other applications, particularly for remote surveillance.

The dawning image
V. K. Zworykin (1889–), a Russian emigrant to the United States, patented the iconoscope in 1923. A similar device was invented independently in Britain. Forerunners of modern television camera tubes, these were electronic devices in each of which a lens focused an optical image on to a screen inside a glass container. The image was scanned by an electron beam that covered the area in a continually repeated series of parallel lines. When the beam struck a bright part of the image the electron current flowing back was greater than when it fell on a darker part of the scene. By using this varying electric current to control the intensity of another electron beam in a cathode ray tube (a beam made to scan the face of the tube in step with the beam in the iconoscope), a replica of the original scene was built up spot by spot and line by line. The scanning covered the entire screen several times a second, the glow caused did not die away instantaneously, and so the human eye could not tell that the image was built up of individual elements.

It was not until 1936 that the world's first high-definition public television service was started by the British Broadcasting Corporation (BBC) in London. British engineers had spent the previous five years developing electronic television [1] as an alternative to the 30-line mechanical system that was invented in 1923–8 by a Scotsman, John Logie Baird (1888–1946).

Television standards
The BBC originally adopted an interlaced scan of 405 lines repeated 25 times a second. Interlacing means that alternate lines are scanned first, followed by those between – like reading lines 1, 3, 5 down this column and then going back to read lines 2, 4, 6.

Other countries set up television services using various standards which made international programme exchange very complex.

Eventually it was agreed that 625 lines, 25 frames a second interlaced, should be standard in most European countries. Today only North and South America and Japan use a different standard.

The sideband theory of radio states that, in order that two radio signals do not interfere with each other, the difference between their frequencies must not be less than the highest frequency of the signal being broadcast. Television signals include very high frequencies, so a bandwidth of about 5MHz (5 million cycles/sec) has largely been accepted as the practical minimum. Because this is equal to the radio space occupied by more than 500 voice channels the entire short-wave radio band would accommodate only five television channels. Thus television is broadcast in the VHF and UHF bands (very and ultra high frequency), which can accommodate up to 80 channels.

Colour television
In theory the range of pure colours (called "hues") is continuous from violet of the shortest visible wavelength to red of the longest.

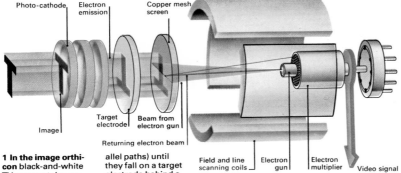

1 In the image orthi-con black-and-white TV camera tube, an optical image of the studio scene is focused on a photo-cathode, which emits electrons in proportion to the amount of light falling on it. These electrons pass through magnetic and electrostatic fields (which keep them moving in par-allel paths) until they fall on a target electrode behind a copper mesh screen. There is a pattern of charges, corresponding to the light and dark areas in the picture. A scanning electron beam from the other side is modulated by the charges, thus forming a varying transmission signal.

2 A cathode ray tube in a television receiver has a screen coated with a material that fluoresces when struck by an electron beam. The beam is produced by a cath-ode that emits elec-trons and is acceler-ated and focused by anodes. The beam is deflected by two scanning coils fed with signals so that it zig-zags across the screen. The vel-ocity of the beam, and thus the intensity of each picture element, is controlled by a grid fed by another signal.

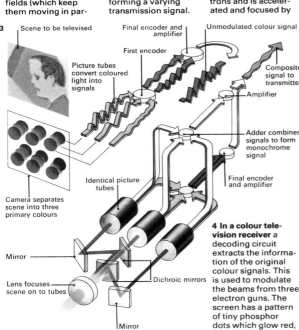

3 A colour television camera first splits the picture into three primary col-ours using colour-separating mirrors. Each beam of col-oured light enters one of three picture tubes, which convert the picture into elec-trical signals. The three signals are combined to form a monochrome signal and processed into a colour signal. This defines the hue (frequency) and sat-uration (intensity) of each colour.

4 In a colour tele-vision receiver a decoding circuit extracts the informa-tion of the original colour signals. This is used to modulate the beams from three electron guns. The screen has a pattern of tiny phosphor dots which glow red, green or blue when struck by electrons. Directly behind the screen is a shadow-mask with thousands of holes arranged so that the beam from each gun can strike only dots of the colour from which its controlling signal was made.

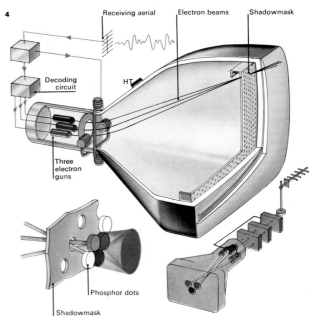

In practice the sensations perceived by the human eye in response to all these hues can be quite accurately matched by simple mixtures of red, green and blue light. Colour television [3] uses this principle by having three camera tubes to convert the red, green and blue light present in each televised scene into three simultaneous but separate electrical signals. In theory these three signals could be transmitted separately, received by three separate circuits in the television receiver and then combined to form a colour picture. But this idea has a major disadvantage. Because the minimum practical bandwidth for one television signal is 5MHz three simultaneous signals would occupy 15MHz – far too much of the already overcrowded radio space available.

Experiments have shown that, provided the black-and-white detail of a picture is sharply defined, the human eye does not require the colour definition to be as high. American engineers devised a clever system to use this information. First the three primary colour signals were added to form a detailed monochrome (black-and-white)

signal for transmission in the usual way. At the same time the three colour signals were converted into a second composite signal that defines the colour mixture in terms of hue and saturation (the amount of white used to dilute the pure hue). Because this colour signal does not need to be of high definition it is possible to sandwich it between the information giving the detail of each monochrome line without interfering with it. In this way a complete colour signal can be transmitted within a 5MHz monochrome signal bandwidth. The television receiver makes a detailed picture from the monochrome signal, extracts the colour information interleaved with it and uses this to deflect the three picture-tube electron beams on to those spots on its screen that will glow with the appropriate primary colours [4].

In the United States this system has been used since 1953 and is known as the National Television Systems Committee (NTSC) system. It works well but has the disadvantage that the colours produced on the receiver screen can be significantly altered by minor changes in the transmitted signal.

Electronic scanning is the basis of television. An electron beam scans the screen in a series of horizontal lines, which are kept in step with the lines scanning an optical image in the television camera in the studio from which the broadcast comes. This synchronization of the scanning process is achieved by a set of timing pulses superimposed on the picture information in the transmitted signal. At the end [1] of each horizontal line [2], there is a pulse that triggers the instant return [3] of the electron beam to the opposite side [4], where it scans the next line [5] below the previous one. After the final line of each picture [6] a different pulse triggers the return of the electron beam to the top of the picture. A saw-tooth current, through electromagnetic deflection coils (round the neck of the tube), controls the beam.

5 The transmission of a live television programme requires the combined efforts of a large, highly skilled team which can be divided into four groups: the studio floor, lighting and colour control, and the sound and direction teams. All are shown in this illustration of the layout of a modern television studio.

1 Director
2 Assistant director
3 Vision mixer
4 Technical manager
5 Timekeeper
6 Lighting supervisor
7 Tariffer (controls iris settings of cameras)
8 Lighting engineer
9 Colour grader
10–12 Sound engineers
13 Floor manager
14 Presenter
15–18 Cameraman
19 Microphone-boom operator
20 Monitor pusher
21 Autocue operator
22 Props man
23–24 Electricians
25 Director's array
26 Microphone to studio
27 Test screen
28 Dimmer bank
29 Switchboard
30 Camera iris controls
31 Lighting display board
32 Picture quality control
33 Sound console
34 Tape decks
35 Amplifiers
36 Cyclorama
37 Hoist control panel
38 Output socket
39 Spotlight
40 Floodlight
41 Cyclorama lights
42 Scenery hoist
43 Soundproof wall
44 Studio speaker
45 Telecine
46 Videotape picture
47 Monochrome final transmission picture
48 Colour final transmission picture
49 General monitor
50 Credit holder
51–54 Pictures from cameras 1–4
55 Monitor bank for extra cameras and outside broadcasts

Sound recording and reproducing

It was possible, at about the turn of the century, to reproduce the expression and "attack" of an actual piano performance on a pianola paper roll. But the story of sound recording and reproduction as the terms are now understood is the story of the gramophone, the "talking" motion picture and the tape recorder.

Thomas Edison (1847–1931) invented a hand-cranked phonograph [1] in 1877. The machine converted the air pressure variations of sound waves into a mechanical record consisting of a groove of varying depth in a sheet of tin-foil wrapped round a cylinder. The foil was soon replaced by a hard wax cylinder and in 1894 Charles Pathé (1863–1957) and his brother Emile (1860–1937) opened a phonograph factory.

The development of early sound systems
Meanwhile Emile Berliner (1851–1929), a German in Washington, DC, patented a "gramophone" in 1887. This used a flat disc instead of a cylinder, the sound groove being cut as a spiral. By 1900 the hill-and-dale recording was replaced by a groove that

made a stylus vibrate from side to side. And with the advent of a shellac disc pressed from a "negative" of the original recording the gramophone, known in the United States as the victrola, became widely popular [Key].

At first methods of recording and playback were entirely mechanical and the quality of reproduction was poor. The invention of the triode valve in 1906 opened the way to electrical recording and by the 1930s music of much improved quality was reproduced electronically from shellac discs running at 78 revolutions per minute (rpm).

In 1948 the American Columbia Company successfully demonstrated an "unbreakable" vinyl plastic disc and high-fidelity microgroove discs playing for 25–30 minutes a side at 33 1/3rpm soon became popular. By 1958 the stereo disc had been introduced. It had separate twin soundtracks in a single groove (each corresponding to the sounds received by a listener's left and right ears) and provided a sense of musical presence hitherto unknown in a recording [2].

Once silent motion pictures had been developed, a system of synchronized sound

followed in the mid-1920s. Early "talkies" used an adaptation of the already popular shellac disc. The most successful system had a 40cm (16in) disc running at 33 1/3rpm, with a motor linked mechanically to the film drive. By 1930, engineers had developed a far better system that recorded the sound optically on one edge of the film in the form of a transparent line varying either in density or in width [3]. A fine beam of light shone through the moving line on to a photoelectric cell, its varying electrical output being amplified and fed to loudspeakers.

Experiments in magnetic recording
The idea of converting the varying pressure waves of sound into a magnetic pattern on a continuous steel wire was developed in the 1920s. The British Broadcasting Corporation, which wanted to transmit the same programmes to different parts of the world at different times, installed an improved machine in 1931. This used a 6mm-wide steel tape running at 1.5m per second to record and replay programmes 20 minutes in length.

In 1929 Fritz Pfleumer patented a

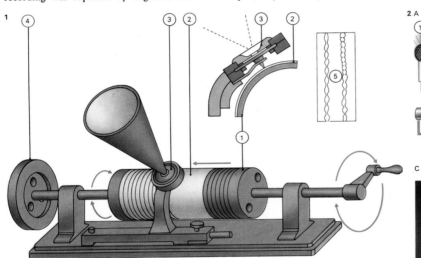

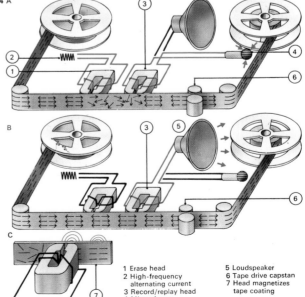

1 Thomas Edison's phonograph consisted of a brass cylinder [1] cut with a spiral groove. Over this was wrapped a sheet of tin foil [2]. A conical funnel focused sound on to a metal diaphragm [3], which touched a steel stylus held by a flat spring. The sharp tip of the stylus pressed on the foil. The stylus was mounted on a screw of the same pitch as the groove, so that when the cylinder was turned the stylus pressed always over the groove. A flywheel [4] helped keep the cylinder speed steady. When sound caused the diaphragm to vibrate, the stylus pressed the foil into the groove in step with the vibrations. The cylinder was wound back to its original position, and the sound reproduced by turning the handle. The stylus and the diaphragm were then vibrated by the indentations [5] in the foil.

3 An optical system is used by the film industry for recording sound, the vibrations being reproduced in the form of a transparent line of varying thickness. In the projector, a light beam passes through the line on to a photoelectric cell. The width of the line controls the amount of light reaching the cell, the resulting electrical signal being amplified to produce sound. In home movie equipment the sound is recorded on a magnetic stripe as on tape.

4 In recording on magnetic tape [A], the tape first passes an erase head which leaves the magnetic particles on the tape in random disarray. Then the record/replay head, energized by a microphone signal, orientates the particles according to the signal's waveform. In playing back [B], the tape again passes the record/replay head. Magnetic variations reproduce in it the currents that formed them [C]. After amplification, the currents drive a loudspeaker.

1 Erase head
2 High-frequency alternating current
3 Record/replay head
4 Microphone
5 Loudspeaker
6 Tape drive capstan
7 Head magnetizes tape coating

2 On a stereo disc, the groove walls are angled at 90° to each other. When a recording is made [A], sound from one microphone [1] produces a hill-and-dale contour on one groove wall. The second channel sound [2] contours the other groove wall. After being pressed between metal moulds [B], the final plastic disc is ready to be played back. The cartridge stylus of the play-back machine vibrates in two planes perpendicular to each other [C]. The movement in each of these planes actuates separate electro magnets, which are wired to two different amplifiers and separate loudspeakers. The sound recorded by the two original microphones is thus reproduced independently.

recording tape that had a flexible insulated base with a magnetic coating. The German AEG company developed this invention and in 1935 in Berlin exhibited the Magnetophone, the first modern tape recorder. But it was not until the end of World War II that the potential of the reel-to-reel tape recorder using 0.5cm (0.25in) plastic-based tape with an iron oxide coating [4], was fully realized. The tape can move at various speeds, with higher speeds providing greater fidelity. The most common speeds for domestic recording are 4.8cm/sec (1.875in/sec), 9.5cm/sec (3.75in/sec) and 19cm/sec (7.5in/sec). For stereo recording two separate soundtracks are recorded side by side using two microphones. Stereo reproduction or playback needs two amplifiers and speakers. The system of dividing the tape into four tracks permits two stereo recordings to be made: tracks 1 and 3 in one direction and 2 and 4 in the reverse.

The cartridge and cassette revolution
The main drawbacks of a reel-to-reel tape recorder are the vulnerability of the tape to damage during threading and the inconvenience of threading and storing it. To eliminate these, the tape cartridge and cassette were invented. The former contains a single loosely wound reel of continuous tape. The tape is fed out from the centre at an angle, guided by rollers to the gate where it touches the playback head of the equipment and then back to the outside of the reel.

A cassette recorder [7] has two spools like those of a reel-to-reel recorder, but much smaller, and is suitable for recording, automatic rewind and playback operations. The tape is only 3.8mm (0.15in) wide and runs at 4.88cm/sec (1.875in/sec). The cassette (plastic case) holds tape for 45, 60, 90 or 120 minutes' playing time and clicks into the cassette player without the need for threading the tape. High-frequency random noise called "tape-hiss" (a consequence of having four tracks recorded on the extremely narrow tape at a slow speed) can lower the quality of the cassette recording. But this blemish on the quality can often virtually be eliminated by using an electronic noise-reduction circuit.

The marvel of sound reproduction was captured in a narrative painting which became a trade-mark as "His Master's Voice". In 1899 the British artist Francis Barraud portrayed a fox terrier called Nipper listening to the voice of his dead master from a phonograph. The Gramophone Co. later EMI, which bought the painting, got Barraud to paint in a gramophone.

6

5 The human hearing process, using a pair of ears, can sense the direction of a sound and so can discriminate between a sound from one direction and background noise from another. This is not so of a microphone, which combines all the sounds it "hears" into one electrical wave. To maintain a high signal-to-noise ratio it is therefore normal to place microphones as near as possible to the desired sound source. To record an orchestra and choir with proper balance between instruments and voices a single microphone would have to be roughly the same distance from each sound source. Except in a perfectly sound-proof studio this would result in an unacceptably low signal-to-noise ratio. The engineer solves this problem by providing separate microphones for each section of the choir and orchestra and for the soloists, the combined outputs then being mixed by electronic means. The balance can then also be adjusted.

5

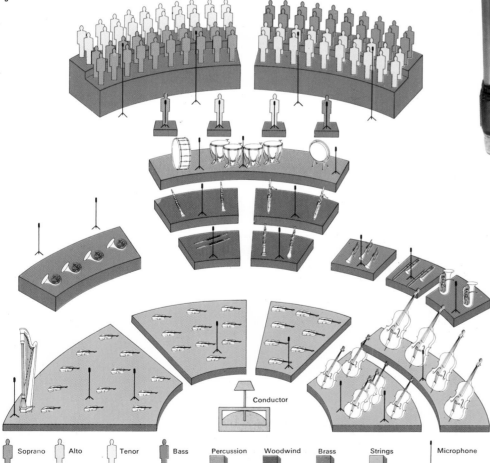

| Soprano | Alto | Tenor | Bass | Percussion | Woodwind | Brass | Strings | Microphone |

Conductor

6 A jukebox, an early example of which is shown here, is a coin-operated machine for selling music. Some modern versions hold 200 or more records and provide stereophonic sound reproduction.

7 A portable cassette player allows tape-recorded music to be heard almost anywhere. With transistorized circuits powered by batteries, it is light and compact and can be used with pre-recorded tapes.

7

Video recording and reproduction

From the earliest days of the television industry, there was a need for a method of recording programmes in such a way that they could be played back immediately. Cinema film is often unsuitable because the delay in processing prevents it being shown immediately, and the fact that film cannot be re-used makes it expensive.

A Scotsman, John Logie Baird (1888–1946), inventor of a mechanical television scanner, was the first man to record a moving picture by other than photographic means. In 1927 he used equipment designed for cutting 10in (25.4cm) 78rpm sound records to record pictures using the output of his 30-line TV scanner.

Magnetic tape recording

The development of magnetic tape recording meant that Baird's aim could be achieved more simply with immediate replay. In the meantime, however, television had progressed from the first 30-line format to a picture having 405 to 819 lines. The best of the early shellac records could reproduce audio (sound) signals up to a frequency of about 4,500 hertz (Hz), and high-fidelity LP records today reproduce musical overtones up to 15,000Hz or more. But a modern TV signal includes frequencies up to 5MHz.

The frequency response of a magnetic tape recorder is limited by the size of the head gap and the speed with which the tape passes the head. The finest equipment operating at 19cm (7.5in) per second cannot reproduce frequencies much above 25,000Hz. An increase in frequency response to 5MHz can be achieved only by increasing the tape-to-head speed to at least 1,270cm (500in) per second.

The earliest videotape recorders (which record pictures on magnetic tape just as a tape recorder records sound) were designed to operate at tape speeds of 254cm (100in) per second or more. They required enormous spools of tape and presented speed control problems; also constant head-to-tape contact was difficult to achieve. Research led to the introduction by Ampex, in 1956, of the first transverse-scan recorder – the system used professionally today. In this system, the tape, normally 5.08cm (2in) wide, moves at either 38cm (15in) or 19cm (7.5in) per second. Four record/replay heads mounted on a drum sweep across the tape producing transverse parallel record tracks for the video signal [2]. Linear tracks at the edges of the tape are used for tape-speed control, picture cueing and sound. Head-to-tape "writing" speeds of 3,810cm (1,500in) per second are achieved.

Helical scanning

The highly sophisticated transverse-scan colour videotape recorder is far too expensive for institutional and domestic use. A cheaper method uses what is termed helical scanning. Here the tape passes in a helix around a rotating drum into which one or more record/replay heads are built [2A]. The drum is rotated rapidly in the opposite direction to the tape (it "slips" round within a loop of tape). As the tape rises by its own width in its journey around the drum, the heads sweep across the tape at an acute angle. Tape-to-head "writing" speeds of up to 2,540cm (1,000in) per second are achieved with helical-scan portable videotape recorders. In domestic applications, this system has the

1 The earliest video recorder worked in exactly the same way as an audio tape recorder, using spools of wide magnetic tape. The video track was recorded by a stationary head and the recording of the high frequencies required to produce pictures of acceptable quality was achieved by using very fast tape speeds. In the most highly developed of these machines, 914.4cm (360in) of tape 5.1cm (2in) wide passed the head each second. Even so, picture quality was poor by modern standards and the machines suitable only for black-and-white pictures.

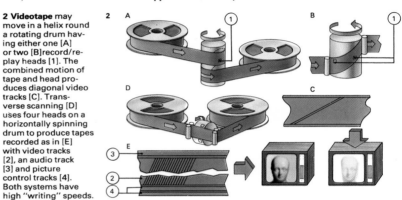

2 Videotape may move in a helix round a rotating drum having either one [A] or two [B] record/replay heads [1]. The combined motion of tape and head produces diagonal video tracks [C]. Transverse scanning [D] uses four heads on a horizontally spinning drum to produce tapes recorded as in [E] with video tracks [2], an audio track [3] and picture control tracks [4]. Both systems have high "writing" speeds.

3 The moving head was invented by a little-known company in Redwood, California, in 1956, as a means of achieving the high tape-to-head speed required for video recording without the tape itself having to move at unmanageably high speed. The latest Ampex machine [A] is a self-contained videotape recorder designed for colour television. The studio videotape recorder is expensive and heavy and many manufacturers, foreseeing a wide market for a cheaper and smaller machine suitable for home, school, police and other work, have made cassette videotape recorders that are compact and efficient, yet not too costly. This cassette player [B] is a typical example of such a model.

advantage that a conventional TV can be used as a combined picture and sound monitor, the signal produced on replay being fed into the aerial socket via a demodulator. Tape 1.27cm (0.5in) wide is commonly used, although some recorders of this kind use 1.9cm (0.75in) or 2.5cm (1.0in) tape. The speed-control mechanisms on these machines are less sophisticated, and generally use servo motors in much the same way as those on studio machines.

Types of video disc
It is possible for magnetic tape to be replaced by magnetic discs in video recording, and the system has the advantage of giving rapid access to any part of the recording for immediate replay, in slow motion if necessary [6]. But a disadvantage is that playing time is short – generally 15-18 seconds using both sides – whereas a modern videotape recorder can accommodate a 90-minute colour TV programme on one spool of tape.

The production of pre-recorded TV tapes is expensive and so some manufacturers have been developing video machines that use discs similar to gramophone records. Such machines are not recorders, but replay programmes on mass-produced discs [4].

There are two kinds of such video discs, which differ in the "pick-up" used to extract the recorded information. In a system developed by Philips, MCA in the United States and other companies there are small elliptical depressions in the disc [6]. The disc revolves at 1,500 revolutions per minute (for a 50Hz electricity supply) or 1,800rpm (for a 60Hz supply). A laster beam scans the lower surface of the disc and the reflected beam becomes modulated and provides video and sound signals for 25 or 30 television pictures per minute (normal rates of transmission). A disc 30.5cm (12in) in diameter records up to half an hour's television. It can provide slow motion or "stationary" pictures.

The video disc developed by Telefunken and Decca is grooved like a long-playing gramophone record [7], although the grooves are finer and spaced much closer together. Playing speeds are again 1,500 or 1,800rpm but such discs record only about seven minutes of television.

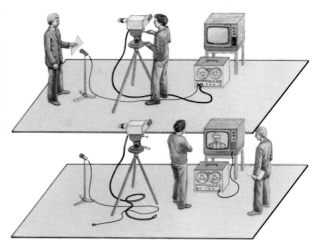

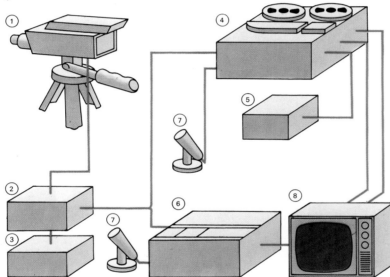

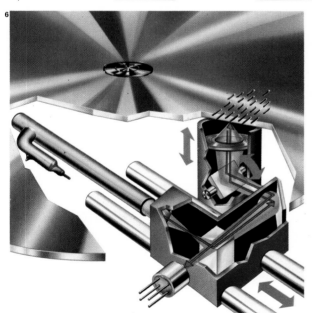

KEY

The recording and reproduction of moving pictures on magnetic tape has been developed to such an extent that reproduced television pictures are nearly indistinguishable from orginals. Even portable video-tape recorders that can be used domestically give remarkably good results.

4 Live video recordings in black-and-white or colour can be made with ease today. The diagram shows a typical set-up using equipment that is readily available on the market. The camera [1] is connected via its control [2] and sync [3] units to the reel-to-reel colour video-tape recorder [4] (with colour pack [5]) or to the video cassette recorder [6]. A microphone is simultaneously connected to record the accompanying sound [7]. The colour TV set [8] has special sockets to enable it to be used as a monitor during recording. Recorded tape can be played back through the TV set immediately after the tape is wound back.

5 The slow-motion "action replay" seen frequently on sports programmes was made possible by the invention of a magnetic disc recorder that provides continuously variable slow-motion forward and reverse as well as stop, "freeze" and natural-speed replay. It records about 7.5 to 9 seconds of television programme material on a series of concentric magnetic tracks on each side of the disc.

6 The non-magnetic video disc is used by several firms as an alternative to magnetic tape systems. Philips and other companies have designed systems that use minute ellipitcal depressions on a 30.5cm (12in) disc to record 30 minutes of colour video signals with sound. It is scanned using a laser beam.

7 Video discs are like sound records. The Teledisc, shown much magnified [right], has about 25 hairline grooves in the space occupied by each groove on a standard LP gramophone record (left). Like the Philips video disc (illustration 6) the Teledisc revolves at 1,500 rpm (UK) or 1,800 rpm (USA).

Radar and sonar

Sonar (formerly known as ASDIC in Britain) is a system of direction finding and rangefinding using sound waves under water. Radar uses the same principles, with radio waves instead of sound waves.

The essentials of the two systems are simple. Acoustic (sound) or electromagnetic (radio) waves are transmitted. When they meet a solid object, some are reflected and return – there is a sound or radio echo. The time that elapses between a wave's transmission and return, multiplied by the speed of the wave, gives the distance travelled. Normally this is twice the distance of the object. Early radars [Key] were mounted on trailers for portability. Some used a "lens" for directing the radar beam.

Development and uses

Sonar (from *SO*und *N*avigation *A*nd *R*anging) was developed principally for the detection of submarines and to act as a submarine commander's "ear" for the detection of other vessels, minefields, submerged ice, wrecks and other underwater hazards. Sonar can be "active" or "passive". In active sonar,

an acoustic wave is transmitted and its echo picked up [1]. In passive sonar, other vessels are detected by listening for noise generated by their engines. Today sonar is also used by fishing vessels seeking shoals of fish and for surveying the ocean bed.

In 1935 a British team headed by Robert Watson-Watt (1892–1973) started a programme of research aimed at adapting and developing radio location for military use [5]. By the outbreak of World War II in 1939, Britain had an aircraft detection system along its east coast. Known as RDF (radio direction finding) it was quickly extended to cover the south coast and was a major factor in Britain's ability to win the war in the air even though its aircraft were outnumbered.

The secrets of RDF were passed to the United States, where further intensive research was conducted and a new name, "radar" (*RA*dio *D*etection *A*nd *R*anging), given to the new technology. German scientists conducted similar research during the early years of the war and achieved similar, although less technically advanced, results.

A radar installation consists of three

separate units: a transmitter that radiates a special form of radio signal; a receiver that picks up and processes any reflected waves; and a presentation unit that gives a visual display from which the operator can immediately read the desired information.

Types of radar aerials

Radar aerials (antennas) vary in design according to their purpose. Many consist of a metal lattice in the shape of a flat dish or rectangular array, which can be steered at any angle to the vertical or horizontal to aim the radar in the required direction. Some can be "locked on" to a target so that they track it automatically. A location radar has a narrow searchlight-type beam, focused by a parabolic reflector, so that the bearing and elevation of reflected waves can be accurately measured. A search radar uses an aerial that radiates waves over a wide arc. The beam is kept relatively flat for a ship's radar, but covers a vertical arc in an aircraft search radar. In both cases, the aerial is sometimes made to revolve horizontally so that the radar sweeps round continuously. Most radar

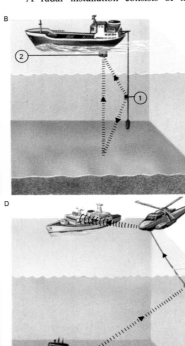

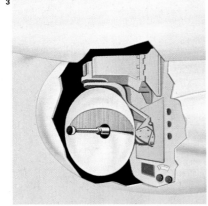

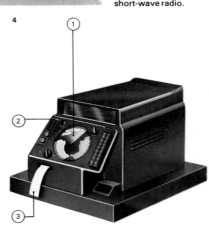

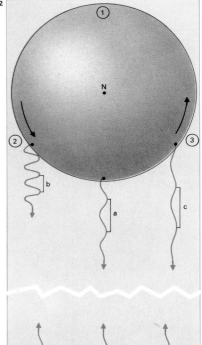

1 The time taken for a sound wave, transmitted from a device in the water under a ship, to travel to the sea-bed and echo back to the ship is measured to calculate the depth of the water. The acoustic generator [1] can be mounted directly under the ship [A], in which case the echo time multiplied by the speed of sound in water gives twice the depth. The signal-to-noise ratio of the echo picked up by the microphone [2] can be improved by lowering the sound generator into the water [B], so reducing the total distance that the sound waves must travel. The surface vessel may confuse an echo from a submarine and so a warship can use a sonar buoy [C]. For maximum mobility a ship makes use of its own helicopter which suspends a "dunking" sonar in the water [D] and transmits signals back to the mother ship by means of a short-wave radio.

2 The reflected wave of a radar beam that hits an approaching or receding object has a frequency greater or less than the transmitted wave. If a wave of a given frequency [a] is reflected from a planet [1] spinning about its pole [N], the frequency [b] of the reflected wave from the approaching side [2] will be more than the frequency [c] from the receding side [3]. The difference between b and c can be used to compute the planet's rate of spin and day length.

3 An aircraft's radar aerial may be hidden beneath a streamlined pod or radome made of a material that protects it from bad weather without seriously affecting the transmissions.

4 The presentation unit of a storm-detector radar has a cathode ray tube [1] that maps signals from a rotating aerial as glowing storm clouds. The function selector [2] controls observed range; paper tape [3] gives a record.

receivers have large aerial arrays designed to receive as much as possible of a reflected signal, which is usually very weak.

Processing and presentation
Radar signals, suitably processed and amplified, are passed to the presentation unit along with the original transmitted signal.

The signal presentation system is generally a cathode ray tube display, based on an oscilloscope, and can give either a simple read-out of range, or of elevation, or both. Alternatively it can display a complete electronic "map" of the position of wave-reflecting objects in all directions [4].

In a simple straight-line display, the direction and elevation of the located object (an aircraft in the sky, for example) are read from dials indicating the direction and elevation of the radar beam. The range is read off a pulsed straight-line oscilloscope trace in which the time between transmission and reception is twice the range.

The "map" or Plan Position Indicator (PPI) display is produced by a straight-line oscilloscope display arranged with a radial

scan that begins at the centre of the tube and ends near its circumference. The scan is then made to rotate, with the start of the scan as the centre of rotation, in step with the rotation of the aerial. The oscilloscope screen is coated with a material having a long afterglow, so that an echo signal (a bright spot) on the screen remains visible during the time taken for one complete revolution of the aerial. The distance of an echo spot from the tube centre represents the range of the object, and its bearing on the screen conforms to its actual bearing.

Most radar installations depend entirely on the weak waves reflected by solid objects, although some systems use a relay receiver-transmitter to receive and retransmit a more powerful return wave. Such a system is known as secondary radar.

When an electromagnetic wave is reflected by an object moving towards or away from the radar installation, the frequency of the reflected wave is altered. This is the Doppler effect, well known in acoustics. The resulting frequency shift can be used to calculate the speed of an object [6].

During World War II, the British developed small, transportable radar sets so that their radar defence system was able to monitor any area of potential air-attack.

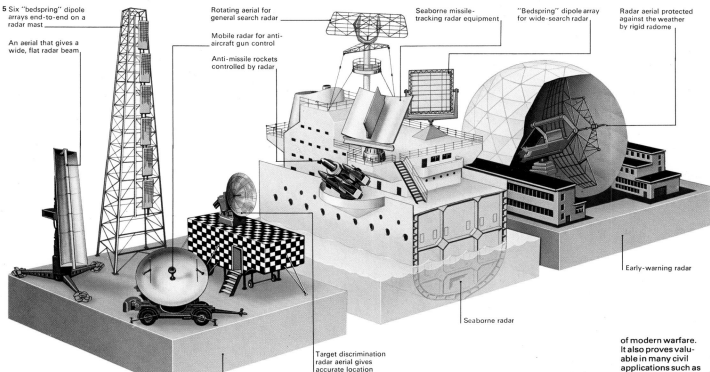

5 Six "bedspring" dipole arrays end-to-end on a radar mast

An aerial that gives a wide, flat radar beam

Rotating aerial for general search radar

Mobile radar for anti-aircraft gun control

Anti-missile rockets controlled by radar

Seaborne missile-tracking radar equipment

"Bedspring" dipole array for wide-search radar

Radar aerial protected against the weather by rigid radome

Early-warning radar

Seaborne radar

Target discrimination radar aerial gives accurate location

German World War II radar

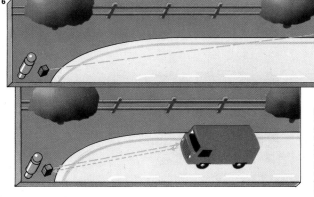

5 Radar was first used in the 1920s to demonstrate the existence and extent of the ionosphere, which was found to reflect radio waves. Research during the 1930s in Britain, USA, Germany and France developed radar for military purposes and it became a vital aid in both defence and attack during World War II when British and German electronic engineers designed and built similar installations. So greatly has its efficiency improved that radar is now regarded as an indispensable tool

of modern warfare. It also proves valuable in many civil applications such as meteorology, navigation, airport traffic control and surveying. Military uses of radar include long-range aircraft and missile warning systems; location radar for automatic control of anti-aircraft guns; airborne radar for use as a night "eye" when attacking enemy bombers in darkness; and naval radar to give information about the presence of enemy shipping in conditions of poor visibility. Radar is also used in sophisticated weaponry as a homing device for steering anti-missile missiles.

6 Radar is used by the police to compute the speed of a passing car. If the speed exceeds the legal limit, the operator can warn a colleague by walkie-talkie radio in time to flag down the offending driver and

charge him with speeding. The radar set continuously measures the car's distance from the set and the change in range is used to compute the speed by electronic means, giving an immediate visual display.

Chemical engineering

Research chemists at their laboratory benches develop new products, generally using only glass apparatus. A new product is made in the laboratory in gramme quantities, but if it is to sell in the open market the makers may well have to sell tonnes – for well-developed commodities such as plastics or fertilizers, up to millions of tonnes a year. It is this transition from laboratory to factory that is the basis of chemical engineering.

The chemical plant

The processes in chemical engineering are often basically only refined versions of activities that take place in the kitchen. But chemical engineering is an exact science devoted to designing, building and operating equipment on an industrial scale in the special kind of factory known as a chemical plant. Yet the process need not literally be a chemical process. Chemical engineers design plant for physical processes such as evaporation, distillation, liquefaction and filtration, as well as for processes that involve a change of chemical composition.

The design of chemical plant and equipment is a separate and distinct discipline, although it overlaps with several others. The task is best done not by a team containing both chemists and mechanical engineers but by individual engineers who have the right combination of skills. They need to know, for instance, how large a vessel should be for "cracking" naphtha from petroleum to change it to ethylene and propylene (which will later be converted to plastics, detergents or dry-cleaning fluids). What should it be made of? How much power is needed for pumping, how much heating and cooling? Will the immediate products be separated by distilling them? It is practical questions such as these that are the stuff of engineering, but in this case they are applied to changing materials in both composition and form.

The foundation of the discipline of chemical engineering lies in chemistry, physics and mathematics. Its operation is based on interweaving these with the knowledge gained from older branches of engineering, as well as assessing the economics of processes. Chemical engineers often use computers for assessing design and operational factors [5].

Many chemical engineers are employed in the chemical industries, where their skills have contributed to an exceptionally high growth record. They also work in oil refining, atomic energy, gas and coal industries.

From theory into practice

The important basic idea of chemical engineering is that processes used for producing quite diverse chemicals – for example, acids, dyes or drugs – can be considered as a series of "unit operations". These unit operations are the same whatever the detailed nature of the material being processed. Consequently a common body of theory can be applied to a range of process industries.

Unit operations include distilling, filtering, mixing, crushing and crystallizing. It is even possible to unify the theory further. Many of the operations can be considered as examples of the study of how fluids flow, how heat is transferred or how material is transferred through surfaces (mass transfer). Thus what is learnt about mass transfer can be applied to unit operations for absorbing gases, leaching soluble materials out of

CONNECTIONS

See also
240 Soaps and detergents
242 Fireworks and explosives
244 Colour chemistry
246 Cosmetics and perfumes

In other volumes
132 Science and The Universe

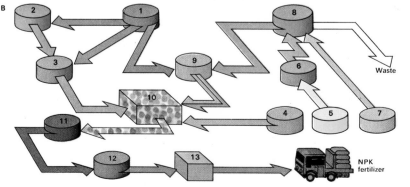

1 The key elements in plant fertilizer are nitrogen, phosphorus and potassium. The basic processes, therefore, in a factory making fertilizers [A] are extracting nitrogen from the air and phosphorus and potassium from natural minerals, and combining these elements (sometimes individually, sometimes jointly) into soluble chemical compounds which are then used to feed plants. On a world scale this demands manufacture of tens of millions of tonnes of fertilizer chemicals. A chemical engineer designs and operates processes [B] to obtain the necessary compounds. He extracts nitrogen from the air and hydrogen from natural gas or water. He then combines the two by "fixing" the nitrogen as ammonia [1]; part of this is converted to nitric acid [2] which reacts with more of the ammonia to form ammonium nitrate [3]. The potassium compound, probably sylvite [4] (potassium chloride), may be used directly. Sulphur [5] is converted into sulphuric acid [6], which is reacted with natural phosphate rock [7] to produce phosphoric acid [8]. With ammonia, this yields diammonium phosphate [9]. The three "NPK" compounds are then mixed and granulated [10], dried [11], coated [12] and packed [13] for market.

solids, or crystallizing dissolved substances.

The chemical engineer must also take into consideration what will happen when he "scales up" the quantities [3]. Sometimes a different reaction takes place in a big tank from that in a small beaker where the reagents have only a short way to travel before meeting. Stirring the contents of the beaker brings them quickly into contact and heat can readily reach all parts or be as rapidly removed. When ammonium di-uranate is produced in the laboratory by reacting the nitrate with ammonia, for example, it precipitates almost immediately. With tonnes of solution in a large tank precipitation may take several hours because the practical speed of reaction depends on the rate at which pumps can operate and stir the liquid. As another example, nitrobenzene is made in the laboratory in a round glass flask which can be immersed in water to remove heat. On a large scale engineers use a different form of vessel to provide enough surface area to remove heat at a reasonable rate, and it is made not of glass but of metal.

Evaporating a liquid in the laboratory may be carried out in a glass vessel over an open flame (if the liquid is not inflammable). But the engineering operation needs large metal vessels with large amounts of surface area to promote efficient transfer of heat between the source and the liquid to be evaporated. Similarly mixing is transformed from stirring with a glass rod in a beaker to a major engineering operation in large metal vessels with motor-driven paddles with elaborate blades. Thus the study of materials for making apparatus (and their fabrication) and of the laws governing change of scale are important parts of this branch of engineering.

Biochemical engineering

Applied to biological processes or materials the discipline of chemical engineering is called biochemical engineering. An important area of this work has been the growing of food proteins on petroleum products to produce animal feed. Biochemical engineering has also been applied to new forms of fermentation processes for making antibiotics and vitamins, including the extraction and concentration of the end products.

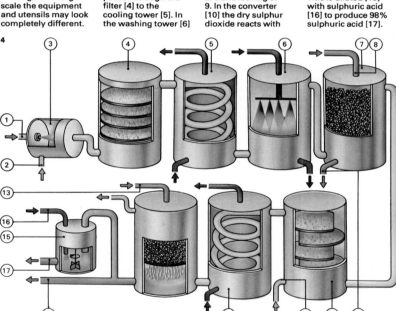

The crowning achievement of a chemical engineer's work is the factory where chemical products such as nylon are made. Chemical engineering serves a modern group of industries rich in challenges because the processes tend to change more quickly than in most other industries. Many of the products help to make life more convenient. Working in close collaboration with chemists and other specialist engineers, chemical engineers choose the process, choose or design the plant, help in its construction and, after starting the process, control its operation to give the most economic results, as either technical or general managers.

2 In one type of evaporator heat is transferred from a heating medium [1] (often steam) to a liquid [2]. The two are kept apart by partitions made of metal or sometimes carbon. The liquid circulates through tubes [3] and is heated to give off a vapour [4]. The loss of solvent concentrates the liquid which is discharged [5] at the bottom of the evaporator. The cooled heating medium [6] is removed.

3 The operations of chemical engineering are essentially scaled-up versions of operations in the chemical laboratory, although on a larger scale the equipment and utensils may look completely different.

4 To make sulphuric acid, liquid sulphur is blown [1] into a burner [3], also fed with air [2], forming sulphur dioxide. This passes through the filter [4] to the cooling tower [5]. In the washing tower [6] it is washed with water then dried in a packed tower [7] of Raschig rings by means of 98% sulphuric acid entering at 8 and leaving at 9. In the converter [10] the dry sulphur dioxide reacts with oxygen [11], is cooled [12] and the resultant sulphur trioxide is absorbed in 98% sulphuric acid [13] to yield oleum [14]. This is diluted [15] with sulphuric acid [16] to produce 98% sulphuric acid [17].

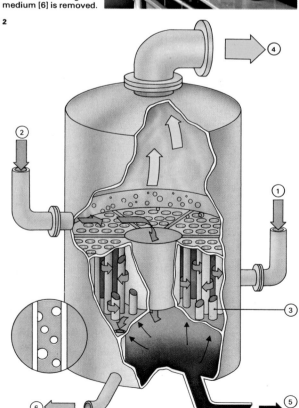

5 Computers are extensively used in chemical engineering for solving design problems and for controlling operations. The basic relationships of a process may be investigated by the computer and corrective action can be applied if some measured quantity in any part of the plant (eg temperature, pressure, viscosity or composition) deviates from requirements and would otherwise result in reduced yield or in poor quality of the final product.

Soaps and detergents

Man has been using various kinds of cleansing agents for thousands of years, because water itself does not readily get rid of dirt and grease, as our ancestors discovered early on. Water is not a good "wetter" because of its high surface tension, which causes it to run off a greasy area or to stay there without penetrating it. Only when some cleansing agent that lowers surface tension is added, can water penetrate to remove grease [5].

The Babylonians added alkaline plant ash (potash) to water. Other primitive cleansers were fuller's earth (a type of fine clay which easily absorbs impurities from oil and fat), soap berries (tree fruits containing a soapy substance called saponin), and the sap of the soapwort plant. Soap itself was probably first made in the Nile valley and about 600 BC Phoenician seamen carried the knowledge to the Mediterranean coasts. In the first century AD the best soap was made from goats' fat and beechwood ashes. Animal fat and wood ash remained the raw materials for centuries. Nowadays, manufacturing materials for keeping clean is a major industry and soaps and detergents are used in most homes [Key].

Soap-making was a small domestic industry until the end of the eighteenth century when a number of changes took place [2]. In 1787 it was discovered that the alkali caustic soda could be made from common salt – a plentiful raw material – so manufacturers were no longer dependent on plant ashes. Vegetable oils such as olive oil had been used in soap-making by the Spaniards as early as AD 700 and now others became more easily available from countries outside Europe. Coconut oil, palm oil, sesame oil and soya-bean oil were imported from Africa, South-East Asia and China and by 1900 were replacing animal fats, which were in short supply. Social changes in the nineteenth century had led to an increased demand for soap and production in Britain rose from 90,000 tonnes in 1853 to 300,000 tonnes by 1900.

Modern soap-making

The treatment of oil or fat with alkali (saponification) is the first stage in the manufacture of all types of "washing" soap [4]. This reaction produces sodium salts of stearic, palmitic and oleic acids. After saponification the soap contains about 30 per cent water – for making the more dense toilet soaps this has to be reduced to about 12 per cent. Then various refinements such as perfumes, preservatives, whiteners or colouring materials and, sometimes, germicides (for medicated soaps) are added and thoroughly mixed. The molten soap is then cooled and cut to size – abrasives are added if scouring soap is wanted.

Soap flakes are produced by spreading the molten soap over water-cooled drums and producing ribbons of soap which are rolled progressively thinner and then broken into flakes. Soap powders normally contain silicates and phosphates which are added to the liquid soap – the resultant slush is superheated under pressure and sprayed into the top of a tower. As the droplets fall they solidify to give the familiar soap powder.

Not all soaps are soda based. Some made with caustic potash are used in liquid soaps and shaving cream and other types are used in textile finishing, as lubricants, in the cosmetics and pharmaceutical industry, in polishes and in emulsion paints.

1 Women in primitive societies still clean their clothes by beating them with rocks in the nearest stream. This method works but shortens the life of the clothes and is hard work for the women.

2 Early soap-making was done in open pans where lye (crude sodium hydroxide) made from plant ash and limestone were heated with animal fats. The soap formed a crust on cooling and could be skimmed off.

3 Synthetic detergents are a mixture of various ingredients, the chief of which is the cleansing agent, or surfactant. A common surfactant has the chemical name sodium dodecylbenzene-sulphonate. It is made from petroleum products and has the molecular structure shown. Much dirt is held on clothing by greasy substances. The long tail (the dodecyl part) of the surfactant molecule "dissolves" in the grease, while the ionic head (the sulphonate part) dissolves in water. The two types of detergent manufacture shown use common hydrocarbon starting materials.

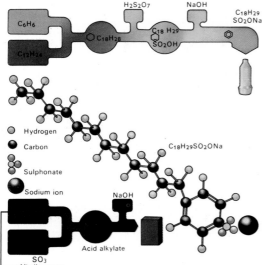

- Hydrogen
- Carbon
- Sulphonate
- Sodium ion

C_6H_6

$C_{12}H_{24}$

$H_2S_2O_7$

$C_{18}H_{28}$

$C_{18}H_{29}SO_2OH$

NaOH

$C_{18}H_{29}SO_2ONa$

$C_{18}H_{29}SO_2ONa$

NaOH

SO_3
Alkylbenzene

Acid alkylate

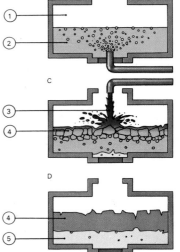

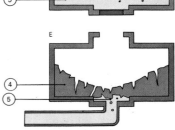

4 Modern soap-making methods [A] follow ancient principles. Fat and alkali are saponified producing soap and glycerol. Today's equipment allows the process to be completed in 15 minutes instead of several days. [B] Fats and oils [2] are reacted with water under high temperature and pressure [1]. [C] Alkali [3] is added to the resulting mixture of fatty acids to produce soap [4]. [D] The soap still contains glycerol [5] which is washed out [E] using brine, and the salt solution is separated from the soap in a centrifugal extractor, which works like a spin dryer. Any fatty acid then left is neutralized with alkali and salt and again separated in a centrifuge. The molten soap is then poured into mixing machines which blend in other ingredients such as perfumes, softeners, germicides and colour. It is then ready for shaping into hard soap, toilet soap, flakes or powder.

However, soaps have a number of disadvantages as cleansing agents. They do not work in even slightly acid water, which is why various alkaline substances – carbonates, phosphates and silicates – are added to household soaps. And, most important, they do not work well in hard water. The soap reacts with calcium and magnesium salts to form the familiar insoluble "scum" which leaves rings on bathtubs or a whitish film on glassware. Moreover, the availability of the raw materials (oils and fat) for making soap varies unpredictably. For these reasons manufacturers began looking, in the late 1940s, for a new type of synthetic detergent that would overcome these problems.

Synthetic detergents
The first synthetic detergents to be made on a large scale were based on products of the distillation of crude oil – at that time a cheap and readily available raw material. By the 1950s detergents based on the synthetic chemical alkylbenzenesulphonic acid (ABS) had captured more than 50 per cent of the fabric washing market.

But early ABS detergents had an important defect. They contained branched chain molecules that made them biologically "hard", or non-degradable, which meant that they were not easily broken down by the bacteria in sewage treatment plants. These detergents were replaced by degradable detergents (linear alkyl sulphonates, LAS).

Threat to the environment
However, even the biodegradable detergents are not free from faults. In addition to the LAS, which is the cleansing agent or surfactant, many other substances are added, such as builders, bleaches, conditioners, optical brighteners and enzymes. A builder prevents the formation of insoluble compounds in hard water. A typical one is sodium tripolyphosphate, which breaks down into phosphates and can lead to an excess growth of algae and other water plants in rivers and lakes [6]. In the 1970s a search began for a replacement for these phosphates. Several alternatives were developed, but all proved to have worse side-effects than the phosphates they were intended to replace.

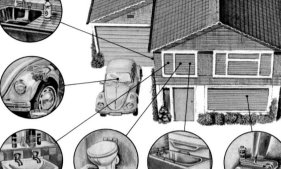

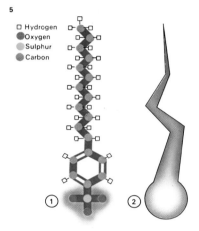

- □ Hydrogen
- ● Oxygen
- ● Sulphur
- ● Carbon

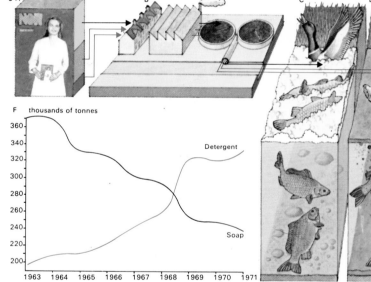

F thousands of tonnes

360
340
320 — Detergent
300
280
260
240 — Soap
220
200

1963 1964 1965 1966 1967 1968 1969 1970 1971

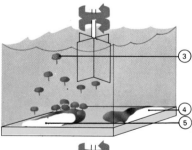

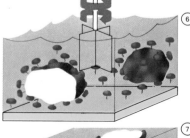

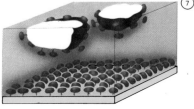

5 Detergents are needed for washing because water is not a good wetter. Detergents solve this problem. One end is water-attracted (hydrophilic); the other dissolves in oils (hydrophobic, or water repellent). Sodium dodecylbenzenesulphonate [1] is a synthetic detergent with a hydrophilic head [2] and hydrophobic tail. The detergent [3] is added to water and the dirty material [4]. The hydrophobic tails stick into the grease [5], while the hydrophilic heads repel each other, forcing dirt and grease up into the water [6]. Dirt particles [7] do not return to the cleaned material because both material and dirt now have the same electric charge and repel each other.

6 The cleanser market has been dominated by detergents since World War II [F]. They are good cleansers but can damage the environment. Surfactants [yellow], phosphates [mauve] and perborates [red] [A] pass unchanged through sewage works [B] to rivers where the surfactant foam [C] kills birds and fish. Phosphates promote algae [D] which absorb the water's oxygen and suffocate fish. [E] Perborates poison sewage bacteria.

7 Foaming rivers in the 1950s were caused by non-biodegradable detergents. The foam prevented oxygen getting to water, and fish died. It also washed the natural oils from birds' feathers, which then became waterlogged, causing drowning.

Fireworks and explosives

An explosive is a substance that can undergo a rapid chemical reaction to produce a large volume of gas. At the instant the gas is formed it occupies the same volume as the explosive and is at an extremely high pressure. The pressure is increased by the generation of heat from the reaction and the rapid expansion which follows moves the surrounding matter. This is an explosion.

There are two main types of explosives: comparatively slow-burning propellants and fast-burning types used for their destructive effect in both military and peaceful operations. In quarrying for building stone, a low-power explosive is necessary to avoid shattering the stone, whereas in mining a high-power explosive is used to break the mineral into the most suitable-sized pieces [3].

Demolition bombs and mines are designed to exert the maximum blast effect and fragmentation bombs and some types of hand grenades have cases that break into many pieces at or above ground level to cause the maximum number of casualties [4]. Some shells contain an explosive charge that is fused to detonate at a predetermined time after the projectile has left the gun, as in "sky bursts". These charges cannot be exploded by the shock of the propellant charge.

Peaceful uses of this power include explosive forming, in which a shock wave from a small charge of high explosive causes a metal sheet to take the shape of a mould.

The manufacture of explosives is under strict government control. "Home-made" explosives can be particularly sensitive to shock and, apart from being illegal, are extremely dangerous.

The development of gunpowder

The first and for hundreds of years the most effective explosive propellant was gunpowder. It was known to the ancient Chinese and was introduced to Europe by the Arabs. In the 1200s it was first used as a propellant for the gun, the principle of which the Arabs had helped to develop. The design and development of firearms was to be closely linked with the quality of the "black powder", as it was then known.

Gunpowder is made by mixing potassium nitrate with carbon and sulphur. The mixture is moistened to prevent spontaneous ignition and the paste is milled to reduce the particle size. The "cake" formed after drying is broken into grains of various sizes. Large grains are slow burning and give a relatively long, slow push to a projectile. This was ideal for a cannon-ball but a shell or bullet in a rifled barrel needed the faster burning properties of small grains.

Cordite, a superior smokeless powder, replaced gunpowder as a military propellant in the late 1800s. It contains nitroglycerine (glyceryl trinitrate) and nitrocellulose (guncotton) with a small amount of mineral jelly. Manufactured as a paste, using acetone as a solvent, it is kneaded into stiff dough and extruded into rods or cords, hence its name. The cords are then cut into suitable lengths and the acetone evaporated.

High explosives are generally based on organic nitro compounds and are detonated ("set off") by a violent shock, such as the explosion of an adjacent charge – a cap or detonator – or a mechanical blow – for example, the firing pin of a pistol or a rifle. High explosives yield stable decomposition

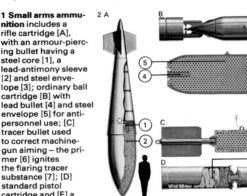
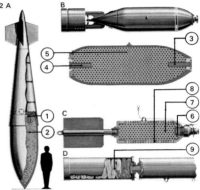

1 **Small arms ammunition** includes a rifle cartridge [A], with an armour-piercing bullet having a steel core [1], a lead-antimony sleeve [2] and steel envelope [3]; ordinary ball cartridge [B] with lead bullet [4] and steel envelope [5] for anti-personnel use; [C] tracer bullet used to correct machine-gun aiming – the primer [6] ignites the flaring tracer substance [7]; [D] standard pistol cartridge and [E] a pistol blank with felt wadding [8] in place of the bullet; [F] large armour-piercing bullet; and [G] primer assembly. The primer [9] is ignited by the cap [10] hitting the anvil [11], which fires the charge [12].

2 **The Grand Slam** [A], a 10-tonne aircraft bomb, was designed to penetrate up to 30m (98.5ft) before exploding. The detonator [1] fired the charge [2]. A general purpose bomb [B] has fuses in the nose and tail [3, 4] to ensure the charge [5] detonates. Fragmentation bombs [C, D] have weakened cases [8] that fragment after the detonator [6] ignites the charge [7]. A built-in parachute [9] may be used to control the rate of descent.

3 **In blasting,** a hole is drilled for the charge, its depth and size depending on the type of rock, the explosive used and the purpose. The detonation causes shock waves to be produced radially, in all directions, with the greatest force at the centre of the explosion. Low-power explosives are used for quarrying building stone to avoid shattering, but for other stones blasting gelatine or dynamite is used. Varying grades of high explosive are available, depending on the hardness of the rock and the depth of hole. Charges are fired by detonator exploded by safety or electric fuse. Large-scale blasting is sometimes used for major excavation work.

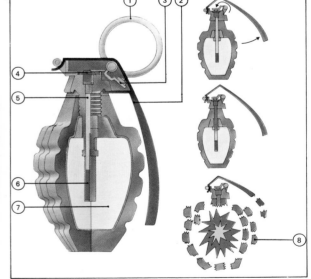

4 **The hand grenade,** of a type used by the United States Army for 50 years, closely resembles the even older British 36 grenade, the "Mills bomb". The thrower crooks his left forefinger through the ring [1], holding the grenade in his right hand. As he pulls the grenade away, the pin is withdrawn and the grenade hurled in an overarm action. The lever [2] flies up, causing the sprung striker [3] to detonate the percussion cap [4], igniting the time fuse [5] made of slow-burning powder. The detonator [6] is fired after a delay of about four seconds, setting off the main charge [7]. Lethal fragments of the cast-iron case [8] are showered in all directions, some of them dangerous for 30m (98.5ft).

products, such as nitrogen, water (in the form of steam) and carbon dioxide.

The development of dynamite

Nitroglycerine, an oily liquid first prepared in 1846, requires only a slight shock to cause detonation. Its use as a high explosive is consequently hazardous. The Swedish chemist and industrialist Alfred Nobel (1833–96) solved the problem in 1866 when he discovered that kieselguhr, a diatomaceous (silica) earth, absorbs up to three times its weight in nitroglycerine, while remaining dry. The granular product, dynamite, retains all the properties of nitroglycerine but is far less sensitive to shock.

Nobel later discovered that nitrocellulose (originally called guncotton) can be gelatinized with nitroglycerine to produce a stiff jelly known as blasting gelatine.

Most military high explosives are organic chemicals such as trinitrotoluene (TNT), ammonium picrate, cyclotrimethylenetrinitramine (RDX or cyclonite) and pentaerythritoltetranitrate (PETN). Bombs and explosive shells contain TNT or amatol (TNT and ammonium nitrate mixture), which are sufficiently insensitive to withstand the shock of the propellant. Armour-piercing shells contain ammonium picrate, which is less sensitive than TNT and can withstand the shock of impact before being detonated. Plastic explosives have military and peaceful uses and consist of RDX blended with wax.

Fuses, detonators and fireworks

A fuse is used to fire an explosive from a distance or after a delay. Safety fuse contains a gunpowder-like substance and is generally used with a detonator. This contains a sensitive primary explosive fired by fuse or by means of an electric current.

Pyrotechny is the use of explosives for signals, display, flares or fireworks. Various colours are obtained by adding a suitable metal salt, usually a sulphide. Antimony gives a white flame, strontium salts red, barium salts green, sodium salts yellow, and a mixture of copper salt and mercury chloride a blue flame. As with all explosives, fireworks are potentially dangerous and should be used only with great care.

KEY

Peaceful uses of explosives include blasting in mines and quarries, the demolition of old buildings and the removal of tree stumps. Only an expert can decide the type and size of explosive charge to be used. Demolishing a tall structure such as the cooling tower of a power station or the chimney of a factory involves the precise placing of small charges that are fired together or in sequence. In this case the space available for falling debris determines the type of explosive and how it is used. A chimney may be made to topple and fall like a tree (by demolishing a wedge shape at the base) or it can be made to fall vertically (by using several charges).

5 The combustion of the propellant in a firework rocket produces its power. The case is wet-rolled and the thrust increased by constricting it near one end (constriction made with a cord before the case is dry). The propellant is packed into the case so that a conical cavity is formed at the burning end, allowing a large surface area of combusion to give the initial push. The cap of the rocket contains flares, or "stars", which are ejected as the propellant burns out.

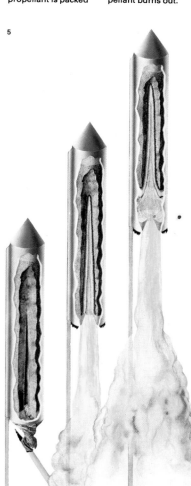

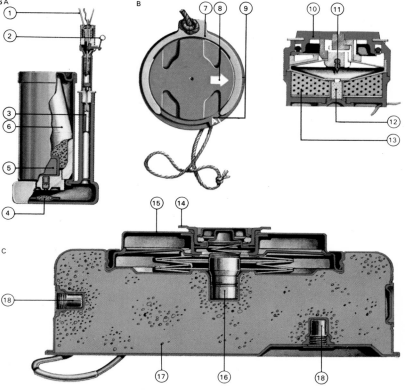

6 Hidden hazards include bounding-type mine [A], buried with the tips of the prongs [1] showing and the safety pin [2] released. When the prongs are touched the firing pin [3] is released, detonating the igniter [4], which in turn detonates the propelling charge [5]. This fires the projectile [6] about 2m (7ft) above the ground before it explodes. [B] is a conventional anti-personnel mine armed by rotating the safety clip [7] from "safe" [8] to the "armed" [9] position. Stepping on the pressure plate [10] pushes the firing pin [11] into the detonator [12], and the charge [13] is exploded. [C] is a heavy anti-tank mine that is armed by inserting the fuse and rotating the arming plug [14]. Heavy pressure on the pressure plate [15] activates the fuse, which ignites the booster [16] and the charge [17]. The activation wells [18] contain fuses to booby-trap it.

7 Firecrackers make the devils jump according to the Chinese, who are said to have developed fireworks for religious festivals. Today various celebrations make use of pyrotechnical displays. Effects are caused by the combustion of a variety of substances; colours are produced by adding various metal salts, sparks by finely ground metal particles. A skilful arrangement of fireworks can produce subtle effects, such as portraits and words spelled in fire.

Colour chemistry

Man has perceived and used colour since the earliest times, as the cave paintings of the Pleistocene in Altamira and Lascaux [5], in Italy and the Urals, and in eastern Siberia and Australia all testify. Early man employed natural materials to produce the colour for these paintings and ever since that time has been experimenting with and perfecting methods of extracting dyes and pigments from natural materials and, comparatively recently, producing them by artificial means.

The composition of dyes and paints

Early man learned how to extract the animal and vegetable dyes with which he daubed his body and, later, impregnated his cloth, but it was not until the nineteenth century that the chemical composition of dyes was fully understood. It was then discovered that dyes are complex organic substances that are chemically bound to the fibres, as opposed to pigments which are larger particles that form a film on the surface. The colour we see depends on the wavelength of the light absorbed by the dye or pigment [6].

In paints, finely ground pigment is bound together in an oily medium. When spread as a thin layer the medium dries to form a hard film binding the pigment particles to the surface. Vegetable and animal pigments were used in antiquity – sepia from cuttlefish, ivory black from burnt ivory, and indigo which was originally a plant extract – although these were less durable than the inorganic pigments derived from minerals. Some inorganic pigments are still used, but most are derived from synthetic organic chemicals.

Plant and animal extracts have been used as dyes [2, 3] and were often mixed to give other colours: woad [1], when combined with madder, gave dark blue; with logwood and nut gall, black; with weld, green; and with kermes, yielded violet and purple.

Classification of dyes

Dyes can be classified in two different ways, either according to how they are used or by their chemical composition [7]. The main dyers' classes are vat dyes, substantive dyes, mordant dyes, sulphur dyes and ingrain dyes. Vat dyes – such as indigo – are insoluble in water and have to be converted to a soluble derivative, which is absorbed by the fabric. Once in the fabric the dye is reconverted to the original form. Substantive dyes impregnate a fabric directly, whereas mordant dyes have to be fixed in the fibres by substances known as mordants, such as alum, which form insoluble complexes with the dye on the fabric. Sulphur dyes are direct dyes for cotton. Ingrain dyes are insoluble azo dyes (organic dyes derived from azo-benzene) produced on the fabric itself; fabric is treated with a chemical that can react with a diazo salt to form the dye.

The nineteenth century saw the end of many of the natural dye industries. The first synthetic dye (aniline purple) was discovered by the British chemist William Perkin (1838–1907) in 1856. This was quickly followed by the discovery of magenta in 1858 and Perkin's tutor August Hofmann (1818–1892) then succeeded in showing that magenta could be converted into violet dyes, a group known as the rosanilines.

At about this time an extremely important chemical discovery, the diazo reaction, was observed by a young German chemist,

1 **The woad plant** sustained the production of blue dye for over 1,000 years. Extracting indigotin from woad produced such a stench that Elizabeth I banned woad mills within five miles of her residences. However, she also protected the industry from the competition of Indian indigo, which was a prohibited import in England until the end of the seventeenth century. In France Henri IV sentenced to death anyone using Indian indigo. The home industry was also protected in Germany and Italy. Despite such protectionism, however, the lower cost and more vivid colour of indigo ultimately prevailed and the woad trade began to decline.

2 **Plants, shellfish and insects** have all been used to produce dyes, sometimes at great trouble and cost. 9,000 of these murex whelks were needed for one gramme of Tyrian purple (Roman imperial purple).

3 **The cochineal insect** is indigenous to Mexico where it lives on prickly pear. It has also been introduced into Spain, the Canary Islands and Central America. A gramme of the red dye requires 2,000 insects.

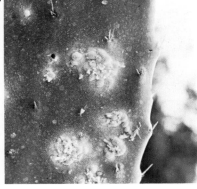

4 **Ancient Persian carpets** dyed with madder, kermes (or cochineal), genista (broom), buckthorn, nut gall, indigo, henna, safflower and tumeric (all vegetable or insect dyes) mellow and improve with age, as in this 400-year-old example.

5 **Cave paintings** in Altamira and Lascaux of 12,000 years ago made use of coloured earths and clays stained with iron or manganese compounds. Natural ochres and lamp-black made from oak charcoal were bound with animal fat, marrow and blood. Australian Aborigines still use the same pigments: black from charcoal; white from pipe clay and gypsum; yellow from limonite, ochre oxide and some fungi; and red from red ochres, laterites, limonites, manganese, iron oxide and various sandstones.

Peter Griess (1829–88). This enabled azo dyes to be prepared from a wide range of intermediates. The discovery of the ring structure of benzene (by another German, Friedrich Kekulé [1829–96]) and other advances in basic chemical understanding meant that the types of molecule responsible for colour could be identified and eventually synthesized. A great commercial industry thus grew up producing synthetic dyes from the organic chemicals found in coal tar.

The modern dyeing industry

Since the nineteenth century the manufacture of both dyes and pigments has been revolutionized by modern techniques. Most are now synthetic compounds derived mainly from organic aromatic chemicals obtained from the distillation of coal tar and crude oil. Using various chemical reactions these compounds are converted into more than a thousand intermediates and these can then be made into the desired dye or pigment.

Azo dyes can be made in many colours and are used on all types of materials and as pigments in printing inks and for plastics.

Anthraquinone dyes give rise to reddish-blue pigments used in paints, stains, enamels, polishes, soaps and plastics as well as fabrics. Indigoid colours (blue and red) are used to dye fabrics. The triphenylmethanes give bright green, blue and purple shades and are used in paper, printing inks, crayons, cosmetics and in processed food. Copper phthalocyanine is a pigment used in printing inks, lacquers, emulsion paints, distempers, rubber and plastics, and for imparting colour to car bodies. Phthalocyanines are also used as dyes; most are bright blue or green.

Dyes or pigments can now be made to suit any purpose. In general, dyes are used to colour textile fabrics whereas pigments are used for paints, printing inks and plastics. Plastics are coloured by mixing the finely divided solid plastic with pigment before moulding. Pigments can be made from dyes by precipitating them with a suitable metal salt to form the "lake" colours. Thousands of different colour chemicals are now manufactured and new compounds are being made all the time to add to the paintbox that man has created for his personal and commercial use.

For centuries the principal way of colouring cloth was by vat dyeing – a method still used today. Bolts of bleached or unbleached cloth are "stewed" in a bath of dye. The invention of synthetic aniline dyes in the 1850s widened the range of colour dyes to choose from. This illustration of about 1870 shows equipment for making aniline dyes.

6 Most mammals are believed to be colour blind. Only man and other primates can distinguish the colours of the visible electromagnetic spectrum. When white light falls upon an object, the chemical composition determines which colours the object absorbs and which it reflects. It is the reflected light that we see and interpret as the different colours; a cloth that absorbs the blue component from white light appears reddish to us.

7 There are four main chemical classes of dyes and pigments in addition to the azo colours that provide 50% of all manufactured dyes and pigments. Alizarin, a natural anthraquinone, was first used as a red dye by the Egyptians. Indigo was first synthesized commercially in 1897. Malachite green was one of the first synthetic dyes to be used. The phthalocyanines were discovered in the 1920s.

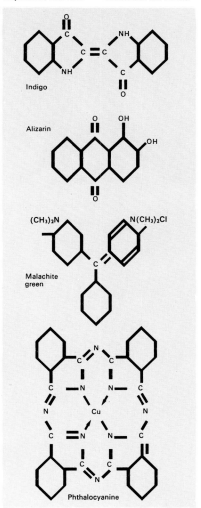

Indigo

Alizarin

Malachite green

Phthalocyanine

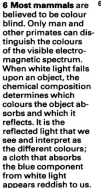

8 Haematite is one of the iron oxides originally used as pigments for yellow, red, brown and black shades. Iron oxides are still widely used because of their durability, inertness and low cost.

9 These printing ink samples show the wide range of colours available. Inorganic pigments (Prussian blue and the lead chromes) and the modern synthetic organic dyes and pigments are used in these inks.

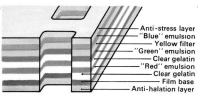

Anti-stress layer
"Blue" emulsion
Yellow filter
"Green" emulsion
Clear gelatin
"Red" emulsion
Clear gelatin
Film base
Anti-halation layer

10 Colour film is made up of three silver halide layers which respond to blue, green and red light. A positive print is obtained from the developed film "negative" by replacing the colours of the "negative" with their complements (as in yellow for blue).

11 Anodized aluminium can be coloured using dyes that are soluble in organic solvents. Both azo and phthalocyanine solvent dyes can be used. By this method, black dyes can be used to "print" letters on notices or, as here, on electronic equipment.

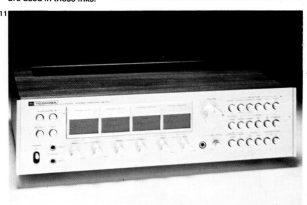

245

Cosmetics and perfumes

Cosmetics have been used since the Stone Age, when men painted their bodies as part of a hunting ritual. Excavations in Egypt have provided evidence that decorating the face and body with oils, aromatics and colour had become a sophisticated art by 5000 BC [Key]. As well as being used for decoration cosmetics are now employed to cleanse, help prevent skin troubles and to disguise minor facial imperfections.

Creams and lotions

Many cosmetic preparations are emulsions of water and oils or waxes. On application the emulsion splits up, water is lost and the oily material remains as a thin film on the skin or hair. Cold cream, often used for removing make-up, is an emulsion consisting of a combination of water, oil and waxes. It gets its name from the cooling effect produced by the evaporation of the water. It can be prepared by mixing about one part by weight of white bees-wax with three parts of liquid paraffin at 70°C (158°F), and adding this to a mixture of two parts water and a sixteenth part of borax. The resulting mixture is stirred as it cools to 35°C (95°F), at which point perfume may be added.

New emulsifiers and better knowledge of emulsion technology have resulted in a large range of different creams. These include foundation creams for use under make-up, cleansing creams that do not de-grease the skin, hand creams to maintain the oil and water balance of the skin and barrier creams.

Cosmetic lotions such as skin tonics or fresheners are mild astringents. They are said to close the "pores" of the skin, but in reality they close the openings of the hair follicles. Such lotions are based on aqueous alcohol and may contain humectants such as glycerol, menthol for its freshening effect and the astringent wych-hazel. Eau-de-Cologne and aftershave lotions may both be regarded as everyday skin fresheners.

Powder, lipstick and eyeshadow

Face powder gives a smooth, even texture to the skin by masking the shine that results from natural secretions. Basic face powder is a blend of various ingredients including zinc oxide for covering power, precipitated chalk for absorbency and bloom, talc for spreading and zinc stearate for adhesion. The colour comes from inorganic pigments and organic lakes. The perfume is either a flowery fragrance or a synthetic bouquet [3]. Cake make-up is applied to the face with a damp sponge and dries to form a water-repellent film of powder. It contains a perfumed powder base and "fillers" mixed with oily and waxy ingredients. The resultant mixture is compressed into cake form.

Lipstick must have a permanent colour, good covering power and an acceptable taste. The base is a mixture of waxes and non-drying oils, and many variations are possible depending on the proportion of oil to wax and the melting-point of the wax. Colour is achieved by blending titanium dioxide (for opacity), inorganic pigments and organic lakes (for intensity and variation of colour) with a staining dye (for indelibility). The most widely used staining dyes are eosin derivatives. A simple lipstick can be made by gently melting together about six parts of ceresin wax, one part olive oil, two parts lanolin, four parts petroleum jelly and one

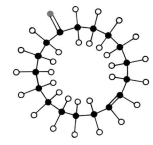

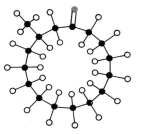

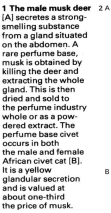

1 **The male musk deer** [A] secretes a strong-smelling substance from a gland situated on the abdomen. A rare perfume base, musk is obtained by killing the deer and extracting the whole gland. This is then dried and sold to the perfume industry whole or as a powdered extract. The perfume base civet occurs in both the male and female African civet cat [B]. It is a yellow glandular secretion and is valued at about one-third the price of musk.

2 **These molecular diagrams** (black = carbon, white = hydrogen, blue = oxygen) show civetone [A] and muscone [B], the active components of civet and musk.

3 **The fragrance of flowers** such as lavender [A] and roses [B] is due to minute traces of essential oils. These are not single substances but complex mixtures of odorous compounds. They are generally volatile liquids and may be extracted for use as raw materials in perfumery. Separation of oil from plants is not easy and the method chosen must ensure that the perfume is not decomposed. Steam distillation may be employed but the high temperature does not suit all oils. The most widely used method is extraction with a low-boiling solvent, which is later removed by distillation. The oils from peels are extracted by crushing and other mechanical means.

part liquid paraffin. When these are completely mixed, the pigments are ground in castor oil and eosin paste is added to achieve the desired shade. The melted lipstick is then poured into a mould and allowed to cool. Eye shadow and eyebrow pencils have a suitable colouring material dispersed in a similar wax base. Eyeshadow base contains an increased proportion of petroleum jelly so that it may be applied with a brush or fingertip, whereas eyebrow pencil contains more high-melting wax to increase its firmness.

Nail varnish, deodorants and antiperspirants
Nail varnish consists of a nitrocellulose base combined with a plasticizer, a modifying resin, volatile solvents and colours. The plasticizer ensures that the film is flexible and does not flake off. The addition of resins improves adhesion, hardness, gloss and resistance to detergent solutions. Solvents are chosen to give even drying in five minutes or less. The colour comes from blending soluble dyestuffs and insoluble lakes with titanium dioxide, which provides opacity. Other materials such as natural pearl (guanine) or coated

micas may be incorporated to give iridescent pearlized or "metallic" finishes.

Unpleasant odours develop with the bacterial decomposition of perspiration on the skin. This can be prevented simply by frequent washing or by the application of an efficient bactericide in a suitable base. Deodorants have no effect on the flow of perspiration and become less effective after a time. Antiperspirants decrease perspiration by a complicated mechanism. It is thought that sweat passes along the sweat duct by the process known as electro-osmosis and that the application of an electropositive material to the electronegative end of the duct inhibits the delivery of sweat. Antiperspirants generally contain an aluminium salt as the active ingredient. Aluminium chloride and sulphate are effective but their acidity causes skin irritation and damage to clothes. Nowadays a buffered form of aluminium chloride in solution is used and this has removed many of the undesirable side-effects. Many liquid formulations are supplied in "spray-on" aerosols, or they may be of the sponge-tipped or ball-ended "roll-on" type.

The Egyptians decorated their eyes by painting the undersides green and (not shown) the lids, lashes and eyebrows black. Green was made from the ore malachite (a copper carbonate) and the black was kohl, a fine black powder produced from stibnite (antimony sulphide) or the ore galena (lead sulphide).

4 A typical manufacturer offers the following selection of cosmetics: mascara in two or three types, eyeshadow in four or more forms, eyeliner (liquid or cake), eyebrow pencils, nail varnish and lipstick (both as pearlized or cream), face powder (loose and cake), liquid make-up and rouge in up to four forms. Each product is available in a wide range of shades that are updated to follow or set new fashions.

5 Cosmetics should be applied to a clean skin [A]. Foundation cream provides a base for colouring the face, to impart a healthy appearance and to modify the shape of the face. Powder hides shine or greasiness and gives a matt bloom to the skin. Eye make-up is used to enhance the natural colour of the eyes, to alter their shape or to draw attention to or detract from the line of the eyes [B]. Lipstick modifies the shape of the mouth and enhances the whiteness of the teeth. A layer of lipstick may also be beneficial in preventing cracked lips – possible sites of infection.

6 Women in North Africa and India use henna (*Lawsonia inermis*). Before the plant flowers its leaves are collected and powdered. This produces a red dye that has long been used. The powder is made into a paste by mixing it with hot water; this is then spread liberally over the part to be dyed. It is generally left overnight and is used to dye fingernails, hands, hair and even the manes of horses.

7 Stage make-up can be used to give a face a natural appearance under the whitening effect of strong light [A]. To age the face [B] shading is applied around the eyes, temples, beneath the cheek bones and at the sides of the nose and mouth. To give an Oriental look [C] white powder and shadow is used to widen and flatten the face and the eyes are elongated with black eyeliner. Grease paint is made in a variety of colours from oil, wax and spermaceti.

Everyday machines and mechanisms: 1

Most everyday products are used without any thought about the history of their invention or development. But modern life would be inconvenient without practical products such as zip fasteners [1], pens [2], door locks [3, 5], water taps [4], lavatory cisterns [6], cigarette lighters [7], aerosols [8] and fire extinguishers [9].

Whitcomb Judson invented the zip fastener in 1891, although the first reliable model was not introduced until 1913 by Gideon Sundback (1880–1954). Before this buttons were used, although they did not become common until the thirteenth century. L. E. Waterman (1837–1901) invented the modern fountain pen in 1884. In postwar years the ball-point pen and fibre-tip pen have become more popular for everyday use.

Locks date back more than 4,000 years and were used by the ancient Egyptians. The lock most commonly used today is the Yale, named after its inventor Linus Yale Jr (1821–68). The modern aerosol has been adapted for many uses. Since the early 1950s it has been used for cosmetics, paints, whipped cream and household cleansers.

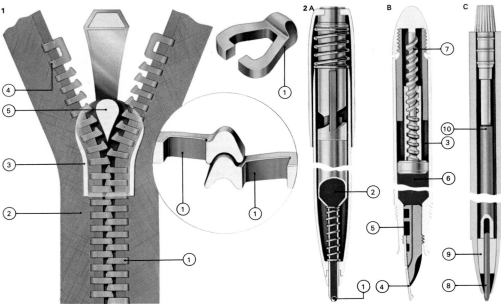

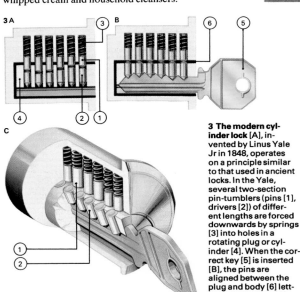

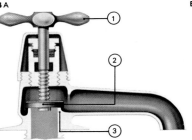

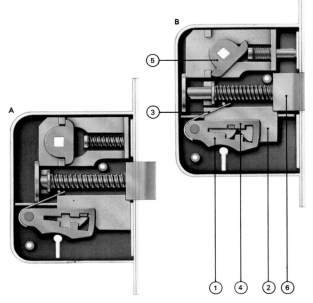

1 The zip fastener, designed as an improved method of fastening garments, comprises two chains of teeth [1] each secured to a length of strong fabric [2], a slide [3], a bottom-end piece and two top-end units [4]. By moving the slide upwards, the teeth are gradually drawn together within the slide, interlocking as shown in the inset diagram. When the slide moves down, the divider within it [5] separates the teeth.

2 Pens are of three main types – ball, fountain and felt. Ball pens [A], developed in 1938 by the Hungarians Laszlo and George Biro, have a ballbearing [1] at the tip of a tube of special ink [2]. Some fountain pens [B] are cartridge-loaded but most have a barrel [3], nib [4], feed [5], an ink reservoir [6] and self-filling mechanism [7]. A felt pen [C] has a fibre tip [8] and "transorb" [9], and an ink reservoir [10].

3 The modern cylinder lock [A], invented by Linus Yale Jr in 1848, operates on a principle similar to that used in ancient locks. In the Yale, several two-section pin-tumblers (pins [1], drivers [2]) of different lengths are forced downwards by springs [3] into holes in a rotating plug or cylinder [4]. When the correct key [5] is inserted [B], the pins are aligned between the plug and body [6] letting the plug turn [C].

4 Taps have not changed their basic design for more than 100 years. Most taps function like the one shown here – closed [A] and open [B]. Turning the handle [1] causes the washer [2] to be screwed upwards away from the valve seat [3].

5 Lever tumbler locks date back to the 18th century. Now usually on internal doors [A], their most important component is the tumbler [1] a simple lever that is securely held on the bolt [2] by a spring [3]. A projection on the tumbler, or stump [4], prevents the bolt from moving back. If a key is inserted, it engages the tumbler and is shaped so that it pushes the tumbler upwards [B] by the right amount, the key is then able to turn enough to engage the bolt at a point and move it back into the lock. Turning the door handle then makes the cam [5] move the latch [6] across. For extra security, a series of different tumblers can be used, with the key lifting each one in turn.

6 The "wash-down" closet, invented in 1889, worked on exactly the same principle as the one used today. Using plastics and superior design, the slim-line cistern is made for use in modern homes. By depressing the top-press device [1], a siphon is formed [2] and the water is sucked from the cistern and flushed down into the WC via a pipe [3]. As the water drains from the cistern, the ballfloat [4] moves down and a lever-system [5] opens the ballvalve [6], allowing the cistern to refill. When it is full, the ballfloat and lever system close the ballvalve. An overflow [7] allows water to drain off in the event of the ballvalve not functioning properly.

7 Cigarette lighters first appeared in 1909 when flint-wheel lighters with Auermetal flints were used. Auermetal, an alloy of iron and magnesium, was invented by Baron Auer von Welsbeck (1858–1929). In a modern flint/ petrol lighter [B], a lever [1] turns the flint-wheel [2]. Petrol [3], generally in wadding, is drawn up the wick [4], and ignited by a spark from the flint [5]. Flint/gas lighters [C] are similar to flint/ petrol models, except that the petrol and wick are replaced by liquid gas [6] and a valve [7]. There are two types of electric cigarette lighters: battery-operated and piezoelectric models. In the battery-

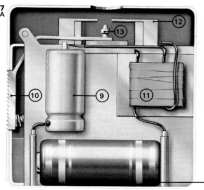

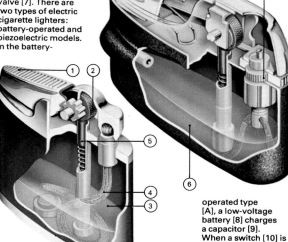

operated type [A], a low-voltage battery [8] charges a capacitor [9]. When a switch [10] is pressed, the capacitor is discharged through a step-up transformer [11] that produces a high-voltage spark across a gap [12], igniting the gas released through a valve [13] when the button was pressed.

In a piezoelectric cigarette lighter [D] the electric current for a spark is generated when a crystal [14] is squeezed, so igniting the gas.

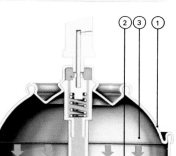

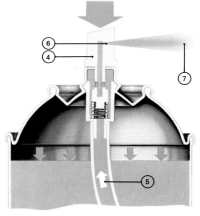

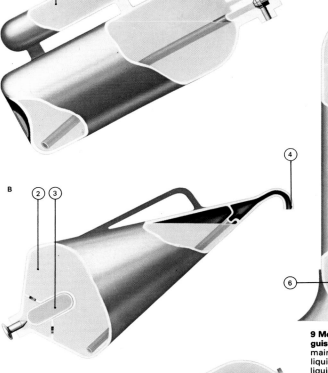

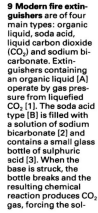

8 Aerosols were patented by L. D. Goodhue and W. N. Sullivan in the USA in 1941 and the aerosol spray has been increasingly used since the early 1950s. First, the can [1] is filled with the product to be sprayed [2] and the propellant [3]. When the pushbutton [4] is pressed, the product is forced up the dip tube [5] and out of the nozzle [6] as a fine spray [7]. Freon is the most common propellant, although it may be a pollutant and scientists are looking for a safer, inert alternative.

9 Modern fire extinguishers are of four main types: organic liquid, soda acid, liquid carbon dioxide (CO_2) and sodium bicarbonate. Extinguishers containing an organic liquid [A] operate by gas pressure from liquefied CO_2 [1]. The soda acid type [B] is filled with a solution of sodium bicarbonate [2] and contains a small glass bottle of sulphuric acid [3]. When the base is struck, the bottle breaks and the resulting chemical reaction produces CO_2 gas, forcing the solution out of the nozzle [4]. The solid sodium bicarbonate unit [C] is "powered" by liquefied CO_2. In the fire the bicarbonate [5] decomposes into soda (to form an air-excluding crust), water vapour and CO_2. The carbon dioxide extinguisher [D] contains 5–6 litres of CO_2 at high pressure [6]. When released, solid CO_2 snow is sprayed on to the fire. In addition to excluding air, it removes heat and lowers the temperature to below the ignition point.

249

Everyday machines and mechanisms: 2

Most of the machines described on these pages are commonplace to anyone who lives in an industrialized country. But some of the inventions that they employ are surprisingly old. Few people realize, for example, that the first gas meters (known as wet meters, because they contained a liquid to make them work) were invented in about 1815. This type of meter is no longer used, but it was more than a hundred years before wet meters were superseded by the modern "dry" type – and even these [1B] are based on a design originally developed between 1830 and 1850. Barometers also measure properties of a gas – the pressure of air in the atmosphere. Early barometers, such as the Fortin barometer [2A] used long tubes filled with mercury. Later barometers, just as gas meters, were superseded by "dry" types, such as the aneroid barometer [2B].

Many modern machines are powered by electricity. Vacuum cleaners [5], electric shavers [6], drills [7] and washing machines [8] all make use of electric motors. A steam iron [3] and the elements in fan heaters [4] use the heating effect of an electric current.

1 Most modern homes have mains electricity and many have a gas supply. Amounts consumed are measured by meters – either coin-operated or "read" periodically by the suppliers, who then send a bill to the consumer. An electricity meter [A] - known technically as an AC watt-hour meter – has a horizontal rotating disc [1] of aluminium with electromagnetic coils above [2] and below [3]. Eddy currents from the coils turn the disc. The magnetic flux of the upper coil is proportional to the supply voltage and that of the lower one depends on the load current. The speed of rotation is proportional to the power passing through the meter. Gears [4] drive a counting mechanism [5]. The positive displacement gas meter [B] has two diaphragm chambers [6] with flexible membranes, which are filled and emptied in turn. Their movements are conveyed by levers [7] to slide valves that control the gas flow to and from the membranes. Other levers work an index drive shaft [8], which actuates a counter [9].

2 The Fortin barometer [A] was designed in the early 1800s. Mercury in a chamois leather bag [1] bears against a screw [2] which is turned until the top of the column touches a pointer [3]. Pressure is read on a scale [4]. The aneroid barometer [B] was invented in 1843. It has an evacuated sealed metal chamber [5] which expands and contracts with shifts in atmospheric pressure. One side is fixed and the other moves a pointer [6].

3 The electric steam iron is based on earlier irons and provides, at the same time, heat and steam for the successful ironing of many fabrics. An electric element [1] heats the solid metal of the sole plate [2] and a container of water [3], which is released by operating the lever [4]. Water turns to steam on touching the sole plate and passes through gulleys on to the material being ironed.

4 Electric heaters with fans are more effective than more conventional electric fires, which rely solely on heat radiation. Recent models [A] are more compact. They incorporate a pair of fans [1] that force air over a grid of electrically heated coils [2]. As long as air passes through the heater, the coils remain at "black" heat. Early models [B] had a pair of bar-type heating elements [3] and a fan [4] to force air over them. In case the airflow stops, both types usually have thermostatic cut-out switches. Many have indicator lamps as well as switches for controlling the fan speed and heat level.

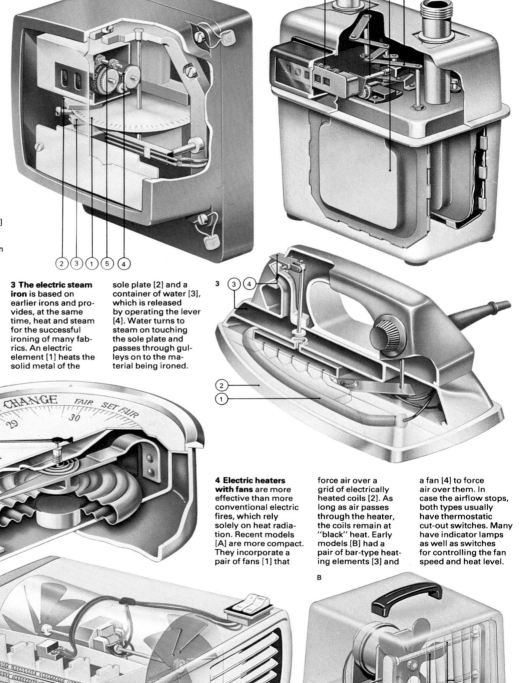

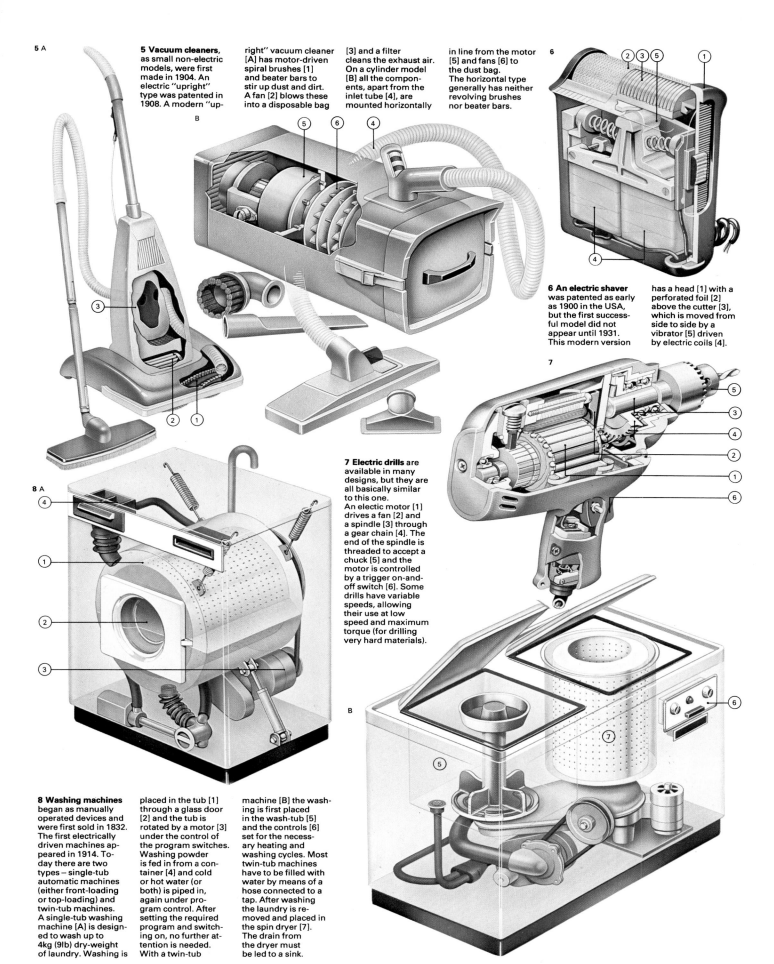

5 A

5 Vacuum cleaners, as small non-electric models, were first made in 1904. An electric "upright" type was patented in 1908. A modern "upright" vacuum cleaner [A] has motor-driven spiral brushes [1] and beater bars to stir up dust and dirt. A fan [2] blows these into a disposable bag [3] and a filter cleans the exhaust air. On a cylinder model [B] all the components, apart from the inlet tube [4], are mounted horizontally in line from the motor [5] and fans [6] to the dust bag. The horizontal type generally has neither revolving brushes nor beater bars.

6 An electric shaver was patented as early as 1900 in the USA, but the first successful model did not appear until 1931. This modern version has a head [1] with a perforated foil [2], above the cutter [3], which is moved from side to side by a vibrator [5] driven by electric coils [4].

7 Electric drills are available in many designs, but they are all basically similar to this one. An electic motor [1] drives a fan [2] and a spindle [3] through a gear chain [4]. The end of the spindle is threaded to accept a chuck [5] and the motor is controlled by a trigger on-and-off switch [6]. Some drills have variable speeds, allowing their use at low speed and maximum torque (for drilling very hard materials).

8 A

8 Washing machines began as manually operated devices and were first sold in 1832. The first electrically driven machines appeared in 1914. Today there are two types – single-tub automatic machines (either front-loading or top-loading) and twin-tub machines. A single-tub washing machine [A] is designed to wash up to 4kg (9lb) dry-weight of laundry. Washing is placed in the tub [1] through a glass door [2] and the tub is rotated by a motor [3] under the control of the program switches. Washing powder is fed in from a container [4] and cold or hot water (or both) is piped in, again under program control. After setting the required program and switching on, no further attention is needed. With a twin-tub machine [B] the washing is first placed in the wash-tub [5] and the controls [6] set for the necessary heating and washing cycles. Most twin-tub machines have to be filled with water by means of a hose connected to a tap. After washing the laundry is removed and placed in the spin dryer [7]. The drain from the dryer must be led to a sink.

Everyday machines and mechanisms: 3

Many modern domestic appliances, including those shown on these pages, are not the product of Space Age technology but owe their existence to nineteenth-century inventors.

The first sewing-machine was patented in 1830. It was wooden, had a hooked needle and was made by the French tailor Barthélemy Thimonnier (1793–1859). In 1841 Thimonnier used 80 such machines to make uniforms for the French army. The sewing-machine [2] did not achieve large-scale factory use until the development of the foot treadle by Isaac Singer (1811–75) in 1851.

In 1834 Britain issued the first refrigerator patent to the American Jacob Perkins (1766–1849), but a practical machine was not developed until the 1850s. The modern electric refrigerator [4] differs little from the one devised by James Harrison (1816–93) in 1851 for freezing meat on cargo ships. The household carpet sweeper [1], commonly named after its inventor Thomas Ewbank (1792–1870), was not made until 1889. The first practical cylinder lawn-mower [3], introduced in 1830, was that of Edwin Beard Budding (1795–1846).

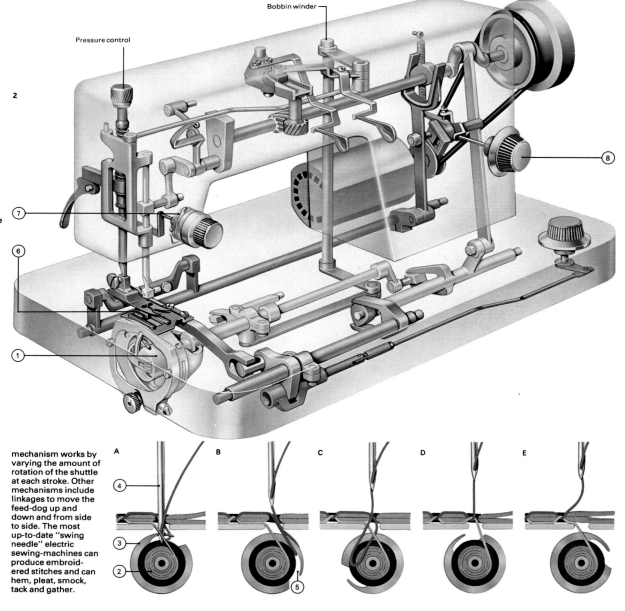

1 The carpet sweeper cleans both carpets and smooth floor surfaces efficiently. Within the strong, lightweight plastic case is a brush [1] with six helical rows of tufts [2]. Alternate rows are designed to suit flat and carpeted surfaces and the brush height is simply adjusted by means of a plastic slider. As the sweeper is moved manually backwards and forwards the brush turns round and round. Dust is collected in pans [3] that can be emptied by raising the lever [4] on top of the sweeper. The unit is completely surrounded by a flexible guard [5] to protect furniture. This type of sweeper is not only labour-saving but has the advantage of using no electricity.

Bobbin winder

Pressure control

2 A modern electric sewing-machine incorporates a rotary or oscillating shuttle [1]. The characteristic machine stitch is created by a pair of threads, one held at the top of the machine on a standard bobbin, which holds a cotton reel, the other wound on a special metal or plastic bobbin [2] in the base of the machine – below the fabric. As the motor revolves (at a speed determined by a foot control) the shuttle [3] turns around the bobbin. The series of events in the sewing of a stitch is shown in the sequence A–E. The needle [4], which is threaded near its point, pierces the fabric [A] and the point of the shuttle moves towards the needle. As the needle moves upwards [B] the hook [5] on the shuttle enters the loop formed in the needle thread. With further movement of the shuttle around the bobbin [C] the loop is drawn over the bobbin, linking it with the bobbin thread. At this point [D] the needle thread loop slips off the shuttle as the take-up lever moves upwards to pull the resulting stitch tight [E]. While the feed-dog [6] moves the material along one stitch length, the shuttle makes one "idle" revolution. The tension of each stitch, the optimum of which varies for different types of fabric, is controlled by tensioning discs [7] and the length is determined by a regulator [8]. The length control

mechanism works by varying the amount of rotation of the shuttle at each stroke. Other mechanisms include linkages to move the feed-dog up and down and from side to side. The most up-to-date "swing needle" electric sewing-machines can produce embroidered stitches and can hem, pleat, smock, tack and gather.

3 Motor mowers
have blades that move either by rotary [A] or cylinder [B] action. In the rotary mower the cutting blade [1] is turned by a petrol engine [2] (or electric motor) at such a high speed that it is not necessary to sharpen the blades. The height of the blades is adjusted using a lever [3] and the engine (and thus the speed of operation) is controlled by a throttle that regulates the flow of petrol/air mixture into the carburettor of the engine. One special type works on the hovercraft principle and rather than having wheels or rollers floats on an air cushion. A disadvantage of most rotary mowers is the absence of a box for grass cuttings – a standard feature of most cylinder mowers. In cylinder machines a series of cutting blades [4] is mounted on a rotating "cylinder" [5]. In the mower illustrated the unit is powered by an electric motor [6] but most cylinder mowers are driven either by petrol engines or pushed across the lawn by hand. The largest types of cylinder mowers are built with seating arrangements for the operator and, when an immaculate finish is required, as on a cricket field, may also incorporate a heavy roller. The first domestic lawnmowers, adaptations of agricultural reaping machines, were drawn by horses whose hoofs were covered with sacking to prevent them damaging the turf.

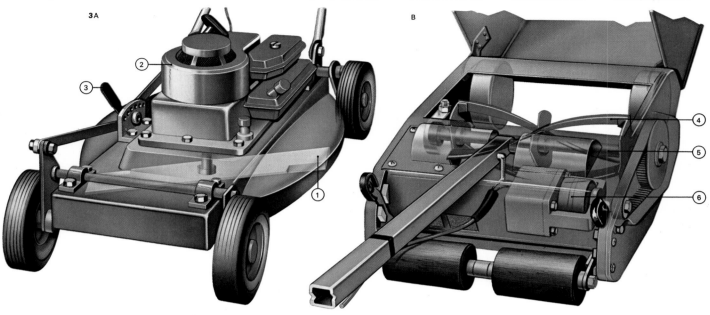

3A

B

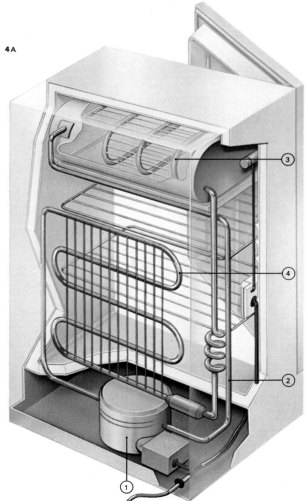

4A

4 Household refrigerators are operated either by electricity [A] or gas [B]. In the electric or compression model an electric motor drives a compressor [1] to circulate a fluid refrigerant [2]. The refrigerant is a liquid cooling agent; it boils at a low temperature and in electric refrigerators is often Freon, a substance containing carbon and fluorine. As the liquid refrigerant changes into a vapour, it absorbs heat from the freezing compartment or evaporator [3]. The vapour is then compressed and passed at high pressure to the condenser [4]. The refrigerant condenses to its liquid form, losing heat to the outside air. The liquid returns to the evaporator after expansion to low pressure at a valve. In the gas or absorption refrigerator a generator [5] is filled with ammonia gas dissolved in water. A gas flame heats this and ammonia vapour [6] is driven off. It is then liquefied in a condenser [7], thus losing heat, and passes into an evaporator [8] where it absorbs heat from the refrigerator and returns to the gaseous state. This gas sinks to the bottom of the evaporator and siphons to the absorber [9] and from there to the generator again.

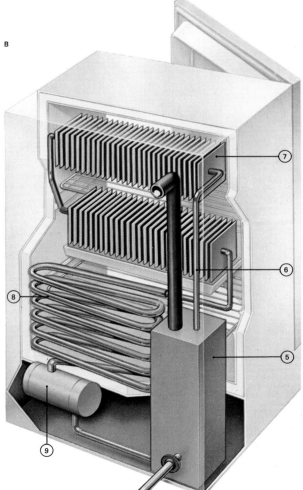

B

Everyday machines and mechanisms: 4

Two of the most useful devices ever invented for commerce – and increasingly for home use – are the typewriter and the electronic calculator. Both are based on the work of past inventors and engineers who designed and built a huge variety of such machines.

The origins of the typewriter [1] can be traced back to a patent granted to the Englishman Henry Mill by Queen Anne in 1714. But the world had to wait until 1874 for the first commercially successful machine – a Remington – which evolved from one made in the USA by Christopher Sholes (1819–90) in 1867. The first electric typewriter was marketed in the mid-1930s.

Pocket calculators [2] owe their existence to the recent rapid development of tiny integrated circuit "chips". These incorporate transistors, invented in 1948 by three American scientists, John Bardeen (1908–), Walter Brattain (1902–) and William Shockley (1910–), who were later awarded a Nobel prize for their work. The same technology permits the present generation of computers to be smaller, more reliable and more powerful than their predecessors.

1 A

B

C

D

1 Modern mechanical typewriters [A] all operate using the same basic principles [B], whether they are portables or desk models. A key [1], labelled with the character to be typed, is tapped and this action moves a chain of linked levers that results in the upward movement of the appropriate type bar [2]. The paper [3] is wound on a cylinder or platen [4], which moves along one character at a time during typing. An inked ribbon is forced against the paper by a metal character at the end of the type bar, so printing a letter on the paper. As the type bar falls back, the carriage moves one character-space to the left. At the end of a line the typist moves a lever that shifts the carriage to the right and at the same time rotates the paper-carrying cylinder round one space line. The electric typewriter [C] was developed to reduce the manual labour of typewriting. In its mechanism [D] a light depression of a key [5] makes a cam [6] contact a drive roller [7] powered by a constant-speed motor [8]. The cam is drawn upwards and the attached cam lever [9] moves back, forcing the type bar [10] upwards against the ribbon and so marking the paper. The force of the typing strokes – unlike a manual machine – is not dependent on the pressure applied by the typist and so even typing results.

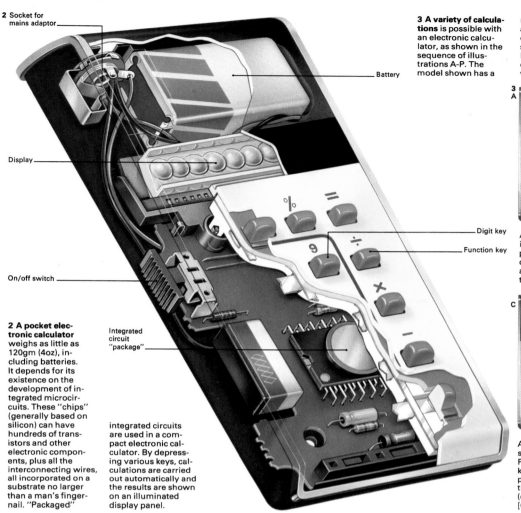

2 Socket for mains adaptor

Battery

Display

Digit key

Function key

9

On/off switch

Integrated circuit "package"

2 A pocket electronic calculator weighs as little as 120gm (4oz), including batteries. It depends for its existence on the development of integrated microcircuits. These "chips" (generally based on silicon) can have hundreds of transistors and other electronic components, plus all the interconnecting wires, all incorporated on a substrate no larger than a man's fingernail. "Packaged"

integrated circuits are used in a compact electronic calculator. By depressing various keys, calculations are carried out automatically and the results are shown on an illuminated display panel.

3 A variety of calculations is possible with an electronic calculator, as shown in the sequence of illustrations A-P. The model shown has a

memory – numbers and the results of part calculations can be stored and called up later as they are required. The CE key, when pressed once,

cancels the existing entry. The same key when pressed twice clears the display, making the calculator ready for the next series of operations.

3

A

B

A simple addition sum is 13 + 92. First press the C key (to clear the registers and display). Press 1 then 3 (display shows

13) [A]. Press + (display remains as 13). Press 9 then 2 (display shows 92) then press = (equals) and display shows 105 [B].

C

D

A simple subtraction sum is 365 – 176. First press the C key (to clear the display). Then press 3, then 6 and then 5 (display shows 365) [C]. Press – (minus)

key (display remains at 365). Press 1, then 7 then 6 (display shows 176). Finally press = (equals) key (display shows 189 [D], the required answer).

A chain calculation makes use of the calculator's memory. To calculate (25 × 92) + (72 × 5), first press the C key, then 2 then 5. Press × (multiply) key and then key in 9 then 2. Press = (equals) key and display shows 2300 [E]. Next press STD – the number is stored in the memory. Then press 7 followed by 2; press × then 5; press = (360 shown) (F). Finally press + then RCL (memory recall) and = . 2660 is displayed [G].

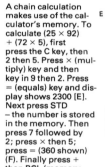

E

F

G

This result could also be stored in the calculator's memory and used as a starting number for additional chain calculations – such as further additions,

subtractions, multiplications or divisions. Pressing C twice clears the memory.

The calculator has a "constant" facility by means of which it can be used for counting. First press C (to clear the display). Press 1 and then = (1 is displayed) [H]. The calculator is now set up as a counter; repeatedly pressing the + key makes it count in sequence – 1, 2, 3, 4 and so on. For example, after five presses of the key it will have counted up to 5 and displayed the number [I]. The constant can also be used for

H

I

carrying out a series of multiplications – for example, converting a set of measurements in inches into centimetres. The conversion factor 2.54 is made the constant.

J

K

L

M

N

O

P

To find 55 per cent of 105, first press the C key; then press 1, then 0 then 5 [J]. Press × key, then 5 then 5 again [K].

Press % key, to get 57.75 – the result. Percentages can be calculated even without a % key. To find 55 per cent of 102,

key in 102, press × (multiply) key, then key in 55. Press = key (display shows 5610). Press ÷ , then 100 and = to give 56.1 [L].

What is the growth of £800 invested at 5% compound interest for two years? Press C; press 1 then · (decimal point), then

0 followed by 5 to display 1.05 [M]. Press × (to make 1.05 the constant). Press = to give 1.1025 [N], the square of 1.05

– the total interest rate at the end of two years. To find the new value of the investment, multiply by 800: press × then

8, then 0 and 0 again [O]. Finally press = to give the result £882 [P]. Simple interest would yield only £880.

INDEX

Héroult, Paul, 34
Hertz, Heinrich, 22, 228
Hevea brasiliensis, 54, *54*
Hildebrand, Henry, *125*
Hildebrand, Wilhelm, *125*
Hill, Rowland, *205*
Hilt's law, 76
Hindenburg, *146*, 147
Hiroshima, *180*
His Master's Voice, *233*
Hispano-Suiza, *128*
Historia Naturalis, 214
History of the Decline and Fall of the Roman Empire, 212
Hitler, Adolf, 184
Hittites, 21, 27, 28, *28*, 30
HMS *Dreadnought*, *174, 175*
HMS *Victory, 173*
Hoe, Richard, *64*
Hoe, 27
Hofmann, August, 22, 244
Hoists, 88, *88, 96*
Holden, Capel, 124
Holland, John P., 118
Homeostasis, 104, *105*
Homo erectus, 24
Homo sapiens, 24, *25*
Hoover Dam, 198
Horatius, 190
Horseshoe, *20*
"Hot spots", 37
Hour-glass, 92
Housing, 52–3
Hovercraft. *See* Air-cushion vehicle
Hovertrain, *144*
Howrah Bridge, *192*
Hughes, David, 225, 226, 228
Humber Bridge, *193*
Huntsman, Benjamin, 31
Huygens, Christiaan, 92, 93
Hydraulic jack, *101*
Hydraulic lever, *86, 87*
Hydraulic lift, *96*
Hydraulic loader (mechanical shovel), *98, 99*
Hydraulic ram, 51
Hydrocarbon, 78, 80–1, *81, 240*
Hydrochloric acid, 32
Hydroelectricity, 70, *71*, 84, 198, *199*
Hydrofoil, 116, *117*
    classes, *116*
    first, *116*
    propulsion, *116*
Hydrogen bonding, 61
Hydrologic cycle, *201*
Hypalon, 54, *55*

Iconoscope, 230
Incas, 190
Inclined plane, 86
    coiled, *87*
    as element of screw, *86*
    mechanical advantage, 87
India, *53*, 70
    brickmaking in, *52*
    metalworking in, *40*
Indian motor cycle, *124*
Indigo, *245*
    Indian, *244*
Indigotin, *244*
Indus civilization, 20, 190
Industrial Revolution, 21, 22, *23*, 82, 194, 201
Information retrieval, 216–17
Insulator, 44
Intaglio printing, 204–6, *205*
Intelsat IV, *229*
Interchange, motorway, *184*
Intercom system, *103*
Internal combustion engine, 68–9
International Federation of Documentation, 217
International Institute of Bibliography, 217
International Practical Temperature Scale, *91*
Ionosphere, *229*

Iron, 22, *23*, 30, 32–3, 64
    early use, 21
    manufacture, 21, 27
    smelting, 21, 40
    workers, 27
Iron, electric steam, *250*
Iron Age, 27–9
Iron oxide, *245*
Irrigation, 50–1, *196*, 197–8
Iso-octane, *81*
Isoprene, 54
Isotope, 74
Isotta-Fraschini A, *128*
Issigonis, Alec, *127*

Jacquard, Joseph, *216*
Jade, Chinese, 26
Jaipur, *40*
Jeep, *136*
Jericho, 188
Jet engine (gas turbine), 36, *63*, 80
Jewellery, 40–1, *41*
Johnson, Samuel, 212, 215
Jordan Canal, *197*
Jordan refiner, *61*
Joule, James, 22
Joy of Knowledge, The, 214–15, *215*
Judson, Whitcomb, 248
Jukebox, *233*
Julius Caesar, 190
Junkers, Hugo, 148, *149*
Junkers F-13, 148
Junkers Ju 52/3m, *149*
Jute, 56
Jutland, 26

Karl-Marx-Stadt, manufacturing at, 95
K-class submarine, *119*
Kekulé, Friedrich, 245
Kelly, William, 31
Keratin, 56
Kerosene, 80
Kiln, *52*
Knitting, 58, *59*
König, Friedrich, 204
Kostienki, *25*

Lama vicugna, 56, *56*
Land reclamation, 198
Lanka Devi, 114
Lanston, Tolbert, *207*
Lascaux cave paintings, 24, 244, *244*
Latex, 54, 61
Lathe, turret, *43*
Latrine, water-seal, 52–3
Lavoisier, Antoine, 22
Lawn-mower, *253*
    robot, *105*
Lawson, H. J., 122
Lawsonia inermis, 247
Lead, 32–3
Lead sulphide, 33
Leakey, Louis, 24
Leather, substitutes for, 37
Lee, William, *59*
Lemercier, Alfred, 205
Leomov, Alexei, 158
Lepanto, Battle of, 172
Lesages, Georges Louis, 224
Letterpress printing, *204, 206*, 207, *207*
Levalloisian culture, 24
Lever, 100
    basic types, *86*
    familiar examples, *86–7*
    mechanical advantage, 86
Lexicon, 214
Lexicon Technicum, 214
Library, 216–17
    security in, *216*
Library of Congress, 217
Licensing Act, 210
Lift, 97
    hydraulic, *96*
Lighthouse, 194

Lignin, 61
Limbs, artificial, *37*
Lime, 51
Linen, 56, 59
Linotype, *205*, 207
Linum, 56
Lipstick, 246–7
Lithography, 204, *205*, 206–7, *207*
Llama, 56
LNG carriers, 79
Load, 86, 100, 193
Lock, 196, *196*, 248
    cylinder, *248*
    lever tumbler, *248*
    pound-, *196*
Lockheed YF12A, *36*
Locomotive,
    diesel, 141
    diesel-electric, *141*
    electric, 140–1
    first commercial, *140*
    steam, *62*, 64–5, *64*, 140
    underground, *140*
    Bavarian Class 53/6, *140*
    Beyer-Garratt, 141
    "Big Boy", *141*
    French Class CC7100, *141*
    General, 140
    High Speed Train, *141*
    Mallard, 140
    Rocket, 140
    Tokaido, *141*
    See also Railway
Lodge, Oliver, 228
London Bridge, *190*, 191
Long-wagon, medieval, 120, *120*
Lovell, James, 159
Lubricants, 36–7
Lunokhod I, *156*
Luppis, G., *118*
Lusitania, 114
Lycurgus cup, 44

McAdam, John, 120, 182
Macchi MC72, *148*
McDonnell-Douglas F-15 Eagle, *178*
Mace, *160*
Machine,
    in common use, 248–55, *248–55*
    earth-moving, 98–9, *98–9*
    lifting, 96–7, *96–7*, 100–1, *100–1*
    measurement,
        distance, 90, *90–1*
        mass, 90, *91*
        pressure, 91, *91*
        radiation, 91, *91*
        speed, *90*, 91
        temperature, 90, *90–1*
        time, 90, 92–3, *92–3*
    metalworking, 42–3, *42–3*
See also specific machines
Machine communication,
    magnetic ink method, *109*
    processing of information, 108
    punched-card method, *107*, *216*, 217
    punched-tape method, *106–7*
Machine control, automated,
    closed-loop (feed-back), 104
    electronic devices, *102–4*
    error signal, *104*
    open loop, 104, *104*
Machine gun. *See* Gun, automatic
Machinery, 36–7, 61–3
    agricultural, 51, *51*
    tunnelling, 188-9, *189*
    typesetting, *211*
Macmillan, Kirkpatrick, 122
Magazine, 210–11
    delivery, *211*
Magdalenian culture, 24
Magenta, discovery of synthetic, 245
Maglemosian culture, 24, 25

Magnet, *75*
Magnetic ink, *109*
Magnetic recording, *232–5*, *233–5*
Magnetite, *28*
Magnetophone, 233
Maiden Castle, *28, 29*
Mail coach, 120
Malachite green, *245*
Manganese, *32*
Marconi, Guglielmo, 228
Marey, Etienne-Jules, *222*
Mark I (British tank), 171
Mass, measurement, 90, *91*
Mass production, 94–5, *94*
Matchlock, *162*
Mathematics, 22
    Egyptian, *27*
Maudsley, Henry, 94
Maxim, Hiram, 164
Maxwell, James Clerk, 22, 104, 228
Mayan people, 21
Maybach, Wilhelm, 124
Me Bf 109G, *176*
Mechanical advantage, 100
    defined, 86
    of inclined plane, 87
    of lever, 86
    of screw, 87
Mechanical shovel (hydraulic loader), *98, 99*
Medway Bridge, *192*
Memory (computer), 106–7, *106*
Menai Bridge, 191
Men-of-war, 172, *173*
Mercedes (1901), *126*
Mercedes (1914), *130*
Mercury, 32–3
Mergenthaler, Ottmar, *205*
Mersey Mole, *189*
Mesolithic period, 221
Mesopotamia, 20–1, 26–7
Metal, 26, 32–3, 36–7, 40–1
    production, 26–7, 40–1
Metal location,
    using electronics, *102*
Metallurgy, 40–1
Metalworking,
    die casting, 42, *42*
    electro-chemical machining, 43, *43*
    forging, 42, *43*
    grinding, *42*
    laser machining, 42, *42*
    machine tooling, 42–3
    milling, *43*
    pressing, *42*
    sand casting, 42, *42*
    spark erosion, 43, *43*
    ultrasonic machining, 43, *43*
Methane, 52, 78
    generation, *51*
Methyl cyclopentane, *81*
Metz, Siege of, 166
Michaux, Ernest, 122, 124
Michaux, Pierre, 122, 124
Microanalyser, electron probe, *36*
Micrometer, *87*, 90, *91*
Microscope, *21*
Micro-switch, 33
Middle East, *32*
    civilizations, 20
MiG-15, *178*
Migration pattern, *37*
Mil, Mikhail, *155*
Military aeroplane,
    development, 176–7
    specialization, 176
    BAC Canberra, *179*
    B-29 Superfortress, *177*
    Camel, *176*
    Handley Page 0/400, *177*
    Junkers Ju 52/3m, *149*
    McDonnell-Douglas F-15 Eagle, *178*
    Me Bf 109G, *176*
    MiG-15, *178*

Picture Credits

Man and Machines
17 Jan Eyerman/T.L.P.A. © Time Inc 1976/Colorific. 18 Daily Telegraph Colour Library. 19 ASEA. 20–1 [8] Ronan Picture Library. 22–3 [Key] Ironbridge Gorge Museum Trust. 24–5 [3] Mark Boulton/Bruce Coleman Ltd; [4] Barnabys Picture Library. 26–7 [2] Michael Holford; [3] Michael Holford; [6A] Basil Booth; [6B] Basil Booth. 28–9 [Key] Aerofilms; [5] Mansell Collection; [6] Mansell Collection. 30–1 [Key] ASEA; [3] ASEA. 32–3 [Key] Picturepoint; [1A] Photoresources; [1B] Paul Brierley; [2] Michael Holford/British Museum; [3] AEC Ltd/Tin Research Institute; [5] Copper Development Association; [6] Mullard Valves Ltd; [7] Kim Sayer/by permission of the Gardening Centre Ltd, Syon Park; [8] Wilmot Breeden Ltd; [9] UKAEA. 34–5 [2] Radio Times Hulton Picture Library; [11] R. Sheridan/ZEFA. 36–7 [Key] Spectrum Colour Library; [1] Photri; [2] Photri; [3] The Cambridge Instrument Co; [4] Michael Francis Wood & Associates; [5] Paul Brierley; [6] Hans Gunter Möuer; [7] F. James. 38–9 [Key] Mansell Collection; [3] Cooper Bridgeman; [4] Trinity College Chapel, Oxford/Leslie Harris; [6] Kim Sayer; [7] Photri; [8] Picturepoint; [9] Richard Cooke. 40–1 [2] Picturepoint; [4] David Strickland; [5] David Strickland. 42–3 [Key] British Steel Corporation. 44–5 [Key] Paul Brierley; [3] Trustees of the British Museum; [6] Paul Brierley; [7] UKAEA. 46–7 [6A] Timber Research & Development Association; [6B] Timber Research and Development Association. 48–9 [Key] Angelo Hornak. 52–3 [2] ZEFA. 54–5 [4] Royal National Lifeboat Institute; [7] GPG Holdings; [9] Source unknown. 56–7 [Key] Courtaulds Ltd; [2] Bruce Coleman/Bruce Coleman Ltd; [3] Jane Burton/Bruce Coleman Ltd; [5] Asbestos & Rubber Co Ltd; [6] MB Copyright; [7] Photri. 58–9 [3] Courtaulds Ltd; [5] Courtaulds Ltd; [6] Courtaulds Ltd; [8] A–Z Botanical Collection. 60–1 [Key] Ronan Picture Library; [1] Ronald Sheridan; [3A] ZEFA; [4] Geoff Goode/by permission of the National Postal Museum. 62–3 [Key] Basil Smith. 64–5 [3] Times Newspapers; [4] Ronan Picture Library. 66–7 [Key] Central Electricity Generating Board; [5] Central Electricity Generating Board. 68–9 [4] Rolls Royce. 70–1 [Key] Barnabys Picture Library. 72–3 [Key] ZEFA; [1] ZEFA; [3] Daily Telegraph Colour Library. 74–5 [7] UKAEA; [8] UKAEA. 76–7 [4] National Coal Board; [5] National Smokeless Fuel Ltd; [6] National Coal Board. 80–1 [Key] R. Halin/ZEFA. 82–3 [5] William MacQuitty. 84–5 [Key] Fabian Acker; [1] Picturepoint; [5] Fabian Acker. 86–7 [Key] Spectrum Colour Library; [2] Picturepoint; [8] Architectural Association; [9] Spectrum Colour Library. 88–9 [4] Phil Sheldon/ZEFA. 92–3 [Key] Mary Evans Picture Library; [7] SSIH [UK] Ltd [Omega Division]. 94–5 [Key] ASEA; [1] The Science Museum; [4] British Leyland; [5A] Ford Motor Co; [5B] Ford Motor Co; [5C] Ford Motor Co; [6] Paul Brierley; [7] ASEA. 96–7 [Key] Ronan Picture Library; [5] Popperfoto; [7] Lancer Boss Ltd. 98–9 [Key] Mansell Collection; [1] E. Webber/ZEFA; [2] F. Park/ZEFA; [3] H. Helbig/ZEFA; [5] Picturepoint. 100–1 [1] Spectrum Colour Library; [2B] By permission of the German Embassy; [3A] ZEFA; [4B] Rolair Systems [UK] Ltd; [4C] Rolair Systems [UK] Ltd; [5A] Hydranautics Inc; [6B] Hydranautics Inc. 102–3 [Key] Toshiba/Michael Turner Associates; [2] Spectrum Colour Library; [4A] David Levin; [4B] David Levin; [6A] Paul Brierley; [6B] Paul Brierley; [7A] Eagle International. 104–5 [6] Queen Mary College. 106–7 [1] Chris Steele-Perkins/Courtesy of the Science Museum; [2] Spectrum Colour Library; [5] IBM. 108–9 [Key] Lloyds Bank Ltd; [1] IBM; [2] Honeywell Ltd; [7] ASEA; [8] David Strickland. 110–1 [Key] Textron's Bell Aerospace

Division, Buffalo, NY. 118–9 [Key] Mansell Collection; [2] Popperfoto; [3] Robert Hunt Library; [4] Photri; [8] Photri. 122–3 [Key] Mansell Collection. 134–5 [Key] Mansell Collection; [2] Fotolink; [3] F. Hackenburg/ZEFA; [4] London Transport Executive; [6] Kim Sayer; [7] P. Phillips/ZEFA; [8] Fotolink. 138–9 [4] Ann Keatley/National Academy of Sciences, Washington; [5] Spectrum Colour Library. 144–5 [2] British Railways Board. 146–7 [7] Radio Times Hulton Picture Library; [8] Times Newspapers. 154–5 [3] Crown Copyright 1973; [5] Smithsonian Institution; [8] Picturepoint. 156–7 [Key] Patrick Moore Collection; [2] Patrick Moore Collection; [4A] Novosti; [4B] NASA; [4C] NASA; [8A] NASA; [8B] NASA. 158–9 [Key] NASA; [2] NASA; [3] NASA; [4] NASA; [5] NASA; [6] NASA; [7] NASA; [8] NASA; [9] NASA; [10] NASA. 164–5 [3] Popperfoto/Imperial War Museum. [5A] Robert Hunt Library; [7] General Electric Aircraft Equipment Division. 166–7 [Key] Crown Copyright, reproduced by permission of the Department of the Environment; [1] Michael Holford/British Museum; [5] Permission by permission of the Governing Body of Christ Church, Oxford; [7A] National Portrait Gallery. 168–9 [Key] John Massey Stewart; [5] Photri; [7] Photri; [8] Photri; [10] Robin Adshead. 170–1 [1] Imperial War Museum; [5] Popperfoto; [7] Imperial War Museum. 180–1 [Key] Robert Hunt Library/Imperial War Museum; [2] Robert Hunt Library; [7] Photri; [8] Photri. 184–5 [Key] Associated Press; [1] Freeman Fox & Partners. 188–9 [Key] Sir Robert McAlpine; [6] Chesapeake Bay Bridge & Tunnel District. 190–1 [2] Mary Evans Picture Library; [3] Loren McIntyre. 192–3 [Key] Douglas Pike; [2] Bill Stirling/Robert Harding Associates; [3] Construction News; [4] Construction News; [5] A. Monk & Co Ltd; [7] Frank Wallis; [8] A. Monk & Co Ltd. 194–5 [Key] J. Allan Cash. 196–7 [Key] Spectrum Colour Library; [7] ZEFA; [9] KLM Aerocarto. 198–9 [Key] J. Allan Cash; [8] Paul Almasy. 202–3 [Key] Mansell Collection; [3] Redland Purle Ltd; [4A] Picturepoint; [4B] Spectrum Colour Library; [5] Redland Purle Ltd. 204–5 [2] Mansell Collection; [3] Mansell Collection; [4] Mansell Collection; [6B] Mansell Collection; [8] Mansell Collection; [9] Mansell Collection; [10] Linotype U.K. 206–7 [Key] G. Sommer/ZEFA; [3E] David Strickland; [4]

Picturepoint. 208–9 [4] Rank Xerox Ltd. 210–1 [Key] Monitor; [2] Picturepoint; [3] Jerry Watcher Colorific; [6] IBM. 212–3 [Key] Kim Sayer; [1] Kim Sayer; [2] BBC Copyright photograph; [3] Mike Peters; [4A] Kim Sayer; [4B] Kim Sayer; [4C] Kim Sayer; [4D] Kim Sayer; [4E] Kim Sayer; [4F] Kim Sayer; [6] Gerfried Brutzer. 214–5 [Key] Kim Sayer; [1] David Levin; [2] David Levin; [3] David Levin; [4A] Graeme French; [4B] Graeme French; [4C] Graeme French; [4D] Graeme French; [4E] Graeme French; [4F] Graeme French; [4G] Graeme French; [4H] Graeme French; [4I] Graeme French. 216–7 [Key] Dupont [UK] Ltd; [1] John Crossley & Sons Ltd; [2A] IBM; [2B] Honeywell; [4A] Angelo Hornak; [5] British Airways. 220–1 [Key] Christ Steele Perkins; [1A] Guy Rycart; [1B] Guy Rycart; [2A] Guy Rycart; [2B] Guy Rycart; [2C] Guy Rycart; [7A] David Strickland; [7B] David Strickland; [7C] David Strickland. 222–3 [7] EMI. 224–5 [Key] By courtesy of the Post Office; [1] Ronan Picture Library; [2] David Strickland; [5] Chris Steele Perkins/The Science Museum. 226–7 [Key] J. Allan Cash. 228–9 [Key] The Science Museum; [1] Mansell Collection; [2A] Chris Steele Perkins/The Science Museum; [3] Radio Times Hulton Picture Library; [6E] BBC; [6B] P. Freytag/ZEFA; [6C] David Levin. 232–3 [Key] EMI. 224–5 [4] Sotheby's, Belgravia; [7] Toshiba/Michael Turner Associates. 234–5 [1] Ampex [Great Britain] Ltd; [3A] Ampex [Great Britain] Ltd; [3B] David Levin; [5] Ampex [Great Britain] Ltd; [7] Paul Brierley/Decca Ltd. 236–7 [Key] Radio Times Hulton Picture Library. 238–9 [Key] ICI Ltd; [1] Aerofilms Ltd; [3] ZEFA; [5] Centre File Ltd. 240–1 [1] Picturepoint; [2] Unilever Ltd; [2] Picturepoint. 242–3 [Key] Popperfoto; [3] Photri; [7] Pains Wessex Ltd/Michael Turner Associates. 244–5 [Key] Ronan Picture Library; [2] David Levin; [3] Bill Holden; [4] Photoresources/Metropolitan Museum of Art; [5] Michael Holford; [8] Basil Booth; [9] David Levin; [11] Toshiba/Michael Turner Associates. 246–7 [Key] Picturepoint; [3A] French Government Tourist Office; [3B] Spectrum Colour Library; [4] Max Factor Ltd; [5A] Educational Products Ltd/Max Factor Ltd; [6] Times Newspapers; [7B] Educational Products Ltd/Max Factor Ltd; [7C] Educational Products Ltd/Max Factor Ltd.

Artwork Credits

Art Editors
Angela Downing; George Glaze; James Marks; Mel Peterson; Ruth Prentice; Bob Scott

Visualizers
David Aston; Javed Badar; Allison Blythe; Angela Braithwaite; Alan Brown; Michael Burke; Alistair Campbell; Terry Collins; Mary Ellis; Judith Escreet; Albert Jackson; Barry Jackson; Ted Kindsey; Kevin Maddison; Erika Mathlow; Paul Mundon; Peter Nielson; Patrick O'Callaghan; John Ridgeway; Peter Saag; Malcolme Smythe; John Stanyon; John Stewart; Justin Todd; Linda Wheeler

Artists
Stephen Adams; Geoffrey Alger; Terry Allen; Jeremy Alsford; Frederick Andenson; John Arnold; Peter Arnold; David Ashby; Michael Badrock; William Baker; John Barber; Norman Barber; Arthur Barvoso; John Batchelor; John Bavosi; David Baxter; Stephen Bernette; John Blagovitch; Michael Blore; Christopher Blow; Roger Bourne; Alistair Bowtell; Robert Brett; Gordon Briggs; Linda Broad; Lee Brooks; Rupert Brown; Marilyn Bruce; Anthony Bryant; Paul Buckle; Sergio Burelli; Dino Bussetti; Patricia Casey; Giovanni Casselli; Nigel Chapman; Chensie Chen; David Chisholm; David Cockcroft; Michael Codd; Michael Cole; Terry Collins; Peter Connelly; Roy Coombs; David Cox; Patrick Cox; Brian Cracker; Gordon Cramp; Gino D'Achille; Terrence Daley; John Davies; Gordon C. Davis; David Day; Graham Dean; Brian Delf; Kevin Diaper; Madeleine Dinkel; Hugh Dixon; Paul Draper; David Dupe; Howard Dyke; Jennifer Eachus; Bill Easter; Peter Edwards; Michael Ellis; Jennifer Embleton; Ronald Embleton; Ian Evans; Ann Evens; Lyn Evens; Peter Fitzjohn; Eugene Flurey; Alexander Forbes; David Carl Forbes; Chris Fosey; John Francis; Linda Francis; Sally Frend; Brian Froud; Gay Galfworthy; Ian Garrard; Jean George; Victoria Goaman; David Godfrey; Miriam Golochoy; Anthea Gray; Harold Green; Penelope Greensmith; Vanna Haggerty; Nicholas Hall; Horgrave Hans; David Hardy; Douglas Harker; Richard Hartwell; Jill Havergale; Peter Hayman; Ron Haywood; Peter Henville; Trevor Hill; Garry Hinks; Peter Hutton; Faith Jacques; Robin Jacques; Lancelot Jones; Anthony Joyce; Pierre Junod; Patrick Kaley; Sarah Kensington; Don Kidman; Harold King; Martin Lambourne; Ivan Lapper; Gordon Lawson; Malcolm Lee-Andrews; Peter Levaffeur; Richard Lewington; Brian Lewis; Ken Lewis; Richard Lewis; Kenneth Lilly; Michael Little; David Lock; Garry Long; John Vernon Lord; Vanessa Luff; John Mac; Lesley MacIntyre; Thomas McArthur; Michael McGuinness; Ed McKenzie; Alan Male; Ben Manchipp; Neville Mardell; Olive Marony; Bob Martin; Gordon Miles; Sean Milne; Peter Mortar; Robert Morton; Trevor Muse; Anthony Nelthorpe; Michael Neugebauer; William Nickless; Eric Norman; Peter North; Michael O'Rourke; Richard Orr; Nigel Osborne; Patrick Oxenham; John Painter; David Palmer; Geoffrey Parr; Allan Penny; David Penny; Charles Pickard; John Pinder; Maurice Pledger; Judith Legh Pope; Michael Pope; Andrew Popkiewicz; Brian Price-Thomas; Josephine Rankin; Collin Rattray; Charles Raymond; Alan Rees; Ellsie Rigley; John Ringnall; Christine Robbins; Ellie Robertson; James Robins; John Ronayne; Collin Rose; Peter Sarson; Michael Saunders; Ann Savage; Dennis Scott; Edward Scott-Jones; Rodney Shackell; Chris Simmonds; Gwendolyn Simson; Cathleen Smith; Les Smith; Stanley Smith; Michael Soundels; Wolf Spoel; Ronald Steiner; Ralph Stobart; Celia Stothard; Peter Sumpter; Rod Sutterby; Allan Suttie; Tony Swift; Michael Terry; John Thirsk; Eric Thomas; George Thompson; Kenneth Thompson; David Thorpe; Harry Titcombe; Peter Town; Michael Trangenza; Joyce Tuhill; Glenn Tutssel; Carol Vaucher; Edward Wade; Geoffrey Wadsley; Mary Waldron; Michael Walker; Dick Ward; Brian Watson; David Watson; Peter Weavers; David Wilkinson; Ted Williams; John Wilson; Roy Wiltshire; Terrence Wingworth; Anne Winterbotham; Albany Wiseman; Vanessa Wiseman; John Wood; Michael Woods; Owen Woods; Sidney Woods; Raymond Woodward; Harold Wright; Julia Wright

Studios
Add Make-up; Alard Design; Anyart; Arka Graphics; Artec; Art Liaison; Art Workshop; Bateson Graphics; Broadway Artists; Dateline Graphics; David Cox Associates; David Levin Photographic; Eric Jewel Associates; George Miller Associates; Gilcrist Studios; Hatton Studio; Jackson Day; Lock Pettersen Ltd; Mitchell Beazley Studio; Negs Photographic; Paul Hemus Associates; Product Support Graphics; Q.E.D. [Campbell Kindsley]; Stobart and Sutterby; Studio Briggs; Technical Graphics; The Diagram Group; Tri Art; Typographics; Venner Artists

Agents
Artist Partners; Freelance Presentations; Garden Studio; Linden Artists; N.E. Middletons; Portman Artists; Saxon Artists; Thompson Artists